Handbook of Experimental Pharmacology

Volume 113

The Pharmacology of Monoclonal Antibodies

Contributors

R. Balint, C.F. Barbas, R.D. Blumenthal, P. Carter, M. Chatterjee
Chen, Y.-C. Jack, R.M. Conry, K.A. Foon, D.M. Goldenberg
E. Haber, M. Hein, A. Hiatt, K. James, K.D. Janda, K. Karjalainen
H. Kohler, J.W. Larrick, A.F. LoBuglio, N. Lonberg, G.E. Mark
E.A. Padlan, S.H. Pincus, A. Plückthun, M.L. Rodrigues
R.G. Rupp, M.N. Saleh, M.R. Shalaby, R.M. Sharkey
A. Traunecker

Editors
Martin Rosenberg and Gordon P. Moore

Springer-Verlag
Berlin Heidelberg New York London Paris
Tokyo Hong Kong Barcelona Budapest

GENENTECH, INC.
460 Pt. San Bruno Blvd.
South San Francisco, CA 94080

MARTIN ROSENBERG, Ph.D.
Vice President and Director
Biopharmaceuticals
Smith Kline Beecham Pharmaceuticals
709 Swedeland Road
King of Prussia, PA 19406-0939
USA

GORDON P. MOORE, Ph.D.
Director
Department of Molecular Genetics
Smith Kline Beecham Pharmaceuticals
709 Swedeland Road
King of Prussia, PA 19406-0939
USA

With 78 Figures and 34 Tables

ISBN 3-540-57123-X Springer-Verlag Berlin Heidelberg New York
ISBN 0-387-57123-X Springer-Verlag New York Berlin Heidelberg

Library of Congress Cataloging-in-Publication Data. The Pharmacology of monoclonal antibodies / con-
tributors, R. Balint . . . [et al.]; editors, Martin Rosenberg and Gordon P. Moore. p. cm. — (Handbook
of experimental pharmacology; v. 113) Includes bibliographical references and index. ISBN 3-540-57123-X:
— ISBN 0-387-57123-X 1. Monoclonal antibodies. I. Balint, Robert Frederick. II. Rosenberg, Martin.
III. Moore, Gordon P., 1950– . IV. Series. [DNLM: 1. Antibodies, Monoclonal—pharmacology.
W1 HA51L v. 113 1994 / QW 575 P536 1994] QP905.H3 vol. 113 [QR186.85] 615′.1 s—dc20 [615′.37]
DNLM/DLC 93-44948

© Springer-Verlag Berlin Heidelberg 1994
Printed in Germany

Typesetting: Best-set Typesetter Ltd., Hong Kong

SPIN: 10077910   27/3130/SPS – 5 4 3 2 1 0 – Printed on acid-free paper

List of Contributors

BALINT, R., Palo Alto Institute of Molecular Medicine, 2462 Wyandotte Street, Mountain View, CA 94043, USA

BARBAS, C.F., III, Department of Molecular Biology, the Scripps Research Institute, 10666 North Torrey Pines Road, La Jolla, CA 92037, USA

BLUMENTHAL, R.D., Center for Molecular Medicine and Immunology, One Bruce Street, Newark, NJ 07103, USA

CARTER, P., Department of Protein Engineering, Genentech Inc., 460 Point San Bruno Boulevard, South San Francisco, CA 94080, USA

CHATTERJEE, M., Department of Microbiology and Immunology, University of Kentucky Medical Center, 800 Rose Street, Lexington, KY 40536, USA

CHEN, Y.-C. JACK, Abbot Laboratories, D-907 AP-20, 1 Abbot Park Rd., Abbot Park, IL 60064, USA

CONRY, R.M., University of Alabama at Birmingham, Comprehensive Cancer Center, L.B. Wallace Tumor Institute—235, 1824 6th Avenue South, Birmingham, AL 35294-3300, USA

FOON, K.A., Lucille Parker Markey Cancer Center and the Department of Medicine, University of Kentucky Medical Center, Lexington, KY 40536, USA

GOLDENBERG, D.M., Center for Molecular Medicine and Immunology, One Bruce Street, Newark, NJ 07103, USA

HABER, E., Center for the Prevention of Cardiovascular Disease and Division of Biological Sciences, Harvard School of Public Health, 677 Huntington Avenue, Boston, MA 02115, USA

HEIN, M., Department of Cell Biology, The Scripps Research Institute, 10666 North Torrey Pines Road, La Jolla, CA 92037, USA

HIATT, A., Department of Cell Biology, The Scripps Research Institute, 10666 North Torrey Pines Road, La Jolla, CA 92037, USA

JAMES, K., Department of Surgery (WGH), The Medical School, Edinburgh University, Edinburgh EH8 9AG, Great Britain

JANDA, K.D., The Scripps Research Institute, Departments of Molecular Biology and Chemistry, 10666 North ⌐orrey Pines Rd., La Jolla, CA 92037, USA

KARJALAINEN, K., Basel Institute for Immunology, Grenzacherstrasse 487, Postfach, CH-4005 Basel, Switzerland

KOHLER, H., Department of Microbiology and Immunology, University of Kentucky Medical Center, 800 Rose Street, Lexington, KY 40536, USA

LARRICK, J.W., Palo Alto Institute for Molecular Medicine, 2462 Wyandotte Street, Mountain View, CA 94043, USA

LOBUGLIO, A.F., University of Alabama at Birmingham, Comprehensive Cancer Center, L.B. Wallace Tumor Institute—235, 1824 6th Avenue South, Birmingham, AL 35294-3300, USA

LONBERG, N., GenPharm International, 297 N. Bernardo Avenue, Mountain View, CA 94043, USA

MARK, G.E., Cellular and Molecular Biology, Merck Research Laboratories, P.O. Box 2000, Rahway, NJ 07065, USA

PADLAN, E.A., Laboratory of Molecular Biology, National Institute of Diabetes and Digestive and Kidney Diseases, National Institutes of Health, Bethesda, MD 20892, USA

PINCUS, S.H., National Institutes of Health, National Institute of Allergy and Infectious Diseases, Laboratory of Microbial Structure and Function, Rocky Mountain Laboratories, Hamilton, MT 59840, USA

PLÜCKTHUN, A., Biochemisches Institut, Universität Zürich, Winterthurerstrasse 190, CH-8057 Zürich, Switzerland

RODRIGUES, M.L., Department of Protein Engineering, Genentech, Inc., 460 Point San Bruno Boulevard, South San Francisco, CA 94080, USA

RUPP, R.G., Process Development and Manufacturing, Regeneron, Inc., 777 Old Saw Mill River Road, Tarrytown, NY 10591, USA

SALEH, M.N., University of Alabama at Birmingham, Comprehensive Cancer Center, L.B. Wallace Tumor Institute—235, 1824 6th Avenue South, Birmingham, AL 35294-3300, USA

SHALABY, M.R., Department of Medicinal and Analytical Chemistry, Genentech Inc., 460 Point San Bruno Boulevard, South San Francisco, CA 94080, USA

SHARKEY, R.M., Center for Molecular Medicine and Immunology, One Bruce Street, Newark, NJ 07103, USA

TRAUNECKER, A., Basel Institute for Immunology, Grenzacherstrasse 487, Postfach, CH-4005 Basel, Switzerland

Preface

It has been almost 20 years since the discovery by Kohler and Milstein of the technology to produce monoclonal antibodies (MAbs), a discovery that promised revolutionary changes in research, clinical diagnosis and human therapy. From today's perspective, it is fair to conclude that this promise has been realized in two areas of the three. As research tools, MAbs have been invaluable: their ability to selectively bind and localize specific antigens, detect and identify new ligands and their receptors, and agonize and/or antagonize specific molecular interactions continues to provide a useful and enabling technology to basic research endeavors. Similarly, MAbs have demonstrated enormous practical impact as diagnostic tools. Recent advances in clinical diagnostic medicine continue to rely heavily on the use of MAb-based reagents for detecting and localizing antigens of clinical import. In contrast, however, MAbs have not proven to have major impact on human disease therapy. With the single exception of an immunosuppressive MAb against the T-cell antigen, CD3, MAbs have as yet found few meaningful applications as therapeutic agents.

During the 1980s, a set of technologies to clone, modify and express genes encoding MAbs was developed. These breakthroughs permitted MAbs to be genetically engineered which consequently gave them the potential to greatly enhance their therapeutic utility as well as significantly expand their research and diagnostic applications. New MAbs, fragments of MAbs, bispecific MAbs, single-chain MAbs, and fusions of MAbs with other gene products became available for study. Of particular significance was the creation of MAbs which strove towards increased immunocompatibility with humans. The engrafting of mouse variable regions onto the gene segments encoding human immunoglobulin constant regions gave rise to the so-called chimeric class of antibodies. This was followed by the more specific (and far less exacting) procedures of engrafting complementarity-determining regions (CDRs), where the DNA encoding the mouse CDR (and often other critical residues of far less obviousness) were engrafted onto human variable region frameworks and constant regions. A third theme has been the recombinant engraftment of the variable regions of primate antibodies onto human constant regions to create "primitized" forms of chimeric antibodies. Concomitantly, technologies for high-level expression of genetically engineered MAbs in a variety of systems emerged.

These expression systems have been scaled up, and industrial purification schemes devised. As a result, a wide variety of engineered MAbs are now entering the clinic. In fact, of the approximately 100 biotech products in clinical trials today, over half are MAbs or MAb derivatives.

More recently, yet another technical revolution in designer antibodies has occurred, ushered in by the discovery of methods to obtain novel MAb variable regions via combinatorial cloning and in vitro mutagenesis. This new technology offers access to a huge variety of natural variable regions, new combinations of heavy and light chains, and wholly novel MAbs created in the test tube. It is of interest that this new technology has its basis in standard recombinant microbial/phage genetics, utilizing expression of antibody fragments both on the surface of filamentous phage and in *E. coli* secretion systems. Clearly, and yet again, basic research in biological systems of no direct or obvious relevance to man has provided the crucial cornerstone for a set of technologies which will have major impact on biomedical research, diagnosis and therapy. When combined with the expression and scale-up methods alluded to above, the potential power of these new methods is astonishing. Thus, it seems only a matter of time before MAbs assume the important place in human therapy which was predicted for them two decades ago.

This book is organized in the following way: Chapters 1 and 2 deal with the creation of human MAbs both by "conventional" cell fusion and the more recently developed polymerase chain reaction (PCR) cloning strategies. Here, the important innovation of mixed primer PCR amplification to clone immunoglobulin genes is described. Chapter 3 introduces the fascinating use of transgenic mice carrying human immunoglobulin genes to make human MAbs. Chapter 4 reports on engineering as opposed to biological approaches to making human MAbs, i.e., genetic adaptation of murine MAbs so as to be immunocompatible in man. Chapter 5 describes genetic fusion of MAbs to create dual specificity and the expression of such constructs in *E. coli*. Chapters 6–8 continue the theme of fusion of genes encoding MAbs (or FAbs) with each other and with non-immunoglobulin genes to create novel therapeutic molecules. Chapter 9 introduces the subject of antibodies as chemical catalysts. Chapter 10 describes the exciting evolution of combinatorial strategies to make MAbs without cell fusion. Chapters 11 and 12 focus on the issue of expression of recombinant MAbs while Chapter 13 deals with industrial scale-up of MAb production. Finally, Chapters 14–16 turn to clinical application of MAbs for imaging, therapy, and prophylaxis of disease.

This volume represents only a set of selected examples from this rapidly evolving field. Nevertheless, the distinguished authors who have contributed to this collection provide important insight into a field of great interest for both basic and applied scientists and clinicians. The potential of the technologies described here is only beginning to be tapped, and another volume five years hence would undoubtedly provide many surprises.

The editors would like to express their gratitude to Springer-Verlag for supporting this venture, and especially to Ms. Doris Walker for her manifold assistance. Most particularly, we thank Ms. Sandi Brown for her excellent editorial and administrative assistance. Finally, we are indebted to the authors for their timely and careful additions to the literature of this important subject, which have made this volume a unique contribution to the field.

Pennsylvania, USA   MARTIN ROSENBERG
April 1994   GORDON P. MOORE

Contents

CHAPTER 3

Transgenic Approaches to Human Monoclonal Antibodies
N. LONBERG. With 11 Figures   49

Section II: Genetically Engineered Monoclonal Antibodies

CHAPTER 4

Humanization of Monoclonal Antibodies

CHAPTER 5

**Applications for *Escherichia coli*-Derived Humanized Fab′
Fragments: Efficient Construction of Bispecific Antibodies**
P. CARTER, M.L. RODRIGUES, and M.R. SHALABY. With 3 Figures 135

Section III: MAb Conjugates and Fusions

CHAPTER 6

Immunotoxins
S.H. PINCUS. With 2 Figures 149

CHAPTER 10

The Combinatorial Approach to Human Antibodies

CHAPTER 11

Antibodies from *Escherichia coli*

Section VI: Medical Applications

CHAPTER 16

Anti-idiotypic Monoclonal Antibodies:
Novel Approach to Immunotherapy
M. CHATTERJEE, K.A. FOON, and H. KOHLER. With 1 Figure . . . .

Section I: Human Monoclonal Antibodies

CHAPTER 1

Human Monoclonal Antibody Technology

K. JAMES

A. Introduction

I. Why Produce Human Monoclonals?

The success of KÖHLER and MILSTEIN (1975) in immortalizing specific antibody secreting cell lines of rodent origin inevitably led to similar attempts to generate human monoclonals. Initially this effort was largely driven by the belief that such products would be clinically (especially therapeutically) superior to their rodent counterparts. In addition to being less immunogenic than rodent antibodies it was also felt that they might recognize more appropriate target antigens and be more efficient at triggering complement and antibody dependent cellular cytotoxic actions in vivo as indeed has proved to be the case. Further impetus was given by the difficulties experienced in producing useful rodent monoclonals against a number of targets of clinical interest including RhD, histocompatability and tumour antigens. This was attributable to the failure of the rodent immune system to recognize the relevant epitopes against a background of immunodominant epitopes. However additional interest in human monoclonal antibody technology has come from an increasing awareness of its potential for investigating the human humoral immune system (e.g., JAMES 1989). As a result it is now being used to probe the genetic basis of antibody diversity, analyze B cell repertoires in health and disease, study antibody responses in microenvironments such as tumours and sites of infection, and identify epitopes on microorganisms which provide both protective and damaging immune responses.

II. Chapter Aims

The aims of this brief introductory chapter are to summarize: (1) the general strategies used in the production of conventional human monoclonal antibodies (2) the range of specificities produced and their application and (3) the limitations of these conventional approaches and how they might be improved.

III. Approach Used

In view of the wide variety of procedures used to date to produce human monoclonal antibodies, general principles will be discussed rather than precise technical details. Such information is to be found in a number of books and reviews dealing in general with this topic (Engelman et al. 1985; Carson and Freimark 1986; James and Bell 1987; Boyd and James 1989; James 1990a) or the development and application of human monoclonal antibodies to specific targets such as microorganisms (Masuho 1988; Boyd and James 1992), blood cells (McCann et al. 1988; Pistillo et al. 1988) and tumours (James 1990b).

B. General Production Strategies

I. Introduction

The basic conventional strategies used to produce human monoclonal antibodies are schematically presented in Fig. 1. As with rodent monoclonal antibody generation the success of the procedure is determined by a number of key factors namely an adequate supply of immune lymphocytes in an appropriate state of differentiation and proliferation and efficient methods for immortalizing, screening and cloning cultures of interest. Unfortunately difficulties may arise with respect to all of these when attempting to produce human monoclonals.

II. Source of Immune Lymphocytes

1. Usual Sources

Generally speaking this does not pose a major problem when producing rodent monoclonal antibodies, for fairly standard immunization schedules can be employed which generate an adequate immune response. Furthermore there is no restriction on the choice of lymphoid tissues. In the human situation however deliberate immunization is restricted to approved immunogens (e.g., bacterial and viral vaccines), adjuvants and routes of injection. Fortunately there are other valuable sources of immune lymphocytes including individuals recovering from infection and patients with active diseases such as cancer and autoimmune disorders. It should also be stressed that many interesting monoclonals, especially against autoantigens, have been generated from lymphocytes obtained from apparently healthy normal individuals (Table 1).

For obvious reasons peripheral blood has been the most widely used source of human lymphocytes for immortalization. Unfortunately comparative studies undertaken with both human and rodent lymphoid tissues reveal

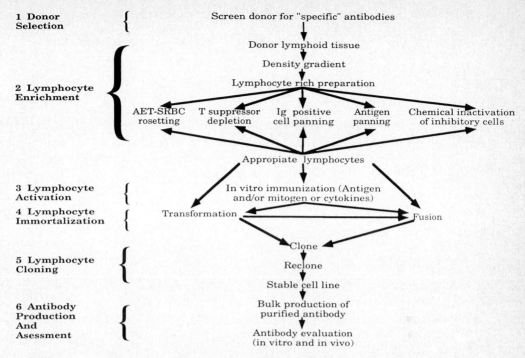

Fig. 1. The key steps involved in the production of conventional human monoclonal antibodies. The initial step is the acquisition of immune lymphocytes. These are generally derived from peripheral blood but on occasion lymph nodes and spleen may be available. The B lymphocyte preparation may be enriched and or stimulated with antigen prior to immortalization. The immortalization is normally achieved by cell fusion, Epstein-Barr virus (EBV) transformation or a combination of these procedures. Dilution cloning or cloning in soft agar is used to isolate stable cell lines. Antibodies of interest are normally produced by bulk culture

Table 1. Sources of immune lymphocytes used for human monoclonal antibody production

Immunization procedure	Examples
Active immunization	Bacterial and viral vaccines
Natural infection	Bacteria and viruses
Active disease	Cancer and autoimmune disease
Inadvertent exposure	Red cell antigens and haptens
No obvious exposure	Autoantigens
Active immunization of human B cell repopulated SCID mice	Tetanus toxoid
In vitro sensitization	Bacterial and viral antigens haptens

For further details see JAMES and BELL 1987; BOYD and JAMES 1989.

that lymphocytes from this source do not perform as well as those derived from spleen and lymph nodes. This inferior performance has been attributed to a number of factors paramount among these being insufficient antigen specific B cells in an appropriate state of differentiation. It is for these and other reasons that investigators attempting to generate antitumour monoclonals have used lymphocytes isolated from tumours themselves or draining lymph nodes.

The above difficulties have led to alternative strategies for obtaining immune lymphocytes. These include in vitro immunization and the immunization of repopulated severe combined immunodeficiency (SCID) mice. Interest is now also being shown in the use of transgenics.

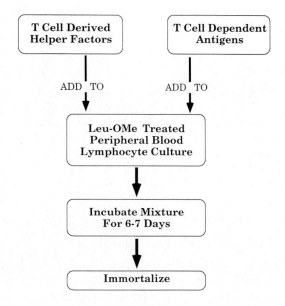

GENERATING HUMAN MONOCLONALS BY
IN VITRO IMMUNIZATION

Fig. 2. Basic steps involved in the in vitro immunization of human peripheral blood lymphocytes. The cell culture contains Leu-OMe treated lymphocytes, the immunizing antigen and T cell derived helper factors. The treatment with Leu-OMe kills or impairs a number of cells which could interfere in the activation, differentiation and proliferation of B cells. These include monocytes, natural killer cells and cytotoxic T lymphocytes. The antigen may be in free form or coupled to a suitable protein carrier and or adjuvant, e.g., lipopeptide, to improve presentation. The T cell derived helper factors may be provided in the form of recombinant derived cytokines, e.g., IL-2 or IFN-γ, or T cell culture supernatants. Specific antibody secreting cell cultures may be immortalized by cell fusion or Epstein Barr virus transformation. Alternatively the RNA may be extracted and antibodies prepared by recombinant techniques

2. In Vitro Immunization

The basic strategy utilized in in vitro immunization has been reviewed elsewhere (BORREBAECK 1988a,b) and is outlined in Fig. 2. It is now apparent that a number of factors determine the outcome of this approach including the source of lymphoid tissue, the removal or inactivation of suppressor and cytotoxic cells, the correct presentation of antigen and the provision of various growth factors. Although this approach has been used to produce human monoclonals, there are still inherent difficulties which need to be resolved. Unfortunately antibodies secreted following the in vitro immunization of truly naive B cells are almost exclusively of the IgM isotype. As a result effective techniques for inducing class switching are urgently needed. On the basis of recent preliminary reports it would appear that this might be achieved by repeated antigen challenge of lymphocyte culture coupled with the addition of fresh feeder cells (BRAMS et al. 1992).

3. Generating Immune Lymphocytes in Severe Combined Immunodeficiency Mice

An approach currently attracting attention is the generation of immune human B cells in SCID mice. These mice lack mature T and B lymphocytes because of defective joining of both T cell receptor and immunoglobulin gene segments. As a consequence they accept grafts of incompatible tissue including human lymphocytes.

To date human antibody responses to a number of antigens, including RhD and tetanus toxoid, have been generated by challenging SCID mice repopulated with human peripheral blood lymphocytes. It would appear that such responses are dependent not only upon the transfer of immune cells but also the dose of antigen injected and are influenced by the presence of other immunocompetent cells in the inoculum. Thus adherent cells may enhance the response while natural killer (NK) cells are believed to be deleterious (LEADER et al. 1992; CARLSSON et al. 1992). Nevertheless human lymphocytes recovered from immunized repopulated SCID mice have been successfully used to generate human monoclonal antibodies and antibody fragments by conventional Epstein-Barr virus (EBV) transformation procedures (CARLSSON et al. 1992) and repertoire cloning techniques respectively (DUCHOSAL et al. 1992).

While this approach is attractive it still remains to be optimized for there are difficulties both with respect to the successful engraftment of transplanted cells and the nature of the antibody response induced. It would appear for example that the generation of primary responses is difficult and that there may also be a skewing of any responses which are evoked particularly with respect to light chain usage. Explanations have been advanced to explain these difficulties including antigen driven proliferation of xenogeneic clones (WILLIAMS et al. 1992) and the aforementioned inhibitory effects of NK cells (CARLSSON et al. 1992).

Table 2. Processing of lymphocytes following density gradient separation

---

T cell removal or impairment
    Rosetting with AET (2 aminoethylisothiouronium bromide hydrobromide) treated
        sheep red blood cells
    Rosetting with anti-T cell[a] coated red blood cells or other particles, e.g., magnetic
    Panning on anti-T cell coated plates
    Lysis with anti-T cell antibody and complement
    Inactivation with cyclosporin or lysosomotrophic agents
Removal or inactivation of other interfering cells[b,c]
    Adherence on plastic, e.g., macrophages
    Inactivation/killing with lysosomotrophic agents, e.g., monocytes, macrophages,
        natural killer cells or cytotoxic T cells
Enrichment of B cells
    Rosetting with anti-immunoglobulin or anti-B cell coated red blood cells or other
        particles
    Panning on anti-immunoglobulin or anti-B cell coated plates
Enrichment of specific B cells
    Rosetting with antigen coated red cells or other particles
    Panning on antigen coated plates
    Incubation with fluorescein coated antigen followed by fluorescence activated cell
        sorting

---

[a] Antibody based separations usually employ monoclonal antibodies of rodent origin.
[b] Removal of interfering cells, e.g., NK cells, Tc cells, is particularly important if in
vitro immunization is being attempted.
[c] On occasions isolated lymphocyte subpopulations are reconstituted to "optimize"
cell mixtures for in vitro immunization.

III. Processing of Lymphoid Tissues

A variety of procedures have been employed to isolate from lymphoid
tissues lymphocytes suitable for immortalization and in vitro immunization.
Lymphocytes are routinely recovered from peripheral blood by routine
density gradient procedures on Ficoll-Hypaque or other such media. Additional processing may also be performed either to enrich B cells (antigen
specific or otherwise) or to remove immunocompetent cells which could
interfere in subsequent in vitro immunization (e.g., lysozyme rich leukocytes)
or EBV immortalization strategies (e.g., cytotoxic T cells). The general
strategies employed to achieve these ends are listed in Table 2.

IV. Immortalization Strategies

1. Introduction

Currently there are three main procedures for immortalizing human B
lymphocytes obtained from the above sources. These involve either fusion
with B cell lines, transformation with EBV or a combination of these
techniques. Some key aspects of these three approaches are detailed below.

2. Cell Fusion

The success of polyethylene-glycol (PEG) induced fusion in the development of rodent monoclonal antibodies has ensured that this approach has been vigorously pursued. However in the human situation progress has been hindered by the lack of suitable human fusion partners. Although a large number of fusion partners have been employed few, if any, of them perform as well as their murine counterparts. The human fusion partners used include plasmacytomas, lymphoblastoid cells and heteromyelomas produced by fusing mouse myelomas with human B cells (Table 3). These hybrid cell lines, which are rapidly gaining in popularity, appear to exhibit the growth and fusion characteristics of the original mouse myeloma while still retaining sufficient human chromosomes to ensure that stable antibody secreting cell lines are generated following fusion with immune human B cells. It is also interesting to note that contrary to expectations a large number of stable antibody secreting cell lines have also been produced by fusing human lymphocytes with mouse myeloma cells.

While the majority of fusions have been performed using PEG, the low efficiency of this process, coupled with its undoubted toxicity, has led to interest in alternate methods of fusion. In this regard particular attention has been devoted to the electrofusion procedure originally developed by Zimmerman and coworkers (ZIMMERMAN and VIENKEN 1982; FOUNG and PERKINS 1989; FOUNG et al. 1990). This process is based on the finding that intimate membrane contact between cells can be established in the presence of an alternating electric field. Subsequent application of a direct current

Table 3. Some fusion partners which have been used in human monoclonal antibody production

Cell type	Examples
Human plasmacytomas	
Myelomas	SKO-007, HFB-1, KMM-1, RPMI-17886TG, U-2030
Lymphomas	RH-L4, NA7-30, Ball-1
Lymphoblastoid cell lines	GM1500-6TG-A12, KR-4, GM672
	GK-5, GMO467.3, GM1500-67G-OB
	H.35.1.1, LICR-LON-HMy2, UC729.6
	WI-L2-729-HF2, WI-L2.727 HF26 TG
	HO-323, TAW-925, HOA 1, LTR228
	M C/MNS-2
Heteromyelomas	
Human/human	KR12, LSM2.7, HM2.0
Mouse/human	SHM-D23, SHM-D33, SBC-H20, Org MHH-1, SP2/SP,
	CB-F7, F3B6, K6H6B, HAB/l, SPA2-4
Mouse/human/human	3HLS, SP2/PT
Rodent	NSO/U, NS1/1. Ag41, SP2/0
	P 3X63 Ag8.653

For further details see JAMES and BELL 1987; BOYD and JAMES 1989.

leads to a disruption of the plasma membrane, creating a cytoplasmic bridge between the cells which promotes cell fusion. The relatively high fusion efficiencies achieved with this approach would seem to make it the method of choice when limited number of B cells are available such as might be the case with respect to intratumour lymphocytes or those isolated from an arthritic joint. In this connection it is interesting to note that the electrofusion approach has recently been employed to generate specific antibody secreting cell lines from single B cells. This was achieved by fusing B cells which had been clonally expanded in vitro in the presence of human T cell supernatant and irradiated murine thymoma helper cells (STEENBAKKERS et al. 1992).

3. Epstein-Barr Virus Fusion

This immortalization strategy depends on the ability of EBV to bind to and transform human B lymphocytes. Following interaction of the virus with C3d receptors on the B cell surface, it is internalized and its genes eventually integrated into the lymphocyte genome. This event leads in turn to the activation of many of the B cells so treated with the subsequent immortalization of a subpopulation. This process is normally achieved by incubating T cell depleted peripheral blood lymphocytes with the culture supernatant from an EBV infected marmoset cell line (designated B95–8). Occasionally however transformation may also occur if the B cells are incubated with EBV infected lymphocytes themselves either intentionally or otherwise. This process is known as cell driven transformation and is reputed to be more efficient than the conventional approach.

Although the EBV immortalization procedure has been used to generate an array of specific antibody secreting cell lines, there are major problems with this approach. In most hands the lines are particularly difficult to clone and they are relatively unstable. The possible reasons for this will be discussed later.

4. Combined Epstein-Barr Virus Transformation and Cell Fusion

An increasingly popular approach to immortalization of human B cells is to transform them with EBV virus and then fuse them several days later, preferably with an heteromyeloma cell line. This combined approach, originally developed by KOZBER and colleagues (1982), appears to offer a number of advantages over either procedure alone. In the first place EBV transformed lines fuse with greater efficiency than normal B cells, while the hybrids clone more readily than transformants. Furthermore antibody secretion may also be improved. In addition antigen specific cells can be selected either before or after virus transformation thus increasing the chances of obtaining specific hybrids. Finally it should be noted that fusion with murine or murine/human partners may lead to the loss of the EBV genome, an added advantage when the antibody is required for clinical use.

5. Novel Approaches

Recently a novel approach for establishing "long-term" human B cell lines free of EBV has been described which could facilitate human monoclonal antibody production in a number of ways (BANCHEREAU et al. 1991). This involves inducing B cells to proliferate by cross-linking the CD40 cell surface antigen. The cross-linking has been achieved by incubating the B cells in anti-CD40 antibody in the presence of irradiated mouse fibroblasts transfected with the human Fc RII. The response is further amplified by the addition of recombinant interleukin-4 (IL-4). This procedure should be of value both in the establishment of antigen specific B cell lines in vitro and in expanding immune B cells for use in conventional cell immortalization or gene cloning strategies.

V. Selection, Cloning and Expansion

1. Introduction

In general the procedures used for selecting, cloning and expanding human monoclonal antibody secreting cell lines are similar to those used in rodent monoclonal antibody production. There are however added difficulties.

2. Selection

Obviously when fusion has been employed the hybrid cells have to be selected by growth in the appropriate selection medium (usually HAT). Rapid primary screening assays are also required to identify those cultures secreting specific antibody, or where this is not feasible, immunoglobulin. In this connection ELISA procedures using microtitre coated plates prove most suitable for many antigens. Wherever possible the primary, and certainly the secondary, screening procedures should be appropriate to the intended use of the antibody. For example if cytotoxic antibodies are required then a lytic assay should be used. In certain instances it may be necessary to perform primary screening on tissue sections as for example is often the case with antitumour antibodies. Under these circumstances endogenous tissue associated immunoglobulins may give rise to high background staining in indirect immunohistochemical procedures. It would appear that problems of this kind can be minimized by the choice of appropriate fixation procedures and by blocking endogenous immunoglobulin with Fab anti-human immunoglobulin (DITZEL et al. 1991).

3. Cloning

Having identified cultures secreting antibodies of interest the next major step is to obtain stable antibody secreting cell lines. Unfortunately it is at this stage that many interesting cultures are lost. At the present time the major cloning strategy is dilution cloning although cloning in soft agar has been applied. A wide variety of feeder cells and media supplements have been used to facilitate cloning, this undoubtedly reflecting the difficulties experienced. They include a range of normal human and murine lymphoid tissues, lymphoblastoid cell lines, endothelial cells and tissues of embryonic origin including fibroblasts, kidney, lung and amnion (see JAMES and BELL 1987). Unfortunately lymphoblastoid cell lines are notoriously difficult to clone, even with feeder layers and growth factor supplements, growth at less than ten cells per well being difficult to achieve. There are however reports of such cells being successfully cloned in soft agar in the presence of human fibroblasts (KOSINSKI and HAMMERLING 1986).

4. Expansion

A variety of in vitro and in vivo procedures have been used to expand selected antibody secreting cell lines. They include growth in ascitic form in nude mice and rats and SCID mice and large scale tissue culture in airlift fermentors and immobilized cell systems. It should be pointed out that while heterohybrids readily grow in nude rodents, growth of hybrid and lymphoblastoid cell lines is difficult. This is presumably due to the fact that human mouse hybrids do not express the HLA antigens which might evoke immune rejection.

5. Additional Evaluation

This may be performed prior to or after limited expansion depending on the amounts of antibody required. In brief it includes isotyping, further analysis of the fine specificity of the antibody and an evaluation of its clinical potential. For example it is vital to screen antitumour monoclonal antibodies against a wide range of normal and malignant targets including frozen tissue sections and established cell lines. Similarly it is also equally important to establish if antimicrobial antibodies are strain as well as species specific.

There are a number of established procedures for ascertaining the therapeutic and prophylactic potential of antibodies of clinical interest. Complement and antibody dependent cellular cytotoxicity assays can be used to determine the capacity of antibodies to lyse bacteria, viruses and tumours. Furthermore there are animal models for a variety of infectious diseases which permit the evaluation of the protective effect of monoclonal antibodies while the potency of antitumour monoclonals can be assessed by determining their effect on the growth of tumour explants in nude rodents or SCID mice. In the future one envisages that the potential of certain antibodies will be assessed in both transgenic and gene knockout animal models of disease.

VI. Bispecific and Trispecific Antibodies

With the exception of IgM and dimeric IgA all immunoglobulin molecules (monoclonal or otherwise) possess two identical antigen combining sites both of which interact with identical epitopes on the target antigen. During recent years however a range of so-called bispecific rodent monoclonal antibodies have been produced which are capable of interacting with two distinct epitopes usually on separate molecules. More recently trispecific monoclonals have been described which bind three distinct epitopes (JUNG et al. 1991; TUTT et al. 1991). The rodent bispecific antibodies have been produced either by fusing two established monoclonal cell lines secreting antibodies of the desired specificity or by chemically linking intact purified monoclonal antibodies or antibody fragments. In view of their potential advantages in targeting both drugs and cells to sites of infection, tumours, etc., it is not surprising that attempts are now being made to produce bispecific human monoclonals (SHINMOTO et al. 1991). In this case the cell fusion approach was used.

At this point it is also worth noting that the chemical linking procedures employed in bispecific antibody production are also being used to link human monoclonals of identical specificity. The so-called human monoclonal antibody homodimers generated by this approach exhibit better avidity, binding and protective capacity than the original monomeric IgG (WOLFF et al. 1992). This has led the authors to suggest that chemical cross-linking of monoclonal antibodies may be a useful strategy for salvaging low affinity IgG monoclonal antibodies that exhibit poor functional activity.

C. Human Monoclonal Targets

I. Antibody Specificities Generated

Despite the difficulties encountered in producing human monoclonals, past successes and the undoubted potential of the technology have combined to ensure that effort in this area has been maintained. This has resulted in the generation of a large number of monoclonal antibodies reacting with a wide range of targets (Table 4). As might be expected many of these are directed against bacterial, viral, red cell and tumour antigens, being developed with diagnosis and therapy in mind. However of increasing theoretical interest at least are those generated against autoantigens (see below).

II. Application of Human Monoclonal Antibody Technology

1. Introduction

Although some of the monoclonals are exhibiting therapeutic potential (see for example ZIEGLER et al. 1991) and others are in phase 1 clinical trials

Table 4. A survey of human monoclonal antibodies which have been produced[a]

Antibodies to	Examples
Bacterial antigens	Cholera toxin, *Clostridium chauvoei*, diphtheria toxin, Gm-ve endotoxin, exotoxin A, *Haemophilus influenzae*, *mycobacterium leprae*, *Neisseria meningitides*, pneumococcus polysaccharide, *Pseudomonas aeruginosa*, tetanus toxoid, *Chlamydia*
Viral antigens	Cytomegalovirus, Epstein-Barr virus, hepatitis A, Bs, Bc and D, herpes simplex, HIV-1, HTLV-1, influenza A, rabies, rubella, T cell leukaemia, varicella zoster
Other infectious agents	*Plasmodium falciparum*
Red and white cell antigens	Blood groups A, i, Kell, rhesus C, D, E and G; Forsmann antigen, chicken and sheep red blood cells; HLA-B and DQ
Tumour antigens	Bladder carcinoma, brain glioma, breast carcinoma, cervical carcinoma, chronic lymphocytic leukaemia, colorectal carcinoma, gastric carcinoma, lung carcinoma, lymphoma, melanoma, myeloid leukaemia, prostatic carcinoma, renal carcinoma, vulva carcinoma
Autoantigens	Astrocytes, cytoskeletal proteins, single- and double-stranded DNA, endothelial cells, Golgi, myelin associated glycoprotein, nuclear and nucleolar antigens, nerve axons, neurones, pancreatic islet cells, pancreatic duct cells, parathyroid, pituitary, platelets, prostatic acid phosphatase, sperm, sperm coating protein, stratified squamous epithelium, thyroglobulin, thyroid follicle, vimentin, rheumatoid factor
Others	Bombesin, 2–4 dinitrochlorobenzene, 2–4 dinitrophenol, 4-hydroxy 3–5 dinitrophenacetic acid, KLH, phosphorylcholine, sperm whale myoglobin

[a] It should be stressed that many of these antibodies have not been adequately characterised and the cell lines producing them are frequently unstable.

(e.g., AULITSKY et al. 1991; DROBYSKI et al. 1991) it can be said that the great breakthrough in this area still remains to be realized. Nevertheless the benefits to date of the technology should not be underestimated. In addition to improving our knowledge of humoral immunity in general it has enabled the study of such responses in distinct microenvironments, facilitated the identification of epitopes evoking protective immunity and permitted the development of new procedures for isolating and assaying tumour and viral antigens. The range of applications of human monoclonal antibodies are summarized in Table 5 while their use in specific clinical situations is outlined in greater detail below.

2. Tumour Field

In the tumour field the immense effort expended in attempting to generate human monoclonal antibodies has not gone entirely unrewarded (JAMES et

Table 5. Possible uses of human monoclonal antibodies

Diagnostic and monitoring	Infectious diseases; malignancy; red cell compatability; tissue typing
Prophylaxis and therapy	
Infectious diseases and malignancy	Passive therapy; identification and purification of candidate antigens for vaccines; as anti-idiotype vaccines; targeting drugs; imaging lesions
Other applications	Autoimmune diseases; renal transplantation; rhesus incompatability; contraception
Investigating the immune system in health and disease	Studying normal B cell repertoire especially in microenvironments; studying the molecular biology of the immunoglobulin locus; analysing B cell response in infection, malignancy, autoimmuinity and allergy; identifying, purifying and characterising antigens of clinical relevance (see above)

For further details of application see JAMES and BELL 1987; MASUHO 1988; BOYD and JAMES 1989; JAMES 1990a,b.

al. 1990). It has provided us with a refined approach to identifying putative tumour antigens and for studying the humoral response to tumours especially in microenvironments such as the tumour itself or in draining lymphnodes. Furthermore the antibodies developed to date have been used as probes for detecting preneoplastic changes, studying differentiation and dedifferentiation in tumours, assaying tumour antigens in sera and urine and for immunoscintigraphy.

3. AIDS Research

More recently the value of human monoclonal antibodies in the study and possible treatment of AIDS has been the subject of increasing interest. The potential applications in this area are summarized in Table 6 and BOYD and JAMES 1992. Perhaps paramount among these is its use as originally advocated (FOLKS 1987) in epitope mapping especially in identifying those structures which elicit protective immunity. On the basis of data accumulated on existing human monoclonals it would appear that the neutralizing anti-

Table 6. Some applications of human monoclonals to HIV

Studying the immune response to HIV
Epitope mapping (conserved, variant, protective, enhancing)
Immunoaffinity purification of HIV proteins
As reagents in competition immunoassays
Targeting of toxins to HIV infected cells
As therapeutics per se
Screening viral isolates

For further details see BOYD and JAMES 1992.

bodies are almost exclusively directed against the gp120 envelope protein. In contrast almost all the enhancing antibodies react with gp41. It should be stressed however that the bulk of anti-gp41 monoclonals neither neutralize or enhance. Additional studies on the gp120 monoclonals exhibiting neutralization indicate that the epitopes eliciting neutralization are in amino acid positions 306–322 or are associated with conformational or carbohydrate epitopes.

4. Autoimmunity

This technology has also contributed significantly to our knowledge of the nature, origins and significance of autoantibodies and will undoubtedly continue to do so in the future (summarised in JAMES 1989). Of particular importance is the information provided on the specificity and variable region gene usage in so-called natural autoantibodies (CAIRNS et al. 1989; SANZ et al. 1989; THOMPSON et al. 1990). More recently it has been employed to investigate the VH repertoire of B cells derived from rheumatoid synovial fluid (BROWN et al. 1992) and to show that human monoclonals against blood group antigens demonstrate the multispecificity characteristics of natural autoantibodies (THOMPSON et al. 1992).

5. Future Targets

A number of recent observations pinpoint another area in which the application of human monoclonal antibody technology could be particularly rewarding. In the first place it appears that circulating antibodies to a number of cytokines and heat shock proteins may be found in normal individuals or in those recovering from infection or suffering autoimmune reactions. Furthermore there are initial reports of the generation of human monoclonals to tumour necrosis factor (TNF) (FRASA et al. 1992). It is obvious that further endeavours in this area should lead to a clearer understanding of the role of cytokines in health and disease and to the development of therapeutically valuable reagents.

D. Limitations of Orthodox Technology

I. Introduction

Reference has already been made to some of the problems associated with the production of human monoclonal antibodies. These include difficulties in obtaining adequate supplies of immune lymphocytes in an appropriate state of differentiation and the poor efficiency of the immortalization procedure itself. Some of the approaches adopted to overcome these major problems have been addressed elsewhere in this chapter.

II. Why Is Antibody Secretion Unsatisfactory?

The aforementioned problems are further compounded by the relatively poor stability and antibody secreting capacity of immortalized human B cells compared with their murine counterparts. A number of interesting hypotheses have been advanced to help explain this unsatisfactory performance and are listed in Table 7. In brief they include overgrowth by nonsecreting lymphocytes, loss of genes coding for the varous immunoglobulin chains, defects in regulatory gene function and imperfections in the synthetic and secretory machinery of the cells themselves.

III. Is Unsatisfactory Secretion Related to Cell Surface Phenotype or Cytokine Secretion?

Unfortunately to date insufficient attention has been devoted to establishing the real causes for the poor performance of such cells and to identifying strategies which might resolve these problems. As we believe that studies in this area are both practically and theoretically important we have closely examined the relationship between cell surface phenotype, cytokine presence and the growth and antibody secretory capacity of a panel of immortalized human B cell lines (JAMES et al. 1990). All the cells were found to exhibit a well differentiated cell surface phenotype irrespective of their growth rates and antibody secreting capacity. Furthermore there was no obvious relationship between the cytokines they produced and antibody secretion. Finally the addition of a wide range of concentrations of IL-2, IL-4 and IL-6 rarely improved antibody secretion. Overall our results seem to suggest that the poor performance of the cell lines we examined was not attributable to defects in cell differentiation (as revealed by cell surface phenotype studies) or in the production of the cytokines we investigated. While these observations proved disappointing we believe that further studies in this area are

Table 7. Explanations advanced for the poor performance of immortalized human B cell lines

---

Inhibitory effects of mycoplasma and other microorganisms
Overgrowth of secreting cells by nonsecretors
Shortage of relevant growth or differentiation factors
Reduction or absence of appropriate receptors for growth or differentiation factors
Loss or inappropriate incorporation of structural genes for immunoglobulin H and
    L chains
Insufficient copies of regulatory and structural genes
Inappropriate H and L chain combinations resulting in hybrid molecules
Failure or loss of relevant regulatory genes or other defects in regulation
Defects in synthetic machinery resulting in impaired transcription, translation or
    assembly
Defects in the secretory machinery of the cell

---

still warranted. These should involve comparable investigations with other cytokines and cytokine mixtures and more detailed phenotypic studies.

E. Conclusion

I. Impact and Potential of Recombinant Technology

1. Chimeric and Humanized Antibodies

There are obviously a number of strategies for improving human monoclonal antibody production and these are summarized in Table 8. Inevitably the difficulties experienced have led to an ever increasing interest in the application of recombinant DNA techniques to the development of therapeutically useful monoclonal antibodies (reviewed by LARRICK and FRY 1991; WINTER and MILSTEIN 1991). Initially this centred on the production of

Table 8. Strategies for improving human monoclonal antibody production

---

Improved immunization procedures
    In vivo, e.g., better immunogens and adjuvants, optimizing bleed times
    SCID mice, e.g., improving take of xenografts, optimizing immunization schedules
    In vitro, e.g., optimizing antigen dose, improving presentation, removal of
        suppressor/cytoxic cells, use of cytokines
Improved selection of B cells
    Use of lymphoid tissues other than blood
    Enrichment of specific B cells
    Selection of cells of appropriate differentiation and proliferation status
Improved immortalization procedures
    Optimisation of existing technologies
    Development of better fusion partners
    Use of more efficient fusion procedures, e.g., electrofusion
    Use of alternative immortalization strategies, e.g., transfection
Improved cloning procedures
    Use of alternative cloning techniques, e.g., micromanipulation, fluorescence
        activated cell sorting
    Use of recombinant growth factors, e.g., IL-2, IL-6
Elucidating factors affecting proliferation and differentiation of B cells in B cell lines
    Expression of receptors for cytokines
    Effects of exogenous cytokines
Studying factors regulating immunoglobulin synthesis and secretion
    Expression of immunoglobulin
    Influence of regulatory genes
    Secretory mechanisms
Application of recombinant techniques
    Cloning and expressing human antibody genes
    Cloning and expressing genes for chimeric (rodent/human) and antibody-
        nonantibody hybrids
    Production of site-directed antibody mutants
    Transfecting cell lines with genes for appropriate growth factors, etc.

---

chimeric antibodies generated by linking the genes coding for the entire variable region of rodent monoclonal antibodies to those coding for human constant region domains. This strategy has permitted the production of therapeutic antibodies of reduced immunogenicity and improved Fc mediated effector functions. Further improvements have resulted from the insertion of gene sequences coding for the hypervariable region of rodent monoclonal antibodies onto an entirely human framework. These strategies will be discussed elsewhere in this volume.

2. Repertoire Cloning

More recently techniques have been developed which enable the direct cloning and expression of human monoclonal antibody fragments (BURTON 1991; WINTER and MILSTEIN 1991). These procedures permit the rapid isolation and expression of specific human antibody genes from small numbers of lymphocytes. As will be seen later these procedures have been made feasible by a number of technical developments. These include the polymerase chain reaction, which permits rapid amplification of nucleic acids; expression systems, which enable the display of functional antibody fragments on the surface of filamentous phage (CHISWELL and McCAFFERTY 1992) and rapid screening techniques.

3. Combining Cell and Gene Cloning Technologies

While these various recombinant approaches will undoubtedly be more widely used in the future it could be argued that the successful development of clinically useful human monoclonal antibodies will be most readily achieved by a combination of cell and gene cloning strategies. This combined approach has already been used to rescue specific antibody genes (LEWIS et al. 1992), enhance their expression (GILLIES et al. 1989; NAKATANI et al. 1989) and to effect an isotype switch (KAMORI et al. 1988). It could obviously be profitably extended to existing human cell lines secreting antibody of proven specificity and to others which might be developed in the future.

References

Aulitsky WE, Schulz TF, Tilg H, Niederwieser D, Larcher K, Östberg L, Scriba M, Martindale J, Stern AC, Grass P, Mach M, Dierich MP, Huber C (1991) Human monoclonal antibodies neutralizing cytomegalovirus (CMV) for prophylaxis of CMV disease: report of a phase 1 trial in bone marrow transplant recipients. J Infect Dis 163:1344–1347
Banchereau J, De Paoli P, Vallé A, Garcia E, Rousset F (1991) Long term human B cell lines dependent on interleukin-4 and antibody to CD40. Science 25:70–72
Borrebaeck CAK (1988a) Critical appraisal of the in vitro immunization technology for the production of mouse and human monoclonal antibodies. Adv Drug Rev 2:143–165
Borrebaeck CAK (ed) (1988b) In vitro immunization in hybridoma technology. Prog Biotechnol 5

Brams P, Royston I, Boerner P (1992) Induction of secondary responses following induction of primary responses on human spleen cells by in vitro immunisation. Presented at the 2nd international conference on human antibodies and hybridomas. Cambridge, England, March 1992

Boyd JE, James K (1989) Human monoclonal antibodies: their potential, problems and prospects. Adv Biotechnol Proc 11:1–43

Boyd JE, James K (1992) B cell responses to HIV and the development of human monoclonal antibodies. Clin Exp Immunol 88:189–202

Brown CMS, Longhurst C, Haynes C, Plater-Zyberk C, Malcolm A, Maini RN (1992) Immunoglobulin heavy chain variable region gene utilization by B cell hybridomas derived from rheumatoid synovial tissue. Clin Exp Immunol 89:230–238

Burton DR (1991) Human and mouse monoclonal antibodies by repertoire cloning. TIBTECH 9:169–175

Cairns E, Kwong PC, Misener V, Ip P, Bell DA, Siminovitch KA (1989) Analysis of variable region genes encoding a human anti-DNA antibody of normal origin. Implications for the molecular basis of human autoimmune responses. J Immunol 143:685–691

Carlsson R, Mårtensson C, Kalliomäki S, Ohlin M, Borrebaeck CAK (1992) Human peripheral blood lymphocytes transplanted into SCID mice constitute an in vivo culture system exhibiting several parameters found in a normal immune response and are a source of immunocytes for the production of human monoclonal antibodies. J Immunol 148:1065–1071

Carson DE, Freimark BD (1986) Human lymphocyte hybridomas and monoclonal antibodies. Adv Immunol 38:275–311

Chiswell DJ, McCafferty J (1992) Phage antibodies: will new "coliclonal" antibodies replace monoclonal antibodies. TIBTECH 10:80–84

Ditzel H, Erb K, Borup-Christensen P, Neilsen B, Jensenius JC (1991) Evaluation of procedures for the fixation and processing of human tissue for immunohistochemical analysis of human monoclonal antibodies. Human Antibod Hybrid 2:135–141

Drobyski WR, Gottlieb M, Carrigan D, Ostberg L, Grebenau M, Schran H, Magid P, Ehrlich P, Nadler PI, Ash RC (1991) Phase 1 study of safety and pharmacokinetics of a human anticytomegalovirus monoclonal antibody in allogeneic bone marrow transplant recipients. Transplantation 51:1190–1196

Duchosal MA, Sabine AE, Fischer P, Leturcq D, Barbas CF, McConahey PJ, Caothien RH, Thornton GB, Dixon FJ, Burton DR (1992) Immunization of hu-PBL-SCID mice and the rescue of human monoclonal Fab fragments through combinatorial libraries. Nature 355:258–262

Engelman EG, Found SKH, Larrick J, Raubitschek A (1985) Human hybridomas and monoclonal antibodies. Plenum, New York

Folks TM (1987) Epitope mapping of human immunodeficiency virus: a monoclonal approach. J Immunol 139:3913–3914

Foung SKH, Perkins S (1989) Electric field induced cell fusion and human monoclonal antibodies. J Immunol Methods 116:117–122

Foung S, Perkins S, Kafader K, Gessner P, Zimmerman U (1990) Development of microfusion techniques to generate human hybridomas. J Immunol Methods 134:35–42

Frasa H, Torensma R, Verhoef J (1992) A human IgM monoclonal antibody against human TNF. Presented at the 2nd internation conference on: human antibodies and hybridomas. Cambridge, England

Gillies SD, Dorai H, Wesolowski J, Majeau G, Young D, Boyd J, Gardner J, James K (1989) Expression of human antitetanus antibody in transfected murine myeloma cells. Biotechnology 7:799–804

James K (1989) Human monoclonal antibody technology – are its achievements, challenges and potential appreciated. Scand J Immunol 29:257–264

James K (1990a) Therapeutic monoclonal antibodies – their production and application. Animal Cell Biotechnol 4:206–255

James K (1990b) Human monoclonal antibodies and engineered antibodies in the management of cancer. Semin Cancer Biol 1:243–253

James K, Bell GT (1987) Human monoclonal antibody production. Current status and future prospects. J Immunol Methods 100:5–40

James K, Gardner J, Skibinski G, McCann M, Thorpe R, Gearing A, Gordon J (1990) Cell surface phenotype, cytokines, and antibody gene expression in immortalized human B cell lines. Human Antibod Hybrid 1:145–153

Jung G, Freimann U, von Marschall Z, Reisfeld RA, Wilmans W (1991) Target cell-induced T cell activation with bi and trispecific antibody fragments. Eur J Immunol 21:2431–2435

Köhler G, Milstein C (1975) Continuous culture of fused cells secreting antibody of predefined specificity. Nature 256:495–497

Komori S, Yamasaki N, Shigeta M, Isojima S, Watanabe T (1988) Production of heavy chain class-switch variants of human monoclonal antibody by recombinant DNA Technology. Clin Exp Immunol 71:508–516

Kosinski S, Hammerling (1986) A new cloning method for antibody-forming lymphoblastoid cells. Increase in cloning efficiency by inclusion of human fibroblasts into semisolid agarose growth layer. J Immunol Methods 94:201–208

Kozbor D, Lagarde AE, Roder JC (1982) Human hybridomas constructed with antigen-specific Epstein virus – transformed cell lines. Proc Natl Acad Sci USA 79:6651–6655

Larrick JW, Fry KE (1991) Recombinant antibodies. Human Antib Hybrid 2:172–189

Leader KA, Macht LM, Steers F, Kumpel BM, Elson CJ (1992) Antibody responses to the blood group antigen D in SCID mice reconstituted with human blood mononuclear cells. Immunology 76:222–234

Lewis AP, Parry N, Peakman TC, Scott Crowe J (1992) Rescue and expression of human immunoglobulin genes to generate functional human monoclonal antibodies. Human Antib Hybrid 3:146–152

Masuho Y (1988) Human monoclonal antibodies: prospects for use as passive immunotherapy. Serodiag Immunother Infect Dis 2:319–340

McCann MC, James K, Kumpel BM (1988) Production and use of human monoclonal anti-D antibodies. J Immunol Methods 115:3–15

Nakatani T, Nomura N, Horigome K, Ohtsuka H, Noguchi H (1989) Functional expression of human monoclonal antibody genes directed against pseudomonal exotoxin A in mouse myeloma cells. Biotechnology 7:805–900

Pistillo MP, Mazzoleni O, Tanigaki N, Hammerling U, Longo A, Frumento G, Ferrara GB (1988) Human anti-HLA monoclonal antibodies: production, characterization and application. Hum Immunol 21:265–278

Sanz I, Casali P, Thomas JW, Notkins AL, Capra JD (1989) Nucleotide sequences of eight human natural autoantibody VH regions reveals apparent restricted use of VH families. J Immunol 142:4054–4061

Shinmoto H, Dosakao S, Tachibana H, Yamada K, Sanetaka S, Murakami H (1991) Generation of hybrid hybridomas secreting human IgM class hybrid antiricin and antidiptheria toxin antibodies. Human Antibod Hybrid 2:39–41

Steenbakkers PGA, van Meel FCM, Olijve W (1992) A new approach to the generation of human and murine antibody producing hybridomas. J Immunol Methods 152:69–77

Strelkauskas AJ (ed) (1987) Human hybridomas. Dekker, New York

Thompson KM, Randen I, Natvig JB, Mageed RA, Jefferis R, Carson DA, Tighe H, Forre O (1990) Human monoclonal rheumatoid factors derived from the polyclonal repertoire of rheumatoid synovial tissue: incidence of cross-reactive idiotopes and expression of VH and VK subgroups. Eur J Immunol 20:863–868

Thompson KM, Sutherland J, Barden G, Melamed MD, Wright MG, Bailey S, Thorpe SJ (1992) Human monoclonal antibodies specific for blood group

antigens demonstrate multispecific properties characteristic of natural autoanti-
    bodies. Immunology 76:146–157
Tutt A, Stevenson GT, Glennie MT (1991) Trispecific F(ab')3 derivatives that use
    cooperative signalling via the TCR/CD3 complex and CD2 to activate and
    redirect resting cytotoxic T cells. J Immunol 147:60–69
Williams SS, Umemoto T, Kida H, Repasky EA, Bankert RB (1992) Engraftment
    of human peripheral blood leukocytes into severe combined immunodeficient
    mice results in the longterm and dynamic production of human xenoreactive
    antibodies. J Immunol 149:2830–2836
Winter G, Milstein C (1991) Man-made antibodies. Nature 349:293–299
Wolff EA, Esselstyn J, Maloney G, Raff HV (1992) Human monoclonal antibody
    homodimers. Effect of valency on in vitro and in vivo antibacterial activity. J
    Immunol 148:2469–2474
Ziegler EJ, Fisher CJ, Sprung CL, Straube RC, Sadoff JC, Foulke GE, Wortel CH,
    Fink MP, Dellinger RP, Teng NNH, Allen IE, Berger HJ, Knatterud GL,
    LoBuglio AF, Smith CR and the HA-1 A Sepsis Study Group (1991) Treatment
    of gram-negative bacteremia and septic shock with HA-1A human monoclonal
    antibody against endotoxin. A randomized, double blind, placebo-controlled
    trial. N Engl J Med 324:429–436
Zimmermann U, Vienken J (1982) Electric field induced cell to cell fusion. J Membr
    Biol 67:165–182

CHAPTER 2

Recombinant Therapeutic Human Monoclonal Antibodies

J.W. LARRICK and R. BALINT

A. Therapeutic Human Monoclonal Antibodies

Cell fusion techniques have made the generation of rodent monoclonal antibodies (MABs) a routine endeavor (KOHLER and MILSTEIN 1975). As early as 1975 the first investigators of this technology recognized the therapeutic and industrial potential of MABs, particularly human monoclonal antibodies (huMABs), yet 18 years later only OKT3 – an anti-T lymphocyte antigen (CD3) murine MAB – is a licensed drug. There are two reasons for this apparent lack of progress: target selection and technical difficulties. Despite this modest beginning a large number of MABs generated primarily by cell fusion are in clinical development (see below). Recent advances using recombinant DNA technology have alleviated many of the technical problems with cell fusion and will accelerate the development of therapeutic huMABs (LARRICK et al. 1987).

What are reasonable therapeutic targets? Table 1 provides a useful classification of therapeutic MABs. A number of first generation targets have been replacement therapy for antisera used for infectious diseases or their toxins. Antibodies have permitted the testing of many novel therapeutic concepts such as inhibition of specific cell adhesion or clotting and complement proteins. Whether monoclonals or other pharmaceuticals are ultimately developed for these targets will depend upon the specific target. For example, although anti-tissue factor antibodies demonstrated the value of this approach to limit lipopolysaccharide (LPS)-induced microvascular coagulation, organic inhibitors of tissue factor have already been identified by Merck Sharp and Dohme laboratories and others. Thus under certain circumstances antibodies have demonstrated a therapeutic concept that is best fulfilled by other types of pharmaceuticals.

Passively administered human antisera against a variety of specific antigens are commonly used (for reviews see LARRICK 1989, 1990; LARRICK and FRY 1991a). huMABs that recognize many of these antigens have been produced, and will probably augment or replace pooled antisera in the near future (see Table 4). Problems related to contamination of pooled globulins by infectious agents, e.g., human immunodeficiency virus (HIV), hepatitis viruses, cytomegalovirus, and various as yet undescribed retroviruses, the diminished availability of serum donors, and the relative ease of reproducible huMAB manufacture favor the development of rMAB technology.

Table 1. Classification of therapeutic antibodies

Antibodies as "antibodies"
  Replace passive immunotherapy
  Toxins, endotoxins
  Infectious agents
    Old agents: influenza, RSV, pneumococcus
    New agents: hepatitis C virus, HIV
Testing of novel therapeutic concepts
  Cell adhesion (CD18/CD11 etc.)
  Initiation of clotting (α tissue factor)
  Cell type-specific (CD3, CD4, etc.)
  Complement activation (αC5a)
Novel uses
  Growth factor mimetics
  Manipulation of sub-subpopulations of cells (e.g., CD28, B7, B cell-specific CD5)
  Bispecific antibodies
  Catalytic antibodies
  Immunoadhesins/immunoconjugates

huMABs are preferred for therapeutic use and will minimize the problems encountered when administering a foreign animal MAB (e.g., anaphylaxis, clinical manifestations of immune complex formation, and reduced efficacy secondary to anti-antibodies). In well over half of the patients treated to date with murine MABs, the human anti-mouse antibody (HAMA) response limited their effectiveness (SCHROFF et al. 1985). Only a fraction of the anti-mouse immune response is directed to the variable region (idiotype) of the rodent immunoglobulins. This suggests that huMABs will be more effective therapeutic molecules than their rodent counterparts. Preliminary pharmacokinetic studies with huMABs demonstrate the superiority of these molecules over foreign (mouse) MABs (McCABE et al. 1989; MASUHO et al. 1990). The first approach to eliminating the immunogenicity problem was to construct recombinant human-mouse chimeric MABs. Recently this approach has been completely replaced by complementarity determining region (CDR)-grafted or humanized antibodies (see below).

huMABs are more likely to have species-specific carbohydrates which may be important in a number of effector functions, such as Fc receptor-mediated antibody-dependent cellular cytotoxicity (ADCC), complement activation, and phagocytosis (NOSE and WIGZELL 1983). Serum half-life and effector functions of immunoglobulin subclasses are very important for designing the optimal anti-infectious disease MAB therapeutic.

rMABs have several attractive advantages over conventional pharmaceuticals (Table 2). These proteins can be generated with exquisite selectivity and specificity in a short period of time. Antibodies have multiple effector functions in addition to their antigen-combining abilities. This permits the design of molecules capable of performing complex tasks involving host cells and other host mediator functions. Antibodies have a long

Table 2. Advantages of recombinant monoclonal antibodies therapeutics over conventional pharmaceuticals

---

Generation of molecules with exquisite selectivity and specificity in a short period of time
Multiple effector functions: design of molecules with complex functions involving host cells and other host mediator functions
Long serum half-life: prophylactic treatment
Specific recombinant monoclonal antibodies generated more rapidly and with far less effort than synthesis of conventional organics: possible to test if neutralization of a given mediator or removing a cell type has therapeutic benefits with less effort than that required to generate organic receptor antagonists
Types of drug toxicity
  Mechanism based
  Nonmechanism based (e.g., metabolites): *not* exhibited by antibodies

---

Table 3. Limitations of recombinant monoclonal antibodies as pharmacological agents

---

Protein composition
Immunogenicity
Tissue distribution, Mr
Parenteral administration
Target heterogeneity
Cost

---

serum half-life. This facilitates prophylactic treatment of at-risk patients. Finally, specific MABs can be generated more rapidly and with far less effort than that required for synthesis of conventional, low molecular weight, organic molecules. This final point is worth emphasizing. As noted above generation of MABs to remove a cell type or neutralize a factor requires much less effort than synthesis of organic receptor antagonists: rMABs permit rapid testing of a therapeutic concept.

As therapeutic agents, antibodies are not without their problems (see Table 3). Antibodies are composed of protein; they are potentially immunogenic (SHAWLER et al. 1985); they must be given parenterally; and they are relatively expensive to produce. Many therapeutic antibodies are aimed at the treatment of acute life-threatening disease. In this case cost may be of minor importance. The construction of rMABs will overcome some of these problems. Production costs of recombinant MABs in some systems (e.g., plants) are substantially below those of hybridomas. Table 4 lists most of the MABs known to be in preclinical and/or in clinical development.

B. Rapid Direct Cloning of Antibody Variable Regions

Despite advances in the in vitro immunization of human B cells (BORREBAECK et al. 1988) and the development of immunodeficient mice (McCUNE et al. 1988) for the reconstitution of the human immune system ex vivo,

Table 4. Targets of therapeutic antibodies

Indication/target	Antibody	Laboratory	Comments
Bacterial sepsis	Centoxin Ha-1A	Centocor	Human MAB-trioma
	E5	Xoma	Mouse/humanized
Anti-ECA		Cetus; Chiron	Human MAB-trioma
	T88		
Anti-*Pseudomonas aeruginosa*		Scotgen; Cetus; Cutter; Miles; Mitsui	
Anti-*P. aeruginosa*		Cetus	Human
exotoxin A	P7E9C7	Takeda	Human
Neonatal sepsis (*E. coli*, Group B streptococcus)	BMY-35037	Bristol Myers Squibb	
Anti-tumor necrosis factor	Cerami	Chiron; Cutter; Miles; Bayer	
	Cen TNF	Centocor	
	CDP571	Celltech	
Anti-integrin (cell adherence) CD18:			
	1B4	Merck	CDR grafted
	60.3	Bristol Myers Squibb	
	H52	Genentech	
Anti-C5a		Chiron; Smith Kline Beecham	Murine Phage display
Other infectious agents Anti-*Pneumocystis carinii*		Research Corporation	
Anti-*Pneumococcus* spp.		Yissum	
Adenocarcinoma	88BV59	Organon Technika	
Glioma EGF		EM Industries	
Solid tumors EGF receptor		Rhone Poulenc Rorer	
Breast cancer HER-2		Genentech	(humanized)
Chimeric L6		Ixsys, Oncogen, Bristol Myers Squibb	
Pancarcinoma	RE-186	NeoRx	Breast, colon, lung, ovarian, pancreas, prostate

Table 4. *Continued*

Indication/target	Antibody	Laboratory	Comments
B cell lymphoma IDs		IDEC	
Acute myelogenous leukemia CD33 M195		SKI Protein Design Labs	
Colorectal antigen 17-1A		Centocor	37 kDa antigen
B72.3 (NIH, Schlom)		Celltech American Cyanamid	Chimeric
Mucin, colon (TAG-72)			
Melanoma ectodermal tumors 14.18 (Scripps, Reisfeld)		Damon Biotech	
CEA		Cytogen; Immuno medics	
	CHA255	Hybritech-Lilly	F(ab)′2, bispecific
CEA-β glucuronidase		Behringwerke AG	
Lymphomas LL2		Immunomedics	
T cell malignancy IL2 receptor (Tac)		Protein Design Labs	
CAMPATH1		Burroughs-Wellcome	
Lewis Y antigen	BR96 SDZ ABL 364	Bristol Myers Squibb Protein Design Labs	
Hepatits B		Sandoz	Human MAB- trioma
		Teijin Takeda	Human MAB-P3 Human MAB- H/H
Cytomegalovirus		Sandoz	Human MAB- trioma
		Miles; Teijin Scotgen	Human MAB Humanized mouse
		Biotechnology General	
Respiratory syncytial virus		Genelabs Protein Design Labs; Scotgen/Smith Kline Beecham	Humanized chimp Humanized mouse
		Scripps	Phage library (human)

Table 4. *Continued*

Indication/target	Antibody	Laboratory	Comments
HIV		Teijin/University of Arizona; Merck Sharpe Dohme/ Medimmune (New York University); Repligen; Bristol Myers Squibb	
Varicella zoster virus		Sandoz; Teijin; Genelabs	
Herpes simplex 1 and 2	Humanized Fd79, gB; Fd138, gD	Protein Design Labs	
		Teijin	
		Sandoz	Human-trioma
Rhinovirus (ICAM-1, etc.)		Merck Sharpe Dohme	
Parainfluenza		Protein Design Labs	
Bone marrow transplantation (graft vs host disease) Immunotoxin	OKT3	Ortho; Johnson and Johnson Xoma; Ortho Johnson and Johnson	
Rheumatoid arthritis, inflammatory bowel, IDDM Immunotoxin		Xoma; Ortho Johnson and Johnson	
CAMPATH1		Burroughs-Wellcome	
CD4		Centocor	
Multiple sclerosis		Centocor	γ-interferon
Transplantation	OKT3	RW Johnson; Ortho Biotech; Johnson and Johnson	
IL2 receptor (Tac)		Protein Design Labs (Roche) Immunex	
	BI-RR1	Boehringer Ingelheim	
CD4		Idec; Becton Dickinson; Boehringer Ingelheim; Centocor	
CD7	SDZ-CHH-380	Sandoz	
	CAMPATH1	Burroughs-Wellcome	

Table 4. *Continued*

Indication/target	Antibody	Laboratory	Comments
Cardiovascular Platelets		Centorex	
(GPIIb/IIIa) Fibrin/pro-urokinase		Takeda	Bispecific

Imaging myocardial
infarctions
  Myosin
  Fibrin
	Myoscint	Centocor	

Cancer imaging CEA		Immunomedics	Colorectal, lung, breast
LL2		Immunomedics	Lymphoma
88BV59		Organon Teknika	Colorectal, aderno- carcinoma
Cancer radio- immunotherapy G103		Cytogen; Sterling	Colorectal cancer
OV103		Cytogen; Sterling	Ovarian cancer

LPS, lipopolysaccharide; ECA, enterobacterial common antigen; CEA, carcino-embryonic antigen; CDR, complementarity determining region; EGF, epidermal growth factor; HIV, human immunodeficiency virus; IDDM, insulin-dependent diabetes mellitus; IL2, interleukin-2

immortalization of antigen-specific human B cells remains the limiting step in the generation of huMABs. Typically this is performed with the aid of Epstein-Barr virus transformation followed by subcloning, confirmation of antigen binding, and hybridization of the B lymphoblasts to suitable fusion partners such as mouse-human heteromyelomas. This general approach is effective and widely used; however, it is time consuming and erratic immortalization occurs. For this reason, we and others have devised methods to directly obtain the variable regions from small numbers of human B cells.

Recent developments suggest that recombinant DNA technology can replace cell fusion as a means of generating MABs (CHISWELL and McCAFFERTY 1992). This type of immunoengineering has been made possible by gene amplification technology, i.e., polymerase chain reaction (PCR). The concept is quite simple: a mixture of oligomer primers in the 5′ leader sequences or framework I region combined with 3′ constant region primers permits the amplification of any immunoglobulin variable region from very small numbers of cells (LARRICK et al. 1989a,b; CHIANG et al. 1989). Primer design and other methodological details can be found in other chapters in this volume and in our previous publications (LARRICK and FRY 1991b). The

PCR fragments can be directly sequenced and/or ligated into expression vectors. The method has been used to obtain variable regions of both heavy and light chains from single human B lymphocytes. It is also possible to obtain sequences from individual B cells deposited by the fluorescence activated cell sorter (FACS) into microtiter wells containing carrier RNA and guanidinium isothiocyanate. Complementary DNA can be synthesized and amplified by PCR for sequencing. Thus the variable region genes of B cells can be obtained from in vitro antigen expanded cultures or from peripheral blood on a suitable day postimmunization. As described in detail below others have reported success using PCR to obtain antibody variable regions for construction of rMABs (Orlandi et al. 1989; Roux and Dhanarajan 1990) directly from libraries of phage combining heavy and light chains together artificially in vitro. Thus, the stage has been set for a new era of rapid progress in understanding and using antibodies.

C. Genetically Engineered Chimeric Monoclonal Antibodies

I. Chimeric Antibodies

Antibodies are composed of disulfide-linked heavy and light chains each comprised of variable (V) and constant (C) domains. It is thought that the most immunogenic portion of antibodies will be the species conserved C regions. For this reason several laboratories have used recombinant DNA technology to construct chimeric rodent-human MABs by attaching human C regions to the rodent V regions (Table 5; Morrison et al. 1984; Boulianne et al. 1984; for reviews see Morrison and Oi 1989). Because the antibody combining site is localized within the V regions these molecules maintain their combining affinity for the antigen and acquire the function of the substituted C regions (Steplewski et al. 1988; Bruggemann et al. 1987). Among the first chimeric constructions was a mouse anti-phosphorylcholine MAB using the S107 myeloma cell line. In another approach the heavy chain V regions of a monoclonal specific for the hapten azophenylarsonate was linked to light chain C regions by Sharon et al. (1984). These chimeric molecules formed hapten-binding light chain dimers.

Although some laboratories have linked variable regions or Fab fragments by biochemical means to human Fc regions (Hamblin et al. 1987) most mouse MABs have been successfully chimerized using recombinant DNA technology (Table 5). When therapeutic use was intended most investigators have used the IgG1 constant regions because of its serum half-life, capacity to fix complement, and bind to Fc receptors. Details of the therapeutic chimeric antibodies summarized in Table 5 can be found in Larrick and Fry (1991a).

Table 5. Therapeutic recombinant chimaeric antibodies

Targets	Reference
Hepatitis surface antigen	LI et al. (1990)
Human immunodeficiency virus antigen gp 120	LIOU et al. (1989); CHIANG et al.
CE7 T lymphocyte antigen	HEINRICH et al. (1989)
CD4	Centocor
Cancer antigens	
Carcinoembryonic antigen (CEA)	BEIDLER et al. (1988); NEUMAIER et al. (1990); KOGA et al. (1990); HARDMAN et al. (1989)
Ganglioside GD2	MUELLER et al. (1990)
Common acute lymphocytic antigen	NISHIMURA et al. (1987); SAGA et al. (1990); YOKOYAMA et al. (1987)
Multiple drug transporter, P170	HAMADA et al. (1990)
Colorectal antigen 17-1A	SUN et al. (1987); SHAW et al. (1987); FOGLER et al. (1989)
Melanoma (Nrml-05)	MARCHITTO et al. (1989)
Tumor associated glycoproteins (e.g., B72.6, L6)	WHITTLE et al. (1987); LIU et al. (1987)
Ovarian cancer	GALLO et al. (1988)
Transferrin receptor	HOOGENBOOM et al. (1990)
Miscellaneous cancer cells	HANK et al. (1990); SAHAGAN et al. (1986)
BR96	Bristol Myers Squibb

1. Summary of Work with Therapeutic Chimeric Monoclonal Antibodies

Many of the above antibodies are in various stages of preclinical development. How successful has this first generation of rMABs been? In all cases the chimeric MAB retained the antigen-binding characteristic of the parental mouse MABs and in most cases the levels of expression were in the middle range for hybridomas. In many cases the chimeric MAB had superior activity in ADCC and other functional activities using human effector cells. At the present time very little is known about the immunogenicity of chimeric MABs, although evidence from rodents (using chimeric rodent MABs) indicates recipients can still recognize these molecules as non-self. BRUGGE-MAN et al. (1989) immunized mice with model xenogeneic (both the VH frameworks and the CH domains of human origin), chimeric (just VH frameworks human), or self MAB, and the anti-antibody responses were dissected. Only the self MAB did not elicit an immune response. A strong response was elicited by the most xenogeneic MAB with approximately 90% against the C domains and approximately 10% against the V domain. The anti-V response was not attenuated in the chimeric antibody, demonstrating that foreign VH frameworks can be sufficient to lead to a strong anti-antibody response. The magnitude of this xenogeneic anti-VH response was similar to that of the allotypic response elicited by immunizing mice of the Igha allotype with an Ighb antibody. Thus, although chimerization can diminish anti-antibody responses, there is reason to believe that chimeric MABs will be immunogenic in immunocompetent human patients. Recent data from LoBuglio et al. (this volume) suggest that some V regions are immunogenic (possibly possessing helper T cell epitopes) whether they are in the original mouse MAB or chimerized or humanized (see below).

II. Recombinant Conjugates and Fusion Proteins

Antibodies provide an appealing method to deliver specific drugs or toxins (Epenetos et al. 1986; Vitetta et al. 1983); however, this large area of investigation is beyond the scope of this discussion. Nonetheless, it should be pointed out that many of the problems encountered with synthesis of these molecules, size of conjugates, and biodistribution may be addressed by rMABs. Construction of antibody fragments for conjugation to toxins or other molecules is a particularly appealing strategy. The generation of bispecific MABs is another important strategy to overcome some of the problems associated with first generation immunoconjugates.

1. Immunotoxins

Chaudhary et al. (1989) constructed and expressed in *Escherichia coli* a single chain antibody toxin fusion protein, anti-Tac(Fv)-PE40, in which the V regions of anti-Tac, were joined in peptide linkage to PE40, a modified form of pseudomonas exotoxin lacking its binding domain. Anti-Tac(Fv)-PE40 was very cytotoxic to two interleukin-2 (IL-2) receptor-bearing human cell lines but was not cytotoxic to receptor-negative cells. This same group developed a strategy using PCR to rapidly clone antibody-toxin conjugates (Chaudhary et al. 1990). Clones encoding recombinant single chain immunotoxins (using MAB OVB3 that recognizes ovarian cancer cells) were expressed in *E. coli* and the protein product bound to and killed cells bearing the OVB3 antigen. Rybak et al. (1992) have constructed immunotoxins with angiogenin.

2. A Recombinant Monoclonal Antibody Linked to Tissue-Type Plasminogen Activator

Schnee et al. (1987) constructed a tissue-type plasminogen activator (t-PA) fusion protein with a MAB specific for the fibrin β chain (anti-fibrin 59D8). This produced a thrombolytic agent that is more specific and more potent that t-PA alone. The rearranged 59D8 heavy chain gene was cloned and combined in the expression vector pSV2gpt with sequence coding for a portion of the γ 2b constant region and the catalytic β chain of t-PA. This construct was transfected into variant cells that lacked heavy chain and which were derived from the 59D8 hybridoma. The cells produced a 65 kDa heavy chain-t-PA fusion protein that was secreted in association with the 59D8 light chain in the form of a 170 kDa disulfide-linked dimer. Chromogenic substrate assays showed the fusion protein to have 70% of the peptidolytic activity of native t-PA and to activate plasminogen as efficiently as t-PA. In a competitive binding assay, reconstituted antibody was shown to have a binding profile similar to that of native 59D8. This hybrid protein was thus capable of high-affinity fibrin binding and plasminogen activation.

3. T Cell Receptor Conjugates

A secreted soluble form of the V domain of a human T cell receptor α chain was constructed from the V α region of the T cell receptor of a diphtheria toxoid-specific human T cell clone fused to a human immunoglobulin κ light chain C region (MARIUZZA and WINTER 1989). Myeloma cells transfected with this chimeric protein secreted a noncovalent homodimer of 65 kDa. The V α C κ protein is extensively glycosylated, and its secretion is glycosylation-dependent. Chimeric genes containing the V β region of this particular T cell receptor linked to immunoglobulin C κ or C γ 2 regions were expressed intracellularly, but the products, although glycosylated, were not secreted nor did they assemble with the V α C κ protein. Thus the chimeric β chain immunoglobulin proteins were incorrectly folded and/or processed due either to the design of the gene fusions themselves or to the absence of vital T cell-specific accessory molecules in the myeloma cells. GASCOIGNE et al. (1987) described a chimeric gene construct containing a T cell receptor α chain V domain and the C region coding sequences of an immunoglobulin γ 2a molecule. Cells transfected with the chimeric gene synthesized a stable protein product that expressed both immunoglobulin and T cell receptor antigenic determinants and protein A binding sites. The T cell receptor V β gene in the same construct was neither assembled nor secreted with the λ light chain and, when expressed with a C κ region, it did not assemble with the chimeric V α C γ 2a protein. Thus not all T cell receptor V regions are similar enough to immunoglobulin V regions for them to be completely interchangeable.

4. Growth Factor Conjugates

SHIN and MORRISON (1990) replaced the C region of a chimeric mouse-human IgG3 anti-dansyl antibody with insulin-like growth factor 1 (IGF1). The chimeric heavy chain was expressed with an anti-dansyl-specific chimeric κ light chain. The IgG3-IGF1 chimeric protein retained its specificity for the dansyl antigen. The chimeric MABs bound to the IGF1 receptors of human lymphoblast IM-9 with reduced affinity and elicited increased glucose and amino acid uptake in human KB cells (with reduced specific activity vs IGF1). It was hypothesized that the reduced affinity and biologic activity resulted from the presence of the unprocessed IGF1 moiety, the large size of the IgG3-IGF1 chimeric protein (160 kDa) compared with IGF1 (7 kDa), or three amino acid substitutions in rat IGF1 compared with human IGF1, which may lead to decreased affinity for the human IGF1 receptor.

5. Other Fusion Proteins

The basic chimeric technology has been applied to the generation of immunoglobulin fusion proteins: immunoglobulin-CD4 (ZETTLMEISSL et al. 1990), photoprotein aequorin and an anti-4-hydroxy-3-nitrophenacetyl MAB (CASADEI et al. 1990), and a MAB-Klenow enzyme conjugate. The photoprotein aequorin and an anti-4-hydroxy-3-nitrophenacetyl antibody gene has permitted the development of a sensitive luminescent immunoassay. The serum half-life of soluble CD4 was substantially prolonged by fusion of this immunoglobulin-like molecule to immunoglublin heavy chains. Different chimeric antibody-like molecules, consisting of the four human CD4 extracellular domains (amino acids 1–369), fused to different parts of human IgG1 and IgM heavy chain C regions lacking the CH1 domain of the heavy chain C region. The chimeric molecules were potent inhibitors of HIV infection and HIV-mediated cytotoxicity. A CD4:IgG1 hinge fusion protein, which was analyzed in more detail, bound efficiently to HIV gp160 and human Fc receptors and showed complement-assisted inhibition of viral propagation in culture. Half-life studies after intravenous application of the latter human fusion protein into mice and monkeys showed significant prolongation of serum survival compared to soluble CD4. An IgG2b murine homologue of the human CD4:IgG1 hinge fusion protein was prepared and evaluated in mice, where it was found to be nontoxic and to have no detectable effect on the humoral response to soluble antigen. WILLIAMS and NEUBEUGER (1986) constructed a hapten-specific MAB/enzyme hybrid molecule, in which the antibody Fc portion was replaced by the Klenow fragment of *E. coli* DNA polymerase I (Pollk). This Fab-Pollk hybrid molecule was secreted in good yield from the myeloma transfectants, could be purified to homogeneity in a single step on hapten-Sepharose columns, and exhibited Pollk activity as judged by its use in dideoxynucleotide sequencing.

6. Antibody-Enzyme Conjugates for Cancer

Chemoimmunoconjugate drug loading, even when using linkers such as dextran or albumin, appears to be limited to less than 100 drug molecules/ antibody (YEH et al. 1992; TROUET et al. 1982; SHEN et al. 1986). Hence insufficient drug may be internalized to totally eradicate the tumor. An alternative approach is to utilize an enzyme conjugate to generate a large number of active drug molecules at the tumor site. Activated prodrugs have a low molecular weight and can diffuse more readily into the tumor mass (JAIN 1990). Chemoimmunoconjugates are also difficult to standardize and require extensive characterization, a disadvantage overcome by genetically engineered MABzymes. A major advantage of enzymatically converting a prodrug to an antineoplastic drug in tumor cells but not in normal tissues is to increase the specificity and lower the toxicity of cancer chemotherapy. Several conditions must be met for this strategy to be feasible: (a) the prodrug should be less toxic than the corresponding parent compound; (b) the prodrug must be converted under defined conditions into the active parent compound; and (c) tumor and normal cells should display sufficient differences in the cellular property used to activate prodrug to parent drug. MAB-enzyme conjugates have been pursued by SENTER et al. (1988, 1989), who demonstrated regression of human lung adenocarcinoma xenografts in nude mice treated with MAB-alkaline phosphatase conjugates followed by mitomycin phosphate administration. Phosphorylated prodrug alone and in combination with a control MAB-alkaline phosphatase also delayed tumor growth. The same group (KERR et al. 1990) also described the activation of a doxorubicin prodrug with a MAB-penicillin-V-amidase conjugate and the conversion of 5-fluorocytosine into the antineoplastic agent 5-fluorouracil by a cytosine deaminase-MAB conjugate (SENTER et al. 1991). BAGSHAWE et al. (1988) have developed glutamic acid prodrugs which can be converted to *bis*-chlorobenzoic acid mustards by carboxypeptidase G2. These prodrugs were able to inhibit or eliminate human choriocarcinoma or colon carcinoma xenografts in nude mice after treatment with antibody-carboxypeptidase G2 conjugates. A cephalosporin-*Vinca* alkaloid prodrug activated by a *β*-lactamase-antibody fragment has also been described (SHEPHERD et al. 1991).

D. Reshaped or Composite Antibodies

The laboratory of Winter has pioneered a more sophisticated approach for construction of human antibodies from rodent monoclonals by splicing the rodent hypervariable CDRs onto human V framework sequences (Table 6). Short of deriving a human MAB from an immune human B cell this is about as "humanized" as a rodent monoclonal can become using rDNA technology. This is feasible because the antibody combining site is constructed from several hypervariable regions held together to form the antigen binding cleft by a *β*-sheet comprised of framework sequences. The first of these "composite" monoclonals was constructed by grafting the CDRs from the heavy chain V region of mouse antibody B1–8, which binds the hapten NP-cap (4-hydroxy-3-nitrophenacetyl caproic acid; KNP-cap = $1.2\,\mu M$), onto a human myeloma protein (JONES et al. 1986). In combination with the B1–8 mouse light chain, the new antibody acquired the hapten affinity of the B1–8 antibody (KNP-cap = $1.9\,\mu M$). The affinity of a second composite MAB was less than the parent murine MAB (VERHOEYEN et al. 1988).

The rat anti-CAMPATH-1 monoclonal (RIECHMANN et al. 1988) recognizes a glycoprotein (CDw52) expressed on virtually all human lymphocytes and monocytes, but is absent from the hematopoietic stem cells. Depletion of cells bearing this antigen appears to be an important therapeutic approach for control of graft vs host disease in bone marrow transplantation, prevention of bone marrow and other organ rejection episodes, and for treatment of various lymphoid malignancies (WALDMANN et al. 1988). The six hyper-

Table 6. Humanized murine monoclonal antibodies

Concept	
NIP	JONES et al. 1986
Lysozyme	VERHOEYEN et al. 1988
Immunomodulatory	
Campath-1 (CDW52)	REICHMANN et al. 1988
CD3 (OKT3)	WOODLE et al. 1992
CD4	GORMAN et al. 1991
IL6	TSUCHIYA et al. 1992
CD18 (60.3)	HSIAO et al. 1992
ICAM-1 (CD54)	MIGLIETTA et al. 1992
IL2 receptor (Tac)	QUEEN et al. 1989; JUNGHANS et al. 1990
Cancer	
CEA	GUSSOW and SEEMANN 1991
HER2 (p185)	CARTER et al. 1992
17-1a	Centocor
B72-3	NIH
EGR receptor	KETTLEBOROUGH et al. 1991
CD33	Co et al. 1992
Lewis Y (SDZ ABL 364)	LOIBNER et al. 1992
Anti-infectious	
E5 (anti-LPS)	Xoma
Hepatitis	Scotgen
Respiratory syncytial virus	TEMPEST et al. 1991
Cytomegalovirus	Scotgen
Herpes simplex virus	Co et al. 1991
HIV	MAEDA et al. 1991
Tetanus toxin	LARRICK et al. 1993

variable regions from the heavy and light chain V region domains of the rat antibody grafted onto the framework regions of a human IgG1 antibody yielded a "reshaped" human MAB with effector functions equal to (complement fixation) or better than (cell-mediated lysis of human lymphocytes) the parent CAMPATH-1 monoclonal. In the inital clinical trials this pioneer reshaped antibody eliminated large numbers of tumor cells, resulting in disease remission for patients with non-Hodgkin's lymphoma. Significantly, there was no antiglobulin response in these patients (HALE et al. 1988).

The Mr 55 000 IL-2 receptor peptide (Tac; CD25) is not expressed by normal resting T cells but is markedly up-regulated in adult T cell leukemia and other malignancies, and on T cells activated in normal immune, auto-immune, allograft, and graft vs host settings. Anti-Tac is a mouse MAB directed against the Tac peptide. This inhibits proliferation of T cells by blocking IL-2 binding. Early attempts to use this MAB in humans for antitumor therapy and immune regulation were limited by weak recruitment of effector functions and neutralization by antibodies to mouse immunoglo-bulins. QUEEN et al. (1989; JUNGHANS et al. 1990) humanized the anti-Tac antibody using human framework and C regions. The human framework regions were chosen to maximize homology with the anti-Tac antibody sequence. A computer model of murine anti-Tac was used to identify several amino acids which, while outside the CDRs, were likely to interact with the CDRs or antigen. These mouse amino acids were also retained in the humanized antibody. The composite anti-Tac antibody was shown to have an affinity for p55 of $3 \times 10^9 M^{-1}$, which is about one third that of murine anti-Tac. Furthermore, the composite Tac rMAB blocked T cell activation and facilitated ADCC with human effector cells.

Respiratory syncytial virus (RSV) is a major cause of acute respiratory morbidity and mortality particularly among young children. When TEMPEST et al. (1991) directly transferred the CDRs of a neutralizing anti-RSV murine MAB into a human IgG1 framework binding activity was lost. Binding activity and neutralizing capacity were restored when murine amino acids 91–94 were used to replace the corresponding human framework amino acids.

Two major approaches to humanization have emerged from this work. The first case, pioneered by Queen et al. at Protein Design Labs, relies on choosing human framework regions most homologous to the murine sequences. Murine amino acids that contact the CDRs are also transferred into the human frameworks. In addition unusual amino acids in the human frameworks are replaced with consensus human amino acids. In an alternative approach, taken by TEMPEST et al. (1991) at Scotgen, a particular human framework is used as the basis to reshape all MABs.

It should be noted that even fully humanized murine MABs may be immunogenic. Although limited studies have demonstrated that chimeric mouse-human antibody 17-1A was less immunogenic in humans than the parent mouse monoclonal (KHAZAELI et al. 1988), more studies will be required to determine how much of a problem the human anti-idiotype response will be. In principle, the idiotype of a reshaped recombinant monoclonal could be changed by altering the CDRs or framework regions (FOOTE and WINTER 1992). However, grafting the CDRs into several cassettes might focus the immune response onto the combining site. This might be one method to potentiate development of effective anti-idiotype vaccines.

E. Immortalization of the Immunoglobulin Repertoire Using rDNA Technology

WARD et al. (1989) used PCR primers flanking the V regions to construct libraries of VH genes from spleen genomic DNA of mice immunized with either lysozyme or keyhole-limpet hemocyanin (KLH). From these libraries, VH domains were expressed and secreted from *E. coli*. Binding activities were detected against both antigens, and two VH domains were characterized with affinities for lysozyme in the 20 nM range. These isolated single domain antibodies were called "dAbs." The immortalization of an entire antibody repertoire laid the groundwork for an enormous technical advance, the construction of whole synthetic antibodies independent of hybridomas.

F. Recombinatorial Antibody Libraries

As noted above, the capacity of PCR to amplify essentially any V region permits the simultaneous amplification and subsequent cloning of an entire library of heavy and/or light chain V regions. This revolutionary finding means that rMABs can be constructed without resort to hybridoma technology. Several groups have produced recombinant libraries in *E. coli* using phage. In one case heavy chains were amplified by PCR primers flanking the V regions and demonstrated to possess antigen-binding activity in the absence of light chains (GUSSOW et al. 1989).

HUSE et al. (1989) described a technique for the generation of recombinant libraries encoding the entire antibody repertoire. PCR primers flanking the V regions are used to amplify V regions combined with expression of Fab fragments in *E. coli*. Heavy and light chains can be expressed in separate vectors and recombined artificially in vitro. The recombinants release Fab fragments into the periplasmic space. Hence the recombinants can be screened directly for antigen-binding fragments in the same manner as a conventional λ gtll library is screened with antibody.

MULLINAX et al. (1990) were the first to apply this technology to a clinically relevant human antibody. They immunized volunteers with tetanus toxoid. mRNA was prepared from lymphocytes harvested 6 days post-immunization. The mRNA was converted to cDNA using light or heavy chain primers. PCR primers were then used to amplify immunoglobulin H or L chain sequences with sets of primers hybridizing to conserved leader sequences in the 5' ends and to the 3' end of the light chain (full length) or just 3' to the first cysteine codon in the hinge exon of the H chain. The product resulted in an Fd fragment of the IgG1 isotype with conservation of the H-L disulfide bond. These fragments were digested with different restriction enzymes and ligated into linearized Lambda Zap vectors. The vectors were constructed to have a ribosome binding site and a pelB leader sequence. The ligated recombinant phage DNA was then packaged. These vectors were constructed to permit coligation and generation of heavy-light chain recombinatorial libraries. Prior to immunization the frequency of B cells producing anti-tetanus antibody was <1/500000. After immunization this rose to as high as 1/3000. In the library that was screened with radioiodine-labeled tetanus toxoid approximately 0.2% of the clones were positive. On further examination several of these showed an apparent affinity of $9 \times 10^8 M^{-1}$ for tetanus. Improvements in this basic technology have been developed by Stratcyte Corporation (La Jolla, CA) in the form of the Surf-ZAP vectors that combine the packaging function of λ phage with phage display (see below).

The immunoexpression approach combined with in vitro recombination of heavy and light chains permits the generation of wholly synthetic antibodies. When the libraries are combined with expression systems a very high number of clones can be screened in a short period of time. Highly conserved antigens, e.g., human antibody fragments, autoantigens or tumor-associated self-antigens and antigens from toxic or dangerous organisms can be used to screen libraries to generate therapeutic human rMABs using this technology. In principle it should be possible to immortalize the entire antigen combining repertoire and many novel recombinants not present in B cells, i.e., heavy and light chains not normally associated in vivo. Libraries can be generated at different time points, from different lymphoid organs, and after different immunization strategies. This important advance will facilitate studies of the antibody network and immunoglobulin development, the immune response, B lymphoma carcinogenesis, etc.

G. Phage Antibody Libraries: Wholly Synthetic Monoclonal Antibodies

Libraries in which antibodies are displayed on the surface of filamentous bacteriophage offer a number of important advantages over the *E. coli* expression libraries discussed above (Table 7). SMITH (1985) and coworkers (PARMLEY and SMITH 1988; SCOTT and SMITH 1990) pioneered the expression of peptide "epitope" libraries on the surface of fd phage by genetically engineering random peptides into the NH_2-terminal domain of the phage gene III protein. Several copies of this protein located at the tip of the phage mediate its attachment to the *E. coli* F pilus, whereupon infection is initiated. PARMLEY and SMITH (1988) showed that fusion of heterologous domains to the NH_2-terminal of the gene III protein does not significantly impair its function and furthermore that such domains are accessible to exogenous ligands. Populations of phage expressing as many as 1×10^8

Table 7. Therapeutic phage-derived monoclonal antibodies

Cencept	
NIP/pOX	HOOGENBOOM and WINTER 1992; CLACKSON et al. 1991;
Lysosome, BSA, etc.	MARKS et al. 1991
Blood group antigens (B,D,E, etc.)	HOOGENBOOM et al. 1992
Immunomodulatory	
C5a	AMES et al. 1992
Tumor necrosis factor	HOOGENBOOM et al. 1992
Cancer	
CEA	HOOGENBOOM et al. 1992
Mucins	HOOGENBOOM et al. 1992
EGF receptor	KETTLEBOROUGH et al. 1993
Anti-infectious	
Tetanus toxin	PERSSON et al. 1991
Hepatitis B virus	ZEBEDEE et al. 1992
Respiratory syncytial virus	BARBAS et al. 1992a
HIV-1	BURTON et al. 1991; BARBAS et al. 1992b
Influenza	CATON and KOPROWSKI 1990

different epitopes can be generated and "panned" against immobilized ligands to enrich for desired binding specificities. Bound phage may be eluted, amplified in *E. coli*, and subjected to successive rounds of panning until maximum affinity epitopes have been identified. Such libraries have been used to identify MAB epitopes (e.g., Cwirla et al. 1990), peptide mimics of nonpeptide epitopes (Devlin et al. 1990; Oldenburg et al. 1992; Scott et al. 1992), alternative ligands for receptors (Scott and Smith 1990), and peptide protease inhibitors (Roberts et al. 1992).

Recently, McCafferty et al. (1990) greatly enlarged the versatility of this technology by demonstrating that antibody fragments encoded at the NH$_2$-terminal of the gene III protein could be displayed on the surface of fd phage with full epitope binding activity. This discovery opened a novel route for the isolation of MABs. Repertoires of antibody V region genes could be amplified by PCR, as discussed above, and cloned into fd or M13 phage to be expressed as gene III fusions, thus creating large libraries of phage, each displaying a specific antibody. By panning such libraries against the antigen of interest binding specificities as rare as one in 10^7 could be isolated. Repertoires from a variety of sources have been used with success. For example, spleens or peripheral blood lymphocytes (PBLs) from immunized or unimmunized donors may be used. Alternatively, specific V region genes may be used as templates for artificially creating diversity by error prone PCR or oligonucleotide-directed randomization. Antibody domains may be expressed as single chain V regions (ScFv) fused to the gene III protein, in which heavy and light chain V regions are tethered together by a flexible linker, or as Fab fragments, in which the heavy chain Fd fragment is fused to the gene III protein (Hoogenboom et al. 1991). For the latter, both chains are expressed separately from the same cistron as signal peptide fusions directing them to the periplasmic space of *E. coli*. Here, they typically accumulate in concentrations exceeding the association constant of the heterodimer, thereby allowing appropriate assembly of the two chain antibody structure on the phage surface. V region repertoire cloning has recently been improved by the introduction of phagemid vectors in place of the original phage genomic vectors (Kang et al. 1991). The greater transforming efficiency of phagemids permits the construction of large libraries and their greater stability insures the production of monospecific phage.

Phage libraries expressing repertoires of antibody V regions obtained from hyperimmunized mouse spleens have yielded MABs with affinities comparable to those of traditional MABs (Clackson et al. 1991), while phage libraries from unimmunized donors have yielded affinities in the submicromolar range, which is comparable to primary response affinities (Marks et al. 1991). The technology has also been used to rescue antibody V regions from immunized Hu-SCID mice (Duchosal et al. 1992). Recently, Winter and coworkers (Marks et al. 1992) described the generation of antibodies with nanomolar affinities using phage libraries constructed from a nonimmunized human repertoire, thus demonstrating the ability of phage technology to produce human antibodies with therapeutically useful affinities without immunization. They constructed a phage library from an unimmunized human PBL repertoire expressed in single chain form, from which they isolated a low-affinity antibody to a hapten. They then recombined the VH domain of this molecule with a VL repertoire from the same donor and then isolated a higher affinity molecule from the resulting library. The increased affinities generated by light chain shuffling were attributed to decreased dissociation rate ("off" rate) rather than increased "on" rates. This was accomplished by preloading phage with biotinylated antigen and then diluting into excess unlabeled antigen for varying times prior to capture on streptavidin-coated paramagnetic beads. The gene fragment encoding VL plus VH CDR3 from the highest affinity phage arising from light chain shuffling was then recombined with a repertoire of VH minus CDR3 from the same donor, since CDR3 usually makes the most extensive contacts with antigen. From this library 90 clones were isolated with higher affinities than the parent, the best of which had a Kd of 1.1 nM, 320-fold lower than that of the initial antibody. Thus, using chain shuffling alone, Winter and coworkers were able to mimic affinity maturation in vitro. However, in view of the known differences in immunogenicity between haptens and proteins, it remains to be shown that such an approach can produce high-affinity antibodies to therapeutic targets.

Random mutagenesis of V regions in vitro has also been used to generate phage libraries from which improved affinities could be isolated. Winter and coworkers (HAWKINS et al. 1992) achieved a fourfold improvement of a hapten-binding antibody by limited randomization (about 1.7 bases per VH) using error prone PCR. Using oligonucleotide-directed randomization, BARBAS et al. (1992c) constructed a Fab phage library containing 5×10^7 heavy chain CDR3 variants of a human anti-tetanus toxoid antibody. This library rivals in size the naive mouse repertoire, which can recognize a seemingly unlimited number of antigens. From this library they isolated fluorescein-binding Fabs with 100- to 1000-fold greater affinity than the parent antibody. Thus, from the successes that have been achieved thus far using phage selection technology along with chain shuffling and random mutagenesis to enhance natural antibody diversity, it is likely that therapeutic huMABs can be produced entirely in vitro from naive human repertoires. This should include huMABs directed against self or other nonimmunogenic antigens. Preliminary work in this direction has been reported by HOOGENBOOM and WINTER (1992). Phage selection technology should also be able to facilitate both humanization of therapeutically promising murine MABs and alterations of specificity and improvement of selectivity of other therapeutically promising MABs.

References

Ames RS, Tornetta MA, Tsui P (1992) Isolation of anti-C5a monoclonal antibodies froma filamentous phage Fab display library. Proceedings of the 3rd international conference on antibody engineering, San Diego

Bagshawe KD, Springer CJ, Searle F, Antoniw P, Sharma SK, Melton RG, Sherwood RF (1988) A cytotoxic agent can be generated selectively at cancer sites. Br J Cancer 58:700–703

Barbas CF, Crowe JE Jr, Cababa D, Jones TM, Zebedee SL, Murphy BR, Chanock RM, Burton DR (1992a) Human monoclonal Fab fragments derived from a combinatorial library bind to respiratory syncytial virus F glycoprotein and neutralize infectivity. Proc Natl Acad Sci USA 89:10164–10168

Barbas CF, Bjorling E, Chiodi F, Dunlop N, Cababa D, Jones TM, Zebedee SL, Persson MA, Nara PL, Norrby E et al. (1992b) Recombinant human Fab fragments neutralize human type 1 immunodeficiency virus in vitro. Proc Natl Acad Sci USA 89:9339–9343

Barbas CF, Bain JD, Hoekstra DM, Lerner RA (1992c) Semisynthetic combinatorial antibody libraries: a chemical solution to the diversity problem. Proc Natl Acad Sci USA 89:4457–4461

Beidler CB, Ludwig JR, Cardenas J, Phelps J, Papworth CG, Melcher E, Sierzega M, Myers LJ, Unger BW, Fisher M et al. (1988) Cloning and high level expression of a chimeric antibody with specificity for human carcinoembryonic antigen. J Immunol 141:4053–4060

Borrebaeck CAK, Danielsson L, Moller SA (1988) Human monoclonal antibodies produced by primary in vitro immunization of peripheral blood lymphocytes. Proc Natl Acad Sci USA 85:3995–3999

Boulianne GL, Hozumi N, Shulman MJ (1984) Production of functional chimaeric mouse/human antibody. Nature 312:644–646

Bruggemann M, Williams GT, Bindon CI et al. (1987) Comparison of the effector functions of human immunoglobulins using a matched set of chimeric antibodies. J Exp Med 166:1351–1361

Bruggemann M, Winter G, Waldmann H, Neuberger MS (1989) The immunogenicity of chimeric antibodies. J Exp Med 170:2153–2157

Burton DR, Barbas CF, Persson MAAA, Koenig S, Chanock RM, Lerner RA (1991) A large array of human monoclonal antibodies to type 1 human immunodeficiency virus from combinatorial libraries of asymptomatic individuals. Proc Natl Acad Sci USA 88:10134–10139

Carter P, Prestal L, Gorman CM, Ridgway JB, Henner D, Wong WL, Rowland AM, Kott C, Shepard HM (1992) Humanization of an anti-p185^{HER2} antibody for human cancer therapy. Proc Natl Acad Sci USA 89:4285–4290

Casadei J, Powell MJ, Kenten JH (1990) Expression and secretion of aequorin as a chimeric antibody by means of a mammalian expression vector. Proc Natl Acad Sci USA 87:2047–2051

Caton AJ, Koprowski H (1990) Influenza virus hemagglutinin-specific antibodies isolated from a combinatorial expression library are closely related to the immune response of the donor. Proc Natl Acad Sci USA 87:6450–6455

Chaudhary VK, Queen C, Junghans RP, Waldmann TA, FitzGerald DJ, Pastan IA (1989) Recombinant immunotoxin consisting of two antibody variable domains fused to Pseudomonas exotoxin. Nature 339:394–397

Chaudhary VK, Batra JK, Gallo MG, Willingham MC, Fitzgerald DJ, Pastan I (1990) A rapid method of cloning functional variable-region antibody genes in E. coli as single-chain immunotoxins. Proc Natl Acad Sci USA 87:1066–1070

Chiang YL, Dong R, Larrick JW (1989) Enzymatic amplification and direct cloning of rearranged immunoglobulin cDNA. Biotechniques 7:360–366

Chiswell DJ, McCafferty J (1992) Phage antibodies: will new "coliclonal" antibodies replace monoclonal antibodies. Trends Biotechnol 10:80–84

Clackson T, Hoogenboom JR, Griffiths AD, Winter G (1991) Making antibody fragments using phage display libraries. Nature 352:624–628

Co MS, Deschamps M, Whitley RJ, Queen C (1991) Humanized antibodies for antiviral therapy. Proc Natl Acad Sci USA 88:2869–2873

Co MS, Audalovic NM, Caron PC, Audalovic MV, Scheinberg DAO, Queen C (1992) Chimaeric and humanized antibodies with specificity for the CD33 antigen. J Immunol 148:1149–1154

Cwirla SW, Peters EA, Barrett RW, Dower WJ (1990) Peptides on phage: a vast library of peptides for identifying ligands. Proc Natl Acad Sci USA 87: 6378–6382

Devlin JJ, Panganiban LC, Devlin PE (1990) Random peptide libraries: a source of specific protein binding molecules. Science 249:404–406

Duchosal MA, Eming SA, Fischer P, Leturcq D, Barbas CF, McConahey PJ, Caothien RH, Thornton GB, Dixon FJ, Burton DR (1992) Immunization of hu-PBL-SCID mice and the rescue of human monoclonal Fab fragments through combinatorial libraries. Nature 355:258–262

Epenetos AA, Snook D, Durbin H et al. (1986) Limitations of radiolabeled monoclonal antibodies for localization of human neoplasms. Cancer Res 46:3183

Fogler WE, Sun LK, Klinger MR, Ghrayeb J, Daddona PE (1989) Biological characterization of a chimeric mouse-human IgM antibody directed against the 17-1A antigen. Cancer Immunol Immunother 30:43–50

Foote J, Winter G (1992) Antibody framework residues affecting the conformation of the hypervariable loops. J Mol Biol 224:487–499

Gallo MG, Chaudhary VK, Fitzgerald DJ, Willingham MC, Pastan I (1988) Cloning and expression of the H chain V region of antibody OVB3 that reacts with human ovarian cancer. J Immunol 141:1034–1040

Gascoigne NR, Goodnow CC, Dudzik KI, Oi VT, Davis MM (1987) Secretion of a chimeric T-cell receptor-immunoglobulin protein. Proc Natl Acad Sci USA 84:2936–2940

Gorman SD, Clark MR, Routledge EG, Cobbold ST, Waldmann H (1991) Reshaping a therapeutic CD4 antibody. Proc Natl Acad Sci USA 88:4181–4185

Gussow D, Seemann G (1991) Humanization of monoclonal antibodies. Methods Enzymol 203:99–121

Gussow D, Ward ES, Griffiths AD, Jones PT, Winter G (1989) Generating binding activities from Escherichia coli by expression of a repertoire of immunoglobulin variable domains. Cold Spring Harb Symp Quant Biol 54:265–272

Hale G, Dyer MJS, Clark MR, Phillips JM, Marcus R, Riechmann L, Winter G, Waldmann H et al. (1988) Remission induction in non-hodgkin lymphoma with reshaped human monoclonal antibody CAMPATH-1H. Lancet ii:1394–1399

Hamada H, Miura K, Ariyoshi K, Heike Y, Sato S, Kameyama K, Kurosawa Y, Tsuruo T (1990) Mouse-human chimeric antibody against the multidrug transporter P-glycoprotein. Cancer Res 50:3167–3171

Hamblin TJ, Cattan AR, Glennie MJ, MacKenzie MR, Stevenson FK, Watts HF, Stevenson GT (1987) Initial experience in treating human lymphoma with a chimeric univalent derivative of monoclonal anti-idiotype antibody. Blood 69: 790–797

Hank JA, Robinson RR, Surfus J, Mueller BM, Reisfeld RA, Cheung NK, Sondel PM (1990) Augmentation of antibody dependent cell mediated cytotoxicity following in vivo therapy with recombinant interleukin 2. Cancer Res 50:5234–5239

Hardman N, Gill LL, De Winter RF, Wagner K, Hollis M, Businger F, Ammaturo D, Buchegger F, Mach JP, Heusser C (1989) Generation of a recombinant mouse-human chimaeric monoclonal antibody directed against human carcinoembryonic antigen. Int J Cancer 44:424–433

Hawkins RE, Russell SJ, Winter G (1992) Selection of phage antibodies by binding affinity. Mimicking affinity maturation. J Mol Biol 226:889–896

Heinrich G, Gram H, Kocher HP, Schreier MH, Ryffel B, Akbar A, Amlot PL, Janossy G (1989) Characterization of a human T cell-specific chimeric antibody (CD7) with human constant and mouse variable regions. J Immunol 143:3589–3597

Hoogenboom HR, Winter G (1992) By-passing immunisation. Human antibodies
    from synthetic repertoires of germline VH gene segments rearranged in vitro. J
    Mol Biol 227:381–388
Hoogenboom HR, Raus JC, Volckaert G (1990) Cloning and expression of a chimeric
    antibody directed against the human transferrin receptor. J Immunol 144:3211–
    3217
Hoogenboom HR, Griffiths AD, Johnson KS, Chiswell DJ, Hudson P, Winter G
    (1991) Multi-subunit proteins on the surface of filamentous phage: metho-
    dologies for displaying antibody (Fab) heavy and light chains. Nucleic Acids Res
    19:4133–4137
Hoogenboom HR, Marks JD, Griffiths AD, Winter G (1992) Building antibodies
    from their genes. Immunol Rev 130:41–68
Hsiao Ku-Chuan, Bajorath J, Harris LJ (1992) Humanization of anti-CD18 mAB
    60.3. Proceedings of the 3rd international conference on antibody engineering,
    San Diego
Huse W, Sastry L, Iverson SA, Kang AS, Alting-Mees M, Burton DR, Benkovic SJ,
    Lerner RA (1989) Generation of a large combinatorial library of the immuno-
    globulin repertoire in phage lambda. Science 246:1275–1281
Jain RK (1990) Physiological barriers to delivery of MABs and other macromolecules
    in tumors. Cancer Res 50:814s–819s
Jones PT, Dear PH, Foote J, Neuberger MS, Winter G (1986) Replacing the
    complementarity-determining regions in a human antibody with those from a
    mouse. Nature 321:522–525
Junghans RP, Waldmann TA, Landolfi NF, Avdalovic NM, Schneider WP, Queen C
    (1990) Anti-Tac-H, a humanized antibody to the interleukin 2 receptor with new
    features for immunotherapy in malignant and immune disorders. Cancer Res
    50:1495–1502
Kang AS, Barbas CF, Janda KD, Benkovic SJ, Lerner RA (1991) Linkage of
    recognition and replication functions by assembling combinatorial antibody Fab
    libraries along phage surfaces. Proc Natl Acad Sci USA 88:4363–4366
Kerr DE, Senter PD, Burnett WV, Hirshberg DL, Hellstrom I, Hellstrom KE
    (1990) Antibody-penicillin-V-amidase conjugate skill antigen positive tumor
    cells when combined with doxorubicin phenoxyacetamide. Cancer Immunol
    Immunother 31:202–206
Kettleborough CA, Saldanha J, Heath VJ, Morrison CJ, Bendig MM (1991)
    Humanization of a mouse monoclonal antibody by CDR-grafting: the import-
    ance of framework residues on loop conformation. Protein Eng 4:773–780
Kettleborough CA, Ansell KH, Allen RW, Gussow D, Bendig MM (1993) Use of
    phage display libraries to isolate novel anti-EGF receptor antibodies. Phar-
    maceutical design using epitope selection technologies, Palo Alto (abstract)
Khazaeli MB, Saleh MN, Wheeler RH et al. (1988) Phase I trial of multiple large
    doses of murine monoclonal antibody CO17-1A. II. Pharmacokinetics and
    immune response. J Natl Cancer Inst 80:937–942
Koga H, Kanda H, Nakashima M, Watanabe Y, Endo K, Watanabe T (1990)
    Mouse-human chimeric monoclonal antibody to carcinoembryonic antigen
    (CEA): in vitro and in vivo activities. Hybridoma 9:43–56
Kohler G, Milstein C (1975) Continuous cultures of fused cells secreting antibody of
    predefined specificity. Nature 256:495–497
Larrick JW (1989) Antibody inhibition of the Immunoinflammatory Cascade. J Crit
    Care 4:211
Larrick J (1990) Potential of monoclonal antibodies as pharmacological agents.
    Pharmacol Rev 41:539–557
Larrick J, Fry KE (1991a) Recombinant antibodies. Human Antibod Hybrid 2:
    172–189
Larrick JW, Fry KE (1991b) PCR amplification of antibody genes. In: Lerner RA,
    Burton DR (eds) METHODS – a companion to methods in enzymology volume:
    new techniques in antibody generation, vol 2. Academic, Orlando, pp 106–110

Larrick JW, Chiang YL, Sheng-Dong R, Senyk G, Casali P (1987) Generation of specific human monoclonal antibodies by in vitro expansion of human B cells. In: International symposium on in vitro immunization in hybridoma technology. Elsevier, Tylösand, Sweden, p 231

Larrick JW, Danielson L, Brenner C, Wallace E, Abrahamson M, Fry KE, Borrebaeck C (1989a) Rapid direct cloning of rearranged immunoglobulin genes from small number of human hybridoma cells. Biotechnology 7:934–938

Larrick JW, Danielsson L, Brenner CA, Abrahamson M, Fry KE, Borrebaeck C (1989b) Cloning of rearranged immunoglobulin genes from small numbers of anti-HIV human hybridoma cells. Biochem Biophys Res Commun 160: 1250–1256

Larrick JW, Wallace EF, Coloma MJ, Bruderer U, Lang AB, Fry KE (1993) Therapeutic human antibodies derived from PCR amplification of B-cell variable regions. Immunol Rev 130:69–85

Li YW, Lawrie DK, Thammana P, Moore GP, Shearman CW (1990) Construction, expression and characterization of a murine/human chimeric antibody with specificity for hepatitis B surface antigen. Mol Immunol 27:303–311

Liou RS, Rosen EM, Fung MS, Sun WN, Sun C, Gordon W, Chang NT, Chang TW (1989) A chimeric mouse-human antibody that retains specificity for HIV gp120 and mediates the lysis of HIV-infected cells. J Immunol 143:3967–3975

Liu AY, Robinson RR, Hellstrom KE, Murray ED Jr, Chang CP, Hellstrom I (1987) Chimeric mouse-human IgG1 antibody that can mediate lysis of cancer cells. Proc Natl Acad Sci USA 84:3439–3443

Loibner H, Baker J, Bednarik K, Janzek E, Neruda W, Plot R, Co MS (1992) Generation and characterization of humanized anti-Lewis Y antibodies. Proceedings of the 3rd international conference on antibody engineering, San Diego

Maeda H, Matsushita S, Eda Y, Kimachi K, Tokiyshi SO, Bendig MM (1991) Construction of reshaped human antibodies with HIV neutralizing activity. Human Antib Hybrid 2:124–135

Marchitto KS, Kindsvogel WR, Beaumier PL, Fine SK, Gilbert T, Levin SD, Woodhouse CS, Morgan AC Jr (1989) Characterization of a human-mouse chimeric antibody reactive with a human melanoma associated antigen. Prog Clin Biol Res 288:101–105

Mariuzza RA, Winter G (1989) Secretion of a homodimeric V alpha C kappa T-cell receptor-immunoglobulin chimeric protein. J Biol Chem 264:7310–7316

Marks JD, Hoogenboom HR, Bonnert TP, McCafferty J, Griffiths AD, Winter G (1991) By-passing immunization. Human antibodies from B-gene libraries displayed on phage. J Mol Biol 222:581–597

Marks JD, Griffiths AD, Malmquist M, Clackson TP, Bye JM, Winter G (1992) Bypassing immunization building high-affinity antibodies by chain shuffling. Biotechnology 10:779–783

Masuho Y, Matsumoto Y-I, Sugano T, Tomiyama T, Sasaki S, Koyama T (1990) Development of a human monoclonal antibody against cytomegalovirus with the aim of passive immunotherapy. In: Borrebaeck CAK, Larrick JW (eds) Therapeutic monoclonal antibodies. Stockton, New York, pp 187–207

McCabe RP, Peters LC, Haspel MV et al. (1989) Preclinical studies on the pharmacokinetic properties of human monoclonal antibodies to colorectal cancer and their use for detection of tumors. Cancer Res 48:4348–4853

McCafferty J, Griffiths AD, Winter G, Chiswell DJ (1990) Phage antibodies: filmentous phage displaying antibody variable domains. Nature 348:552–554

McCune JM, Namikawa R, Kaneshima H et al. (1988) The SCID-hu mouse: murine model for the analysis of human hematolymphoid differentiation and function. Science 241:1632–1639

Miglietta J, Shrutkowski A, Farrell T, Kishimoto K, Brown M, Kehry M, Morrison S, Griffin J (1992) Alteration of framework residues modulate binding of a CDR-grafted anti-human ICAM-1 (CD54). Proceedings of the 3rd international conference on antibody engineering, San Diego

Morrison SL, Oi VT (1989) Genetically engineered antibody molecules. Adv Immunol 44:65–92

Morrison SL, Johnson MJ, Herzenberg LA, Oi VT (1984) Chimeric human antibody molecules: mouse antigen-binding domains with human constant region domains. Proc Natl Acad Sci USA 81:6851–6855

Mueller BM, Romerdahl CA, Gillies SD, Reisfeld RA (1990) Enhancement of antibody-dependent cytotoxicity with a chimeric anti-GD2 antibody. J Immunol 144:1382–1386

Mullinax RL, Gross EA, Amberg JR, Hay BN, Hogrefe HN, Kubitz MM, Greener A, Alting-Mees M, Ardourel D, Short JM, Sorge JA, Shopes B (1990) Identification of human antibody fragment clones specific for tetanus toxoid in a bacteriophage γ immunoexpression library. Proc Natl Acad Sci USA 87:8095–8099

Neumaier M, Shively L, Chen FS, Gaida FJ, Ilgen C, Paxton RJ, Shively JE, Riggs AD (1990) Cloning of the genes for T84.66, an antibody that has a high specificity and affinity for carcinoembryonic antigen, and expression of chimeric human/mouse T84.66 genes in myeloma and Chinese hamster ovary cells. Cancer Res 50:2128–2134

Nishimura Y, Yokoyama M, Araki K, Ueda R, Kudo A, Watanabe T (1987) Recombinant human-mouse chimeric monoclonal antibody specific for common acute lymphocytic leukemia antigen. Cancer Res 47:999–1005

Nose M, Wigzell H (1983) Biological significance of carbohydrate chains on monoclonal antibodies. Proc Natl Acad Sci USA 80:6632–6636

Oldenburg KR, Loganathan D, Goldstein IJ, Schultz PG, Gallop MA (1992) Peptide ligands for a sugar-binding protein isolated from a random peptide library. Proc Natl Acad Sci USA 89:5393–5397

Orlandi R, Gussow DH, Jones PT, Winter G (1989) Cloning immunoglobulin variable domains for expression by the polymerase chain reaction. Proc Natl Acad Sci USA 86:3833–3837

Parmley SF, Smith GP (1988) Antibody-selectable filamentous fd phage vectors: affinity purification of target genes. Gene 73:305–318

Persson MA, Caothien RH, Burton DR (1991) Generation of diverse high-affinity human monoclonal antibodies by repertoire cloning. Proc Natl Acad Sci USA 88:2432–2436

Queen C, Schneider WP, Selick HE, Payne PW, Landolfi NF, Duncan JF, Avdalovic NM, Levitt M, Junghans RP, Waldmann TA (1989) A humanized antibody that binds to the interleukin 2 receptor. Proc Natl Acad Sci USA 86:10029–10033

Riechmann L, Clark M, Waldmann H, Winter G (1988) Reshaping human antibodies for therapy. Nature 332:323–327

Roberts BL, Markland W, Ley AC, Kent RB, White DW, Guterman SK, Ladner RC (1992) Directed evolution of a protein: selection of potent neutrophil elastase inhibitors displayed on M13 fusion phage. Proc Natl Acad Sci USA 89:2429–2433

Roux KH, Dhanarajan P (1990) A strategy for single site PCR amplification of dsDNA: priming digested cloned or genomic DNA from an anchor-modified restriction site and a short internal sequence. Biotechniques 8:48–57

Rybak SM, Hoogenboom HR, Meade HM, Raus JCM, Schwatz D, Youle RJ (1992) Humanization of immunotoxins. Proc Natl Acad Sci USA 89:3165–3170

Saga T, Endo K, Koizumi M, Kawamura Y, Watanabe Y, Konishi J, Ueda R, Nishimura Y, Yokoyama M, Watanabe T (1990) In vitro and in vivo properties of human/mouse chimeric monoclonal antibody specific for common acute lymphocytic leukemia antigen. J Nucleic Med 31:1077–1083

Sahagan BG, Dorai H, Saltzgaber-Muller J, Toneguzzo F, Guindon CA, Lilly SP, McDonald KW, Morrissey DV, Stone BA, Davis GL et al. (1986) A genetically engineered murine/human chimeric antibody retains specificity for human tumor-associated antigen. J Immunol 137:1066–1074

Schnee JM, Runge MS, Matsueda GR, Hudson NW, Seidman JG, Haber E, Quertermous T (1987) Construction and expression of a recombinant antibody-targeted plasminogen activator. Proc Natl Acad Sci USA 84:6904–6908

Schroff R, Foon K, Beatty S, Oldham R, Morgan A Jr (1985) Human anti-murine immunoglobulin responses in patients receiving monoclonal antibody therapy. Cancer Res 45:879–885

Scott JK, Smith GP (1990) Searching for peptide ligands with an epitope library. Science 249:386–390

Scott JK, Loganathan D, Easley RB, Gong X, Goldstein IJ (1992) A family of concanavalin A-binding peptides from a hexapeptide epitope library. Proc Natl Acad Sci USA 89:5398–5402

Senter PD, Saulnier MG, Schreiber GJ, Hirshberg DL, Brown JP, Hellstrom I, Hellstrom KE (1988) Anti-tumor effects of antibody-alkaline phosphatase conjugates in combination with etoposide phosphate. Proc Natl Acad Sci USA 85:4842–4846

Senter PD, Schreiber GJ, Hirshberg DL, Ashe SA, Hellstrom I, Hellstrom KE (1989) Enhancement of the in vitro and in vivo antitumor activities of phosphorylated mitomycin C and etoposide derivatives by monoclonal antibody-alkaline phosphatase conjugates. Cancer Res 49:5789–5792

Senter PD, Su PCD, Katsuragi T, Sakai T, Coasand WL, Hellstrom I, Hellstrom KE (1991) Generation of 5-fluorouracil from 5-fluorocytosineby monoclonal antibody-cytosine deaminase conjugates. Bioconjugate Chem 2:447–451

Sharon J, Gefter ML, Manser T, Morrison SL, Oi VT, Ptashne M (1984) Expression of a VHC kappa chimaeric protein in mouse myeloma cells. Nature 309:364–367

Shaw DR, Khazaeli MB, Sun LK, Ghrayeb J, Daddona PE, McKinney S, LoBuglio AF (1987) Characterization of a mouse/human chimeric monoclonal antibody (17-1A) to colon cancer tumor-associated antigen. J Immunol 138:4534–4538

Shawler DL, Bartholomew RM, Smith LM, Dillman RO (1985) Human immune response to multiple injections of murine monoclonal IgG. J Immunol 135:1530–1535

Shen W-C, Ballou B, Ryser, HJ-P, Hakala TR (1986) Targeting, internalization, and cytotoxicity of methotrexate-monoclonal anti-stage specific embryonic antigen-1 antibody conjugatesin cultured F-9 teratocarcinoma cells. Cancer Res 46:3912–3916

Shepherd TA, Jungheim LN, Meyer DL, Starling JJ (1991) A novel targeted delivery system utilizing a cephalosporin-oncolytic prodrug activated by an antibody β-lactamase conjugate for the treatment of cancer. J Biomed Chem Lett 1:21–26

Shin SU, Morrison SL (1990) Expression and characterization of an antibody binding specificity joined to insulin-like growth factor 1: potential applications for cellulartargeting. Proc Natl Acad Sci USA 87:5322–5326

Smith GP (1985) Filamentous fusion phage: novel expression vectors that display cloned antigens on the virion surface. Science 228:1315–1317

Steplewski Z, Sun LK, Shearman CW et al. (1988) Biological activity of human-mouse IgG1, IgG2, IgG3 and IgG4 chimeric monoclonal antibodies with anti-tumor specificity. Proc Natl Acad Sci USA 85:4852–4856

Sun LK, Curtis P, Rakowicz-Szulczynska E, Ghrayeb J, Chang N, Morrison SL, Koprowski H (1987) Chimeric antibody with human constant regions and mouse variable regions directed against carcinoma-associated antigen 17-1A. Proc Natl Acad Sci USA 84:214–218

Tempest PR, Bremner P, Lambert M, Taylor G, Furze JM, Carr FJ, Harris WJ (1991) Reshaping a human monoclonal antibody to inhibit human respiratory syncytial virus infection in vivo. BioTechnology 9:266–271

Trouet A, Masquelier M, Baurain R, Deprez-De Campeneere D (1982) A covalent linkage between daunorubicin and proteins that is stable in serum and reversible by lysosomal hydrolases, as required for a lysosomotropic drug-carrier conjugate: in vitro and in vivo studies. Proc Natl Acad Sci USA 79:626–629

Tsuchiya M, Sato K, Saldanha J, Tsunenari T, Koishihara Y, Ohsugi Y, Kishimoto T, Bendig MM (1992) The humanization of two mouse antibodies that inhibit IL-6-dependent tumor cell growth. Proceedings of the 3rd international conference on antibody engineering, San Diego

Verhoeyen M, Milstein C, Winter G (1988) Reshaping human antibodies: grafting an antilysozyme activity. Science 239:1534–1536

Vitetta ES, Krolick KA, Miyama-Inaba M et al. (1983) Immunotoxins: a new approach to cancer therapy. Science 219:644

Waldmann H, Hale G, Clark M et al. (1988) Monoclonal antibodies for immunosuppression. Prog Allergy 45:16–30

Ward ES, Gussow D, Griffiths AD, Jones PT, Winter G (1989) Binding activities of a repertoire of single immunoglobulin variable domains secreted from *Escherichia coli*. Nature 341:544–546

Whittle N, Adair J, Lloyd C, Jenkins L, Devine J, Schlom J, Raubitschek A, Colcher D, Bodmer M (1987) Expression in COS cells of a mouse-human chimaeric B72.3 antibody. Protein Eng 1:499–505

Williams GT, Neuberger MS (1986) Production of antibody-tagged enzymes by myeloma cells: application to DNA polymerase I Klenow frament. Gene 43: 319–324

Woodle ES, Thistlewaite JR, Jolliffee LK, Zivin RA, Coltins A, Adair JR, Bodmer M, Athwal D, Alegre ML, Bluestone JA (1992) Humanized OKT3 antibodies: successful transfer of immune modulating properties and idiotype expression. J Immunol 148:2756–2763

Yeh M-Y, Roffler SR, Yu M-H (1992) Doxorubicin-monoclonal antibody conjugate for therapy of human cervical carcinoma. Int J Cancer 51:274–282

Yokoyama M, Nishimura Y, Watanabe T (1987) Suppression of tumor growth by in vivo administration of a recombinant human-mouse chimeric monoclonal antibody. Jpn J Cancer Res 78(11):1251–1257

Zebedee SL, Barbas CF, Hom Y-L, Caothien RH, Graff R, Degaw J, Pyati J, LaPolla R, Burton DR, Lerner RA, Thornthon GA (1992) Human combinatorial antibody libraries to hepatitis B surface antigen. Proc Natl Acad Sci USA 89:3175–3180

Zettlmeissl G, Gregersen JP, Duport JM, Mehdi S, Reiner G, Seed B (1990) Expression and characterization of human CD4: immunoglobulin fusion proteins. DNA Cell Biol 9:347–353

CHAPTER 3

Transgenic Approaches to Human Monoclonal Antibodies

N. LONBERG

A. Introduction

Monoclonal antibodies (MoAbs), which combine high specificity and low toxicity, would seem to be ideal candidates for Ehrlich's "magic bullet." However, despite the widespread use of one MoAb product (GOLDSTEIN et al. 1985), antibodies in general have not lived up to their initial expectations. This is in part due to the intrinsic immunogenicity of nonhuman antibodies. The most commonly used technique for generating MoAbs employs rodent B cells, and patients respond to therapeutic doses of rodent monoclonals by making antibodies against the rodent immunoglobulin sequences. These human anti-mouse antibodies (HAMAs) can neutralize the therapeutic antibodies, leading to a shorter in vivo half life and reduced effectiveness (TJANDRA et al. 1990), thus motivating a search for ways to generate human MoAbs. One potential route involves manipulating the mouse genome to create mice with transplanted human immunoglobulin genes and a human antibody repertoire. This is made possible by techniques developed over the last decade. In 1980 Gordon et al. reported a method for the incorporation of cloned DNA sequences into mouse embryos; the resulting transgenic mouse carries the foreign DNA within its own genetic material and passes it on to its offspring. The procedure was quickly used to generate strains of mice expressing light (BRINSTER et al. 1983) and heavy (GROSSCHEDL et al. 1984) chain immunoglobulins encoded by cloned genes. By 1985 Alt et al. proposed that it might be possible to use this technology to engineer a mouse that would respond to antigen with human sequence antibodies. The authors conceded that this scheme was "conceptually outlandish", but that it "may actually be realized in the not too distant future". Eight years later we appear to be close to that realization; however, because such a mouse has not yet been used to generate human MoAbs for therapeutic use, it remains a largely theoretical construct. I will therefore review some of the technological considerations necessary for an evaluation of the feasibility of this scheme.

B. Competing Technologies for the Generation of Therapeutic Antibodies

What are the currently available methods for generating human MoAbs, and is there a need for new technologies such as a human antibody transgenic mouse? Present technology for generating MoAbs involves preexposing an animal (usually a rat or mouse) with antigen. This preexposure leads to the formation of splenic B cells that secrete immunoglobulin molecules with high affinity for the antigen. The spleen cells are then fused with myeloma cells to form immortal, antibody secreting, hybridoma cell lines (KÖHLER and MILSTEIN 1975). Individual hybridoma clones are screened to identify those cells producing immunoglobulins directed against a particular antigen. Ideally, any new technology for generating human MoAbs should be no more difficult to use than the current methodology for producing rodent monoclonals. However, obtaining human sequence MoAbs has proved to be far more difficult than obtaining rodent MoAbs. Human hybridomas are more difficult to generate and maintain than rodent hybridomas, human subjects cannot be immunized and manipulated like laboratory mice, and humans are intrinsically tolerant to many of the human antigens that represent potential therapeutic targets. As a result no single technology has emerged for making human antibodies, and several competing systems are under development. These technologies can be grouped into four basic strategies: (1) find a human B cell that produces the right antibody; (2) start with a nonhuman antibody and make it look as human as possible; (3) start from scratch and use the powerful screening methods developed by molecular biologists to select the right antibody from an essentially random collection of synthetic sequences; and (4) transplant human immunoglobulin genes into a mouse, let the mouse immune system generate high affinity antibodies, and isolate human, antibody secreting, mouse hybridomas by conventional methods.

All of these strategies have potential disadvantages. The first strategy, direct isolation from human B cells, includes several different methods of obtaining cells and several different methods of immortalizing the cells or isolating the antibody genes. Antibody producing cells can be directly isolated from human subjects or from immunodeficient mice that have been xenografted with human immune tissue and then exposed to antigen (LUBIN et al. 1991), or generated by in vitro affinity maturation (BORREBAECK 1988; BANCHEREAU et al. 1991; BANCHEREAU and ROUSSET 1991). These human B cells can then be directly immortalized with Epstein-Barr virus or fused with a myeloma line to form a hybridoma (JAMES and BELL 1987; THOMPSON 1988), or the antibody genes isolated by molecular cloning (DUCHOSOL et al. 1992). Recent advances have been made in molecular cloning with the development of new expression vectors for antibody variable region cDNA sequences (ORLANDI et al. 1989; HUSE et al. 1989; CLACKSON et al. 1991; KANG et al. 1991). Random heavy and light chain variable region sequences derived from donor RNA can be expressed together on the surface of a single bacterium or bacteriophage, and the resulting specificity of these cloned sequences used to isolate individual clones. It appears that the repertoire expressed by these so-called combinatorial libraries is very similar to the original repertoire of the RNA donor (CATON and KOPROWSKI 1990), which makes them useful for isolating human sequences encoding pre-existing specificities from human B cells. Using this technique, DUCHOSOL et al. (1992) were able to obtain tetanus toxoid specific, human antibody Fab fragments with apparent affinities in the 10^{-8}–10^{-9} range from human peripheral blood lymphocyte (PBL) xenografted immunodeficient mice. All of the steps involved in isolating human sequence antibodies directly from human B cells are technically challenging. This strategy may also be particularly ineffective for generating antibodies against human immune cell surface proteins, which are unlikely to be recognized by human B cells.

The second strategy – engineering of existing nonhuman antibodies by grafting antigen recognition sequences from a characterized monoclonal onto framework sequences from a human antibody – is being widely pursued and has generated several different molecules that are now in preclinical and clinical trials (HALE et al. 1988; QUEEN et al. 1989). Drawbacks of this approach include potential loss of affinity, retention of antigenic nonhuman sequences, and the time involved in individually "humanizing" candidate rodent antibodies. A promising variation on this theme involves starting with an "almost-human" antibody. Immunoglobulins from nonhuman primates are very close in sequence to human immunoglobulins and are presumably less immunogenic than rodent proteins. NEWMAN et al. (1992) used this technique to generate a chimeric antibody consisting of cynomolgus monkey heavy and light chain variable region sequences fused to human λ and IgG1 constant region sequences. The resulting monoclonal is similar to consensus human variable region framework sequences at greater than 90% of its amino acid residues and binds to human CD4 with an affinity of 3 × 10^{-11}. One potential problem with this method is the difficulty of isolating nonrodent hybridomas. A second potential problem involves the underlying logic of this approach. If immunoglobulins from a given species are so similar to human proteins that they do not illicit a human immune response, this particular species is in turn likely to be tolerant to many human proteins (or at least individual epitopes) that represent potential therapeutic targets.

The third approach – construction and screening of recombinant libraries comprising synthetic, or mutated, human variable region sequences – is derived from the combinatorial library approach discussed above. This method promises to deliver new antibody specificities by introducing synthetic sequences into cloned variable regions and screening for high affinity antibodies (BARBAS et al. 1992; GRAM et al. 1992a; LERNER et al. 1992; HAWKINS et al. 1992; HOOGENBOOM and WINTER 1992). It is basically a molecular biology analogue of the natural process of B cell affinity maturation that occurs within germinal centers upon immunization. This strategy may be limited by the nature of the screening process, which requires the antigen to be purified, or abundant, and to be stable under the conditions of the screen. It also remains to be seen whether or not in vitro affinity maturation will lead to antibodies with affinities as high as those generated in vivo.

The last strategy – obtaining human sequence MoAbs from a transgenic mouse – is the newest and least tested. However, it could solve many of the problems associated with the methods discussed above. It would be possible to make human MoAbs using methodologies that are now routinely employed for producing mouse monoclonals. Furthermore, the transgenic mouse is not tolerized against human antigens, providing access to this otherwise unavailable portion of the human antibody repertoire.

C. Origins of Antibody Diversity

To build a transgenic mouse useful for generating human MoAbs it may be necessary for the inserted transgenes to recapitulate the natural human immunoglobulin loci's ability to direct B cell diversity. The next two sections will discuss the molecular origins of this diversity and the structure of the human loci that must be mimicked within the mouse.

I. Functional Requirements for a Human Immunoglobulin Transgene

An IgG antibody molecule consists of four polypeptide chains, two identical light chains and two identical heavy chains, folded into 12 homologous structural domains (Fig. 1). Each of these 12 domains consists of approximately 110 amino acids folded into two anti-parallel β-sheets. The antigen combining site is formed at the interface between the NH_2-terminal heavy chain domain and the NH_2-terminal light chain domain. For each of the

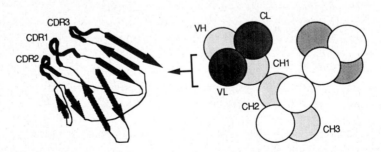

Fig. 1. Three dimensional structure of an IgG antibody. The antibody consists of 12 structural domains, each of which comprises a similar folding pattern. This fold is depicted for one of the variable domains which contacts antigen. Most of the antigen contacts are contained in the three complimentarity determining regions *CDR1, 2,* and *3*

Unrearranged light chain gene:

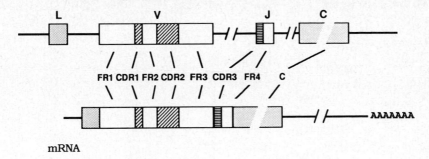

mRNA

Unrearranged heavy chain gene:

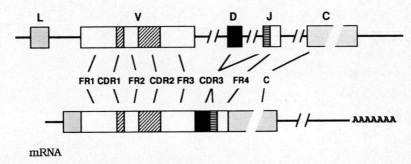

mRNA

Fig. 2. Genomic organization of unrearranged and rearranged immunoglobulin genes. Gene segments encoding different portions of the heavy and light chain variable regions are joined during B cell development to generate a primary repertoire

two domains, the residues that interact with antigen are located on three loops connecting β-strands. These three loops, termed complimentarity determining regions 1, 2, and 3 (CDR1, 2, and 3), are also the regions of greatest sequence diversity between different antibody molecules recognizing different antigens. Thus, the antibody repertoire is determined by sequence diversity at CDR1, 2, and 3. This diversity is derived from three sources: recombinational diversity, junctional diversity, and somatic mutation. Recombinational diversity at CDR1 and 2 comes from the choice of different V segments containing different CDR1 and 2 sequences. Recombinational diversity at CDR3 comes from the choice of different D and J segments. Light chain CDR3 sequences are formed by gene rearrangements that bring together light chain V and J segments, while heavy chain CDR3s are formed by the fusion of heavy chain V, D, and J gene segments (Fig. 2). The size of the recombinational component is a function of the number of possible V, D, and J combinations and the number of different possible

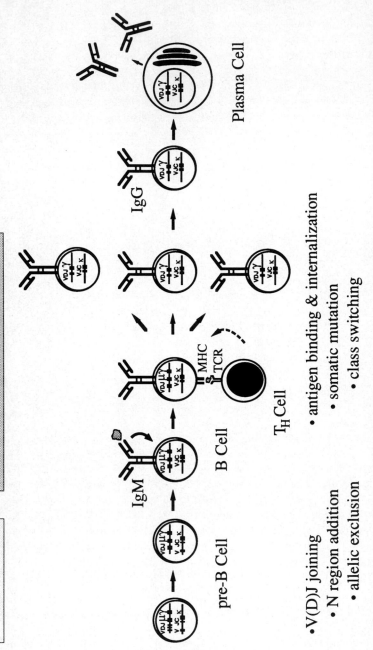

Fig. 3. T cell dependent pathway of B cell development. Rearrangement of heavy and light chain gene segments within the fetal liver or adult bone marrow generates a set of virgin B cells displaying cell surface receptors with different specificities. This constitutes the primary repertoire. Virgin B cells bind antigen with low affinity. Bound antigen is internalized, processed, and presented on the cell surface by class II MHC molecules. The MHC/antigen complex is recognized by germinal center T cells, which stimulate the B cells to undergo somatic mutation accompanied by affinity maturation. Class switched, high affinity, memory B cells can then give rise to IgG secreting plasma cells

heavy and light chain combinations that produce functional antibodies. Junctional diversity contributes only to CDR3 diversity, while somatic mutation, acting across the entire V region, contributes to diversity at all three CDRs. Recombinational and junctional diversity together constitute the diversity of the primary repertoire. Thus VDJ joining generates a set of IgM expressing primary B cells. Any B cell that expresses a cell surface IgM molecule with a certain minimal affinity for a foreign antigen will internalize that antigen as IgM is cycled off the cell surface. The antigen will then be processed and associated peptides will be presented on the cell surface by class II MHC molecules (LANZAVECCHIA 1985). If enough foreign antigen is presented at the cell surface this will trigger a T cell response that will in turn trigger the T cell dependent maturation of the B cell (Fig. 3). This is the so-called secondary response. Part of this response involves the hypermutation of the variable portion of the immunoglobulin genes. Thus a B cell clone undergoing a secondary response will constantly be giving rise to new clones with altered immunoglobulin molecules. Those clones with higher affinities for the foreign antigen will be selectively expanded by helper T cells, giving rise to affinity maturation of the expressed antibody.

Ideally, a transgenic mouse that responds to antigen stimulation with a human antibody repertoire would contain inserted human genes that function correctly throughout the pathway of B cell development described above. To accomplish this, the transgenes will have to satisfy a number of different criteria. These include: (1) high level and cell type specific expression; (2) functional gene rearrangement; (3) activation of, and response to, allelic exclusion; (4) expression of a sufficient primary repertoire; (5) signal transduction; (6) class switching; (7) somatic hypermutation; and (8) domination by the transgene antibody locus during the immune response. The immunoglobulin molecule, which is central to the entire process of B cell development, is not only the repository of diversity, but is also a component of the B cell receptor complex responsible for antigen internalization and signal transduction. To satisfy the criteria listed above, it is therefore necessary to build transgenes that can reconstitute a functional *mouse* B cell receptor using *human* sequences. The structure of the human immunoglobulin loci, from which these transgenes must be assembled, is described in the next section.

II. Structure of the Human Immunoglobulin Loci

In humans the primary repertoire is encoded by a single heavy chain locus and two light chain loci, κ and λ (Fig. 4). The three unlinked loci are located on chromosomes 2, 22, and 14, respectively. The loci are quite large relative to the sizes of DNA fragments which are usually cloned and manipulated by standard molecular biological techniques. The structures of the loci are discussed below, and the available techniques for transplanting these loci into the mouse germline are discussed in the following section.

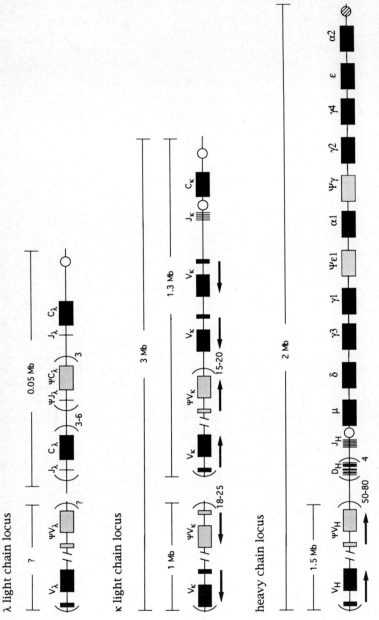

Fig. 4. Organization of the human immunoglobulin gene loci. *Solid rectangles* indicate functional protein coding gene segments and *shaded rectangles* indicate pseudogenes; *circles* indicate transcriptional enhancer elements. The 3′ heavy chain enhancer has been identified in mice and rats but has not yet been found in humans. Relative orientation of V gene segments is given by *arrows*. *Parentheses* indicate regions with multiple gene segments; the approximate number of segments within each of these regions is given (pseudogenes and functional genes are interspersed, not paired, within these regions). The diagram is not drawn to scale; references are given in the text

1. The Human λ Light Chain Locus

The λ light chain locus, which is the least well characterized, accounts for approximately 40% of the expressed immunoglobulins. This is an order of magnitude higher than the expression levels of λ light chains in the mouse and may be a consequence of the greater relative diversity of the human λ locus. While the mouse λ locus consists of three functional Vλ segments and three functional Jλ segments, each of which is paired with a functional Cλ segment (SELSING et al. 1989), the human λ locus consists of at least seven families of variable segments, each of which appears to contain multiple members (ANDERSON et al. 1984), and at least four functional Jλ-Cλ pairs (VASICEK and LEDER 1990). The number of V segments is estimated by Southern hybridization to be between 50 and 100; however, the size of this region is unknown. The exact number of joining and constant regions appears to vary between alleles (TAUB et al. 1983). All of the functional Jλ-Cλ pairs are clustered in a 50 kb region that includes 3 Jλ-Cλ pseudogenes and a transcriptional enhancer element located 12 kb downstream of the last constant segment (BLOMBERG et al. 1991).

2. The Human κ Light Chain Locus

The other 60% of the expressed human light chain repertoire is encoded by the κ locus. This locus consists of approximately 80–90 Vκ gene segments, five Jκ segments and a single Cκ segment spread out over approximately 3 Mb (LORENZ et al. 1987; STRAUBINGER et al. 1988; PARGENT et al. 1991). Only 36 of the 67 Vκ germline gene segments that have been sequenced appear to be functional, suggesting that there are fewer than 50 functional segments in the entire locus (MEINDL et al. 1990).

Two transcriptional enhancers have been identified in the human κ locus: one located within the J-C intro (EMORINE et al. 1983) and one located 12 kb downstream from the constant region (MÜLLER et al. 1990; JUDDE and MAX 1992).

3. The Human Heavy Chain Locus

The heavy chain locus encodes considerably greater diversity because it includes both D and J segments. There are approximately 100–150 human VH gene segments, 20–30 D region segments, six JH segments, and nine functional constant region genes (PASCUAL and CAPRA 1991; HOFKER et al. 1989; WALTER et al. 1990, 1991; VAN DIJK et al. 1992; MATSUDA et al. 1993); however, a significant fraction of the V and D segments may be either nonfunctional or rarely expressed. MATSUDA et al. (1993) have isolated clones spanning the D proximal half of the 1.5 Mb human heavy chain variable region. They determined the sequence of each of the 64 VH segments within this contig and found that, like the κ locus, almost half (31/64) of the heavy chain V segments are pseudogenes. If this ratio can be applied to the rest of the heavy chain variable region, there are probably less than 75 functional human VH segments. Using a completely different approach Tomlinson et al. (1992) used universal PCR primers to compile a complete set of germline heavy chain variable regions from a single individual. Fifty-one of these sequences contain an open reading frame. When the authors compared their sequences to 32 other germline sequences reported in the literature they found very little allelic polymorphism. They estimated from the set of all 83 published sequences that there are only about 50 functional VH genes with distinct CDR sequences.

The expressed VH repertoire could be derived from only a subset of the 50 structurally distinct gene segments encoded in the germline. It is not yet possible to estimate the frequency with which each of these segments are incorporated into functional genes; however, deletion maps of the heavy chain locus provide a preliminary indication that the frequency varies between VH segments. Two such deletion maps of the VH locus (Walter et al. 1991; Van Dijk et al. 1992) have been constructed by analyzing the structure of rearranged B cell chromosomes. Surprisingly, only two of the 42 V to DJ rearrangements appear to fall outside the region mapped by Matsuda et al. (1993), suggesting that the distal half of the VH locus does not contribute equally to the expressed primary repertoire, which consequently may be encoded by fewer than 50 VH gene segments.

The heavy chain locus appears to contain two transcriptional enhancers. An intronic enhancer located between the J and μ gene segments has been well characterized (Hayday et al. 1984), and an additional 3′ enhancer has been located 25 kb and 12 kb downstream of the last constant region gene (Cα) in rat and mouse, respectively (Pettersson et al. 1990; Lieberson et al. 1991; Dariavach et al. 1991). An analogous human heavy chain 3′ enhancer has not yet been reported; however, the relative positions of enhancers elsewhere within the immunoglobulin heavy and light chain loci are conserved. If the position of the 3′ heavy chain enhancer is conserved between mouse and human, it is not clear exactly where the human sequence would be located because the human constant region represents a duplication relative to the mouse region (Honjo et al. 1989). It may turn out that the 3′ enhancer was included in this duplication event and that there is an enhancer downstream of each of the two human Cα genes.

D. Transgenic Technology

What tools are available to manipulate the DNA sequences described above to that they can be inserted into the mouse genome and function correctly? The following sections describe technologies for making transgenic mice.

I. Pronuclear Microinjection

The most commonly used technique for generating transgenic animals is pronuclear microinjection (GORDON et al. 1980; BRINSTER et al. 1985; HOGAN et al. 1986). A dilute solution of a linear DNA fragment ($1-5\,\mu$g/ml) is injected into one of the two pronuclei of a one-half day embryo using a drawn glass capillary as a needle. The injected embryos are then reimplanted into pseudopregnant females. Approximately 20% of the resulting newborn animals will contain the injected DNA within their own genetic material and pass it on to their offspring. The foreign DNA usually integrates as a multicopy tandem array at a single random chromosomal site (PALMITER and BRINSTER 1986). This tandem array structure, which consists largely of head-to-tail fused inserts, is probably due to rapid ligation and homologous recombination between injected fragments prior to integration. Similar integration structures are obtained when DNA is microinjected into cultured cells (FOLGER et al. 1982). This phenomenon can be exploited for generating animals that contain two different DNA fragments integrated at the same site. Coinjected fragments tend to cointegrate (SMALL et al. 1985; STORB et al. 1986a), and if the two fragments share sequences in common, they frequently undergo homologous recombination prior to microinjection. This makes it possible to build transgene inserts by in vivo repair of individually cloned overlapping fragments (PALMITER et al. 1985; PIEPER et al. 1992).

II. Embryonic Stem Cells

A more recently developed technology for generating transgenic animals involves the use of pluripotent stem cell lines. These so-called embryonic stem (ES) cells, which are derived from cultured blastocysts, can be passaged, transfected, and selected like most other cell lines; however, they can also be used to generate transgenic animals containing cell line derived chromosomes (EVANS and KAUFMAN 1981; MARTIN 1981; ROBERTSON 1987). The ES cells are injected into the blastocoel cavity of blastocyst stage embryos, which are then transferred into the uterus of a pseudopregnant female (BRADLEY et al. 1984; BRADLEY 1987). Usually a high percentage of the resulting newborns are chimeric; the somatic and germ tissues are formed from a mix of ES cell line and host blastocyst derived cells. Animals with a high ES cell line derived contribution to their germ cells can be bred to generate stable lines of transgenic animals. This technology has many advantages over pronuclear microinjection. Perhaps the major advantage is that ES cells can be selected for drug resistance markers in culture. This makes it possible to select for rare events such as homologous recombination between transfected DNA sequences and chromosomal sequences (THOMAS and CAPPECHI 1987; MANSOUR et al. 1988; THOMPSON et al. 1989). A DNA fragment consisting of a positively selectable marker, such as the neomycin resistance (neoR) gene, flanked by target specific genomic sequences is transfected into ES cells. Most of the resulting neomycin resistant clones include randomly integrated DNA fragments; however a fraction of the clones result from double homologous recombination events between chromosomal sequences and the cloned sequences flanking the neoR gene. These targeted clones, which can be identified by Southern blot hybridization, contain specific sequence insertions and deletions at selected chromosomal locations. The frequency of targeted clones can be increased by including a negatively selectable marker, such as the herpesvirus thyrmidine kinase gene, at one or both ends of the DNA fragment. Targeted insertions will have lost these flanking sequences during the double crossover events; however, a high proportion of the random insertions will include them. Most of the neomycin resistant clones which resulted from random insertions of the transfected fragment can therefore be selected against. ES cell lines resulting from homologous recombination events are then used to generate chimeric mice. Subsequent breeding generates animals containing the targeted chromosomal modification within every somatic and germline cell. This technique has been used to engineer mouse strains in which specific genes have been disrupted (SCHWARTZBERG et al. 1989; ZIJLSTRA et al. 1989) or even subtly modified (HASTY et al. 1991).

III. Transgene Constructs

As discussed above, the human immunoglobulin loci are spread out over very long stretches of DNA. The loci are so large that they probably cannot be manipulated and inserted intact into the mouse genome. Even relatively modest fractions of these loci are difficult to isolate and maintain using the standard plasmid cloning methods of molecular biology. Furthermore, extreme care must be taken when paring down the loci to construct transgenes, because elimination of seemingly neutral sequences such as introns can adversely effect expression in transgenic animals (BRINSTER et al. 1988). The following sections discuss methods for manipulating large DNA sequences and inserting them into the mouse genome.

1. Bacteriophage Cloning Vectors

The first vectors used for cloning DNA sequences larger than about 20 kb were the cosmid vectors (ISH-HOROWICZ and BURKE 1981). These vectors use plasmid origins of replication but take advantage of the size selection offered by packaging in phage lambda. The problem with this technology is that insert size is restricted to the range of 35–45 kb because of the mechanism of phage λ packaging. One solution involves the use of other bacteriophage systems, such as P1, which do not package discreet lengths of DNA. This type of phage based vector has been developed by STERNBERG (1990; STERNBERG et al. 1990) for isolating fragments in the 75–100 kb range.

2. Plasmid Cloning Vectors

Plasmid vectors are more flexible than phage vectors with regards to insert size, and multicopy plasmids make it possible to isolate large quantities of cloned DNA. However, large inserts cloned into multicopy plasmids tend to be unstable and undergo deletions. For this reason a number of investigators have used low copy plasmids for cloning large inserts. F factor based vectors appear to be particularly well suited for cloning large DNA fragments because they are low copy and they have a partition system that ensures that each daughter cell maintains the plasmid. Hosoda et al. (1990) used such a vector to directly clone a 103 kb fragment from human genomic DNA. O'Connor et al. (1989) used an F factor based vector to build a 125 kb insert of *Drosophila* DNA sequences by repeated rounds of homologous recombination with smaller "shuttle" plasmids. Non-F factor based low copy cloning vectors have also been generated (Wang and Kushner 1991; Stoker

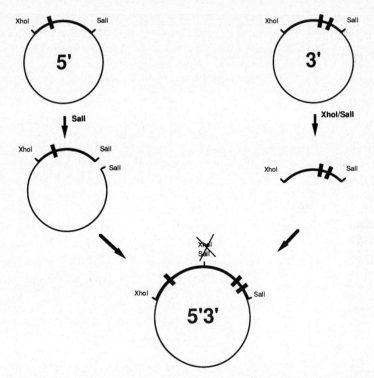

Fig. 5. Strategy for building large transgenes in plasmid vectors. A simplified scheme for combining multiple plasmid inserts to build large DNA constructs is shown. Individual fragments are first cloned into a vector with *Xho*I and *Sal*I sites flanking the polylinker. *Xho*I/*Sal*I fragments from one clone can be inserted at either the 5′ *Xho*I or 3′ *Sal*I site of another clone. The internal *Xho*I and *Sal*I sites are destroyed so that the resulting plasmid still has unique *Xho*I and *Sal*I sites for subsequent cloning steps

et al. 1982). We have designed a series of medium copy, pBR322 derived
vectors that offer a compromise between the stability of low copy plasmids
and the ease of DNA isolation from high copy plasmids (TAYLOR et al.
1992). A simplified version of our cloning strategy is outlined below. First, a
set of phage clone inserts that together span the desired final construct are
subcloned into the polylinker region of one of our plasmid vectors such that
all of the inserts are arrayed in identical 5' to 3' orientations relative to the
polylinker. The polylinker is flanked by unique *Xho*I and *Sal*I restriction
sites which are in turn flanked by a pair of *Not*I sites. All of the internal
*Xho*I and *Sal*I sites are destroyed by partial digestion and Klenow fill-in.
Neighboring clones are then concatenated by isolating the *Xho*I/*Sal*I fragment from one clone and ligating it to the second clone, which has been
linearized with either *Xho*I or *Sal*I depending on the relative 5' to 3' order
of the inserts. The resulting fusion destroys the internal *Xho*I and *Sal*I sites,
maintaining unique flanking *Xho*I and *Sal*I sites (Fig. 5). This process is
repeated until all of the fragments have been assembled into a single clone.
Prior to microinjection or transfection, the plasmid is digested with *Not*I
and the insert DNA isolated from vector sequences which may adversely
influence transgene expression (CHADA et al. 1985; TOWNES et al. 1985a,b;
WIDERA et al. 1987). We have used this strategy to build several human
immunoglobulin gene minilocus constructs in the 50–80 kb range (Fig. 6).

3. Yeast Artificial Chromosome Vectors

Yeast artificial chromosome (YAC) vectors consist of cloned telomere,
centromere, and replication sequences that allow megabase sized foreign
inserts to be propagated as stable chromosomes in yeast (BURKE et al. 1987).
These vectors make it theoretically possible to isolate substantial portions
of the human immunoglobulin loci. An additional advantage of working
with YAC clones is that individually isolated clones can be combined by
homologous recombination in yeast to generate larger clones (DEN DUNNEN
et al. 1992). However, it is still a technical challenge to generate a transgenic
animal with an intact megabase sized DNA fragment (HUXLEY and GNIRKE
1991). The primary source of difficulty is the fragility of fragments larger
than 1–200 kb.

Manipulations required for purification and microinjection or transfection can shear the DNA. This is further complicated by the fact that YAC

Fig. 6. Human minilocus transgenes. Human immunoglobulin minilocus transgene
constructs. The four transgene inserts, *KC1*, *KC2*, *HC1*, and *HC2*, are depicted as
they appear prior to microinjection (after linearization with the restriction enzyme
*Not*I and isolation from vector sequences). *Open triangles* indicate discontinuities
between the structure of the transgene and the natural chromosomal structural of the
intact human gene loci. The start site of the human γ1 pre-switch sterile transcript is
indicated by the *wavy arrow* below HC1. *V*, variable segment; *D*, diversity segment;
J, joining segment, *C*, constant region gene; *S*, switch region; *E*, enhancer. (From
TAYLOR et al. 1992 and TAYLOR and LONBERG 1993)

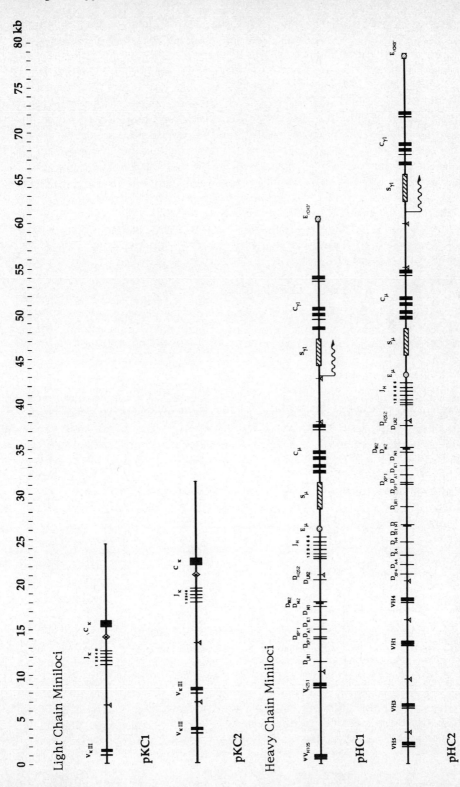

vectors are maintained at one copy per cell, making it difficult to obtain large quantities of DNA. One solution to this problem is presented by the recent development of amplifiable YAC vectors (SMITH et al. 1990). SCHEDL et al. (1992) used an amplifiable vector to isolate a 35 kb YAC containing a tyrosinase gene and generated transgenic mice by pronuclear microinjection. While this technique may not be practical for larger YAC clones, the result is significant because the YAC vector derived telomere sequences integrated into the mouse chromosome without eliminating expression or affecting the stability of the transgene. It is particularly encouraging because cloned human telomere sequences have been found to cause fragmentation of the chromosome upon integration into a mammalian cell line (FARR et al. 1992). Other investigators have used lipofection (STRAUSS and JAENISCH 1992) and spheroplast fusion (PACHNIS et al. 1990; GNIRKE et al. 1991) to transfer YAC clones as large as 680 kb (HUXLEY et al. 1991) into a variety of cell lines, including ES cells (DAVIES et al. 1992).

E. Immunoglobulin Transgenics

Given the various technologies discussed in the preceding section, what success have researchers had generating functional immunoglobulin transgenes? In the discussion of the origins of antibody diversity, I listed eight criteria that may have to be met by a successful immunoglobulin transgene: high level and cell type specific expression; functional gene rearrangement; activation of, and response to, allelic exclusion; expression of a sufficient primary repertoire; signal transduction; class switching; somatic hypermutation; and domination by the transgene antibody locus during the immune response. The following sections describe experimental results related to each of these criteria.

I. High Level and Cell Type Specific Expression

1. *Cis*-acting Regulatory Sequences

The first requirement for correct transgene function is high level and cell type specific expression. The transgene must be actively transcribed in primary B cells in order to make enough surface IgM to trigger T cell dependent maturation. The expression of secreted antibody must increase in IgG secreting plasma cells, and this level of expression must be sustained after fusion with myeloma cells to generate efficient MoAb producing hybridomas. For a transgene to fulfill these requirements it must contain all of the *cis*-acting transcriptional regulatory elements required for cell type specific expression. A number of these *cis*-acting elements have been identified. Both the κ and heavy chain variable region promoters have been extensively characterized in tissue culture transfection experiments and have been shown to include related protein recognition sequences that confer some B cell specificity (SEN and BALTIMORE 1985). As described above, the immunoglobulin λ, κ, and heavy chain loci each include identified transcriptional enhancer sequences. The λ locus contains a 3' enhancer (BLOMBERG et al. 1991) and the κ and heavy chain loci contain intronic and 3' enhancers (JUDDE and MAX 1992; MEYER and NEUBERGER 1989; MEYER et al. 1990; HOLE et al. 1991; PETTERSSON et al. 1990; LIEBERSON et al. 1991). Experiments using cultured cells have demonstrated that each of these enhancers confer some degree of cell type specificity. The 3' heavy chain enhancer appears to be the most restricted: it is active only in plasma cell lines, and may only be important for immunoglobulin expression at this late stage of B cell development (DARIAVACH et al. 1991). At the other extreme, the heavy chain intronic enhancer is active in both B and T cells. While this enhancer shows some lymphoid specificity, it also has activity in a number of other tissue types such as brain and kidney, but not liver (JENUWEIN and GROSSCHEDL 1991). B cell specific expression of the heavy chain may be a product of B cell specific rearrangement (discussed below). Additional regulatory elements may be located between the heavy chain intronic enhancer and the CH1 exon. JENUWEIN and GROSSCHEDL (1991) found a sequence immediately 3' of the μ enhancer that drives expression of a sterile μ transcript in skeletal muscle, and GRAM et al. (1992b) found that sequences within or near the 3' half of the human $\gamma 1$ switch region were required for high level expression of a transgene consisting of mouse VDJ and enhancer sequences fused to human $\gamma 1$ coding exons.

2. Transgene Expression

With the exception of the 3' enhancers, all of these *cis*-acting sequences are closely linked to those coding regions which would typically be included in transgene constructs. Extensive work has been published describing the expression of rearranged heavy and light chain immunoglobulin genes in transgenic mice (STORB 1989). Rearranged mouse transgenes have generally been found to express at high levels on the cell surface and in the serum. The transgenes are also expressed with the correct tissue specificity, with the exception of heavy chain μ and δ genes, which are transcribed in skeletal muscle and both T cells and B cells (TSANG et al. 1988; JENUWEIN and GROSSCHEDL 1991), consistent with the idea that heavy chain cell type specificity is conferred by restricted VDJ joining. Abnormal cell type specific transcription does not, however, lead to cell surface expression because membrane bound heavy chain is part of a complex that includes light chain (or light chain surrogate) as well as Ig-α and Ig-β (discussed below).

3. Human Transgene Constructs

Most of the experiments involving immunoglobulin transgenes have involved mouse sequences; however, several reports describe human constructs. Rearranged human μ (NUSSENZWEIG et al. 1987) and $\gamma 1$ (YAMAMURA et al. 1986) heavy chain genes express specifically in B cells in transgenic mice. We (TAYLOR et al. 1992) have compared the cell surface immunoglobulin

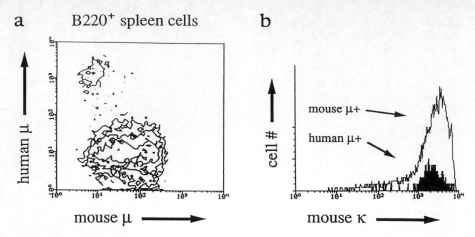

Fig. 7a,b. Cell surface expression of human heavy chain on mouse B cells. Spleen cells from a human heavy chain minilocus transgenic animal (animal #1287, line HC1–26; Taylor et al. 1992) were stained with fluorescent antibodies to mouse B220, mouse IgM, mouse Igκ, and human IgM. **a** Two dimensional flow cytometric analysis of human and mouse μ expression on surface of B220 gated lymphocyte population (B cells). Human and mouse μ expressing populations are distinct, indicating xenotypic exclusion (*x* and *y axes*: log fluorescence intensity). **b** One dimensional flow cytometric analysis of mouse κ light chain expression on surface of human and mouse μ expressing populations shown in panel **a** (*x axis*, log fluorescence intensity; *y axis*, linear cell number). The distribution of mouse κ expression is similar for the two populations, indicating similar B cell receptor levels

levels of transgenic B cells expressing human heavy chains to those of cells expressing endogenous mouse chains, and found them to be similar (Fig. 7). However, we and others (BRÜGGEMANN et al. 1989; TAYLOR et al. 1992) have found serum levels of human κ, μ, and γ1 immunoglobulins in transgenic mice containing unrearranged transgenes to be in the μg/ml rather than the normal mg/ml range. This low level of serum expression cannot be accounted for by the overall frequency of B cells that express the human transgenes and may be an indication that either the B cell compartment responsible for serum expression is not being occupied by transgene expressing cells or the secreted forms of the human chains are inefficiently transcribed or processed. BRÜGGEMANN et al. (1989) found that hybridomas from these animals expressed only 10% the normal level of human heavy chain, supporting the latter explanation.

II. Rearrangement

1. Target Sequences

Experiments with transfected cell lines indicate that in this type of a test system very little specific DNA sequence is required as a target for the enzymes involved in immunoglobulin gene rearrangement (LEWIS and

GELLERT 1989; LIEBER 1992). These experiments have defined the target sequences and suggest that the accessibility of these sequences to the recombinase is modulated chromatin structure, which may be coupled to transcription. The necessary and sufficient sequences for V(D)J joining are a highly conserved, near-palindromic heptamer and a less well conserved AT-rich nonamer separated by a spacer of either 12 or 23 bp (TONEGAWA 1983; HESSE et al. 1989). Efficient recombination occurs only between sites containing recombination signal sequences with different length spacer regions. This so called 12–23 joining rule prevents heavy chain V, D, and J regions from being incorrectly linked. It insures that D regions are not linked to each other, or skipped over entirely, by direct V-to-J joining.

The recombination signals and enzymatic machinery appear to be identical in B and T cells and cannot, therefore, be responsible for the restriction of immunoglobulin gene rearrangement to B cells and T cell receptor (TCR) gene rearrangement to T cells (YANCOPOULOS et al. 1986). Instead, the cell type specificity of receptor gene recombination may be regulated by the accessibility of these DNA sequences to the recombinases. This accessibility may be determined by the transcriptional activity of the region (YANCOPOULOS and ALT 1985). Transcription initiating from the unrearranged V regions and from within the D region and within the JC intron may serve to open up the locus to the recombinase system. In the heavy chain locus, D-to-J joining and subsequent V-to-DJ joining may be regulated, respectively, by JC intron and/or D region transcription and by V region transcription. This interpretation was used by FERRIER et al. (1990) to explain the cell type specific gene rearrangements they observed in a transgenic mouse system. They generated transgenic animals containing a hybrid test construct consisting of unrearranged TCR β-chain V, D, and J region elements linked to an immunoglobulin μ heavy chain constant region with or without the μ enhancer. The enhancerless constructs showed no rearrangements, while the enhancer containing constructs showed D-to-J rearrangements in both T and B cells, and V-to-DJ rearrangements in T cells only. Apparently, the μ enhancer, which has been shown to function in both B and T cells in transgenic mice, is able to activate lymphoid cell specific transcription in the JC intron and/or the nearby D region, thus triggering D-to-J joining. However, the μ enhancer is only capable of activating transcription from the TCR Vβ region promoter in the thymus, thus restricting complete VDJ joining to T cells. Natural immunoglobulin heavy chain VDJ joining is restricted to B cells in an analogous fashion. D-to-J joining occurs in T and B cells but the B cell specificity of the IgH variable region promoters prevents V-to-DJ joining in T cells.

The actual relationship between recombination and transcription is not clear. While HSIEH et al. (1992) found that transcription alone did not affect recombination in a minichromosome system, this group demonstrated that CpG methylation of the substrate DNA directs the assembly of a chromatin complex that is inaccessible to the recombinase (HSIEH and LIEBER 1992). A relationship between methylation and rearrangement was also observed by ENGLER et al. (1991) in transgenic mice containing a hybrid rearrangement construct that included mouse light chain V and J rearrangement recognition sequences, a heavy chain enhancer, a metallothionein promoter, a bacterial gpt gene, and other sequences. Rearrangement of this test construct was found to be inversely correlated with the degree of methylation of the transgene. Cytosine methylation is inversely correlated with transcription (CEDER 1988); however, it is difficult to distinguish cause from effect.

2. Immunoglobulin Gene Rearrangements in Transgenic Mice

There have been several reports of immunoglobulin minilocus transgenes expressed in mice. BUCCHINI et al. (1987) reported that an unrearranged chicken λ gene transgene was expressed and that approximately 1%–2% of the transgenes in the spleen were rearranged. In one of the four transgenic lines generated, the chicken gene was also rearranged (at lower levels) in the thymus. GOODHARDT et al. (1987) introduced an unrearranged rabbit κ gene into transgenic mice and observed rearrangement in both the spleen and the thymus. BRÜGGEMANN et al. (1989) used a hybrid mouse/human heavy chain minigene construct. These authors observed a high frequency (approximately 50%) of rearrangement in both thymus and spleen; however, the majority of those rearrangement events involved only DJ joining, without V region joining. Approximately 3%–4% of the B cells appeared to be synthesizing antibodies encoded by the rearranged transgenes. The same group has also reported the generation of transgenic animals containing two nonoverlapping cosmid clones that together span almost the entire human D region and include VH6, all of the J segments and all of μ. The human sequence transgenes in these mice appear to undergo VDJ joining; however, as described below, no recognizable human D segments were found to be incorporated (BRÜGGEMANN et al. 1991). TAYLOR et al. (1992) generated transgenic mice with κ light and heavy chain minilocus transgenes containing only human coding sequences (Fig. 4). Both human heavy and light chain epitopes were observed in the serum and the surface of the transgenic B cells. Sequence analysis of the transgene encoded transcripts demonstrated that the transgenes undergo V(D)J joining and N region addition to generate a repertoire of different CDR3 sequences. The structure of these V(D)J joints is described in detail below.

3. Light Chain Junctions

The human light chain minilocus encoded transcripts reported by TAYLOR et al. (1992) are diverse and incorporate all five human Jκ segments (Fig. 8). Approximately one quarter of the Vκ-Jκ joints include nongermline encoded sequences. This addition of junctional random nucleotides is commonly associated with heavy chain N regions (LIEBER et al. 1988; YANCOPOULOS and ALT 1986). The large number of naturally occurring Vκ segments makes it

V$_\kappa$65.8   J$_\kappa$

Jκ1	CAG CAG TAT GGT AGC TCA CCT CC		G TGG ACG	TTC GGC CAA GGG
883-1	--- --- --- --- --- --- --		- --- ---	--- --- --- ---
883-3	--- --- --- --- --- --- --G A-		- --- ---	--- --- --- ---
883-4	--- --- --- --- --- --- ---		--- ---	--- --- --- ---
883-11	--- --- --- --- --- --- -		-- ---	--- --- --- ---
883-12	--- --- --- --- ---		---	--- --- --- ---
883-13	--- --- --- --- --- --- --- -		-A ---	--- --- --- ---
883-15	--- --- --- --- ---		--- ---	--- --- --- ---
883-17	--- --- --- --- --- --- -		--- ---	--- --- --- ---
878-19	--- --- --- --- --- --- --		- ---	--- --- --- ---
878-20	--- --- --- --- --- ---		--- ---	--- --- --- ---
878-25	--- --- --- --- --- --- --		- ---	--- --- --- ---
878-28	--- --- --- --- --- --- -		- ---	--- --- --- ---
878-30	--- --- --- --- --- --- --		- --- ---	--- --- --- ---

Jκ2	CAG CAG TAT GGT AGC TCA CCT CC		G TAC ACT	TTT GGC CAG GGG
883-9*	--- --- --- --- --- --- --- -		--- ---	--- --- --- ---
883-16	--- --- --- --- --- --- --C AT		- --- ---	--- --- --- ---
878-26	--- --- --- --- --- --- T-- -A		- AG- ---	--- --- --- ---

Jκ3	CAG CAG TAT GGT AGC TCA CCT CC		A TTC ACT	TTC GGC CCT GGG
883-7	--- --- --- --- --- --- --		- --- ---	--- --- --- ---
883-10	--- --- --- --- --- --- --		- --- ---	--- --- --- ---
883-53	--- --- --- --- --- --- --		- --- ---	--- --- --- ---
878-21	--- --- --- ---⌐ --- --- ---		--- ---	--- --- --- ---
878-29	--- --- --- --- --- --- ---		--- ---	--- --- --- ---
878-31	--- --- --- --- --- --- ---		--- ---	--- --- --- ---

Jκ4	CAG CAG TAT GGT AGC TCA CCT CC		G CTC ACT	TTC GGC GGA GGG
883-2	--- --- --- --- --- --- --		- --- ---	--- --- --- ---
878-27	--- --- --- --- --- --- ---		--- ---	--- --- --- ---
878-32	--- --- --- --- --- --- --- GG		- GC- ---	--- --- --- ---
878-34	--- --- --- --- --- --- --- --		- AG- ---	--- --- --- ---

Jk5	CAG CAG TAT GGT AGC TCA CCT CC		G ATC ACC	TTC GGC CAA GGG
883-14*	--- --- --- --- --- --- T		- --- ---	--- --- --- ---
883-18	--- --- --- --- --- --- ---		--- ---	--- --- --- ---
883-71	--- --- --- --- --- --- --- --		- ---	--- --- --- ---

Fig. 8. Transgene encoded human light chain CDR3 sequence diversity. The nucleotide sequence of the junctional region of 29 independent cDNA clones is shown. Out of frame VJ joints are indicated by *asterisks*. Sequences are divided into categories based on J segment use. The germline encoded sequence is depicted above each category. A *dash* indicates no divergence from the germline sequence and a *blank space* or a *letter* indicates a missing or substituted nucleotide. Each clone is identified by two numbers separated by a dash; the first number indicates the identification number of the animal that provided the RNA, and the second number specifies the clone. Animal #883 was a double (heavy and light chain minilocus) transgenic derived from lines HC1–26 and KC1–665 (heavy chain sequences from this animal are shown in Fig. 4). Animal #878 contained only the light chain minilocus (line KC1–665). (From Taylor et al. 1992)

difficult to determine whether or not N region addition is a normal component of κ light chain VJ joining (HELLER et al. 1987; DERSIMONIAN et al. 1989; MEINDL et al. 1990; MARKS et al. 1991); however, because the KC1 minilocus construct contains only a single variable segment, the transgenic result is unambiguous. Similar N region additions have been reported previously in light chain transgene rearrangements. GOODHARDT et al. (1989) introduced a chimeric mouse VκJκ/rabbit Cκ construct and analyzed the structure of cDNA clones derived from transcripts of the rearranged transgene. They found that the VJ joints included deletions and N region type additions. The authors interpreted this finding as an indication that transgene VJ joining was occurring at a developmental time point during which endogenous heavy chains normally rearrange. During this developmental window endonucleases and terminal transferase might be more active, leading to heavy chain-like joints. It is possible that the abnormal chromosomal location of the transgene or the concatenated structure of the integrated locus could lead to premature rearrangement. Alternatively, limited N region addition may be a normal component of light chain rearrangement that is difficult to recognize beneath the usual divesity of κ variable segments and somatic mutations. Whether or not the observed light chain N regions are an artifact of the transgenic system, they do not lead to abnormally long CDR3 sequences because the additions are compensated for by exonucleolytic reduction of the V and J segments. Six of the seven transcripts with N region additions result from in-frame VJ joints. Of these, five produce a ten amino acid CDR3 (the expected length given exact V-J joining with no exonucleolytic activity) and the sixth generates a nine residue CDR3. Furthermore, out of all of the 27 in frame transcripts we analyzed, 15% have eight residue CDR3 sequences while 52% have nine residue and 19% have ten residue CDR3's. In comparison, analysis of the 34 naturally occurring Vκ-III nucleotide sequences reported by KABAT et al. (1991) shows that 12%, 71%, and 15% have eight, nine, and ten residue CDR3s, respectively. Therefore, N region addition does not appear to skew the size distribution of the light chain CDR3s away from that of an authentic human repertoire.

4. Heavy Chain Junctions

Both BRÜGGEMANN et al. (1991) and TAYLOR et al. (1992) reported the structure of human heavy chain VDJ joints formed in transgenic mice. Although the mice generated by Brüggeman et al. contained almost the entire human D region, none of the CDR3s included sequences that could be unambiguously assigned to known human D segments. The sequences were obtained using a JH4 specific PCR primer, so that no information on J segment distribution could be obtained. The reported VDJ joints are all unique; however, only three out of 11 sequences were derived from in-frame rearrangements, indicating a selection bias against transgene expressing B cells. The human heavy chain minilocus encoded transcripts reported by

TAYLOR et al. (1992) are also diverse and incorporate all six human JH segments, at least eight of the ten transgene encoded human D segments, and both transgene encoded heavy chain isotypes (Fig. 9). A comparison of the distribution of JH segments in the transgenic system with the distribution found by YAMADA et al. (1991) in authentic human PBLs is shown in Fig. 10. The transgenic mice preferentially use JH4 (47%) followed by JH6 (22%). Yamada et al. found a similar pattern; 53% of the authentic human joints incorporate JH4 and 22% incorporate JH6. It is more difficult to compare D segment usage between the transgenic mice and human PBLs because the transgene minilocus does not include all of the human D region. Of the 75 in-frame clones analyzed by Yamada et al., 48% could be assigned to D segments included in the HC1 transgene, and a further 11% could not be assigned to any known human D segment. These CDR3s either consist almost entirely of N region additions flanking very short D segment remnants or incorporate previously unrecognized D genes. Given these constraints two observations can be made. First, the DXP family is the most heavily used in both the transgenic animals and in human PBL, accounting for 31% and 29%, respectively, of the in-frame sequences. The second observation is that while only one of the in-frame human PBL sequences used DHQ52, 33% of the in-frame transgenic sequences (25% of all transgenic sequences) used DHQ52. However, this frequency varied from 0% to 46% between individual animals (Fig. 9).

The average length of the CDR3 sequences encoded by the 36 in-frame transcripts from the transgenic animals is 10.6 amino acids. This is similar to the average CDR3 length of 10.3 residues found for adult PBL sequences by SANZ (1991). However, the transgenic sequences are considerably shorter than the 14.5 residue average found by YAMADA et al. (1991) for adult PBL sequences. As with D segment distribution, there are differences in the average heavy chain CDR3 length between individual animals. The length difference between the average naturally occurring heavy chain CDR3 and the sequences found in the transgenic animals is predominantly due to differences in N region addition. The average number of N region nucleotides per CDR3 sequence (excluding from analysis those sequences for which no D segment could be assigned and thus the N-D border could not be established) is 5.7 for the transgenic sequences and 14.3 for the adult human sequences reported by YAMADA et al. (1991). This average increase in N nucleotides adds approximately three amino acids to the authentic human sequences.

5. Repercussions of Mouse B Cell Environment on Human VDJ Joints

The over representation of DHQ52, the shorter average CDR3 length, and the shorter N regions that characterize the transgenic heavy chain transcripts are all reminiscent of human fetal VDJ junctions. SCHROEDER and WANG (1990) reported a similar high frequency of DHQ52 in a small sample (six of

Clone	D	J		V	n-D-n	J	C
215-1	DHQ52	J3	γ1	AGA	cgg - CTAACTGGGG - ttgat	GCTTTGATATCTGGGGCCAAGGGACAATGGTCACCGTCTCTTCAG	CC
215-2	DN1	J4	γ1	AGA	cacc - GTATAGCAGCAGCTGG	CTTTGACTACTGGGGCCAAGGGACCACGGTCACCGTCTCCTCAG	CC
215-3	D?	J6	γ1	AGA	t	ATTACTACTACTACGGTATGGACGTCTGGGGCCAAGGGACCACGGTCACCGTCTCCTCAG	CC
215-4	DXP1	J6	γ1	AGA	c - ATTACGATATTTGACTGGT - c	CTACTACTACTACGGTATGGACGTCTGGGGCCAAGGGACCACGGTCACCGTCTCCTCAG	CC
215-5, 17	DXP1	J4	γ1	AGA	cggagg - TACTATGGTTCGGGGAGTTATTATAAC - gt	CTTTGACTACTGGGGCCAAGGGACCAATGGTCACCGTCTCCTCAG	CC
215-6	D?	J3	γ1	AGA	cggggggtctgat	GCTTTGATATCTGGGGCCAAGGGACAATGGTCACCGTCTTTCAG	CC
215-7	DHQ52	J6	γ1	AGA	gc - AACTGG - c	GCTTTGATATCTGGGGCCAAGGGACAATGGTCACCGTCTCTTCAG	GG
215-8	DHQ52	J1	μ	AGA	tcgg - CTAACTGGGGA - tc	CTACTACTACGGTATGGACGTCTGGGGCCAAGGGCACCCTGGTCACCGTCTCCTCAG	GG
215-9		J1	μ	AGA		TACTTCCAGCACTGGGGCCAGGGCACCCTGGTCACCGTCTCCTCAG	GG
215-10	DLR2	J4	μ	AGA	cac - GTAGCTA - actct	TTTGACTACTGGGGCCAAGGGACCCTGGTCACCGTCTCCTCAG	GG
215-11	DXP1	J4	μ	AGA	caa - ATTACTATGGTTCGGGGAGTT - cc	CTTTGACTACTGGGGCCAAGGGACCCTGGTCACCGTCTCCTCAG	GG
215-12	D?	J1	μ	AGA	c	AATACTTCCAGCACTGGGGCCAGGGCACCCTGGTCACCGTCTCCTCAG	GG
215-13	DHQ52	J6	γ1	AGA	ca - AACTGGGG	ACTACTACTACTACGGTATGGACGTCTGGGGCCAAGGGACCACGGTCACCGTCTCCTCAG	CC
215-14	DXP1	J6	γ1	AGA	ca - TTACTATGGTTCGGGGAGTTAT - g	ACTACTACTACTACGGTATGGACGTCTGGGGCCAAGGGACCACGGTCACCGTCTCCTCAG	CC
215-15	DXP1	J4	γ1	AGA	ca - GGGAG	TGGGGCCAGGGAACCCTGGTCACCGTCTCCTCAG	CC
215-18	DHQ52	J4	γ1	AGA	caaac - CTGGGGA - gga	GACTACTGGGGCCAAGGGACCACGGTCACCGTCTCCTCAG	CC
215-19	DK1	J6	γ1	AGA	GGATATAGTGGCTACGAT - a	ACTACTACTACTACGGTATGGACGTCTGGGGCCAAGGGACCACGGTCACCGTCTCCTCAG	GG
215-20	DHQ52	J4	γ1	AGA	CAAACTGGGGA - gg	ACTTTGACTACTGGGGCCAAGGGACCACGGTCACCGTCTCCTCAG	CC
215-21	DK1	J2	γ1	AGA	TATAGTTGGCTACGATTAC	CTACTGGTACTTCGATCTCTGGGGCCGTGGCACCGTCACTGTCTCCTCAG	GG
215-22	DHQ52	J6	γ1	AGA	gca - TCCCTCCCC - tcctttg	ACTACTACGGTATGGACGTCTGGGGCCAAGGGACCACGGTCACCGTCTCCTCAG	GG
215-23	DIR2	J1	μ	AGA	cg - GGGTGGGGG	TTTGACTACTGGGGCCAAGGGACCCTGGTCACCGTCTCCTCAG	GG
215-26	DN1	J3	μ	AGA	cat - GGTATAGCAGCAGCTGGTAC	TGCTTTGATATCTGGGGCCAAGGGACAATGGTCACCGTCTTCAG	GG
215-28	DN1	J4	μ	AGA	gaa - ATAGCAGCAGCTG - cc	CTACTTTGACTACTGGGGCCAAGGGACCCTGGTCACCGTCTCCTCAG	GG
215-30	DHQ52	J4	μ	AGA	caa - AACTGGGG	TGACTACTGGGGCCAAGGGACCCTGGTCACCGTCTCCTCAG	GG
883-5.2	DXP1	J5	μ	AG	TATTACTATGGTTCGGGGAGTT - c	GGTTCGACCCCTGGGGCCAGGGAACCCTGGTCACCGTCTCCTCAG	GG
883-8.2	DHQ52	J4	μ	AGA	g - CTAACTGGGGA - aac	TACTTTGACTACTGGGGCCAAGGGACCCTGGTCACCGTCTCCTCAG	GG
883-75	DIR2	J4	μ	AGA	GGCTGGGC - ctac	TACTTTGACTACTGGGGCCAAGGGACCCTGGTCACCGTCTCCTCAG	GG
883-84	DHQ52	J2	μ	AGA	c - AACTGGGG - at	CTGGTACTTCGATCTCTGGGGCCGTGGCACCGTCACTGTCTCCTCAG	GG
883-87	DM2	J6	γ1	AG	ggc - TAGCAGCAGCTGG	TACTACTACTACTACGGTATGGACGTCTGGGGCCAAGGGACCACGGTCACCGTCTCCTCAG	GG
883-90	DHQ52	J3	μ	AGA	cggg - ACTGGGGA - cccgat	GCTTTGATATCTGGGGCCAAGGGACAATGGTCACCGTCTCCTCAG	GG
883-91	DHQ52	J4	μ	AGA	catg - AACTGGGG	TTTGACTACTGGGGCCAAGGGACCCTGGTCACCGTCTCCTCAG	GG
883-92	D?	J3	μ	AGA	caactcctgat	GCTTTGATATCTGGGGCCAAGGGACAATGGTCACCGTCTCCTCAG	GG
883-93	DIR2	J2	μ	AGA	cttgccca - GTTCCGG - caga	ACTGGTACTTCGATCTCTGGGGCCGTGGCACCGTCACTGTCTCCTCAG	GG
883-94	DXP1	J1	μ	AGA	CGATATTTGACTGGT - ccgatgact	ACTGGTACTTCGATCTCTGGGGCCGTGGCACCGTCACTGTCTCCTCAG	GG
883-95	DHQ52	J4	μ	AGA	gg - CTGGGG - cctac	TACTTTGACTACTGGGGCCAAGGGACCCTGGTCACCGTCTCCTCAG	GG
640-2	DXP1	J4		AGA	c - ATTACTATGGTTCGGGGAGTTATT - gt	CTACTGGGGCCAAGGGACCCTGGTCACCGTCTCCTCAG	GG
640-5	DXP1	J5		AGA	caatttccctg - TTTTGACTGGT - ccctc	AACTGGTTCGACTCCTGGGGCCAGGGAACCCTGGTCACCGTCTCCTCAG	GG
640-8, 3.2	DXP1	J6		AGA	caggggggct - TATTACGATATTTGACTGGT - g	ACTACTACTACTACGGTATGGACGTCTGGGGCCAAGGGACCACGGTCACCGTCTCCTCAG	GG
640-9	DM2	J4		AGA	cagag - AGCAGCTGG - aa	GACTACTGGGGCCAGGGAACCCTGGTCACCGTCTCCTCAG	GG
640-13	DXP1	J4		AGA	caagg - ACTATGGTTCGGGGAGTTATTAT - cct	CTTTGACTACTGGGGCCAAGGGACCACGGTCACCGTCTCCTCAG	GG
640-14	DK1	J4		AGA	catgggg - TAGTGGCTACGAT - cttt	GACTACTGGGGCCAGGGAACCCTGGTCACCGTCTCCTCAG	GG
640-15	DXP1	J4		AGA	CGATATTTGACTGGTATTA - gagagg	TTTGACTACTGGGGCCAGGGAACCCTGGTCACCGTCTCCTCAG	GG
640-1.2	DK1	J4		AGA	c - ATAGTGGCTACGA - cct	GACTACTGGGGCCAGGGAACCCTGGTCACCGTCTCCTCAG	GG
640-7.2	DXP1	J6		AGA	ccaagggg - TTACGATATTTGACTGGTTAT - cg	TACTACTACTACGGTATGGACGTCTGGGGCCAAGGGACCACGGTCACCGTCTCCTCAG	GG
640-13.2	DXP1	J4		AGA	c - ATTACGATATTTGACTGGTTATTA	TGACTACTGGGGCCAGGGAACCCTGGTCACCGTCTCCTCAG	GG
640-14.2	DXP1	J4		AGA	ccgg - TATGGTTCGGGGAGTTATTATT - cgagg	CTACTTTGACTACTGGGGCCAGGGAACCCTGGTCACCGTCTCCTCAG	GG
640-16.2	DXP1	J1		AGA	gag - ACGATATTTGACTGGTTATTAT - tc	GCTGAATACTTCCAGCACTGGGGCCACCCTGGTCACTGTCTCCTCAG	GG

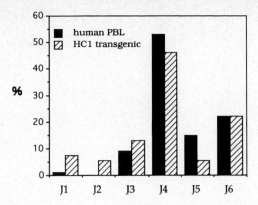

Fig. 10. Distribution of human J segment incorporation in natural and transgenic repertoires. The frequency with which different human J segments are found in rearranged clones isolated from adult human peripheral blood lymphocytes (YAMADA et al. 1991) is compared to the frequency observed in clones from transgenic mice containing the HC1 minilocus. (From TAYLOR et al. 1992)

13 sequences) from human fetal liver. In a much larger sample of 120 fetal sequences, SANZ (1991) found that 14% incorporated DHQ52. SANZ (1991) also reported an average fetal CDR3 length of 7.3 amino acids (71% of the average length of adult sequences). However, a closer examination of the structure of the N regions suggests that this apparent fetal-like structure of the transgenic heavy chains is coincidental and that it is simply a consequence of the rearrangements taking place in mouse rather than human pre-B cells (Table 1). Sanz found that the average length of each N region is not reduced in fetal joints as compared to adult joints, but rather that the frequency of occurrence of these N regions is reduced. It appears that 93% of the V-D and D-J junctions in the 88 D containing adult human PBL genomic DNA clones reported by YAMADA et al. (1991) include N regions and that the average length of these individual N regions is 7.7 bp. This is

◀──────────────────────────────────────

Fig. 9. Transgene encoded human heavy chain CDR3 sequence diversity. The nucleotide sequence of the junctional region of 49 independent cDNA clones is shown. Two of the D segment assignments (clones 215–15 and 883–95) are based on only five nucleotides of homology and therefore represent possible assignments. All other assignments are based on greater than five nucleotides of homology. Nucleotides assigned to N regions are in *lower case letters*. Each clone is identified by two numbers separated by a dash; the first number indicates the identification number of the animal that provided the RNA, and the second number specifies the clone. Animal #215 was a heavy chain minilocus transgenic derived from line HC1–57. Animal #883 was a double (heavy and light chain minilocus) transgenic derived from lines HC1–26 and KC1–665 (light chain sequences from this animal are shown in Fig. 4). Animal #640 was a heavy chain minilocus transgenic derived from the single copy line HC1–112. Clones 215–5 and 215–17, and clones 640–8 and 640–3.2 are identical. (From TAYLOR et al. 1992)

Table 1. N regions from natural human and mouse heavy chain rearrangements compared to human transgene N regions

	N regions/ junction	bp/N region	N-nucleotides/ VDJ	Average CDR3 length
Adult human	0.93	7.7	14.3 (4.8 aa)	14.5 aa
Fetal human	0.77	7.7	11.9	
Adult mouse	0.70	3.0	4.2	
Newborn mouse	0.06	1.4	0.2	
Adult transgenic	0.77	3.8	5.7 (1.9 aa)	10.6 aa

Adult human data from YAMADA et al. (1991), fetal human data from SCHROEDER and WANG (1990), mouse data from FEENEY (1990), transgenic data from TAYLOR et al. (1992).

identical to the average length of the individual N regions in the eight fetal liver clones for which D segments could be identified reported by SCHROEDER and WANG (1990). In contrast, the average length of an N region in the transgenic mouse cDNA clones is 3.8 bp. This is close to the average length of 3 bp/N region for the 63 adult mouse cDNA clones published by FEENEY (1990). The transgenic sequences are certainly not reflective of mouse fetal gene rearrangements which are even more dramatically reduced in N region addition than human fetal rearrangements. Feeney found that the frequency of N region containing genomic clones fell from 83% in the adult to below 2% in the fetal liver. Therefore, N region addition characteristic of adult mouse B cells accounts for the smaller average size of the transgenic heavy chain CDR3 sequences. Although the heavy chain CDR3 sequences appear superficially like a human fetal liver repertoire because of the overuse of DHQ52 and the shorter average size of the N regions, the high frequency of N regions indicates that the cDNA clones do not derive from B cells that developed in the mouse fetal liver. Instead, the fetal character of the CDR3s is a consequence of mouse B cell N nucleotide addition (which is less extensive than human) coupled with an increase in DHQ52 incorporation that may be peculiar to the transgene and that varies between individual animals or transgenic lines. While the aggregate characteristics of transgenic VDJ joints differ from human VDJ joints, individual transgenic sequences are indistinguishable from authentic human sequences.

III. Allelic Exclusion

1. Background

Any given immunoglobulin secreting plasma B cell normally produces only a single type of antibody. This monospecificity serves two functions: first, it may prevent serum autoimmunity; B cell clones selected by specific T helper clones produce only those antibodies for which they were selected. At the same time monospecificity prevents clones expressing potentially useful immunoglobulin specificities from being anergized because they co-express immunoglobulin molecules that react with self antigens. The process by which diploid B cells maintain monospecificity is termed allelic exclusion.

Both active and passive processes have been proposed as mechanisms for allelic exclusion. Passive allelic exclusion occurs by default because of the high error rate of VDJ joining, which makes it improbable that two successful rearrangement events will occur in the same cell. The high frequency of incorrect VDJ joining can be seen from the analysis of immortalized B cell lines: at least 90% of immunoglobulin expressing B cell clones contain one functionally rearranged gene and one nonfunctionally rearranged gene, rather than one functional and one unrearranged copy. In addition to this stochastic mechanism of allelic exclusion, there is good evidence for the existence of an active mechanism, whereby the successful rearrangement of one immunoglobulin gene allele activates a feedback loop that turns off the process of gene rearrangement, thus preventing the rearrangeeement and expression of the second copy of the gene.

2. Induction of Allelic Exclusion by Rearranged Transgenes

The transgenic mouse system has generated considerable data in support of the feedback inhibition model for allelic exclusion. Rearranged mouse μ and δ heavy chain, and κ and λ light chain, transgenes are capable of excluding the rearrangement of endogenous immunoglobulin genes (WEAVER et al. 1985; IGLESIAS et al. 1987; RITCHIE et al. 1984; NEUBERGER et al. 1989; HAGMAN et al. 1989). The ability of a transgene to activate allelic exclusion appears to be dependent on the expression of its protein product on the cell surface; μ transgenes lacking transmembrane exons did not prevent rearrangement of the endogenous genes (MANZ et al. 1988). Data from κ transgenic mice indicate that the extent of allelic exclusion may be a function of the ability of the transgene encoded products to form stable heavy/ light chain complexes. The rearranged κ gene from the hybridoma MOPC-21 encodes a light chain that is known to form complexes with a large number of different heavy chains. Transgenic mice expressing this gene only rearrange their endogenous κ genes in B cells that fail to correctly rearrange an endogenous heavy chain gene (RITCHIE et al. 1984; STORB et al. 1986b). However, transgenic mice containing MOPC-167 derived κ genes, which encode a light chain that is not capable of combining with most heavy chains, showed only a low level of allelic exclusion (MANZ et al. 1988), and transgenic mice containing another, poorly expressed, κ gene showed no evidence of allelic exclusion.

It may be that the mechanism of allelic exclusion operating in these transgenic animals is very different from that operating in normal mice. The expression of high levels of a rearranged transgene in a pre-B cell that would not normally have begun to even rearrange its immunoglobulin genes may accelerate that cell through those stages of differentiation in which rearrangement can take place. This abnormally rapid rate of B cell maturation has been postulated to explain the protective effects of a rearranged human μ transgene from pre-B cell lymphomas in an otherwise susceptible mouse strain (NUSSENZWEIG et al. 1988). On the other hand, accelerating the differentiation of the pre-B cell may be the normal mechanism by which feedback inhibition occurs. The presence of cell surface immunoglobulin could cause the cell to exit from the bone marrow. Outside of the bone marrow, the cell would no longer be exposed to the particular cocktail of cytokines required to maintain the immature developmental state during which the recombinase genes are expressed. Protein turnover would then nonspecifically shut off all V(D)J joining.

Allelic exclusion has also been demonstrated using a rearranged human μ transgene (NUSSENZWEIG et al. 1987), thus demonstrating that a human heavy chain can mediate this process in mouse cells. This phenomenon might more properly be referred to as xenotypic exclusion. TAYLOR et al. (1992) also observed xenotypic exclusion with an unrearranged human heavy chain transgene. Figure 7 shows a flow cytometric analysis of transgenic splenic B cells stained with reagents specific for mouse and human IgM. The human and mouse IgM expressing populations are distinct.

3. Response to Allelic Exclusion by Unrearranged Transgenes

None of the published transgenic experiments have fully addressed the issue of the response of an unrearranged transgene to feedback inhibition. The unrearranged rabbit κ transgene described above (GOODHARDT et al. 1987) was found to be arranged and co-expressed in cells with rearranged endogenous κ genes; however, without knowing the history of an individual cell, it is impossible to determine whether the breakdown of allelic exclusion was due to the inability of the transgene to respond to feedback inhibition or to trigger it. In fact, this particular transgene was poorly expressed and, because it coded for a xenogeneic light chain, it might be expected to form unstable heavy/light complexes, thus, based on the experiments discussed above using rearranged transgene constructs, it would not be expected to block the rearrangement of the endogenous mouse κ genes. Similarly, it is impossible to interpret the results of BRÜGGEMANN et al. (1989) with regard to the response of transgenes to rearrangement feedback inhibition. Several lines of mice containing an unrearranged human/mouse hybrid μ gene construct did not show evidence of allelic exclusion of the transgene; however, the level of expression of the transgene was considerably less than that of the endogenous genes, and the observed lack of allelic exclusion may have been due to the inability of the transgene to activate, rather than respond to, a feedback mechanism.

4. Alternatives to Direct Feedback Allelic Exclusion

If allelic exclusion proves difficult to obtain, there are several alternatives. An attractive approach involves breeding a single-copy transgenic mouse line with a line of mice that has its endogenous immunoglobulin genes inactivated by homologous recombination. An hemizygous transgenic mouse in an Ig$^-$ background would have individually monospecific B cells by default.

IV. Primary Repertoire

An important requirement for correct transgene function is the generation of a primary repertoire that is diverse enough to trigger a secondary immune response to a wide range of different antigens. Because of the large size of the immunoglobulin loci, and the technical difficulties involved in manipulating large pieces of DNA, an important issue in the design of the transgenes is what fraction of the coding sequence within the heavy and light chain loci is actually necessary to provide an adequate primary repertoire? Most of these sequences consist of multiple V segments interspersed with large intergenic regions and perhaps half of these V segments are pseudogenes. Furthermore, many of the remaining V segments may be rarely used or may never contribute to the expressed repertoire. It could, therefore, be possible to design an immunoglobulin transgene that is considerably smaller than the natural immunoglobulin locus, yet still provides an adequate primary repertoire. BRÜGGEMANN and NEUBERGER (1991) reported the production – from a human heavy chain minilocus transgene comprising only two VH segments and four D segments – of partially human antibodies that bind to sheep red blood cells. However, sheep red blood cells represent a very large collection of epitopes, and it cannot be determined whether the antigen recognition is encoded by the transgene encoded heavy chain repertoire or the endogenous mouse light chain repertoire.

A large enough collection of V, D, and J segments will be needed to generate a primary repertoire that contains at least one B cell clone with a minimum threshold affinity for each antigen. This minimum threshold affinity is just high enough to allow the B cell to internalize and process enough of the antigen to trigger T cell dependent B cell maturation. Because T cell dependent maturation is accompanied by somatic hypermutation across the entire V region, the existence within the primary repertoire of B cell clones with affinities higher than the minimum threshold is presumably redundant. A high affinity clone will eventually be produced during the secondary response. Diversity at CDR3, created by VJ and VDJ joining, could be enough to provide minimal affinities for a large number of different antigens. This minimal affinity would then trigger the T cell dependent maturation that would give rise to high affinity antibodies for each of the antigens. Considerable support for this hypothesis comes from the structure

of the rabbit heavy chain locus (KNIGHT and BECKER 1990; KNIGHT 1992).
Rabbits use only three to four germline VH segments for their entire heavy
chain repertoire, with between 70% and 90% of the antibodies encoded
by a single VH gene segment. While rabbits appear to have an extremely
restricted primary repertoire, this may not be directly applicable to the
behavior of immunoglobulin transgenes in mice because of differences in the
modes of somatic diversification between mice and rabbits. Analysis of
rearranged rabbit gene sequences shows that somatic diversifications are
frequently in the form of clusters of nucleotide changes that correlate with
the sequences of unused V segments. This suggests that gene conversion, a
mechanism used by chickens for primary diversification but not observed in
mouse B cell differentiation, is responsible for much of the somatic diversity
in rabbits. However, of most importance is the limited nature of the primary
repertoire of the rabbit heavy chain locus. Unlike chicken B cells, which
undergo gene conversion in the bursa prior to hatching (REYNAUD et al.
1989), splenic B cells in 3 week old rabbits have still not undergone diversification (SHORT et al. 1991). While it is still unclear whether later somatic
diversification is spontaneous or is triggered by weak binding of antigens to
the limited primary repertoire, the rabbit example still strongly supports the
concept that a single V segment can give rise to a considerable range of
different specificities. There presumably will be some constraints on the
choice of V segments that could give rise to a complete repertoire. CHOTHIA
et al. (1992) analyzed the predicted three dimensional structures of CDR1
and 2 from a large set of DNA sequences that may encompass the entire VH
repertoire. They were able to group all of the CDR1 sequences into three
structural classes and all of the CDR2 sequences into five classes. One
possible approach to achieving a broad range of antigen reactivities is
to include variable segments that include different combinations of these
structural classes. An actual assessment of the number and type of V region
segments required to produce the minimum primary repertoire will involve
work with transgene constructs containing different numbers of V segments.

V. Intracellular Signaling

1. Background

Aside from its role as an antigen binding protein, the transgene encoded
immunoglobulin will have to function correctly as part of a multimolecular
cell surface receptor complex. This complex appears to transmit a signal
from the cell surface indicating the occupancy status of the receptor. This
signals is important for triggering allelic exclusion, initiating the pathway
of T cell dependent maturation, and activating or inactivating (tolerizing)
mature B cells.

   The critical role of the cell surface receptor complex in B cell development is demonstrated by analysis of mice that lack μ gene transmembrane

exons (KITAMURA et al. 1991). The mice were generated from ES cells in which the μ gene had been altered by homologous recombination. While heterozygous mice are normal, B cell development is completely arrested at the pre-B stage in animals homozygous for the transmembrane deletion.

It is well established that antigen binding to cell surface immunoglobulin can induce an increase in intracellular calcium and stimulate phosphoinositol turnover (COGGESHALL and CAMBIER 1984; BIJSTERBOSCH et al. 1985; CAMBIER and RANSOM 1987). Evidence suggesting a role for cell surface immunoglobulin in signal transduction also comes from studies of B cell tolerance. B and T cell tolerance is induced at two levels: self reacting cells are either destroyed (clonal deletion) or converted to an unresponsive state (clonal anergy). When mature, IgM positive B cells encounter antigen in the presence of antigen specific helper T cells, they are activated. When the same B cells encounter antigen in the absence of such T cells, the level of surface IgM is reduced ten-fold. Furthermore, the anergized B cells are rendered incapable of differentiating into immunoglobulin secreting plasma cells (GOODNOW et al. 1989; GOODNOW 1992). However, if the antigen level is reduced so that the number of occupied receptors drops from 50% to 5%, the B cells remain in a responsive state, even in the absence of antigen specific T cells. This suggests that two signals are required for B cell activation, and one signal is required for anergization. Activation requires one signal from a helper T cell and a second signal from an occupied cell surface antigen receptor. When the receptor is occupied in the absence of T cell help, the B cell becomes unresponsive. Presumably, the cell is "aware" of the occupancy status of its surface receptors by virtue of a signal transduced through the membrane via a complex of proteins associated with surface immunoglobulin.

2. B Cell Receptor Complex

The B cell receptor complex appears to be analogous to the well characterized TCR complex. The TCR consists of two antigen binding immunoglobulin-like chains, α and β, associated with a set of transmembrane accessory proteins (CD3-γ, -δ, -ε, and -ζ) that are involved in signal transduction and are required for the assembly of the α and β chains on the cell surface (CLEVERS et al. 1988). Like the TCR, B cell immunoglobulin chains also require accessory molecules to reach the cell surface (HOMBACH et al. 1988). Two of these accessory molecules, Ig-α (encoded by the mb-1 gene), and Ig-β (encoded by the B29 gene), have been characterized and shown to be structurally related to the T cell CD3-γ, δ, and ζ proteins (SAKAGUCHI et al. 1988; RETH 1989; CAMPBELL et al. 1991). Association with these two molecules is sufficient to bring IgM to the surface of transfected fibroblasts (VENKITARAMAN et al. 1991; MATSUUCHI et al. 1992). Both Ig-α and Ig-β are associated with all five of the heavy chain isotypes, although there does appear to be differential dependence on α and

β for cell surface expression (Venkitaraman et al. 1991; Van Noesel et al. 1992). IgM is absolutely dependent on both α and β, while IgG2b and IgD may not always require both chains. The α/β independent form of IgD has been shown to be a phosphatidylinositol linked rather than transmembrane bound molecule (Wienands and Reth 1992).

Immunoglobulin heavy chain probably interacts with the Ig-α and β chains via its transmembrane domain. The various heavy chain transmembrane domains are unusually hydrophilic; for example, the mouse μ sequence includes six threonine residues and three serine residues clustered into two short sequences, TTAST and SLFYSTTVI. Replacement of the entire IgM transmembrane sequence with a more conventional class I MHC transmembrane domain, or even substitution of the first four of the threonine and serine residues with valine and alanine, leads to constitutive cell surface expression in plasmacytoma lines which do not express Ig-α (Williams et al. 1990). Replacement of the YS sequence in the second cluster with two valine residues abrogates signal transduction and antigen presentation (Shaw et al. 1990) while replacement of the first cluster with the equivalent segment of an MHC class II transmembrane domain eliminates antigen presentation but not signaling (Parikh et al. 1992). These polar residues within the IgM transmembrane may interact with polar glutamic acid and glutamine residues within the respective transmembrane domains of Ig-α and Ig-β (Reth et al. 1991).

The significance of these details for the success of a human antibody transgenic mouse is that such a mouse will rely on hybrid B cell receptor complexes assembled from human heavy chains and mouse α/β accessory proteins. Will sequence divergence between mouse and human disrupt B cell development and function? Figure 11 presents the amino acid sequences of the COOH-terminal regions of several different mouse and human heavy chains. The only differences within the last two exons of mouse and human

```
                     *  *               *
mouse IgM    EGEVNAEEEGFE  NLWTTASTFIVLFLLSLFYSTTVTLF  KVK
human IgM    EGEVSADEEGFE  NLWATASTFIVLFLLSLFYSTTVTLF  KVK


mouse IgD    QSDSYMDLEEEN  GLWPTMCTFVALFLLTLLYSGFVTFI  KVK
human IgD    NSDDYTTFDDVG  SLWTTLSTFVALFILTLLYSGIVTFI  KVK


mouse IgG1   ETCAEAQDGELD  GLWTTITIFISLFLLSVCYSAAVTLF  KVKWIFSSVVELKQTLVPEYKNMIGQAP
mouse IgG2a  DVCAEAQDGELD  GLWTTITIFISLFLLSVCYSASVTLF  KVKWIFSSVVELKQTISPDYRNMIGQGA
mouse IgG2b  DICAEAKDGELD  GLWTTITIFISLFLLSVCYSASVTLF  KVKWIFSSVVELKQKISPDYRNMIGQGA
mouse IgG3   ETCAEAQDGELD  GLWTTITIFISLFLLSVCYSASVTLF  KVKWIFSSVVQVKQTAIPDYRNMIGQGA


mouse IgA    SQDILEEEAPGA  SLWPTTVTFLTLFLLSLFYSTALTVTTV  RGPFGSKEVPQY
human IgA    PQETLEEETPGA  NLWPTTITFLTLFLLSLFYSTALTVTSV  RGPSGNREGPQY


             extracellular      transmembrane            cytoplasmic
```

Fig. 11. COOH-terminal sequences of membrane bound forms of mouse and human heavy chains. *Asterisks* indicate differences between mouse and human IgM. (From Kabat et al. 1991 and Yu et al. 1990)

IgM are two relatively conservative changes outside the membrane, and a single substitution within the membrane. The extracellular spacer sequence that includes the first two changes is not conserved between different heavy chain classes. Furthermore, both residues are divergent between mouse IgM and IgD, each of which has been shown to be capable of mediating B cell activation, deletion, and anergy in transgenic mice (BRINK et al. 1992). Replacement of the μ transmembrane exons with the equivalent δ exons in transfected cell lines also did not disrupt signaling (WEBB et al. 1989). It is, therefore, unlikely that these two differences between mouse and human IgM will disrupt B cell function. The single residue substitution within the transmembrane region is more problematic. This threonine to alanine change (TTAST to ATAST) falls within the first polar cluster thought to be important for receptor complex assembly and function; however, it is not conserved between mouse μ and δ. Additional a priori support for equivalence of the mouse and human heavy chains is provided by the complete sequence identity of the mouse and human Ig-α genes (YU and CHANG 1992).

Experimental evidence that a human/mouse hybrid receptor might be functional also comes from the observation that completely human IgM transgenes can xenotypically exclude rearrangement of endogenous mouse IgM (NUSSENZWEIG et al. 1987; TAYLOR et al. 1992). A possible solution to potential problems arising from hybrid receptors involves substituting the penultimate exon of the human gene for the equivalent exon of the mouse gene. This exon encodes the transmembrane region and is not part of the message encoding the secreted (and potentially therapeutic) antibody. Such a transgene, in which the last two mouse exons were spliced to the remaining portion of the human μ gene, was used by BRÜGGEMANN et al. (1989) in their minilocus experiments.

3. Pre-B Cell Complex

In pre-B cells IgM associates with surrogate light chains to form a transitional complex that may be important for allelic exclusion of heavy chain rearrangement and further maturation within the bone marrow (CHERAYIL and PILLAI 1991). The transitional complex is found prior to and immediately following κ gene rearrangement and expression. This represents another cross-species protein complex that the human heavy chain will presumably have to form in transgenic B cells. The surrogate light chain is composed of two proteins, ω and ι, encoded by the genes $\lambda5$ and VpreB, respectively (SAKAGUCHI et al. 1986; PILLAI and BALTIMORE 1987; KUDO and MELCHERS 1987). Neither of these genes requires rearrangement to be expressed. An analysis of B cell maturation in human heavy chain transgenic mice that have bred with mice that do not express endogenous μ (KITAMURA et al. 1991; CHEN et al. 1993b) will be necessary to demonstrate that a mouse/human hybrid transitional complex can function.

VI. Class Switching

1. Background

Because use of the μ or δ constant regions is largely determined by alternate splicing, IgM and IgD can be coexpressed in a single cell. However, the other heavy chain isotypes (γ, α, and ε) are only expressed after a gene rearrangement event deletes the Cμ and Cδ exons. This gene rearrangement process, termed class switching, occurs by recombination between so called switch segments located immediately upstream of each heavy chain constant region gene segment (except δ). The individual switch segments are between 2 and 10 kb in length and consist primarily of short repeated sequences (MOWATT and DUNNICK 1986; MILLS et al. 1990). The exact point of recombination differs for individual class switching events, and switch products can also be substrates for further switching (MILLS et al. 1992). Induction of class switching appears to be associated with sterile transcripts that initiate upstream of the switch segments (LUTZKER and ALT 1988; STAVNEZER et al. 1988; ESSER and RADBRUCH 1989; BERTON et al. 1989; ROTHMAN et al. 1990). For example, the observed induction of the $\gamma 1$ sterile transcript by IL-4 and inhibition by IFN-γ correlates with the observation that IL-4 promotes class switching to $\gamma 1$ in B cells in culture, while IFN-γ inhibits $\gamma 1$ expression. Ideally then, transgene constructs that are intended to undergo class switching should include all of the *cis*-acting sequences necessary to regulate these sterile transcripts. A possible alternative method for obtaining class switching in transgenic mice involves the inclusion of the 400 bp direct repeat sequences that flank the human μ gene (YASUI et al. 1989; WHITE et al. 1990). Homologous recombination between these two sequences deletes the μ gene in IgD-only B cells.

2. Class Switching in Transgenic Mice

The ability of a transgene construct to switch isotypes by autonomous recombinational deletion during B cell maturation has not been directly demonstrated in transgenic mice. However, two laboratories have published descriptions of relevant model systems. The first system involves only the 5′ half of the switch apparatus, while the second system may be an example of transgene autonomous switching. DURDIK et al. (1989) microinjected a rearranged mouse μ heavy chain gene construct and found that in four independent mouse lines a high proportion of the transgenic B cells expressed the transgene encoded variable region associated with IgG rather than IgM. They then showed that an IgG expressing hybridoma contained a chromosomal translocation between the site of insertion of the transgene and the endogenous heavy chain locus (GERSTEIN et al. 1990). Thus, isotype switching appears to have taken place between the transgene and the endogenous γ constant region on another chromosome. In this case the 3′ switch acceptor sequences are derived from the endogenous locus; however,

TAYLOR et al. (1992) built a transgene that included two heavy chain genes, human μ and $\gamma1$, together with the respective switch regions. The transgene also included the start site of the $\gamma1$ sterile transcript implicated in class switching. Mice generated with this construct expressed both human μ and human $\gamma1$ transcripts and protein. The human $\gamma1$ expressing B cells in these mice may have undergone class switching using only transgene derived sequences. However, because the genomic structure of individual IgG expressing hybridomas was not analyzed, alternative mechanisms cannot be ruled out. The transgene included the two 400 bp repeats that flank the human μ gene, and these could be involved in deletional recombination. The close proximity of the μ and $\gamma1$ segments in the transgene could also lead to the synthesis of a single transcript that would express both isotypes by alternate splicing. In addition, the transgene inserts consisted of multicopy tandem arrays, and some of the sequences could have been scrambled to allow for direct expression of $\gamma1$. Expression of $\gamma1$ could also be explained by *trans*-splicing. Another group (SHIMIZU et al. 1989, 1991) has identified a small population of cells in human μ transgenic mice that simultaneously express both human μ and mouse γ heavy chain sequences associated with the same transgene encoded VDJ sequence. The authors propose that *trans*-splicing of human VDJ sequences to germline γ sterile transcripts could be responsible fot the double isotype expression. *Trans*-splicing models have also been used to explain double isotype expression in BCL1 leukemia cells (NOLAN-WILLARD et al. 1992). These observations are open to alternative explanations; however, *trans*-splicing of transfected trypanosome and nematode sequences has been directly demonstrated in mammalian cell lines (BRUZIK and MANIATIS 1992).

3. Importance of Class Switching for a Human Antibody Mouse

Which constant region gene segments need to be included in a functional heavy chain transgene? This is a two part question: which constant region segments are important for B cell development, and which segments are useful for generating a therapeutic product? Because IgM is the heavy chain class that is expressed during early stages of development it is presumably important to include the μ gene. The importance of membrane bound IgM is supported by the observation that targeted deletion of the μ transmembrane exons results in a B cell deficient mouse (KITAMURA et al. 1991); however, this deletion also removed the first half of the δ gene, and it is not clear that enough class switch recombination would occur in early B cell development to give other heavy chain isotypes the opportunity to rescue the B cell compartment. It has not been demonstrated that a different segment, such as one of the γ genes, would not function throughout B cell development if it were substituted for μ.

IgD is less likely to be important for early B cell maturation or for functioning of mature B cells. It is not expressed in early B cell development, and it is deleted in IgG, IgA, and IgE expressing cells (YUAN and VITETTA 1978). Furthermore, disruption of the IgD gene, by homologous recombination in ES cells, does not prevent the development of mature functional B cells (ROES and RAJEWSKY 1991). However, the resulting IgD deficient B cells do not develop a secondary response to antigen as rapidly as normal B cells, suggesting a possible role for this receptor in the initiation of affinity maturation (ROES and RAJEWSKY 1993). The authors propose that the flexibility of the IgD molecule, conferred by its extended hinge region, facilitates binding to – and internalization of – large multimeric antigens. Thus IgD may have evolved as a specific receptor for recruiting B cells into the affinity maturation pathway within germinal centers. Inclusin of an IgD gene could conceivably improve the functionality of an immunoglobulin transgene.

The individual heavy chain classes that are expressed following switch recombination are also unlikely to be required for B cell function. Although heavy chain class switching is associated with affinity maturation, somatically mutated IgM antibodies are common and switching can be induced in vitro without somatic mutation (BEREK and MILSTEIN 1988; MANSER 1987; WYSOCKI et al. 1992). There is also in vivo evidence that switching can take place in the absence of somatic mutation. During a T cell dependent immune response, class switching takes place in spleen foci both outside of and within germinal centers, while somatic mutation is restricted to germinal center B cells (JACOB et al. 1991a). Class switching could help to trigger affinity maturation; however, it is more likely that class switching and somatic mutation are associated because they occur at the same stage of B cell development and are stimulated by overlapping sets of cytokines and cell-cell interactions (RAJEWSKY et al. 1987; SIEKOVITZ et al. 1987). This may be a good reason to build class switch substrates into human immunoglobulin transgenes: the appearance of non-IgM isotypes following class switching would not only provide a good indication that an immunized mouse is undergoing a secondary immune response, it could also provide an additional screening tool for selecting hybridomas expressing high affinity antibodies. The therapeutic characteristics of different heavy chain classes provides an additional reason for engineering transgenes that can undergo class switching. IgM pentamers have less access to some tissues because of their size, and the molecules expressed in transgenic animals are not completely human because they include the mouse encoded J chain. The IgG subclasses are probably more useful therapeutically. Each of these classes differ in their ability to activate complement lysis and antibody dependent cytotoxicity (BINDON et al. 1988). An IgG1 antibody might be more useful for applications involving cell killing, while an IgG4 antibody could be better for receptor blocking. It might therefore be advantageous to build different strains of transgenic mice that class switch to different constant regions. Alternatively the mice could be used simply to generate human

VDJ segments that can be cloned and manipulated in vitro to generate therapeutic targets.

VII. Substrate for Somatic Mutation

During the initial stages of a T cell dependent response a subset of antigen reactive B cells will undergo a process of affinity maturation during which the rearranged immunoglobulin loci collect random point mutations (KOCKS and RAJEWSKY 1989; WEISS and RAJEWSKY 1990). These point mutations, which accumulate over the heavy and light chain V(D)J regions at a rate of 10^{-3}/bp/cell division, increase the diversity of the available repertoire (ALLEN et al. 1987). From this repertoire high affinity clones are expanded and loss of function clones are deleted in an evolutionary process leading to a pool of memory cells that can give rise to plasma cells expressing immunoglobulins with affinities greater than $10^8 M^{-1}$. This process appears to take place only within specific spleen, lymph node, and Peyers patch structures (germinal centers) (BEREK et al. 1991; JACOB et al. 1991b, 1992).

It is a fundamental requirement for a correctly functioning transgene immunoglobulin locus that it serve as a substrate for somatic hypermutation during affinity maturation. Affinity maturation is particularly important for obtaining useful antibodies derived from a minilocus because virtually any conceivable transgene construct will have a smaller primary repertoire than the natural human immunoglobulin loci. An efficient process of affinity maturation could overcome the limited primary repertoire of a transgenic system and might be the key feature that distinguishes such a system from the other methods of obtaining low immunogenicity antibodies.

Somatic mutation of rearranged mouse κ light chain transgenes has been reported (O'BRIEN et al. 1987; HACKETT et al. 1990; SHARPE et al. 1991). The observed frequency of mutations within the transgene variable regions ranged from 0.1% to 1%, compared to frequencies of 1%–2% found in naturally occurring affinity selected κ genes (STEELE et al. 1992). Somatic mutations of κ transgenes were found to be dependent on hyperimmunization and may be dependent on sequences 3' of the constant region gene segment. Both of the transgene constructs that were found to be substrates for hypermutation included almost 10 kb of 3' flanking sequences, while two other constructs that did not undergo somatic mutation included only 1 kb of 3' flanking sequences (SHARPE et al. 1990, 1991; CARMACK et al. 1991). The flanking sequences left off of the nonmutating transgenes include the 3' transcriptional enhancer sequence (MEYER and NEUBERGER 1989), leading to speculation that the enhancer may be important for targeting somatic mutation (SHARPE et al. 1991). However, this conclusion may be contradicted by data from transgenic animals containing a tRNA reporter gene immediately downstream of a rearranged VJκ gene (UMAR et al. 1991). The authors observed a high frequency of tRNA mutations despite the fact that their construct only included 1.2 kb downstream of the Cκ exon. Although the exact sequences involved in targeting light chain somatic mutation have not been identified, the important point is that they lie within the region immediately flanking the gene and are able to function when removed from the intact κ locus.

In addition to sequence requirements for somatic mutation, it has been proposed that the orientation of the transgene relative to the direction of DNA replication may be important. ROGERSON et al. (1991) found that in a κ transgenic line containing a three copy tandem array of the injected fragment, only one of the three copies underwent somatic mutation, and this copy was integrated in the opposite orientation as the other two. Whether or not the orientation of the transgene turns out to be important, this result highlights the difficulty of interpreting transgenic somatic mutation data. Only a fraction of the integrated transgenes may function as a hypermutation substrate, making it difficult to collect data from a statistically significant number of independent lines to draw firm conclusions regarding differences in the structure of the transgenes. The authors also carefully analyzed the structure of the integration site and showed that all three of the transgenes did not integrate intact. It cannot therefore be assumed a priori that sequences included in the injected fragment will be included in the transgenic mouse.

The sequences that define the substrate for hypermutation within the heavy chain locus have not been found. There has been one report of a rearranged μ transgene undergoing somatic mutation; however, the mutated copy of the transgene VH segment had recombined into the endogenous immunoglobulin locus via class switching from the transgene encoded μ to an endogenous γ gene (DURDIK et al. 1989). The sequences that directed these mutations could have been derived from the endogenous immunoglobulin locus rather than the transgene. It is unlikely that the heavy chain enhancer alone is sufficient because a hybrid B cell expressing transgene consisting of a TCR coding sequence and an immunoglobulin heavy chain intronic enhancer was not mutated in B cells that had undergone hypermutation at the endogenous immunoglobulin loci (HACKETT et al. 1992).

VIII. Domination of the Immune Response

This final requirement, that the transplanted human immunoglobulin genes dominate over the endogenous mouse immunoglobulin genes, requires first that all of the other requirements for correct transgene function are fulfilled. If all of these requirements are fulfilled, it is still essential for the transgene locus to be dominant so that a high proportion of the hybridomas obtained will be producing therapeutically useful human antibodies, and not mouse or mouse/human hybrid antibodies.

There are several basic strategies for generating a dominant transgene response. The first strategy involves only the transgenes themselves. If the transgene loci undergo V(D)J joining with high efficiency at a point in B cell development when the endogenous loci have not yet begun to rearrange, then it is possible for the transgenic response to dominate by xenotypic exclusion. A problem with this approach is that over time escape clones expressing endogenous immunoglobulins could be selected for if the primary repertoire of the transgenes is insufficient. It is probably necessary to actively interfere with the expression of the endogenous immunoglobulin genes to get an efficient human antibody response. Three possible strategies are discussed below: antibody depletion, anti-sense transgenes, and gene targeting by homologous recombination.

1. Antibody Depletion

Intravenous injection of anti-immunoglobulin antibodies has become almost a standard research tool in immunology. This procedure was used over 20 years ago in the first convincing demonstration that class switching is responsible for the appearance different immunoglobulin isotypes (KEARNEY et al. 1970). Anti-μ antibody was injected into developing chickens and found to suppress not only IgM but IgG, showing that IgG expressing cells developed from IgM expressing cells. The same experiment was also performed in mice (MANNING and JUTILA 1972). Antibodies against mouse μ, κ, and λ could be used to suppress B cells expressing endogenous immunoglobulins in transgenic mice. This procedure would be analogous to experiments reported by WEISS et al. (1984) in which mice were generated that completely lacked κ light chain expression because of anti-κ antibody injections. These mice had normal numbers of B cells and normal serum immunoglobulin levels because γ expression compensated for the lack of κ. When the mice were immunized with antigens that normally elicit a κ response, they showed a reduced primary response (presumably due to the limited diversity of the λ repertoire in mice) but a fairly normal secondary response. In a similar fashion, a limited transgene encoded repertoire could be selected for if the endogenous response were suppressed by antibody injections.

2. Anti-sense Transgenes

Coexpression of cross-hybridizing transcripts represents another potential method of suppressing gene expression. It was first recognized as a mechanism used in prokaryotes to regulate DNA replication (TOMIZAWA et al. 1981) and protein synthesis (SIMONS and KLECKNER 1983). IZANT and WEINTRAUB (1984) showed that antisense transcripts introduced into mammalian cells reduced the expression of targeted genes. While this technique has worked well in tissue culture and transgenic plants (HAMILTON et al. 1990), it has been only moderately successful as a method of suppressing gene expression in transgenic animals (MUNIR et al. 1990). However, there have been several reports of altered phenotypes in transgenic mice expressing antisense constructs. KATSUKI et al. (1988) obtained a shiverer-like phenotype in mice containing an antisense myelin basic protein construct; HAN et al. (1991) found reduced levels of Moloney murine leukemia virus-induced leukemias in transgenic mice expressing an antisense viral packaging sequence; and PEPIN et al. (1992) created an obese strain of mice by inserting an antisense glucocorticoid receptor transgene. The constant region RNA sequences of mouse and human immunoglobulins are different enough that antisense transgenes could theoretically be used to specifically reduce expression of mouse μ, κ, and λ without affecting the expression of human transgene encoded antibodies.

3. Gene Targeting

Perhaps the most effective means of reducing the expression of the endogenous mouse immunoglobulins is to inactivate the genes by homologous recombination in ES cells. The technology involved is briefly outlined above in the section on ES cells. Homologous recombination has been used to generate a variety of mice containing different germline alterations since it was first successfully employed in 1989 to correct a mutant HPRT gene (THOMPSON et al. 1989). This was perhaps the simplest gene targeting event to engineer because the functional HPRT gene could be selected for under relatively innocuous conditions; however, other methods had been developed for targeting nonselectable loci (THOMAS and CAPPECHI 1987). Once it was demonstrated that these methods could be used to target specific chromosomal sequences regardless of their function (MANSOUR et al. 1988) or transcriptional state (JOHNSON et al. 1989) and that the drug selection protocols employed do not prevent the targeted cells from populating the germline (SCHWARTZBERG et al. 1989; ZIJLSTRA et al. 1989), it was apparent that the immunoglobulin loci could be specifically altered to reduce their expression. Both the mouse μ heavy chain (KITAMURA et al. 1991; CHEN et al. 1993b) and κ light chain (CHEN et al. 1993a; ZOU et al. 1993) genes have now been disrupted by this process. Disruption of these two loci is relatively straightforward: functional rearrangement at each locus is dependent upon a unique small cluster of J segments that can be entirely deleted by a single gene targeting event. However, the remaining mouse λ locus will be more difficult to knock out because it consists of three functional constant region genes, each paired with a single J segment. Two of the Jλ-Cλ are adjacent, but the third gene is located over 100 kb upstream (STORB et al. 1989; CARSON and WU 1989). The genes could be inactivated by two independent targeting steps, or it might be possible to delete the entire 100 kb locus using a targeting vector that selects for recombination events at each end. MOMBAERTS et al. (1991, 1992) created a 15 kb deletion in the T cell receptor β locus using this approach. If the λ locus is eliminated by gene targeting it will require a great deal of mouse breeding to obtain a mouse containing heavy and light chain transgenes in a μ, κ, λ background. The ideal mouse would be hemizygous at each of the two transgene loci and homozygous null for each of the three endogenous alleles. The total number of possible genotypes for mice with variations at five different loci is 243. It will be much simpler to avoid eliminating the λ locus, in which case there are only 81 different possible genotypes to deal with. Expression of λ, which contributes only 5% of the normal mouse light chain, may not be a problem. The mouse λ locus is not very diverse, consisting of only three variable segments, and might not compete with the transgene locus. In addition, λ appears to rearrange late in B cell development, and does not completely restore B cell numbers in κ knockout mice (CHEN et al. 1993b; ZOU et al. 1993).

F. Perspective

There are several questions that still have to be answered before it can be determined if a transgenic animal can be used as an efficient source of human sequence MoAbs. The first major question involves the functionality of a hybrid mouse/human B cell receptor. This question will be answered by an analysis of mice resulting from crosses between endogenous immunoglobulin knockout animals and minilocus transgenic animals. If the transgene encoded immunoglobulin chains can rescue the B cell compartment with cells that can respond to antigen stimulation, then the hybrid receptors should be functional. The second big question involves the size of the primary repertoire. Will a limited minilocus encoded repertoire be sufficient to initiate affinity maturation with a wide variety of antigens? It is possible to speculate about the relative importance of the different CDR loops for the development of a primary repertoire and on the plasticity of that repertoire for the development of a secondary repertoire; however, only experimental results from immunized transgenic animals can answer the question. The third major unresolved question is whether heavy chain trans-genes can be a substrate for somatic mutation (as has been observed for light chain transgenes) and if the resulting transgenic B cells will undergo affinity maturation to generate useful therapeutic antibodies? This question can also only be resolved by experimental results. If it turns out that transgenic B cells do not undergo affinity maturation, then transgenic mice will be no more useful for generating human MoAbs than the combinatorial library approach. In addition to these three major unresolved questions there are a number of questions regarding the design of an optimal trans-genic mouse. For example, can a transgene lacking the δ heavy chain gene function efficiently? Will the expression of the endogenous mouse λ light chain complicate the isolation of hybridomas secreting completely human antibodies? Again these and other questions will only be resolved by actual experience using transgenic mice to generate human MoAbs.

References

Allen D, Cumano A, Dildrop R, Kocks C, Rajewsky K, Rajewsky N, Roes J, Sablitzky F, Siekevitz M (1987) Timing, genetic requirements and functional consequences of somatic hypermutation during B-cell development. Immunol Rev 96:5–22

Alt FW, Blackwell TK, Yancopoulos GD (1985) Immunoglobulin genes in transgenic mice. Trends Genet 1:231–236

Anderson MLM, Szajnert MF, Kaplan JC, McColl L, Young BD (1984) The isolation of a human Ig Vλ gene from a recombinant library of chromosome 22 and estimation of its copy number. Nucleic Acids Res 12:6647–6661

Banchereu J, Rousset F (1991) Growing human B lymphocytes in the CD40 system. Nature 353:679–679

Banchereau J, de Paoli P, Vallé A, Garcia E, Rousset F (1991) Long-term human B cell lines dependent on interleukin-4 and antibody to CD40. Science 251:70–72

Barbas CF, Bain JD, Hoekstra DM, Lerner RA (1992) Semisynthetic combinatorial antibody libraries: a chemical solution to the diversity problem. Proc Natl Acad Sci USA 89:4457–4461

Berek C, Milstein C (1988) The dynamic nature of the antibody repertoire. Immunol Rev 105:5–26

Berek C, Berger A, Apel M (1991) Maturation of the immune response in germinal centers. Cell 67:1121–1129

Berton MT, Uhr JT, Vitetta ES (1989) Synthesis of germ-line γ1 immunoglobulin heavy-chain transcripts in resting B cells: induction by interleukin 4 and inhibition by interferon γ. Proc Natl Acad Sci USA 86:2829–2833

Bijsterbosch MK, Meade CJ, Turner GA, Klaus GGB (1985) B lymphocyte receptors and polyphosphoinositide degradation. Cell 41:999–1006

Bindon CI, Hale G, Brüggemann M, Waldmann H (1988) Human IgG isotypes differ in complement activating function at the level of C4 as well as C1q. J Exp Med 168:127–142

Blomberg BB, Rudin CM, Storb U (1991) Identification of an enhancer for the human λ 1 chain Ig gene complex. J Immunol 147:2354–2358

Borrebaeck CAK (1988) Human mAbs produced by primary in-vitro immunization. Immunol Today 9:355–359

Bradley A (1987) Production and analysis of chimaeric mice. In: Robertson EJ (ed) Teratocarcinomas and embryonic stem cells, a practical approach. IRL Press, Oxford, pp 113–151

Bradley A, Evans M, Kaufman MH, Robertson E (1984) Formation of germ-line chimeras from embryo derived teratocarcinoma cells. Nature 309:225–256

Brink R, Goodnow CC, Crosbie J, Adams E, Eris J, Mason DY, Hartley SB, Basten A (1992) Immunoglobulin M and D antigen receptor are both capable of mediating B lymphocyte activation, deletion, or anergy after interaction with specific antigen. J Exp Med 176:991–1005

Brinster RL, Ritchie KA, Hammer RE, O'Brien RL, Arp B, Storb U (1983) Expression of a microinjected immunoglobulin gene in the spleen of transgenic mice. Nature 306:332–336

Brinster RL, Chen HY, Trumbauer ME, Yagle MK, Palmiter RD (1985) Factors affecting the efficiency of introducing foreign DNA into mice by microinjecting eggs. Proc Natl Acad Sci USA 82:4438–4442

Brinster RL, Allen JM, Behringer RR, Gelinas RE, Palmiter RD (1988) Introns increase transcriptional efficiency in transgenic mice. Proc Natl Acad Sci USA 85:836–840

Brüggemann M, Neuberger MS (1991) Generation of antibody repertoires in transgenic mice. Methods 2:159–165

Brüggemann M, Caskey HM, Teale C, Waldmann H, Williams GT, Surani MA, Neuberger MS (1989) A repertoire of monoclonal antibodies with human heavy chains from transgenic mice. Proc Natl Acad Sci USA 86:6709–6713

Brüggemann M, Spicer C, Buluwela L, Rosewell I, Barton S, Surani MA, Rabbits TH (1991) Human antibody production in transgenic mice: expression from 100 kb of the human IgH locus. Eur J Immunol 21:1323–1326

Bruzik JP, Maniatis T (1992) Spliced leader RNAs from lower eukaryotes are trans-spliced in mammalian cells. Nature 360:692–695

Bucchini D, Reynaud CA, Ripoche M-A, Grimal H, Jami J, Weill J-C (1987) Rearrangement of a chicken immunoglobulin gene occurs in the lymphoid lineage of transgenic mice. Nature 326:409–411

Burke DT, Carle GF, Olson MV (1987) Cloning of large segments of exogenous DNA into yeast artificial chromosome vectors. Science 236:806–812

Cambier JC, Ransom JT (1987) Molecular mechanisms of transmembrane signaling in B lymphocytes. Annu Rev Immunol 5:175–199

Campbell KS, Hager EJ, Friedrich RJ, Cambier JC (1991) IgM antigen receptor complex contains phosphoprotein products of B29 and mb-1 genes. Proc Natl Acad Sci USA 88:3982–3986

Carmack CE, Camper SA, Mackle JJ, Gerhard WU, Weigert MG (1991) Influence of a Vκ8 L chain transgene on endogenous rearrangements of the immune response to the HA(SB) determinant on influenza virus. J Immunol 147:2024–2033

Carson S, Wu GE (1989) A linkage map of the mouse immunoglobulin lambda light chain locus. Immunogenetics 29:173–179

Caton AJ, Koprowski H (1990) Influenza virus hemagglutinin-specific antibodies isolated from a combinatorial library are closely related to the immune response of the donor. Proc Natl Acad Sci USA 87:6450–6454

Ceder H (1988) DNA methylation and gene activity. Cell 53:3–4

Chada K, Magram J, Raphael K, Radice G, Lacy E, Costantini F (1985) Specific expression of a foreign beta-globin gene in erythroid cells of transgenic mice. Nature 314:377–380

Chen J, Trounstine M, Kurahara C, Young F, Kuo CC, Xu Y, Loring JF, Alt FW, Huszar D (1993a) B cell development in mice that lack one or both immunoglobulin κ light chain genes. EMBO J 12:821–830

Chen J, Trounstine M, Alt FW, Kurahara C, Young F, Loring JF, Huszar D (1993b) Immunoglobulin gene rearrangements in B cell deficient mice generated by targeted deletion of the JH locus. Int Immunol 5:647–656

Cherayil BJ, Pillai S (1991) The ω/λ5 surrogate immunoglobulin light chain is expressed on the surface of transitional B lymphocytes in the murine bone marrow. J Exp Med 173:111–116

Chothia C, Lesk AM, Gherardi E, Tomlinson IM, Walter G, Marks JD, Llewelyn MB, Winter G (1992) Structural repertoire of the human VH segments. J Mol Biol 227:799–817

Clackson T, Hoogenboom HR, Griffiths AD, Winter G (1991) Making antibody fragments using phage display libraries. Nature 352:624–628

Clevers H, Alarcon B, Wileman T, Terhorst C (1988) The T cell receptor/CD3 complex: a dynamic protein ensemble. Annu Rev Immunol 6:629–662

Coggeshall KM, Cambier JC (1984) B cell activation. VIII. Membrane immunoglobulins transduce signals via activation of phosphatidylinositol hydrolysis. J Immunol 133:3382–3386

Dariavach P, Williams GT, Campbell K, Pettersson S, Neuberger MS (1991) The mouse IgH 3′-enhancer. Eur J Immunol 21:1499–1504

Davies NP, Rosewell IR, Brüggemann M (1992) Targeted alterations in yeast artificial chromosomes for inter-species gene transfer. Nucleic Acids Res 20:2693–2698

den Dunnen JTD, Grootscholten PM, Dauwerse JG, Walker AP, Monoco AP, Butler R, Anand R, Coffey AJ, Bentley DR, Steensma HY, van Ommen GJB (1992) Reconstruction of the 2.4 Mb human DMD-gene by homologous YAC recombination. Hum Mol Genet 1:19–28

Dersimonian H, McAdam KPWJ, Mackworth-Young C, Stollar BD (1989) The recurrent expression of variable region segments in human anti-DNA autoantibodies. J Immunol 142:4027–4033

Duchosal MA, Eming SA, Fischer P, Leturcq D, Barbas CF, McConahey PJ, Caothien RH, Thornton GB, Dixon FJ, Burton DR (1992) Immunization of hu-PBL-SCID mice and the rescue of human monoclonal Fab fragments through combinatorial libraries. Nature 355:258–262

Durdik J, Gerstein RM, Rath S, Robbins PF, Nisonoff A, Selsing E (1989) Isotype switching by a microinjected μ immunoglobulin heavy chain gene in transgenic mice. Proc Natl Acad Sci USA 86:2346–2350

Emorine L, Kueh M, Weir L, Leder P, Max EE (1983) A conserved sequence in the immunoglobulin J kappa-C kappa intron: possible enhancer element. Nature 304:447–449

Engler P, Haasch D, Pinkert CA, Doglio L, Glymour M, Brinster R, Storb U (1991) A strain specific modifier on mouse chromosome 4 controls the methylation of independent transgene loci. Cell 65:939–947

Esser C, Radbruch A (1989) Rapid induction of transcription of unrearranged $S\gamma 1$ switch regions in activated murine B cells by interleukin 4. EMBO J 8:483–488

Evans MJ, Kaufman MH (1981) Establishment in culture of pluripotential cells from mouse embryos. Nature 292:154–156

Farr CJ, Stevanovic M, Thomson EJ, Goodfellow PN, Cooke HJ (1992) Telomere-associated chromosome fragmentation: applications in genome manipulation and analysis. Nature Genet 2:275–282

Feeney AJ (1990) Lack of N regions in fetal and neonatal mouse immunoglobulin V-D-J junctional sequences. J Exp Med 172:1377–1390

Ferrier P, Kripple B, Blackwell TK, Furley AJW, Suh H, Winoto A, Cook WD, Hood L, Costantini F, Alt FW (1990) Separate elements control DJ and VDJ rearrangement in a transgenic recombination substrate. EMBO J 9:117–125

Folger KR, Wong EA, Wahl G, Capecchi MR (1982) Patterns of integration of DNA microinjected into cultured mammalian cells: evidence for homologous recombination between injected plasmid DNA molecules. Mol Cell Biol 11:1372–1387

Gerstein RM, Frankel WN, Hsieh CL, Durdik JM, Rath S, Coffin JM, Nisonoff A, Selsing E (1990) Isotype switching of an immunoglobulin heavy chain transgene occurs by DNA recombination between different chromosomes. Cell 63:537–584

Gnirke A, Barnes TS, Patterson D, Schild D, Featherstone T, Olson MV (1991) Cloning and in vivo expression of the human GART gene using yeast artificial chromosomes. EMBO J 10:1629–1634

Goldstein G et al. (1985) A randomized clinical trial of OKT3 monoclonal antibody for acute rejection of cadaveric renal transplants. N Engl J Med 313:337–342

Goodhardt M, Cavelier P, Akimenko MA, Lutfalla G, Babinet C, Rougeon F (1987) Rearrangement and expression of rabbit immunoglobulin κ light chain gene in transgenic mice. Proc Natl Acad Sci USA 84:4229–4233

Goodhardt M, Babinet C, Lutfalla G, Kallenbach S, Cavelier P, Rougeon F (1989) Immunoglobulin κ light chain gene promoter and enhancer are not responsible for B-cell restricted gene rearrangement. Nucleic Acids Res 17:7403–7415

Goodnow CC (1992) Transgenic mice and analysis of B-cell tolerance. Annu Rev Immunol 10:489–518

Goodnow CC, Crosbie J, Jorgensen H, Brink RA, Basten A (1989) Induction of self-tolerance in mature peripheral B lymphocytes. Nature 342:385–391

Gordon JW, Scangos GA, Plotkin DJ, Barbosa JA, Ruddle FH (1980) Genetic transformation of mouse embryos by microinjection of purified DNA. Proc Natl Acad Sci USA 77:7380–7384

Gram H, Marconi L-A, Barbas CF, Collet TA, Lerner RA, Kang AS (1992a) In vitro selection and affinity maturation of antibodies from a naive combinatorial immunoglobulin library. Proc Natl Acad Sci USA 89:3576–3580

Gram H, Zenke G, Geisse S, Kleuser B, Bürki K (1992b) High-level expression of a human γ1 transgene depends on switch region sequences. Eur J Immunol 22:1185–1191

Grosschedl R, Weaver D, Baltimore D, Costantini F (1984) Introduction of a μ immunoglobulin gene into the mouse germ line: specific expression in lymphoid cells and synthesis of functional antibody. Cell 38:647–658

Hackett J, Rogerson B, O'Brien R, Storb U (1990) Analysis of somatic mutations in κ transgenes. J Exp Med 172:131–137

Hackett J, Stebbins C, Rogerson B, Davis MM, Storb U (1992) Analysis of a T-cell receptor gene as a target of the somatic hypermtation mechanism. J Exp Med 176:225–231

Hagman J, Lo D, Doglio LT, Hackett J, Rudin CM, Haasch D, Brinster R, Storb U (1989) Inhibition of immunoglobulin gene rearrangement by the expression of a λ2 transgene. J Exp Med 169:1911–1929

Hale G, Dyer MJS, Clark MR, Phillips JM, Marcus R, Reichmann L, Winter G, Waldmann H (1988) Remission induction in non-Hodgkin lymphoma with re-shaped human monoclonal antibody CAMPATH-1H. Lancet 2:1394–1399

Hamilton AJ, Lycett GW, Grierson D (1990) Antisense gene that inhibits synthesis of the hormone ethylene in transgenic plants. Nature 346:284–287

Han L, Yun JS, Wagner TE (1991) Inhibition of maloney murine leukemia virus-induced leukemia in transgenic mice expressing antisense RNA complementary to the retroviral packaging sequences. Proc Natl Acad Sci USA 88:4313–4317

Hasty P, Ramirez-Solis R, Krumlauf R, Bradley A (1991) Introduction of a subtle mutation into the Hox-2.6 locus in embryonic stem cells. Nature 350:243–246

Hawkins RE, Russell SJ, Winter G (1992) Selection of phage antibodies by binding affinity: mimicking affinity maturation. J Mol Biol 226:889–896

Hayday AC, Gillies SD, Saito H, Wood C, Wiman K, Hayward WS, Tonegawa S (1984) Activation of a translocated c-myc gene by an enhancer in the immunoglobulin heavy-chain locus. Nature 307:334–340

Hesse JE, Lieber MR, Mizuuchi K, Gellert M (1989) V(D)J recombination: a functional definition of the joining signals. Genes Dev 3:1053–1061

Heller M, Owens JD, Mushinski JF, Rudikoff S (1987) Amino acids at the site of Vκ-Jκ recombination not encoded by germline sequences. J Exp Med 166:637–646

Hofker MH, Walter MA, Cox DW (1989) Complete physical map of the human immunoglobulin heavy chain constant region gene complex. Proc Natl Acad Sci USA 86:5567–5571

Hogan B, Costantini F, Lacy E (1986) Methods of manipulating the mouse embryo. Cold Spring Harbor Laboratory, New York

Hole NJK, Harindranath N, Young-Cooper GO, Garcia R, Mage RG (1991) Identification of enhancer sequences 3' of the rabbit Ig κL chain loci. J Immunol 146:4377–4384

Hombach J, Leclercq L, Radbruch A, Rajewsky K, Reth M (1988) A novel 34-kd protein co-isolated with the IgM molecule in surface IgM-expressing cells. EMBO J 7:3451–3456

Honjo T, Shimizu A, Yaota Y (1989) Constant-region genes of the immunoglobulin heavy chain and the molecular mechanism of class switching. In: Honjo T, Alt FW, Rabbits TH (eds) Immunoglobulin genes. Academic, New York, pp 123–150

Hoogenboom HR, Winter G (1992) By-passing immunisation: human antibodies from synthetic repertoires of germine VH gene segments rearranged in vitro. J Mol Biol 227:381–388

Hosoda F, Nishimura S, Uchida H, Ohki M (1990) An F factor based cloning system for large DNA fragments. Nucleic Acids Res 18:3863–3869

Hsieh C-L, Lieber MR (1992) CpG methylated minichromosomes become inaccessible for V(D)J recombination after undergoing replication. EMBO J 11:315–325

Hsieh C-L, McCloskey RP, Lieber MR (1992) V(D)J recombination on mini-chromosomes is not affected by transcription. J Biol Chem 267:15613–15619

Huse WD, Sastry L, Iverson SA, Kang AS, Alting-Mees M, Burton DR, Benkovic SJ, Lerner RA (1989) Generation of a large combinatorial library of the immunogobulin repertoire in phage lambda. Science 246:1275–1281

Huxley C, Gnirke A (1991) Transfer of yeast artificial chromosomes from yeast to mammalian cells. BioEssays 13:545–550

Huxley C, Hagino Y, Schlessinger D, Olson MV (1991) The human HPRT gene on a yeast artificial chromosome is functional when transferred to mouse cells by cell fusion. Genomics 9:742–750

Iglesias A, Lamers M, Köhler G (1987) Expression of immunoglobulin delta chain causes allelic exclusion in transgenic mice. Nature 330:482–484

Ish-Horowicz D, Burke JF (1981) Rapid and efficient cosmid cloning. Nucleic Acids Res 9:2989–2998

Izant JG, Weintraub H (1984) Inhibition of thymidine kinase gene expression by anti-sense RNA: a molecular approach to genetic analysis. Cell 36:1007–1015

Jacob J, Kassir R, Kelsoe G (1991a) In situ studies of the primary immune response to (4-hydroxy-3-nitrophenyl)acetyl. I. The architecture and dynamics of responding cell populations. J Exp Med 173:1165–1175

Jacob J, Kelsoe G, Rajewski K, Weiss U (1991b) Intraclonal generation of mutants in germinal centres. Nature 354:352–353

Jacob J, Miller C, Kelsoe G (1992) In situ studies of the antigen-driven somatic hypermutation of immunoglobulin genes. Immunol Cell Biol 70:145–152

James K, Bell GT (1987) Human monoclonal antibody production: current status and future prospects. J Immunol Methods 100:5–40

Jenuwein T, Grosschedl R (1991) Complex pattern of immunoglobulin μ gene expression in normal and transgenic: nonoverlapping regulatory sequences govern distinct tissue specificities. Genes Dev 5:932–943

Johnson RS, Sheng M, Greenberg ME, Kolodner RD, Papaioannou VE, Spiegelman BM (1989) Targeting of nonexpressed genes in embryonic stem cells via homologous recombination. Science 245:1234–1236

Judde J-G, Max EE (1992) Characterization of the human immunoglobulin kappa gene 3' enhancer: functional importance of three motifs that demonstrate B-cell-specific in vivo footprints. Mol Cell Biol 12:5206–5216

Kabat EA, Wu TT, Perry HM, Gotteman KS, Foeller C (1991) Sequences of proteins of immunological interest, 5th edn. US Department of Health and Human Services, NIH Publ no 91-3242

Kang AS, Barbas CF, Janda KD, Benkovic SJ, Lerner RA (1991) Linkage of recognition and replication functions by assembling combinatorial antibody Fab libraries along phage surfaces. Proc Natl Acad Sci USA 88:4363–4366

Katsuki M, Sato M, Kimura M, Yokoyama M, Kobayashi K, Nomura T (1988) Conversion of normal behavior to shiverer by myelin basic protein antisense cDNA in transgenic mice. Nature 241:593–595

Kearney JF, Lawton AR, Bockman DE, Cooper MD (1970) Suppression of immunoglobulin G synthesis as a result of antibody-mediated suppression of immunoglobulin M synthesis in chickens. Proc Natl Acad Sci USA 67:1918–1922

Kitamura D, Roes J, Kühn R, Rajewsky K (1991) A B cell-deficient mouse by targeted disruption of the membrane exon of the immunoglobulin μ chain gene. Nature 350:423–426

Knight KL (1992) Restricted VH gene usage and generation of antibody diversity in rabbit. Annu Rev Immunol 10:593–616

Knight KL, Becker RS (1990) Molecular basis of the allelic inheritance of rabbit immunoglobulin VH allotypes: implications for the generation of antibody diversity. Cell 60:963–970

Kocks C, Rajewsky K (1989) Stable expression and somatic hypermutation of antibody V regions in B-cell developmental pathways. Annu Rev Immunol 7:537–559

Kudo A, Melchers F (1987) A second gene VpreB in the λ5 locus of the mouse, which appears to be selectively expressed in pre-B lymphocytes. EMBO J 6:2267–2272

Köhler G, Milstein C (1975) Continuous cultures of fused cells secreting antibody of predenfined specificity. Nature 256:495–497

Lanzavecchia A (1985) Antigen-specific interaction between T and B cells. Nature 314:537–539

Lerner RA, Kang AS, Bain JD, Burton DR, Barbas CF (1992) Antibodies without immunization. Science 258:1313–1314

Lewis S, Gellert M (1989) The mechanism of antigen receptor gene assembly. Cell 59:585–588

Lieber MR (1992) The mechanism of V(D)J recombination: a balance of diversity, specificity, and stability. Cell 70:873–876

Lieber MR, Hesse JE, Mizuuchi K, Gellert M (1988) Lymphoid V(D)J recombination: nucleotide insertion at signal joints as well as coding joints. Proc Natl Acad Sci USA 85:8588–8592

Lieberson R, Giannini SL, Birshtein BK, Eckhardt LA (1991) An enhancer at the 3′ end of the mouse immunoglobulin heavy chain locus. Nucleic Acids Res 19:933–937

Lorenz W, Straubinger B, Zachau HG (1987) Physical map of the human immunoglobulin κ locus and its implications for the mechanism of Vκ-Jκ rearrangement. Nucleic Acids Res 15:9667–9676

Lubin I, Faktorowich Y, Lapidot T, Gan Y, Eshhar Z, Gazit E, Levite M, Reisner Y (1991) Engraftment and development of human T and B cells in mice after bone marrow transplantation. Science 252:427–431

Lutzker S, Alt FW (1988) Structure and expression of germ line immunoglobulin γ2b transcripts. Mol Cell Biol 8:1849–1852

Manning DD, Jutila JW (1972) Immunosuppression in mice injected with heterologous anti-immunoglobulin anti-sera. J Immunol 108:282–285

Manser T (1987) Mitogen-driven B cell proliferation and differentiation are not accompanied by hypermutation of immunoglobulin variable region genes. J Immunol 139:234–238

Mansour SL, Thomas KR, Capecchi MR (1988) Disruption of the proto-oncogene int-2 in mouse embryo-derived stem cells: a general strategy for targeting mutations to no-selectable genes. Nature 336:348–352

Manz J, Denis K, Witte O, Brinster R, Storb U (1988) Feedback inhibition of immunoglobulin gene rearrangement by membrane μ, but not by secreted μ heavy chains. J Exp Med 168:1363–1381

Martin GR (1981) Establishment of pluripotential cell lines from embryos cultured in medium conditioned by teratocarcinoma stem cells. Proc Natl Acad Sci USA 78:7634–7638

Marks JD, Tristem M, Karpas A, Winter G (1991) Oligonucleotide primers for polymerase chain reaction amplification of human immunoglobulin variable genes and design of family-specific oligonucleotide probes. Eur J Immunol 21:985–991

Matsuda F, Shin EK, Nagaoka H, Matsumura R, Haino M, Fukita Y, Taka-ishi S, Imai T, Riley JH, Anand R, Soeda E, Honjo T (1993) Structure and physical map of 64 variable segments in the 3′ 0.8 megabase region of the human immunoglobulin heavy-chain locus. Nature Genet 3:88–94

Matsuuchi L, Gold MR, Travis A, Grosschedl R, DeFranco AL, Kelly RB (1992) The membrane IgM-associated proteins MB-1 and Ig-β are sufficient to promote surface expression of a partially functional B-cell antigen receptor in a nonlymphoid cell line. Proc Natl Acad Sci USA 89:3404–3408

Meindl A, Klobeck H-G, Ohnheiser R, Zachau HG (1990) The Vκ gene repertoire in the human germ line. Eur J Immunol 20:1855–1863

Meyer KB, Neuberger MS (1989) The immunoglobulin κ locus contains a second, stronger B-cell-specific enhancer which is located downstream of the constant region. EMBO J 8:1959–1964

Meyer KB, Sharpe MJ, Surani MA, Neuberger MS (1990) The importance of the 3'-enhancer region in immunoglobulin κ gene expression. Nucleic Acids Res 18:5609–5615

Mills FC, Brooker JS, Camerini-Otero RD (1990) Sequences of human immunoglobulin switch regions: implications for recombination and transcription. Nucleic Acids Res 18:7305–7316

Mills FC, Thyphronitis G, Finkelman FD, Max EE (1992) Ig υ-ε isotype switch in IL-4 treated human B lymphoblastoid cells. J Immunol 149:1075–1085

Mombaerts P, Clarke AR, Hooper ML, Tonegawa S (1991) Creation of a large genomic deletion at the T cell antigen receptor β-subunit locus in mouse embryonic stem cells by gene targeting. Proc Natl Acad Sci USA 88:3084–3087

Mombaerts P, Clarke AR, Rudnicki MA, Iacomini J, Itohara S, Lafaille JJ, Wang L, Ichikawa Y, Jaenisch R, Hooper ML, Tonegawa S (1992) Mutations in the T-cell antigen receptor genes α and β block thymocyte development at different stages. Nature 360:225–231

Mowatt MR, Dunnick WA (1986) DNA sequence of the murine γ1 switch segment reveals novel structural elements. J Immunol 136:2674–2683

Müller B, Stappert H, Reth M (1990) A physical map and analysis of the murine Cκ-RS region show the presence of a conserved element. Eur J Immunol 20:1409–1411

Munir MI, Rossiter BJF, Caskey CT (1990) Antisense RNA production in transgenic mice. Somat Cell Mol Genet 16:383–394

Neuberger MS, Caskey HM, Pettersson S, Williams GT, Surani MA (1989) Isotype exclusion and transgene down-regulation in immunoglobulin-λ transgenic mice. Nature 338:350–352

Newman R, Alberts J, Anderson D, Carner K, Heard C, Norton F, Raab R, Reff M, Shuey S, Hanna N (1992) "Primatization" of recombinant antibodies for immunotherapy of human diseases: a macaque/human chimeric antibody against human CD4. Biotechnology 10:1455–1460

Nolan-Willard M, Berton MT, Tucker P (1992) Coexpression of μ and γ1 heavy chains can occur by a discontinuous transcription mechanism from the same unrearranged chromosome. Proc Natl Acad Sci USA 89:1234–1238

Nussenzweig MC, Shaw AC, Sinn E, Danner DB, Holmes KL, Morse HC, Leder P (1987) Allelic exclusion in transgenic mice that express the membrane form of immunoglobulin μ. Science 236:816–819

Nussenzweig MC, Schmidt EV, Shaw AC, Sinn E, Campos-Torres J, Mathey-Prevot B, Pattingale PK, Leder P (1988) A human immunoglobu, in gene reduces the incidence of lymphomas in c-Myc-bearing transgenic mice. Nature 336: 446–450

O'Brien RL, Brinster RL, Storb U (1987) Somatic hypermutation of an immunoglobulin transgene in κ transgenic mice. Nature 326:405–409

O'Connor M, Peifer M, Bender W (1989) Construction of large DNA segments in Escherichia coli. Science 244:1307–1312

Orlandi R, Gussow DH, Jones PT, Winter G (1989) Cloning immunoglobulin variable domains for expression by the polymerase chain reaction. Proc Natl Acad Sci USA 86:3833–3837

Pachnis V, Pevney L, Rothstein R, Costantini F (1990) Transfer of a yeast artificial chromosome carrying human DNA from Saccharomyces cerevisia into human cells. Proc Natl Acad Sci USA 87:5109–5113

Palmiter RD, Brinster RL (1986) Germ-line transformation of mice. Annu Rev Genet 20:465–499

Palmiter RD, Hammer RE, Brinster RL (1985) Expression of growth hormone genes in transgenic mice. In: Costantini F, Jaenisch R (eds) Genetic manipulation of the early mammalian embryo. Cold Spring Harbor Laboratory, Cold Spring Harbor, pp 123–132

Pargent W, Schäble KF, Zachau HG (1991) Polymorphisms and haplotypes in the human immunoglobulin κ locus. Eur J Immunol 21:1829–1835

Parikh VS, Bishop GA, Liu K-J, Do BT, Ghosh MR, Kim BS, Tucker PW (1992) Differential structure-function requirements of the transmembranal domain of the B cell antigen receptor. J Exp Med 176:1025–1031

Pascual V, Capra JD (1991) Human immunoglobulin heavy-chain variable region genes: organization, polymorphism, and expression. Adv Immunol 49:1–74

Pepin M-C, Pothier F, Barden N (1992) Impaired type II glucocorticoid-receptor function in mice bearing antisense RNA transgene. Nature 355:725–728

Pettersson S, Cook GP, Brüggemann M, Williams GT, Neuberger MS (1990) A second B cell-specific enhancer 3' of the immunoglobulin heavy-chain locus. Nature 344:165–168

Pieper FR, deWit CM, Pronk ACJ, Kooiman PM, Strijker R, Krimpenfort PJA, Nuyens JH, deBoer HA (1992) Efficient generation of functional transgenes by homologous recombination in murine zygotes. Nucleic Acids Res 20:1259–1264

Pillai S, Baltimore D (1987) Formation of disulfide linked $\mu2\omega2$ tetramers in pre-B cells by the 18K ω-immunoglobulin light chain. Nature 329:172–174

Queen C, Schneider WP, Selick HE, Payne PW, Landolfi NF, Duncan JF, Avdalovic NM, Levitt M, Junghans RP, Waldmann TA (1989) A humanized antibody that binds to the interleukin 2 receptor. Proc Natl Acad Sci USA 86:10029–10033

Rajewsky K, Förster I, Cuamano A (1987) Evolutionary and somatic selection of the antibody repertoire in the mouse. Science 238:1088–1094

Reth M (1989) Antigen receptor tail clue. Nature 338:383–384

Reth M, Hombach J, Weinands J, Campbell KS, Chein N, Justement LB, Cambier JC (1991) The B-cell antigen receptor complex. Immunol Today 12:196–201

Reynaud C-A, Dahan A, Anquez V, Weill J-C (1989) Development of the chicken antibody repertoire. In: Honjo T, Alt FW, Rabbits TH (eds) Immunoglobulin genes. Academic, New York, pp 151–162

Ritchie KA, Brinster RL, Storb U (1984) Allelic exclusion and control of endogenous immunoglobulin gene rearrangement in κ transgenic mice. Nature 312:517–520

Robertson EJ (1987) Embryo-derived stem cell lines. In: Robertson EJ (ed) Teratocarcinomas and embryonic stem cells, a practical approach. IRL Press, Oxford, pp 71–112

Roes J, Rajewsky K (1991) Cell autonomous expression of IgD is not essential for the maturation of conventional B cells. Int Immunol 3:1367–1371

Roes J, Rajewsky K (1993) Immunoglobulin D (IgD)-deficient mice reveal an auxiliary function for IgD in antigen-mediated recruitment of B cells. J Exp Med 177:45–55

Rogerson B, Peters A, Hackett J, Haasch D, Storb U (1991) Mutation pattern of immunoglobulin transgenes is compatible with a model of somatic hypermutation in which targeting of the mutator is linked to the direction of DNA replication. EMBO J 10:4331–4341

Rothman P, Lutzker S, Gorham B, Stewart V, Coffman R, Alt FW (1990) Structure and expression of germline immunoglobulin $\gamma3$ heavy chain gene transcripts: implications for mitogen and lymphokine directed class-switching. Int Immunol 2:621–627

Sakaguchi N, Berger CN, Melchers F (1986) Isolation of a cDNA copy of an RNA species expressed in murine pr-B cells. EMBO J 5:2139–2147

Sakaguchi N, Kashiwamura S, Kimoto M, Thalmann P, Melchers F (1988) B lymphocyte lineage-restricted expression of mb-1, a gene with CD3-like properties. EMBO J 7:3457–3464

Sanz I (1991) Multiple mechanisms participate in the generation of diversity of human H chain CDR3 regions. J Immunol 147:1720–1729

Schedl A, Beermann F, Theis E, Montoliu L, Kelsey G, Schütz G (1992) Transgenic mice generated by pronuclear injection of a yeast artificial chromosome. Nucleic Acids Res 20:3073–3077

Schroeder HW, Wang JY (1990) Preferential utilization of conserved immunoglobulin heavy chain variable gene segments during human fetal life. Proc Natl Acad Sci USA 87:6146–6150

Schwartzberg PL, Goff SP, Robertson EJ (1989) Germ-line transmission of a c-*able* mutation produced by targeted gene disruption in ES cells. Science 246:799–803

Selsing E, Durdik J, Moore MW, Persiani DM (1989) Immunoglobulin λ genes. In: Honjo T, Alt FW, Rabbits TH (eds) Immunoglobulin genes. Academic, New York

Sen R, Baltimore D (1989) Factors regulating immunoglobulin gene transcription. In: Honjo T, Alt FW, Rabbits TH (eds) Immunoglobulin genes. Academic, New York, pp 327–342

Sharpe MJ, Neuberger M, Pannell R, Surani MA, Milstein C (1990) Lack of somatic mutation in a κ light chain transgene. Eur J Immunol 20:1379–1385

Sharpe MJ, Milstein C, Jarvis JM, Neuberger MS (1991) Somatic hypermutation of immunoglobulin κ may deoend on sequences 3′ of Cκ and occurs on passenger transgenes. EMBO J 10:2139–2145

Shaw AC, Mitchell RN, Weaver YK, Campos-Torres J, Abbas AK, Leder P (1990) Mutations of immunoglobulin transmembrane and cytoplasmic domains: effects on intracellular signaling and antigen presentation. Cell 63:381–392

Shimizu A, Nussenzweig MC, Mizuta T-R, Leder P, Honjo T (1989) Immunoglobulin double isotype expression by trans-mRNA in a human immunoglobulin transgenic mouse. Proc Natl Acad Sci USA 86:8020–8023

Shimizu A, Nussenzweig MC, Han H, Sanchez M, Honjo T (1991) Trans-splicing as a possible mechanism for the multiple isotype expression of the immunoglobulin gene. J Exp Med 173:1385–1393

Short JA, Sethupathi P, Zhai SK, Knight KL (1991) VDJ genes in VHa2 allotype suppressed rabbits: limited germline VH gene usage and accumulation of somatic mutations in D regions. J Immunol 147:4014–4018

Siekovitz M, Kocks C, Rajewsky K (1987) Analysis of somatic mutation and class switching in naive and memory B cells generating adoptive primary and secondary responses. Cell 48:757–770

Simons RW, Kleckner N (1983) Translational control of IS10 transposition. Cell 34:683–691

Small JA, Blair DG, Showalter SD, Scangos GA (1985) Analysis of a transgenic mouse containing simian virus 40 and v-myc sequences. Mol Cell Biol 5:642–648

Smith DR, Smyth AP, Moir DT (1990) Amplification of large artificial chromosomes. Proc Natl Acad Sci USA 87:8242–8246

Stavnezer J, Radcliff G, Lin YC, Berggren L, Sitia R, Severinson E (1988) Immunoglobulin heavy-chain switching may be directed by prior induction of transcripts from constant-region genes. Proc Natl Acad Sci USA 85:7704–7708

Steele EJ, Rothenfluh HS, Both GW (1992) Defining the nucleic acid substrate for somatic hypermutation. Immunol Cell Biol 70:129–144

Sternberg N (1990) A bacteriophage P1 cloning system for the isolation, amplification and recovery of DNA fragments as large as 100 kbp. Proc Natl Acad Sci USA 87:103–107

Sternberg N, Reuther J, DeReil K (1990) Generation of qa 50,000-member human DNA library with an average DNA insert size of 75–100 kbp in a bacteriophage P1 cloning vector. New Biol 2:1551–162

Stoker NG, Fairwather NF, Spratt BG (1982) Versitile low-copy-number vectors for cloning in *Escherichia coli*. Gene 18:335–341

Storb U (1989) Immunoglobulin gene analysis in transgenic mice. In: Honjo T, Alt FW, Rabbits TH (eds) Immunoglobulin genes. Academic, New York, pp 303–326

Storb U, Pinkert C, Arp B, Engler P, Gollahon K, Manz J, Brady W, Brinster RL (1986a) Transgenic mice with μ and κ genes encoding antiphosphorylcholine antibodies. J Exp Med 164:627–641

Storb U, Ritchie KA, O'Brien R, Arp B, Brinster R (1986b) Expression, allelic exclusion and somatic mutation of mouse immunoglobulin kappa genes. Immunol Rev 89:85–102

Storb U, Haasch D, Arp B, Sanchez P, Cazenave P-A, Miller J (1989) Physical linkage of the mouse λ genes by pulsed field gel electrophoresis suggests that the rearrangement process favors proximate target sequences. Mol Cell Biol 9:711–718

Straubinger B, Huber E, Lorenz W, Osterholzer E, Pargent W, Pech M, Pohlenz H-D, Zimmer F-J, Zachau HG (1988) The human Vκ locus, characterization of a duplicated region encoding 28 different immunoglobulin genes. J Mol Biol 199:23–34

Strauss WM, Jaenisch R (1992) Molecular complementation of a collagen mutation in mammalian cells using yeast artificial chromosomes. EMBO J 11:417–422

Taub RA, Hollis GF, Heiter PA, Korsmeyer SJ, Waldmann TA, Leder P (1983) Variable amplification of immunoglobulin λ light-chain genes in human populations. Nature 304:172–174

Taylor LD, Lonberg N (1993) A plasmid vector for building large DNA constructs. (submitted)

Taylor LD, Carmack C, Schramm SR, Mashayekh R, Higgins KM, Kuo C-C, Woodhouse C, Kay RM, Lonberg N (1992) A transgenic mouse that expresses a diversity of human sequence heavy and light chain immunoglobulins. Nucleic Acids Res 20:6287–6295

Thomas KR, Cappechi MR (1987) Site directed mutagenesis by gene targeting in mouse embryo-derived stem cells. Cell 51:503–512

Thompson KM (1988) Human monoclonal antibodies. Immunol Today 9:113–116

Thompson S, Clarke AR, Pow AM, Hooper ML, Melton DW (1989) Germ line transmission and expression of a corrected HPRT gene produced by gene targeting in embryonic stem cells. Cell 56:313–321

Tjandra JJ, Ramadi L, McKenzie IFC (1990) Development of human anti-murine antibody (HAMA) response in patients. Immunol Cell Biol 68:367–376

Tomizawa J, Itoh T, Selzer G, Som T (1981) Inhibition of ColE1 RNA primer formation by a plasmid-specified small RNA. Proc Natl Acad Sci USA 78:1421–1425

Tomlinson I, Walter G, Marks JD, Llewelyn MB, Winter G (1992) The repertoire of human germline VH sequences reveals about fifty groups of VH segments with different hypervariable loops. J Mol Biol 227:776–798

Tonegawa S (1983) Somatic generation of antibody diversity. Nature 302:575–581

Townes TM, Chen HY, Lingrel JB, Palmiter RD, Brinster RL (1985a) Expression of human beta-globin genes in transgenic mice: effects of a flanking metallothionein-human growth hormone fusion gene. Mol Cell Biol 5:1977–1983

Towners TM, Lingrel JB, Palmiter RD, Brinster RL (1985b) Erythroid-specific expression of human beta-globin genes in transgenic mice. EMBO J 4:1715–1723

Tsang H, Pinkert C, Hagman J, Lostrum M, Brinster RL, Storb U (1988) Cloning of a γ2b gene encoding anti *pseudomonas aeruginosa* H chains and its introduction into the germ line of transgenic mice. J Immunol 141:308–314

Umar A, Schweitzer PA, Levy NS, Gearhart JD, Gearhart PJ (1991) Mutation in a reporter gene depends on proximity to and transcription of immunoglobulin variable transgenes. Proc Natl Acad Sci USA 88:4902–4906

van Dijk KW, Milner LA, Sasso EH, Milner ECB (1992) Chromosomal organization of the heavy chain variable region gene segments comprising the human fetal antibody repertoire. Proc Natl Acad Sci USA 89:10430–10434

van Noesel CJM, Brouns GS, van Schijndel GMW, Bende RJ, Mason DY, Borst J, van Lier RAW (1992) Comparison of human B cell antigen receptor complexes: membrane-expressed forms of immunoglobulin (Ig)M, IgD, and IgG are associated with structurally related heterodimers. J Exp Med 175:1511–1519

Vasicek TJ, Leder P (1990) Structure of the human immunoglobulin λ genes. J Exp Med 172:609–620

Venkitaraman AR, Williams GT, Dariavach P, Neuberger MS (1991) The B-cell antigen receptor of the five immunoglobulin classes. Nature 352:777–781

Walter MA, Surti U, Hofker MH, Cox DW (1990) The physical organization of the human immunoglobulin heavy chain gene complex. EMBO J 9:3303–3313

Walter MA, Dosch HM, Cox DW (1991) A deletion map of the human immunoglobulin heavy chain variable region. J Exp Med 174:335–349

Wang FR, Kushner SR (1991) Construction of versitile low-copy-number vectors for cloning, sequencing and gene expression in *Escherichia coli*. Gene 100:195–199

Weaver D, Costantini F, Imanishi-Kari T, Baltimore D (1985) A transgenic immunoglobulin mu gene prevents rearrangement of endogenous genes. Cell 42: 117–127

Webb CF, Nakai C, Tucker PW (1989) Immunoglobulin receptor signaling depends on the carboxyl terminus but not the heavy-chain class. Proc Natl Acad Sci USA 86:1977–1981

Weiss S, Lehmann K, Raschke WC, Cohn M (1984) Mice completely suppressed for the expression of immunoglobulin κ light chain. Proc Natl Acad Sci USA 81:211–215

Weiss U, Rajewsky K (1990) The repertoire of somatic antibody mutants accumulating in the memory compartment after primary immunization is restricted through affinity maturation and mirrors that expressed in the secondary response. J Exp Med 172:1681–1689

White MB, Word CJ, Humphreies CG, Blattner FR, Tucker PW (1990) Immunoglobulin D switching can occur through homologous recombination in human B cells. Mol Cell Biol 10:3690–3699

Widera G, Burkly LC, Pinkert CA, Bottger EC, Cowing C, Palmiter RD, Brinster RL, Flavell RA (1987) Transgenic mice selectively lacking MHC class II (I-E) antigen expression on B cells: an in vivo approach to investigate Ia gene function. Cell 51:175–187

Wienands J, Reth M (1992) Glycosyl-phosphatidylinositol linkage as a mechanism for cell surface expression of immunoglobulin D. Nature 356:246–248

Williams GT, Venkitaraman AR, Gilmore DJ, Neuberger MS (1990) The sequence of the μ transmembrane segment determines the tissue specificity of the transport of immunoglobulin M to the cell surface. J Exp Med 171:947–952

Wysocki LJ, Creadon G, Lehmann KR, Cambier JC (1992) B-cell proliferation initiated by Ia cross-linking and sustained by interleukins leads to class switching but not somatic mutation in vitro. Immunology 75:116–121

Yamada M, Wasserman R, Reichard BA, Shane S, Caton AJ, Rovera G (1991) Preferential utilization of specific immunoglobulin heavy chain diversity and joining segments in adult human peripheral blood B lymphocytes. J Exp Med 173:395–407

Yamamura K-I, Kudo A, Ebihara T, Kamino K, Araki K, Kumahara Y, Watanabe T (1986) Cell-type-soecific and regulated expression of a human γ1 heavy-chain immunoglobulin gene in transgenic mice. Proc Natl Acad Sci USA 83:2151–2156

Yancopoulos GD, Alt FW (1985) Developmentally controlled and tissue-specific expression of unrearranged V(H) gene segments. Cell 40:271–281

Yancopoulos GD, Alt FW (1986) Regulation of the assembly and expression of variable-region genes. Annu Rev Immunol 4:339–368

Yancopoulos GD, Blakwell TK, Such H, Hood L, Alt FW (1986) Introduced T cell receptor variable region gene segments recombine in pre-B cells: evidence that B and T cells use a common recombinase. Cell 44:251–259

Yasui H, Akahori Y, Hirano M, Yamada K, Kurosawa Y (1989) Class switch from υ to δ is mediated by homologous recombination between $\sigma\mu$ and $\Sigma\mu$ sequences in human immunoglobulin gene loci. Eur J Immunol 19:1399–1403

Yu L, Peng C, Starnes SM, Liou RS, Chang TW (1990) Two isoforms of human membrane-bound α Ig resulting from alternative mRNA splicing in the membrane segment. J Immunol 145:3932–3936

Yu L-M, Chang TW (1992) Human mb-1 gene: complete cDNA sequence and its expression in B cells bearing membrane Ig of various isotypes. J Immunol 148:633–637

Yuan D, Vitetta ES (1978) Cell surface immunoglobulin. XXI. Appearance of IgD on murine lymphocytes during differentiation. J Immunol 120:353–356

Zijlstra ME, Li E, Sajjadi F, Subramani S, Jaenisch R (1989) Germ line transmission of a disrupted β2-microglobulin gene produced by homologous recombination in embryonic stem cells. Nature 342:435–438

Zou Y-R, Takeda S, Rajewsky K (1993) Gene targeting in the Igκ locus: efficient generation of λ chain expressing B cells, independent of gene rearrangements in Igκ. EMBO J 12:811–820

Section II: Genetically Engineered Monoclonal Antibodies

CHAPTER 4

Humanization of Monoclonal Antibodies

G.E. Mark and E.A. Padlan

A. Introduction

Antibodies of predefined specificity have many potential uses in human therapy and diagnosis, and hybridoma technology (Koehler and Milstein 1975) has made possible the generation of virtually limitless amounts of such antibodies. Unfortunately, hybridoma proteins are more easily obtained from nonhuman, usually rodent, sources and the use of those antibodies in human subjects will be hindered by the patient's immune system. The reduction of the immunogenicity in humans of xenogenic antibodies will make those molecules more efficacious reagents and various procedures for "humanizing" antibodies have been developed with this objective in mind.

The goal of humanization is to make an antibody appear as human-like as possible to a patient's immune system. Two general procedures have been proposed to achieve this goal: the first, by the construction of human/ nonhuman antibody chimeras (Morrison and Oi 1988; Jones et al. 1986; Verhoeyen et al. 1988), with as few nonhuman parts as possible; the second, by "veneering" or "cloaking" an antibody with a human-like surface (Padlan 1991). A humanized antibody should retain all the antigen-binding properties of the original molecule, and that requires the faithful reproduction of the combining site structure. In this regard, humanization has been aided by the structural knowledge that is available on antibodies.

In terms of structure, antibodies are probably the most studied of all proteins, with amino acid and nucleotide sequence data available for thousands of different chains (Kabat et al. 1991) and three-dimensional structures available for both whole antibodies and for a variety of fragments. Here, we review the relevant structural data, demonstrate how these data have been used in the design of humanization protocols, and exemplify two alternative humanization procedures.

B. Structure of Antibodies

I. General

An antibody molecule is traditionally characterized as being composed of three fragments: the two Fabs, which are identical and each of which contains the light chain and the first two domains of the heavy chain, and the Fc, which contains the COOH-terminal constant domains of the two heavy chains. The Fabs are connected to the Fc by the hinge region, which varies in length and flexibility in the different antibody isotypes. The combining sites are located at the tips of the Fabs.

X-ray analysis has shown that the variable and constant domains of antibodies form compact globular structures with a characteristic fold – the immunoglobulin fold (POLJAK et al. 1973). Each domain consists of a stable arrangement of hydrogen-bonded β-strands which form a bilayer structure, further stabilized by a disulfide bond between the two layers. The variable domains of the light and the heavy chains, the V_L and the V_H, are similar to each other in three-dimensional structure, as are the constant domains. Homologous domains from different species are very similar, so that the V_H domains of human and murine antibodies, for example, are superposable, except in the hypervariable or complementarity determining regions (CDRs) and usually only if length differences exist in those regions (PADLAN and DAVIES 1975).

In the Fab (Fig. 1), V_L and V_H associate closely to form a compact module, the Fv, and are related by a pseudodyad. The contact between the

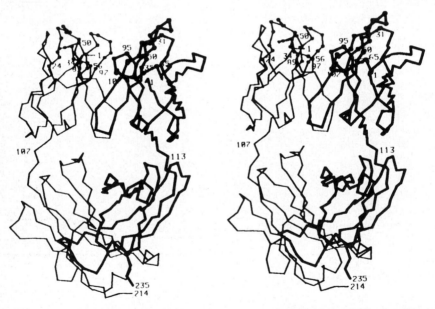

Fig. 1. Stereodrawing of the α-carbon trace of the Fab of antibody HyHEL-10 (PADLAN et al. 1989). The light chain is drawn with *thinner lines* on the *left* and the heavy chain with *thicker lines* on the *right*. The variable domains are on *top* and the constant domains are at the *bottom*. The complementarity determining regions (CDR) residues are indicated by *circles*. The beginning and end of each CDR are labeled, as are the NH$_2$- and COOH-terminals of both chains and the ends of the variable domains

variable domains involves both framework and CDR residues. The involvement of CDR residues in the $V_L : V_H$ contact contributes to the variation seen in the pseudodyad. The constant domains in the Fab, the C_L and the C_H1, likewise form a compact module and are also related by a pseudodyad. In both light and heavy chains, the variable and constant domains are linked by a short segment of polypeptide chain, the switch, which is flexible, allowing variation in the relative disposition of the variable and constant modules.

II. The Antibody Combining Site

The CDRs are seen as loops mainly situated at the NH_2-terminal tip of the Fab, where they form a continuous surface. Crystallographic analysis of several antibody/antigen complexes and other studies have shown that antigen binding mainly involves this surface (although some framework residues have been found to be involved also in the interaction with antigen). The CDR surface is therefore usually equated with the combining site of the antibody. Thus, the antigen-binding specificity of an antibody is defined by the topography of its CDR surface and by the chemical characteristics of this surface; these in turn are determined by the conformation of the individual CDRs, by the relative disposition of the CDRs, and by the nature and disposition of the side chains of the amino acids comprising the CDRs.

Crystallographic analysis of antibody structures, with and without bound ligand, has shown that, in some cases, the combining site structure may change on binding, in keeping with the hypothesis of "induced fit" (EDMUNDSON et al. 1974; BHAT et al. 1990; STANFIELD et al. 1990; HERRON et al. 1991; RINI et al. 1992; DAVIES and PADLAN 1992). The conformation of the individual CDRs may change, as could the mode of quarternary association of the variable domains (COLMAN et al. 1987; BHAT et al. 1990; STANFIELD et al. 1990; HERRON et al. 1991; RINI et al. 1992). These results suggest that the combining site structure is not rigid, rather, it is plastic and may assume different conformations depending on circumstance.

III. Complementarity Determining Regions

There are three hypervariable regions in each of the light and heavy chains of an antibody. On the basis of sequence variation, the CDRs of the light chain are comprised of residues 24–34 (CDR1-L), 50–56 (CDR2-L), and 89–97 (CDR3-L), and those of the heavy chain contain residues 31–35 (CDR1-H), 50–65 (CDR2-H), and 95–102 (CDR3-H) (KABAT et al. 1991) (Fig. 1). Within each CDR, there are residue positions that are more hypervariable than others; these positions are presumably more involved in the determination of antigen-binding specificity and in the diversification of specificities, while the more conserved residues in the CDRs probably play a

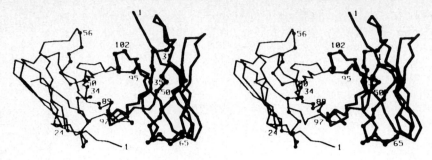

Fig. 2. End on view of the Fv of antibody HyHEL-10. The figure is rotated 90° relative to Fig. 1

more structural role and serve to stabilize the combining site structure (Kabat et al. 1977; Padlan 1977).

Variations in length accompany the variability in sequence in these CDRs, with CDR3-H displaying particularly large length variations. The longer CDRs seem to be more flexible and have been found to display greater structural variability. Some of the longer CDRs have been observed to assume different conformations when subject to different crystal environments (Schiffer et al. 1973) and to suffer the larger deformations upon ligand binding (e.g., Stanfield et al. 1990; Herron et al. 1991; Rini et al. 1992).

The six CDRs are disposed (Fig. 2) such that the NH_2-terminal part of CDR1-L and the COOH-terminal parts of CDR2-L and CDR2-H are farther from the center of the CDR surface, while CDR1-H, CDR3-H, CDR3-L, the COOH-terminal part of CDR1-L, and the NH2-terminal parts of CDR2-L and CDR2-H are closer to the center and may play a more prominent role in antigen binding. There is close contact among the CDRs of the same domain, and among the CDRs from opposite domains in the Fv.

IV. Influence of Framework Residues on Combining Site Structure

The framework regions, by and large, show conserved amino acid substitutions and very similar three-dimensional structures. The different antibody combining sites, therefore, can be pictured as being constructed with CDRs of varied shapes and sizes, which are grafted onto a scaffolding of conserved structure. The framework, on account of the intradomain disulfide bond, a largely hydrogen-bonded backbone structure, a hydrophobic interior with several large aromatic residues, and sharp turns frequently involving prolines, provides a strong foundation for the highly variable CDRs.

Many residues from the framework are in contact with the CDRs in the same domain and, across the $V_L : V_H$ interface, with those in the opposite domain. These framework/CDR contacts significantly affect the conformation of the CDRs. Indeed, canonical structures have been observed for most CDRs and those structures are determined by the nature of a small number of framework residues that contact the CDRs (CHOTHIA and LESK 1987; CHOTHIA et al. 1989; TRAMONTANO et al. 1990).

C. Strategies for the Humanization of Antibodies

Ideally, humanization should result in a molecule that is totally nonimmunogenic and with all the antigen-binding properties of the original antibody. In order to preserve the fine specificity of an antibody, its CDRs (and probably also some of the neighboring framework residues), their interaction with each other, and their interaction with the rest of the variable domains must be strictly maintained. Further, if the interaction with antigen involves induced fit, the structural elements which permit the conformational change must also be preserved in the humanized molecule.

I. Transplanting a Nonhuman Combining Site onto a Human Framework

The easiest way to preserve the ligand-binding properties of an antibody is to keep the entire Fv. This has been achieved by the construction of chimeric antibodies in which the variable domains of the xenogenic proteins are fused with the constant regions of human molecules (reviewed by MORRISON and OI 1988). The close similarity among human and nonhuman antibodies, even in the switch regions, makes this procedure feasible. The chimeric molecules have been found to possess all the antigen-binding characteristics of the original xenogenic antibodies. Unfortunately, the chimeras retain the immunogenicity of the nonhuman variable domains.

The greatest reduction in immunogenicity is achieved by the procedure of Winter and coworkers (JONES et al. 1986; VERHOEYEN et al. 1988), who graft only the CDRs of the xenogeneic antibodies onto human framework and constant regions. Antibodies humanized in this manner would be expected to be essentially nonimmunogenic (assuming allotypic or idiotypic diversities go unnoticed). However, CDR grafting per se may not result in the complete retention of antigen-binding properties. Indeed, it is frequently found that some framework residues from the original antibody need to be preserved in the humanized molecule if significant antigen-binding affinity is to be recovered (see, for example, RIECHMANN et al. 1988; QUEEN et al. 1989; TEMPEST et al. 1991; Co and QUEEN 1991; Co et al. 1991; FOOTE and WINTER 1992).

In theory, all the framework residues which could influence the structure of the combining site should be preserved. These include those which are in contact with the CDRs, since they provide the primary support for the combining site structure, and those which are involved in the $V_L : V_H$

contact, since they influence the relative disposition of the CDRs. It may be necessary to keep also those framework residues which are buried in the domain interior. Structural studies of the effect of the mutation of interior residues, in which changes in side chain volume are involved, have shown that the resulting local deformations are accommodated by shifts in side chain positions that are propagated to distant parts of the molecular interior (see, for example, Alber et al. 1987). This suggests that a humanization protocol, in which an interior residue is replaced by one of a different size, could result in a significant modification of the combining site structure.

The important framework residues can be identified if the three-dimensional structure of the antibody, or, at least, of the Fv, is known. The structures of murine Fabs, for which atomic coordinates are available, have been analyzed and those framework residues which play a role in maintaining the combining site structure are presented in Tables 1–5. The framework residues in the V_L domains whose side chains contact CDR residues are listed in Table 1; those in the V_H are listed in Table 2. The framework residues which contact framework residues in the opposite domain and potentially influence the quaternary structure of the Fv are listed in Table 3. The buried, inward-pointing, framework residues in the V_L (i.e., those which are located in the domain interior) are listed in Table 4; similarly, those in the V_H are listed in Table 5. These results are summarized in Fig. 3 for V_L and in Fig. 4 for V_H.

It is seen in Tables 1–5 that: (a) there are many framework residues which either contact the CDRs, contact the opposite domain, or are found in the domain interior and (b) these framework residues, which could influence the structure of the combining site and antigen-binding characteristics of an antibody, are different from antibody to antibody.

If an antibody was to be humanized, it would probably be wise to retain all of the framework residues that are listed in Tables 1–5. At first glance, it would appear that there would be too many nonhuman residues to keep; however, searching through the tabulation of immunoglobulin sequences (Kabat et al. 1991), one finds that human variable domain sequences are known that already have most of the framework residues that need to be preserved. Hence, many mutations would not be needed to transform a human framework sequence to one that will accomodate the nonhuman CDRs with retention of antigen-binding properties. For example, the humanization of the murine antibody HyHEL-5 would require keeping nine murine framework residues in the V_L using the human V_κ sequence BI (or Rei and a few others) as template, and 13 framework residues in the V_H, using the human V_H sequence AND (or 21/28′CL and a few others). Humanization of the murine antibody B13i2 would require the retention of only three murine framework residues in the V_L, using the human $V\kappa$ sequence CUM (or NIM) as template, and only two framework residues in the V_H, using the human V_H sequence M72 (or M74 and a few others). It is possible that there exist other human sequences that are known but are not

Table 1. V_L framework residues that contact CDR residues in murine Fabs of known three-dimensional structure

Position	Antibody 1	2	3	4	5	6	7	8	9	10	11	12
1	E(2)[a]	D(5)	D(10)	D(3)	D(5)	D(8)		D(4)	D(5)	D(11)	D(5)	Q(5)
2	I(11)	I(15)	I(18)	I(18)	I(20)	V(12)	V(9)	I(21)	V(7)	I(12)	I(25)	I(17)
3		V(3)	V(2)	V(3)	Q(2)	V(2)		Q(2)	L(1)		Q(2)	V(2)
4	L(13)	M(11)	L(12)	L(17)	M(14)	M(18)	M(12)	M(9)	M(6)	M(12)	M(9)	L(14)
5		T(3)	T(1)		T(1)	T(3)	T(2)		T(1)	T(1)	T(1)	T(2)
7					T(3)							
23	C(3)	C(3)	C(4)	C(4)	C(3)	C(3)	C(2)	C(3)	C(3)	C(1)	C(3)	C(5)
35	W(8)	W(6)	W(7)	W(8)	W(8)	W(5)	W(6)	W(10)	W(7)	W(10)	W(9)	W(6)
36	Y(17)	Y(21)	Y(12)	Y(10)	Y(19)	Y(13)	Y(10)	Y(19)	Y(14)	Y(18)	Y(8)	Y(17)
45					K(5)	K(5)						
46	P(8)	L(9)	L(4)	R(21)	L(8)	V(15)	L(12)	L(8)	L(12)	L(9)	L(7)	L(9)
48	I(10)	L(11)	I(11)	I(12)	I(11)	I(9)	I(10)	I(13)	I(10)	V(9)	I(13)	I(13)
49	Y(36)	Y(33)	K(17)	Y(18)	Y(31)	Y(29)	Y(31)	Y(30)	Y(28)	Y(29)	Y(32)	Y(27)
58	V(8)	V(6)	I(6)	V(11)	V(7)	V(9)	V(12)	V(4)	V(8)	V(6)	V(9)	V(9)
60		D(1)				D(2)			D(1)			V(1)
62				F(1)	F(1)	F(1)	F(1)	F(1)				F(1)
63									S(1)			
67	S(3)	S(2)								S(1)		
69	T(5)	T(9)	T(8)	T(2)	T(8)	T(14)	T(7)	T(8)	T(4)	T(9)	T(9)	T(10)
70		D(2)			D(3)		D(2)	D(6)	D(6)		D(9)	
71	Y(14)	F(23)	F(23)	Y(23)	Y(21)	F(17)	F(28)	Y(21)	F(21)	Y(25)	Y(25)	Y(12)
88	C(4)	C(3)	C(3)	C(2)	C(3)	C(3)	C(2)	C(2)	C(2)	C(5)	C(4)	C(4)
98	F(11)	F(15)	F(13)	F(7)	F(14)	F(12)	F(14)	F(12)	F(17)	F(17)	F(14)	F(17)

CDR, complementarity determining region.

[a] The number in parentheses after each residue name corresponds to the number of atomic contacts in which the amino acid is involved. Only the contacts involving side chain atoms are presented. Atoms are designated as being in contact if they are within the sum of their van der Waals' radii plus 0.5 Å. The numbering scheme of KABAT et al. (1991) is used throughout.

Table 2. V_H framework residues that contact CDR residues in murine Fabs of known three-dimensional structure

Position	Antibody											
	1	2	3	4	5	6	7	8	9	10	11	12
1	V(11)	V(3)	V(8)		V(3)		E(3)	E(3)	V(13)	V(12)	E(2)	L(5)
2	L(3)	L(6)	L(5)		L(1)			V(7)	L(3)	L(4)	V(3)	V(5)
4		T(2)	V(6)	L(2)		L(3)		L(3)			L(7)	Y(16)
24						A(1)	A(1)				T(1)	S(1)
27	F(3)	F(2)		Y(14)	Y(7)	F(26)	F(2)	Y(4)	F(2)	F(4)	F(8)	I(7)
28	D(9)	T(6)		T(3)	T(4)	T(6)	S(1)	T(2)	T(7)	L(1)	T(8)	T(2)
29	F(4)	F(4)		F(10)	F(6)	F(13)	F(4)	F(6)	F(5)		F(4)	W(3)
30	S(2)	S(1)	T(4)	S(3)	T(9)	S(9)	N(1)	T(4)	S(1)	T(2)	T(10)	I(4)
36	W(4)	W(3)	W(6)	W(5)	W(6)	W(7)	W(4)	W(8)	W(4)	W(6)	W(4)	
37		V(1)			V(1)		V(1)		V(4)	V(1)		
38	R(1)	R(2)	R(4)	K(2)	K(2)	R(4)	R(1)	K(2)	R(1)			R(1)
40				R(1)								
46	E(3)	E(4)	E(1)	E(27)	E(7)	E(4)	E(2)	E(9)		E(1)		E(2)
47	W(34)	W(46)	Y(22)	W(22)	W(36)	W(31)	W(34)	W(36)	W(25)	W(36)	W(28)	W(37)
48	I(1)	I(1)	M(6)	I(12)	I(8)	V(1)	V(4)	I(9)	V(2)	L(1)	L(1)	M(3)
49	A(4)	A(4)				A(3)	A(5)		A(3)			
66	K(2)	R(3)	R(316)	K(1)	K(2)	R(8)	R(8)	K(1)	R(5)	R(3)	R(2)	R(3)
67	F(6)	F(10)	I(9)	A(1)	T(7)	F(13)	F(5)	T(8)	F(10)	L(10)	F(9)	I(10)
68		I(1)				T(11)					T(1)	S(1)
69	I(9)	V(7)	I(12)	F(16)	L(12)	I(25)	I(19)	L(9)	I(12)	I(9)	I(9)	I(10)
71	R(27)	R(28)	R(7)	A(4)	V(2)	R(13)	R(28)	V(7)	R(21)	K(9)	R(23)	R(23)
73	N(3)	T(3)		A(1)	R(3)	D(3)	D(1)	K(2)	N(1)		N(7)	
78	L(5)	L(7)	Y(15)	N(1)	A(1)	V(2)	L(3)	A(1)	L(5)	V(4)	L(3)	F(11)
80						L(1)						
82					R(2)	D(2)				M(1)		K(1)
82a			L(2)									
86						T(6)						
92	C(1)	C(1)	C(1)	C(2)	C(1)			C(2)		C(2)		C(1)
93	A(5)	A(6)	A(1)	L(6)	A(4)		V(8)	A(2)	T(6)	A(5)	T(3)	A(2)
94	R(39)	R(26)	N(18)	H(9)	R(35)		R(14)	R(29)	R(27)	R(34)	R(36)	R(39)
103	W(13)	W(14)	W(15)	W(3)	W(16)	W(7)	W(10)	W(14)	W(7)	W(7)	W(2)	W(12)

For explanation of table, see Table 1.

Table 3. Framework residues that contact framework residues in the opposite domain in murine Fabs of known three-dimensional structure

Position	Antibody 1	2	3	4	5	6	7	8	9	10	11	12
Position in V$_L$												
36	Y(3)	Y(3)	Y(3)	Y(5)	Y(2)	Y(11)	Y(7)	Y(7)	Y(1)	Y(7)	Y(3)	Y(2)
38	Q(10)	Q(7)	Q(9)	Q(5)	Q(9)	Q(3)	Q(2)	Q(6)	Q(8)	Q(6)	Q(6)	Q(7)
43	S(12)	P(4)	S(10)	S(9)	T(4)	S(6)	S(8)		S(3)	S(5)		S(5)
44	P(10)	P(16)	P(8)	P(11)	V(11)	P(7)	P(7)	I(20)	P(7)	P(16)	V(14)	P(11)
46	P(3)											
85			M(2)				F(4)					
87	Y(10)	Y(8)	F(8)	Y(6)	F(8)		F(18)	F(6)	Y(8)	Y(12)	I(1)	Y(10)
98	F(12)	F(11)	F(10)	F(20)	F(11)	F(13)	A(2)	F(10)	F(12)	F(17)	F(17)	F(7)
100		A(1)										V(2)
Position in V$_H$												
37	V(4)		I(2)	V(1)	V(2)	V(2)		V(1)	V(2)	V(4)	V(2)	I(1)
39	Q(10)	Q(7)	K(8)	Q(5)	Q(9)	Q(3)	Q(1)	Q(6)	Q(8)	Q(6)	Q(5)	Q(8)
43			N(4)				K(4)		K(1)			N(7)
44		R(7)									A(1)	K(1)
45	L(13)	L(12)	L(8)	L(14)	L(11)	L(8)	L(17)	L(11)	L(7)	L(14)	L(8)	L(13)
47	W(1)		Y(12)							W(3)	W(1)	
91	Y(8)	Y(5)	Y(4)	Y(13)	F(1)	Y(2)	Y(6)	F(6)	Y(2)	Y(6)	Y(4)	F(4)
103	W(13)	W(17)	W(16)	W(15)	W(17)	W(18)	W(19)	W(26)	W(13)	W(22)	W(20)	W(14)
105	Q(6)											Q(2)

For explanation of table, see Table 1.

Table 4. Inward-pointing,[a] buried framework residues in the V_L of murine Fabs of known three-dimesional structure

Position	Antibody											
	1	2	3	4	5	6	7	8	9	10	11	12
2	I	I	I	I	I	V	V	I	V	I	I	I
4	L	M	L	L	M	M	M	M	M	M	M	L
6	Q	Q	Q	Q	Q	Q	Q	Q	Q	Q	Q	Q
11	T	L	L	M	L	L	L	L	L	L	L	M
13	A	V	V	A	A	V	V	A	V	A	A	A
19	V	V	V	V	V	A	A	V	A	V	V	V
21	I	M	L	M	I	I	I	I	I	I	I	M
23	C	C	C	C	C	C	C	C	C	C	C	C
35	W	W	W	W	W	W	W	W	W	W	W	W
37	Q	Q	Q	Q	Q	L	L	Q	L	Q	Q	Q
47	W	L	L	W	L	L	L	L	L	L	L	L
48	I	I	I	I	I	I	I	I	I	V	I	I
58	V	V	I	V	V	V	V	V	V	V	V	V
61	R	R	R	R	R	R	R	R	R	R	R	R
62	F	F	F	F	F	F	F	F	F	F	F	F
71	Y	F	F	Y	Y	F	F	Y	F	Y	Y	Y
73	L	L	L	L	L	L	L	L	L	L	L	L
75	I	I	I	I	I	I	I	I	I	I	I	I
78	M	V	V	M	L	V	V	L	V	L	L	M
82	D	D	D	D	D	D	D	D	D	D	D	D
83									F			
84	A	A		A	A			A			A	A
86	Y	Y	Y	Y	Y	Y	Y	Y	Y	Y	Y	Y
88	C	C	C	C	C	C	C	C	C	C	C	C
102	T	T	T	T	T	T	T	T	T	T	T	T
104	L	L	L	L	L	L	L	L	L	L	L	L
106	L	I	I						I		I	L

[a] An inward pointing residue is designated as buried if at least 50% of its side chain is inaccessible to solvent. Solvent accessibilities were computed as described previously (Padlan 1990); residue exposure is defined in the context of an isolated domain.

included in the compilation of Kabat et al. (1991) which are even more similar to these murine domains.

In the absence of three-dimensional structure, the identification of the important framework residues could be accomplished by other means, for example, by modelling of the combining site structure (e.g., Queen et al. 1989; Co and Queen 1991; Co et al. 1991) or by studying the effect of site-specific mutations on the ligand-binding properties of the molecule (e.g., Tempest et al. 1991; Foote and Winter 1992). Some useful hints are provided by the results compiled in Figs. 3 and 4.

It is seen from Figs. 3 and 4 that many of the important framework residues flank the CDRs. Among these flanking positions are most of the framework residues that are involved in the contact with the opposite

Table 5. Inward-pointing[a], buried framework residues in the V_H of murine Fabs of known three-dimesional structure

Position	Antibody											
	1	2	3	4	5	6	7	8	9	10	11	12
2	V	V	V		V			V	V	V	V	V
4	L	L	L	L	L	L	P	L	L	L	L	L
6	E	E	E	Q	Q	E	E	Q	E	E	E	E
12	V	V	V	M	V	V	V	V	V	V	V	V
18	L	L	L	V	V	M	L	V	L	L	L	Q
20	L	L	L	I	M	L	L	M	L	I	L	L
22	C	C	C	C	C	C	C	C	C	C	C	C
24	A	T	V	A	A	A	A	A	A	A	T	A
27	F	F		Y	Y	F		Y	F	F	F	Y
29	F	F	I	F	F	F	F	F	F	L	F	I
36	W	W	W	W	W	W	W	W	W	W	W	W
38	R	R	R	K	K	R	R	K	R	R	R	R
40						S						
46			E	E								E
48	I	I	M	I	I	V	V	I	V	L	L	M
49		A				A	A		A			
66		R	R			R	R		R	R	R	R
67	F	F	I	A	T	F	F	T	F	L	F	I
69	I	V	I	F	L	I	I	L	I	I	I	I
71	R	R	R	A	V	R	R	V	R	K	R	R
76						S	D			N		
78	L	L	Y	A	A	V	L	A	L	V	L	F
80	L	L	L	M	M	L	L	M	L	L	L	L
82	M	M	L	L	L	M	M	L	M	M	M	L
82c	V	L	V	L	L	L	L	L	L	L	L	V
86	D	D	D	D	D	D	D	D	D	D	D	D
88	A	A	A		A		A	A	A	A	A	A
90	Y	Y	Y	Y	Y	Y	Y	Y	Y	Y	Y	Y
92	C	C	C	C	C	C		C	C	C	C	C
94	R	R	N	H	R		R	R	R	R	R	R
107	T	T	T	T	T	T	T	T	T	T	T	T
109	V	V	V	L	L	V	V	L	L	L	V	V
111	V	V	V	V	V	V	V	V	V	V	V	V

[a] An inward-pointing residue is designated as buried if at least 50% of its side chain is inaccessible to solvent. Solvent accessibilities were computed as described previously (PADLAN 1990); residue exposure is defined in the context of an isolated domain.

domain (Table 3) and many of those which are in contact with the CDRs (Tables 1, 2). Moreover, all of the framework residues which have been observed to participate in the binding to antigen (SHERIFF et al. 1987; PADLAN et al. 1989; BENTLEY et al. 1990; FISCHMANN et al. 1991; TULIP et al. 1992) are in these flanking regions. These results suggest that if during humanization not just the CDRs are transplanted, but also some of the residues immediately adjacent to the CDRs, there would be a good chance of retaining the ligand-binding properties of the original antibody. The

```
                    10        20        35   40      49
J539       EI.L.Q....T.A.....V.I.C      WYQQ....SP.PWIY
McPC603    DIVMTQ....L.V.....V.M.C      WYQQ....PP.LLIY
HyHEL-10   DIVLTQ....L.V.....V.L.C   C  WYQQ....SP.LLIK
HyHEL-5    DIVL.Q....M.A.....V.M.C      WYQQ....SP.RWIY
R19.9      DIQMTQT...L.A.....V.I.C   D  WYQQ....TVKLLIY
4-4-20     DVVMTQ....L.V.....A.I.C      WYLQ....SPKVLIY
YST9-1     DIQMTQ....L.A.....V.I.C   R  WYQQ.....V.LLIY
36-71      DIQM.Q....L.A.....V.I.C      WYQQ.....I.LLIY
B13I2      DVLM.QT...L.V.....A.I.C      WYLQ....SP.LLIY
D1.3       DI.MTQ....L.A.....V.I.C   1  WYQQ....SP.LLVY
BV04-01    .V.MTQ....L.V.....A.I.C      WYLQ....SP.LIIY
AN02       QIVLTQ....M.A.....V.M.C      WYQQ....SP.LLIY


                    60        70        80     88    98       107
J539         .V..RF......T.Y.L.I..M...D.A.YYC     F...T.L.L.
McPC603      .V.DRF....S.TDF.L.I..V...D.A.YYC     F.A.T.L.I.
HyHEL-10   C .I..RF....S.T.F.L.I..V...D..MYFC   C F...T.L.I.
HyHEL-5      .V..RF......T.Y.L.I..M...D.A.YYC     F...T.L...
R19.9      D .V..RF......TDY.L.I..L...D.A.YFC   D F...T.L...
4-4-20       .V.DRF......T.F.L.I..V...D...Y.C     F...T.L...
YST9-1     R .V..RF......TDY.L.I..L...D.A.YIC   R F...T.L.I.
36-71        .V..RF......TDY.L.I..L...D.A.YFC     F...T.L...
B13I2        .V.DRFS....TDF.L.I..V...D...YYC     F...T.L.I.
D1.3       2 .V..RF....S.T.Y.L.I..L...DF..YYC   3 F...T.L...
BV04-01      .V..RF......TDF.L.I..V...D...YFC     F.A.T.L...
AN02         .V.VRF......T.Y.L.I..M...D.A.YYC     F.V.T.L.L.
```

Fig. 3. The framework residues in V_L, the side chains of which are in contact with CDRs or with V_H and those which are inward-pointing

likelihood will be even greater if the first few amino acids in the NH$_2$-terminals of both chains are transplanted also, since some of them are found to be in contact with CDRs. In fact, the NH$_2$-terminals are contiguous with the CDR surface and are in position to be involved in ligand binding (Fig. 2).

II. Recombinant Methodology of Complementarity Determining Region Transfer

1. Polymerase Chain Reaction-Mediated Complementarity Determining Region Transfer

Several approaches have been employed to construct CDR-grafted variable regions, the most common of which involves the preparation of a single-stranded DNA template from an M13 vector containing the human variable region chosen for CDR engraftment. An oligodeoxynucleotide-based in vitro mutagenesis protocol with the single-stranded DNA template and DNA polymerase may be employed to replace the existing CDRs with those of the xenogenic monoclonal antibody (MAb) (Taylor et al. 1985;

```
                 10           20          30          40        49
J539        .V.L.E.....V.....L.L.C.A..FDFS      WVRQ.....LEWI.
McPC603     .V.L.E.....V.....L.L.C.T..FTFS      WVRQ....RLEWIA
HyHEL-10    .V.L.E.....V.....L.L.C.V....IT   C  WIRK...N.LEYM.
HyHEL-5     ...L.Q.....M.....V.I.C.A..YTFS      WVKQR....LEWI.
R19.9       .V.L.Q.....V.....V.M.C.A..YTFT   D  WVKQ.....L.WI.
4-4-20      ...L.E.....V.....M.L.C.A..FTFS      WVRQS....LEWVA
YST9-1      EV.L.E.....V.....L.L.C.T..FTFT   R  WVRQ....AL.WL.
36-71       EV.L.Q.....V.....V.M.C.A..YTFT      WVKQ.....LEWI.
B13I2       .V.L.E.....V.....L.L.C.A..FTFS      WVRQ...K.L.WVA
D1.3        .V.L.E.....V.....L.I.C.V..F.LT   1  WVRQ.....LEWL.
BV04-01     E..P.E.....V.....L.L.C.A..FSFN      WVRQ...K.LEWVA
AN02        .V.L.E.....V.....Q.L.C.V..YSIT      WIRQ...NKLEWM.


                 70           82abc        90        102    110
J539           KF.I.R.N....L.L.M..V...D.A.YYCAR     W.Q.T.V.V..
McPC603        RFIV.R.T....L.L.M..L...D.A.YYCAR     W...T.V.V..
HyHEL-10    C  RI.I.R......Y.L.L..V...D.A.YYCAN   C W.....V.V..
HyHEL-5        KA.F.A......A.M.LN.L...D...YYCLH     W...T.L.V..
R19.9       D  KT.L.V.R....A.M.LR.L...D.A.YFCAR   D W...T.L.V..
4-4-20         RFTI.R.D..S.V.L.M..L...D...YYCT.     W...T.V.V..
YST9-1      R  RFTI.R.N....L.L.M..L...D.A.YYCTR   R W...T.V.V..
36-71          KT.L.V.K....A.M.L..L...D.A.YFCAR     W...T.L.V..
B13I2          RF.I.R.N....L.L.M..L...D.A.YYCTR     W...T.L.V..
D1.3        2  RL.I.K......V.L.M..L...D.A.YYCAR   3 W...T.L.V..
BV04-01        RF.I.R.D....L.L.M..L...D.A.YYCVR     W...T.V.V..
AN02           RISI.R......F.L.LK.V...D.A.YFCAR     W.Q.T.V.V..
```

Fig. 4. The framework residues in V_H, the side chains of which are in contact with CDRs or with V_L and those which are inward-pointing

NAKAMAYE and ECKSTEIN 1986; REICHMANN et al. 1988; VERHOEYEN et al. 1988; TEMPEST et al. 1991; KETTLEBOROUGH et al. 1991). Alternatively, PCR-based methodology provides this process with flexibility, ease, rapidity, and results in high frequency generation of the desired CDR-grafted variable region fragment (DAUGHERTY et al. 1991; DEMARTINO et al. 1991). Two general strategies involving PCR recombination may be employed to graft the xenogenic CDRs into their appropriate human frameworks. When the human framework is available as a cDNA sequence, short oligodeoxynucleotide primers containing terminal complementarity and all or part of the xenogenic CDR sequence may be synthesized. Following annealing of each primer pair to the human template V region cDNA and subsequent PCR amplification the resultant fragments are themselves combined by PCR to generate the humanized CDR-grafted variable region (Fig. 5). When the human framework template is not available, long oligodeoxynucleotides (80–100 bases in length) of alternating polarity may be synthesized to contain both the human framework residues (FRs) and the xenogenic CDR sequences, and terminal regions of complementarity (Fig. 6). PCR amplification of the combined oligodeoxynucleotides with two short terminal amplifying primers result in a DNA fragment encoding the xenogenic CDR-

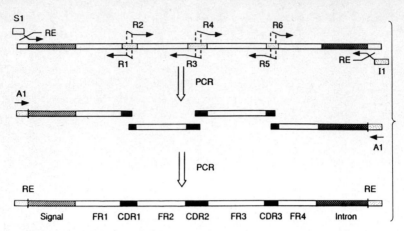

Fig. 5. PCR amplification and recombination for the construction of CDR-grafted V regions: Rei light chain grafting using short oligodeoxynucleotides

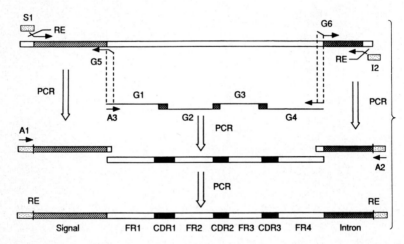

Fig. 6. PCR amplification and recombination for the construction of CDR-grafted V regions: Gal heavy chain grafting using long oligodeoxynucleotides

grafted human V region framework. In either case, the CDR-grafted FRs are subsequently combined with additional PCR-generated fragments representing, in part, the immunoglobulin signal peptide and a portion of the intron 3′ of the human heavy or light chain J regions. When the total number of PCR amplification cycles are kept below 45 approximately 90% of the V region clones are found to be error-free. Finally, these molecules are cloned into expression vectors containing an insert encoding either the human light chain constant region or the human heavy chain constant region.

2. Humanization of the Murine Monoclonal Antibody IB4

To identify human framework sequences which would be compatible with the CDRs of murine IB4 MAb (mIB4), human frameworks with a high degree of sequence similarity to those of the mIB4 were identified. Sequence similarity was measured using identical residues and evolutionarily conservative amino acid substitutions (SCHWARTZ and DAYHOFF 1979; RISLER et al. 1988). Similarity searches (DEVEREUX et al. 1984) were performed using the mIB4 framework sequence from which the CDR sequences had been removed. This sequence was used to query a database of human immunoglobulin sequences that had been culled from multiple nucleic acid and

```
TEMPLATES USED AS HEAVY AND LIGHT CHAIN VARIABLE REGION FRAMEWORKS


Heavy Chain



    mIB4:  DVKLVESGGDLVKLGGSLKLSCAASGFTFS  [DYYMS]  WVRQTP
     Jon:  DVQLVESGGGLVKPGGSLRLSCAASGFTFS  [TAWMK]  WVRQAP
     Gal:  EVQLVESGGDLVQPGRSLRLSCAASGFTFS  [BLGMT]  WVRQAP
    mGal:                      G



    EKRLELVA  [AIDNDGGSISYPDTVKG]  RFTISRDNAKNTLYLQM
    GKGLEWVV  [WRVEQVVEKAFANSVNG]  RFTISRNDSKNTLYLQM
    GKGLEWVA  [NIKZBGSZZBYVDSVKG]  RFTISRDNAKNSLYLQM
         L
                                                         %Id

    SSLRSEDTALYYCAR  [-QGRLRRDYFDY]WGQGTTLTVSS...
    ISVTPEDTAVYYCAR  [VPLYGBYRAFNY]WGQGTPVTVSS...      78
    NSLRVEDTALYYCAR  [-----GWGGGD-]WGQGTLVTVST...      82
                                          L            85

Light Chain


    mIB4:  DIVLTQSPASLAVSLGQRATISC  [RASESVDSYGNSFMH--]WY
     REI:  DIQLTQSPSSLSASVGDRVTITC  [RASGNIHNYLA------]WY
     Len:  DIVMTQSPNSLAVSLGERATINC  [KSSQSVLYSSNSKNYLA]WY



    QQKPGQPPKLLIY  [RASNLES]  GIPARFSGSGSRTDFTLTINPV
    QQKPGKAPKLLIY  [YTTTLAD]  GVPSRFSGSGSGTDFTFTISSL
    QQKPGQPPKLLIY  [WASTRES]  GVPDRFSGSGSGTDFTLTISSL


                                                  %Id
    EADDVATYYC  [QQSNEDPLT]  EGAGTKLELKR...
    QPEDIATYYC  [QHFWSTPRT]  FGQGTKVVIKR...      69
    QAEDVAVYYC  [QQYYSTPYS]  FGQGTKLEIKR...      81

%Id: percent identity to mIB4 FRs
Packing residues are underlined
```

Fig. 7. Sequence comparison of heavy and light chain frameworks

protein sources (George et al. 1986; Kabat et al. 1991). Human variable region sequences with a high degree of identity to the murine sequences were then examined individually for their potential as humanizing framework sequences. This involved focusing the comparison upon those amino acid residues deemed important in maintaining the combining site structure. Two heavy and light chain frameworks were selected for CDR grafting; Gal and Jon were chosen to represent the murine heavy chain framework and Rei and Len were chosen to represent the murine light chain framework (Fig. 7). This approach is similar to that utilized by Queen et al. (1989), but varies in two important points. First, any combination of human heavy and light chain frameworks may be chosen to receive the murine CDRs even though they may never have existed previously as a heterodimer. Second, molecular model construction of the antibodies involved need not be undertaken since choices are based upon primary sequence information and identification of the relevant interactive residues. In this way, the human homologues providing the murine CDRs with the structural support most similar to their native murine framework were selected for subsequent construction of the humanized variable regions.

In order to systematically evaluate the process by which the mIB4 MAb was humanized its conversion was separated into three steps. First, a CDR-grafted humanized version of the human light chain was expressed with a chimeric version of the mIB4 heavy chain so as to determine the relative importance of the light chain variable region. Then the grafted light chain was coexpressed with CDR-grafted humanized heavy chain variable regions derived from several human frameworks. This would allow for the selection of the best scaffolding for the murine CDRs and assist in our understanding of those variable region elements which contribute to the successful transfer of CDRs. Finally, mutagenesis of the FRs was undertaken to clarify the role of these residues in the positioning of the transposed CDRs.

The CDR-grafted human light chain V region was constructed from a cDNA encoding the Rei light chain framework (Reichmann et al. 1988) by substituting the mIB4 CDR sequences for its resident sequences. These manipulations were accomplished by PCR amplification of this cDNA template with primers which incorporated the desired murine CDR sequences. The DNA sequence of the CDR-grafted V region was confirmed prior to its insertion into an expression vector containing the human κ constant region. A CDR-grafted version employing the human Len light chain FR was constructed from long oligodeoxynucleotides and placed into an identical expression vector.

The murine heavy chain V region required for the construction of a chimeric murine/human γ-4 heavy chain was obtained through PCR amplification of the murine FR1–FR4 sequences, followed by the PCR-mediated attachment to this product of DNA fragments containing the signal peptide (including its intron) and 3' J-C intronic sequences (Fig. 8). This V region was inserted into the expression vector containing the coding region of the

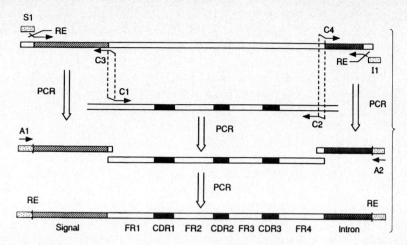

Fig. 8. PCR amplification and recombination for the construction of a hemichimeric heavy chain V region

human γ-4 constant domain. To obtain sufficient quantities of recombinantly expressed antibody for measurement of its avidity, monkey kidney cells (CV1P) were cotransfected with heavy and light chain vectors and the protein secreted into serum-free medium was collected. The hemichimeric grafted antibody (chimeric heavy chain and grafted light chain) was assayed for it ability to compete with ^{125}I-mIB4 for their ligand, the CD18 receptors on activated polymorphonuclear (PMN) leukocytes (Table 6). No loss in avidity was measurable upon grafting the murine CDRs onto the human framework Rei.

The human heavy chain frameworks were chosen in accordance with the hypothesis stated above. The CDR-grafted versions of the Gal and Jon heavy chain domains were constructed using long oligodeoxynucleotides and PCR amplification (DAUGHERTY et al. 1991; DeMARTINO et al. 1991). The sequence-verified heavy chain variable regions were inserted into the heavy chain expression vector in place of the chimeric V region and these plasmids

Table 6. Summary of competitive binding activity of murine IB4 and recombinant human IB4 antibodies

Heavy chain	Light chain	IC_{50} (nM)
Murine IB4	Murine IB4	0.52 ± 0.20
Murine IB4	Grafted IB4/Rei	0.46 ± 0.08
Grafted IB4/mGal	Grafted IB4/Rei	0.67 ± 0.08
Grafted IB4/Gal	Grafted IB4/Rei	1.68 ± 0.26
Grafted IB4/Gal	Grafted IB4/Len	2.80 ± 1.04
Grafted IB4/Jon	Grafted IB4/Rei	5.88 ± 0.13

were cotransfected into CV1P cells with the CDR-grafted Rei κ light chain vector and the secreted antibody was purified from the serum-free conditioned medium. Competitive binding curves were again used to evaluate the avidities of the various antibodies for the CD18 ligand on the activated PMNs (Table 6). Although each heterodimeric antibody contains the same six CDRs, they do not exhibit the same avidity for the CD18 ligand. Thus, we can see that the biological properties of an antibody molecule (i.e., its avidity) rely significantly on the variable region framework structure which supports the CDR loops. In the case of this murine MAb, the choice of light chain framework appears to be less critical than the choice of the heavy chain structure. The human Gal framework was the best of those chosen for the humanization and resulted in the synthesis of a fully grafted antibody whose avidity is not markedly diminished from that of its parent MAb.

Although the Len light chain V region framework sequences, relative to the Rei light chain frameworks, show more identical residues and more similar residues when aligned to mIB4 framework, this has little, if any, impact on the measured antibody/antigen interactions. Comparison of the presumed three dimensional structure of these two light chain V regions indicates that the α carbon trace of the IB4 CDRs residing within these FRs are superimposable, again suggesting that both FRs identically support these CDRs.

3. Antibody Reshaping

Individual FR residues may be altered to increase the avidity of the humanized antibody for its ligand. Through a comparison of the murine and human heavy chain FR packing residues, mismatches between the Gal and murine FRs were noted. A mutated form of the CDR-grafted Gal variable region was constructed and the avidity of the recombinant antibody secreted by CV1P cells transformed by mutant Gal (mGal) and Rei expression vectors, when measured by competitive binding, was found to be equivalent to that of the parent murine antibody (Table 6). This improvement of binding reveals that subtle changes in FR packing residues may substantially impact the way in which the CDRs are displayed within the antigen binding site.

III. Replacing Surface Residues To Humanize (Veneering)

It may be possible to reduce the immunogenicity of a nonhuman Fv, while preserving its antigen-binding properties, by simply replacing those exposed residues in its framework regions which differ from those usually found in human antibodies (Padlan 1991). This would humanize the *surface* of the xenogenic antibody while retaining the interior and contacting residues which influence its antigen-binding characteristics. The judicious replacement of exterior residues should have little, or no, effect on the interior of the domains, or on the interdomain contacts.

The solvent accessibility patterns of the Fvs of J539 (a murine IgA$_\kappa$) and of KOL (a human IgG1$_\lambda$) have been analyzed (PADLAN 1991) and are reproduced here. Among all the Fab structures currently available from the Protein Data Bank (BERNSTEIN et al. 1977), those of J539 (Protein Data Bank entry: 2FBJ) and KOL (Protein Data Bank entry: 2FB4) have been studied to the highest resolution and the most extensively refined. The fractional accessibility values for the framework residues in the J539 and KOL V$_H$ are compared in Table 7 and those for the framework residues in the J539 and KOL V$_L$ are presented in Table 8.

Examination of Table 7 reveals a very close similarity in the exposure patterns of the V$_H$ of KOL and J539. Only at positions 88 and 104 are the two patterns very different and at these positions one or both antibodies have glycine. The fractional solvent accessibility values for the individual residues were computed as described by PADLAN (1990); residues whose side chains have fractional accessibility values between 0.00 and 0.20 are designated as being completely buried, between 0.20 and 0.40 as mostly buried, between 0.40 and 0.60 as partly buried/partly exposed, between 0.60 and 0.80 as mostly exposed, and at least 0.80 as completely exposed. In the special case of glycine, the residue is considered completely exposed if its α-carbon atom is accessible to solvent, otherwise it is considered completely buried (Table 7). The exposure patterns of the V$_L$ domains (Table 8) likewise are very similar, with large differences only at positions 2, 13, 66, 99, and 101. The conformation of the NH$_2$-terminal segments of the KOL and J539 light chains are slightly different, and this results in the difference observed at position 2; at the other positions, one or both molecules again have glycine.

The very close similarity of the exposure patterns for the variable domains of KOL and J539 points to the close correspondence of the tertiary structures of the homologous domains and of the dispositions of the individual residues in these proteins. This is particularly remarkable since: (a) these antibodies are from different species, (b) their light chains are of different types (J539 has a κ light chain, while KOL has a λ light chain), (c) half of their CDRs, specifically CDR1-L, CDR3-L, and CDR3-H, have very different lengths and backbone conformations, and (d) KOL and J539 have only 44 identical residues out of the 79 corresponding positions in the V$_L$ framework and 60 out of the 87 in the V$_H$ framework. An even closer similarity in overall structure and in the exposure patterns might be expected for two molecules that are more similar in sequence than this pair. These results suggest that the solvent exposure of a residue can be more easily predicted than perhaps its involvement in the maintenance of the combining site structure.

The procedure that was proposed (PADLAN 1991) for reducing the antigenicity of a xenogenic variable domain, while preserving its ligand-binding properties, would replace only the exposed FRs which differ from those of the host with the corresponding residues in the most similar host

Table 7. Solvent exposure of side chains of framework residues in KOL and J539 V_H

Position	Fractional accessibility				
	KOL			J539	
	Residue	Exposure		Residue	Exposure
1	GLU	1.00	Ex	GLU	1.00	Ex
2	VAL	0.23	mB	VAL	0.37	mB
3	GLN	0.82	Ex	LYS	0.82	Ex
4	LEU	0.00	Bu	LEU	0.10	Bu
5	VAL	0.87	Ex	LEU	1.00	Ex
6	GLN	0.00	Bu	GLU	0.09	Bu
7	SER	0.94	Ex	SER	0.94	Ex
8	GLY	1.00	Ex	GLY	1.00	Ex
9	GLY	0.00	Bu	GLY	0.00	Bu
10	GLY	1.00	Ex	GLY	1.00	Ex
11	VAL	0.90	Ex	LEU	0.81	Ex
12	VAL	0.25	mB	VAL	0.25	mB
13	GLN	0.71	mE	GLN	0.87	Ex
14	PRO	0.59	pB	PRO	0.64	mE
15	GLY	1.00	Ex	GLY	1.00	Ex
16	ARG	0.73	mE	GLY	1.00	Ex
17	SER	0.66	mE	SER	0.75	mE
18	LEU	0.28	mB	LEU	0.26	mB
19	ARG	0.66	mE	LYS	0.75	mE
20	LEU	0.00	Bu	LEU	0.00	Bu
21	SER	0.71	mE	SER	0.82	Ex
22	CYS	0.00	Bu	CYS	0.00	Bu
23	SER	1.00	Ex	ALA	1.00	Ex
24	SER	0.00	Bu	ALA	0.00	Bu
25	SER	0.87	Ex	SER	1.00	Ex
26	GLY	1.00	Ex	GLY	1.00	Ex
27	PHE	0.10	Bu	PHE	0.10	Bu
28	ILE	0.85	Ex	ASP	0.72	mE
29	PHE	0.00	Bu	PHE	0.00	Bu
30	SER	0.74	mE	SER	0.83	Ex
36	TRP	0.00	Bu	TRP	0.00	Bu
37	VAL	0.00	Bu	VAL	0.00	Bu
38	ARG	0.10	Bu	ARG	0.31	mB
39	GLN	0.15	Bu	GLN	0.28	mB
40	ALA	0.95	Ex	ALA	0.75	mE
41	PRO	0.90	Ex	PRO	0.73	mE
42	GLY	1.00	Ex	GLY	1.00	Ex
43	LYS	0.86	Ex	LYS	0.86	Ex
44	GLY	1.00	Ex	GLY	1.00	Ex
45	LEU	0.00	Bu	LEU	0.00	Bu
46	GLU	0.75	mE	GLU	0.73	mE
47	TRP	0.10	Bu	TRP	0.04	Bu
48	VAL	0.00	Bu	ILE	0.00	Bu
49	ALA	0.00	Bu	GLY	0.00	Bu
66	ARG	0.36	mB	LYS	0.51	pB

Table 7. *Continued*

Position	Fractional accessibility					
	KOL			J539		
	Residue	Exposure		Residue	Exposure	
67	PHE	0.00	Bu	PHE	0.00	Bu
68	THR	0.87	Ex	ILE	0.88	Ex
69	ILE	0.00	Bu	ILE	0.00	Bu
70	SER	0.78	mE	SER	0.79	mE
71	ARG	0.11	Bu	ARG	0.00	Bu
72	ASN	0.61	mE	ASP	0.55	pB
73	ASP	0.44	pB	ASN	0.43	pB
74	SER	0.85	Ex	ALA	0.97	Ex
75	LYS	0.88	Ex	LYS	0.77	mE
76	ASN	0.69	mE	ASN	0.68	mE
77	THR	0.41	pB	SER	0.33	mB
78	LEU	0.00	Bu	LEU	0.00	Bu
79	PHE	0.45	pB	TYR	0.35	mB
80	LEU	0.00	Bu	LEU	0.00	Bu
81	GLN	0.53	pB	GLN	0.69	mE
82	MET	0.00	Bu	MET	0.00	Bu
82a	ASP	0.73	mE	SER	0.58	pB
82b	SER	0.98	Ex	LYS	0.96	Ex
82c	LEU	0.00	Bu	VAL	0.00	Bu
83	ARG	0.73	mE	ARG	0.83	Ex
84	PRO	0.75	mE	SER	0.90	Ex
85	GLU	0.82	Ex	GLU	0.90	Ex
86	ASP	0.00	Bu	ASP	0.11	Bu
87	THR	0.54	pB	THR	0.47	pB
88	GLY	1.00	Ex	ALA	0.00	Bu
89	VAL	0.58	pB	LEU	0.63	mE
90	TYR	0.00	Bu	TYR	0.00	Bu
91	PHE	0.00	Bu	TYR	0.08	Bu
92	CYS	0.00	Bu	CYS	0.00	Bu
93	ALA	0.00	Bu	ALA	0.00	Bu
94	ARG	0.17	Bu	ARG	0.15	Bu
103	TRP	0.09	Bu	TRP	0.07	Bu
104	GLY	0.00	Bu	GLY	1.00	Ex
105	GLN	0.93	Ex	GLN	0.99	Ex
106	GLY	0.00	Bu	GLY	0.00	Bu
107	THR	0.22	mB	THR	0.26	mB
108	PRO	0.99	Ex	LEU	0.67	mE
109	VAL	0.00	Bu	VAL	0.00	Bu
110	THR	0.76	mE	THR	0.69	mE
111	VAL	0.00	Bu	VAL	0.00	Bu
112	SER	0.98	Ex	SER	0.74	mE
113	SER	0.94	Ex	ALA	0.84	Ex

Bu, completely buried; mB, mostly buried; pB, partly buried/partly exposed; mE, mostly exposed; Ex, completely exposed.

Table 8. Solvent exposure of side chains of framework residues in KOL and J539 V_L

Position	Fractional accessibility					
	KOL			J539		
	Residue	Exposure		Residue	Exposure	
1	GLN	1.00	Ex	GLU	0.99	Ex
2	SER	1.00	Ex	ILE	0.16	Bu
3	VAL	0.77	mE	VAL	0.87	Ex
4	LEU	0.00	Bu	LEU	0.00	Bu
5	THR	0.92	Ex	THR	0.80	mE
6	GLN	0.00	Bu	GLN	0.00	Bu
7	PRO	0.62	mE	SER	0.89	Ex
8	PRO	1.00	Ex	PRO	0.67	mE
9	SER	1.00	Ex	ALA	1.00	Ex
10	–	–	–	ILE	0.94	Ex
11	ALA	0.34	mB	THR	0.30	mB
12	SER	0.71	mE	ALA	0.59	pB
13	GLY	1.00	Ex	ALA	0.00	Bu
14	THR	0.73	mE	SER	0.78	mE
15	PRO	0.75	mE	LEU	0.79	mE
16	GLY	1.00	Ex	GLY	1.00	Ex
17	GLN	0.69	mE	GLN	0.64	mE
18	ARG	0.79	mE	LYS	0.74	mE
19	VAL	0.21	mB	VAL	0.22	mB
20	THR	0.62	mE	THR	0.65	mE
21	ILE	0.00	Bu	ILE	0.00	Bu
22	SER	0.92	Ex	THR	0.69	mE
23	CYS	0.00	Bu	CYS	0.00	Bu
35	TRP	0.00	Bu	TRP	0.00	Bu
36	TYR	0.00	Bu	TYR	0.00	Bu
37	GLN	0.46	pB	GLN	0.14	Bu
38	GLN	0.00	Bu	GLN	0.24	mB
39	LEU	0.75	mE	LYS	0.69	mE
40	PRO	0.91	Ex	SER	1.00	Ex
41	GLY	1.00	Ex	GLY	1.00	Ex
42	MET	0.74	mE	THR	0.90	Ex
43	ALA	0.62	mE	SER	0.30	mB
44	PRO	0.00	Bu	PRO	0.00	Bu
45	LYS	0.95	Ex	LYS	0.90	Ex
46	LEU	0.23	mB	PRO	0.43	pB
47	LEU	0.15	Bu	TRP	0.16	Bu
48	ILE	0.00	Bu	ILE	0.00	Bu
49	TYR	0.39	mB	TYR	0.42	pB
57	GLY	1.00	Ex	GLY	1.00	Ex
58	VAL	0.14	Bu	VAL	0.13	Bu
59	PRO	0.70	mE	PRO	0.61	mE

Table 8. *Continued*

Position	Fractional accessibility					
	KOL			J539		
	Residue	Exposure		Residue	Exposure	
60	ASP	0.95	Ex	ALA	1.00	Ex
61	ARG	0.31	mB	ARG	0.36	mB
62	PHE	0.12	Bu	PHE	0.00	Bu
63	SER	0.85	Ex	SER	0.94	Ex
64	GLY	0.00	Bu	GLY	0.00	Bu
65	SER	1.00	Ex	SER	1.00	Ex
66	LYS	0.41	pB	GLY	1.00	Ex
67	SER	1.00	Ex	SER	1.00	Ex
68	GLY	1.00	Ex	GLY	1.00	Ex
69	ALA	0.71	mE	THR	0.75	mE
70	SER	1.00	Ex	SER	0.98	Ex
71	ALA	0.00	Bu	TYR	0.09	Bu
72	SER	1.00	Ex	SER	0.70	mE
73	LEU	0.00	Bu	LEU	0.00	Bu
74	ALA	0.74	mE	THR	0.43	pB
75	ILE	0.00	Bu	ILE	0.00	Bu
76	GLY	1.00	Ex	ASN	0.83	Ex
77	GLY	1.00	Ex	THR	0.83	Ex
78	LEU	0.00	Bu	MET	0.00	Bu
79	GLN	0.76	mE	GLU	0.63	mE
80	SER	1.00	Ex	ALA	0.96	Ex
81	GLU	0.78	mE	GLU	0.91	Ex
82	ASP	0.09	Bu	ASP	0.13	Bu
83	GLU	0.64	mE	ALA	0.55	pB
84	THR	0.34	mB	ALA	0.00	Bu
85	ASP	0.30	mB	ILE	0.58	pB
86	TYR	0.00	Bu	TYR	0.00	Bu
87	TYR	0.16	Bu	TYR	0.11	Bu
88	CYS	0.00	Bu	CYS	0.00	Bu
98	PHE	0.04	Bu	PHE	0.00	Bu
99	GLY	0.00	Bu	GLY	1.00	Ex
100	THR	0.59	pB	ALA	1.00	Ex
101	GLY	1.00	Ex	GLY	0.00	Bu
102	THR	0.00	Bu	THR	0.00	Bu
103	LYS	0.82	Ex	LYS	0.79	mE
104	VAL	0.00	Bu	LEU	0.00	Bu
105	THR	0.86	Ex	GLU	0.89	Ex
106	VAL	0.19	Bu	LEU	0.44	pB
106a	LEU	0.70	mE	–	–	–
107	GLY	1.00	Ex	LYS	0.77	mE

Bu, completely buried; mB, mostly buried; pB, partly buried/partly exposed; mE, mostly exposed; Ex, completely exposed.

sequence. Thus, the FRs which are at least partly exposed in the corresponding domains of KOL or J539 (those with mE, or Ex designations in Tables 7 and 8) would be replaced, while the FRs corresponding to those which in KOL and J539 are completely or mostly buried would be retained. With this procedure also, the number of FRs in xenogenic domains that would be needed to be replaced by human residues was found to be not very large in the cases examined (PADLAN 1991).

The judicious replacement of exterior residues should have little, or no, effect on the interior domains or on the interdomain contacts. Thus, ligand binding properties should be unaffected as a consequence of alterations which are limited to the surface exposed variable region FRs. We refer to this procedure of humanization as veneering since only the outer surface of the antibody is altered, the supporting residues remain undisturbed.

There are two steps in the process of veneering. First, the framework of the mouse variable domains are compared with the human variable region database. The most homologous human variable regions are identified and subsequently compared residue for residue to the corresponding murine regions. Second, those residues in the mouse framework which differ from its human homologue are replaced by the residues present in the human homologue. This switching occurs only with those residues which are at least partially exposed (PADLAN 1991). One retains in the veneered mouse antibody: its CDRs, the residues neighboring the CDRs, those residues defined as buried or mostly buried, and those residues believed to be involved with interdomain contacts (PADLAN 1991). Attention is also paid to the NH2-terminals of the heavy and light chains since they are often contiguous with the CDR surface and are in a position to be involved in ligand binding. Care should likewise be exercised in the placement of proline, glycine, and charged amino acids since they may have significant effects on tertiary structure and electrostatic interactions of the variable region domains.

An appropriate human framework is determined utilizing the criteria discussed above. In practice, it was found that the human light chain variable region framework with significant homology to the mIB4 framework was determined to be the human Len framework (a similarity of 90% and an identity of 81%). For the purposes of exemplifying the veneering process a IB4 CDR-grafted version of the Len light chain variable region was used as the template into which mutations were placed so as to easily create the veneered framework sequence. Specific amino acid residues within the human Len framework were replaced with residues found in the murine IB4 framework so that the final light chain V-region appeared as it should were the murine V-region the starting material for the veneering process (Fig. 9). The veneered heavy chain portion of the recombinant antibody was derived by mutating the murine IB4 heavy chain variable region so that it contained only human surface exposed residues. In this case the human Gal framework was used as the template for surface residue comparisons (Fig. 7). The

HEAVY AND LIGHT CHAIN VARIABLE REGION FRAMEWORKS
USED FOR PREPARATION OF THE VENEERED IB4

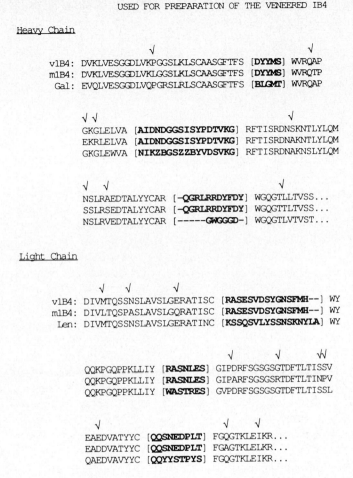

<u>Heavy Chain</u>

```
                      √                              √
    vIB4:  DVKLVESGGDLVKPGGSLKLSCAASGFTFS [DYYMS] WVRQAP
    mIB4:  DVKLVESGGDLVKLGGSLKLSCAASGFTFS [DYYMS] WVRQTP
     Gal:  EVQLVESGGDLVQPGRSLRLSCAASGFTFS [BLGMT] WVRQAP


           √ √                            √
           GKGLELVA [AIDNDGGSISYPDTVKG] RFTISRDNSKNTLYLQM
           EKRLELVA [AIDNDGGSISYPDTVKG] RFTISRDNAKNTLYLQM
           GKGLEWVA [NIKZBGSZZBYVDSVKG] RFTISRDNAKNSLYLQM


           √  √                               √
           NSLRAEDTALYYCAR [-QGRLRRDYFDY] WGQGTLLTVSS...
           SSLRSEDTALYYCAR [-QGRLRRDYFDY] WGQGTTLTVSS...
           NSLRVEDTALYYCAR [-----GWGGGD-] WGQGTLVTVST...
```

<u>Light Chain</u>

```
                      √      √       √
    vIB4:  DIVMTQSSNSLAVSLGERATISC [RASESVDSYGNSFMH--] WY
    mIB4:  DIVLTQSPASLAVSLGQRATISC [RASESVDSYGNSFMH--] WY
     Len:  DIVMTQSSNSLAVSLGERATINC [KSSQSVLYSSNSKNYLA] WY


                              √       √      √√
           QQKPGQPPKLLIY [RASNLES] GIPDRFSGSGSGTDFTLTISSV
           QQKPGQPPKLLIY [RASNLES] GIPARFSGSGSRTDFTLTINPV
           QQKPGQPPKLLIY [WASTRES] GVPDRFSGSGSGTDFTLTISSL


           √                    √      √
           EAEDVATYYC [QQSNEDPLT] FGQGTKLEIKR...
           EADDVATYYC [QQSNEDPLT] FGAGTKLELKR...
           QAEDVAVYYC [QQYYSTPYS] FGQGTKLEIKR...
```

Fig. 9. Heavy and light chain V region sequence alterations performed to create veneered surfaces; *vIB4*, veneered heavy and light chain V regions; *checkmarks*, location of point mutations placed so as to convert an exposed murine residue to a human-appearing residue

sequences of the veneered heavy and light chain variable regions are shown in Fig. 9. In most instances the corresponding residues in the human templates (Gal or Len) are used to substitute for undesired murine residues at analogous positions. The two exceptions are residues 74 and 84 in the heavy chain, where a more preferred amino acid is used. The veneered recombinant antibody secreted into the culture supernatants was purified by protein A chromatography, and its avidity determined, as described previously. The results of the binding assays indicated that the avidity of the veneered recombinant IB4 antibody was equal to that of the mIB4 MAb

(Table 6). This result suggests that an antibody with presumptive human allotype may be recombinantly constructed from the murine MAb by the introduction of numerous point mutations into its FRs, followed by expression of these V regions fused to human κ and γ-4 constant domains, without loss in avidity for the antigen. It can be inferred from this result that the point mutations within the framework regions do not alter the presentation of the mIB4 light chain and heavy chain CDRs.

D. Immunogenicity of Humanized Antibodies

Irrespective of the skill and approach of the molecular alchemist, success will ultimately require demonstration of the metamorphosis of the original murine antibody into one which is immunologically acceptable to the human recipient. Ideally, demonstration of this outcome would best be evaluated in a preclinical model (Hakimi et al. 1991). Human MAbs have been found to have the same pharmacokinetics in Rhesus monkeys as they do in humans (Ehrlich et al. 1987; Jonker et al. 1991). After repeated dosing of human MAbs into these monkeys they are well tolerated and rarely result in immune recognition. Groups of three Rhesus monkeys were injected, at weekly intervals, for 5 weeks with 1 mg MAb (either murine, CDR-grafted, or veneered) per kilogram body weight. At various times following each injection the level of circulating MAb and the development of anti-MAb antibodies were assayed by ELISAs. The IB4 MAbs all bound their CD18 target on Rhesus PMNs. Differences between the humanized versions of

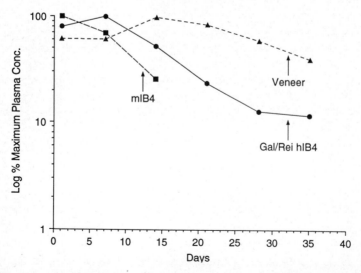

Fig. 10. Alteration in Rhesus peak serum levels of recombinant antibodies during the course of weekly dosing at 1 mg/kg

IB4 and its murine predecessor were appreciated by the third dose of antibody when two of the three monkeys receiving the murine antibody displayed moderate anaphlactic symptoms. This response was never seen in animals treated with the other forms of the IB4 MAb during the 6 weeks of this study. Distinguishing differences between the recombinant IB4 MAbs were most evident following their fourth dose, at which time peak plasma levels for the CDR-grafted MAb was significantly reduced relative to the veneered MAb. This trend continued for the remainder of the observation period (Fig. 10) and was attributable to the progressively higher levels of anti-IB4 antibodies in these animals. These findings suggest that a veneering approach to humanization may not only result in recombinant antibodies which retain all of their affinity and potency, but these antibodies may also be less immunogenic than those humanized by CDR grafting and reshaping procedures.

E. Conclusion

Antibodies, by virtue of their exquisite specificity and high potency, have long been considered therapeutics of tomorrow. They have tremendous potential in the management of immune responsiveness, the detection and treatment of cancer, and the prophylaxis and treatment of viral and bacterial infections. While their long biologic half-life would suggest that MAbs would be well tolerated for long-term therapy, the specter of possible anti-antibody responses has focused their use to solely short-term situations. The recent advent of the combinatorial library approach for the identification and construction of human MAbs and the development of transgenic and reconstituted mice with the capability of producing a human antibody in response to an antigen will supersede the need to reshape rodent MAbs. However, until these technologies are well proven, human therapy will be the providence of the humanized antibody.

The ideal humanization procedure would completely eliminate immunogenicity, while strictly preserving the fine specificity of the antibody. No such procedure exists to date. The two procedures described above both seek to achieve this end, albeit from different directions. The procedure proposed by Winter and co-workers (JONES et al. 1986; VERHOEYEN et al. 1988) tries to keep human as much of the antibody surface as possible, while modifying its interior to recover the ligand-binding properties of the original molecule. The veneering procedure proposed by PADLAN (1991) preserves the interior and thereby the structure responsible for the antigen-binding properties, while modifying the surface to make it look human-like. Independently designed humanization protocols based on these two procedures are already predicting very similar humanizing sequences (PADLAN, unpublished results), and the predicted sequences will become even more similar as we learn more about the structural requirements for humanization.

References

Alber T, Sun DP, Wilson K, Wozniak JA, Matthews BW (1987) Contributions of hydrogen bonds of Thr 157 to the thermodynamic stability of phage T4 lysozyme. Nature 330:41–46

Bentley GA, Boulot G, Riottot MM, Poljak RJ (1990) Three-dimensional structure of an idiotope-anti-idiotope complex. Nature 348:254–257

Bernstein FC, Koetzle TF, Williams GJB, Meyer EF Jr, Brice MD, Rodgers JR, Kennard O, Shimanouchi T, Tasumi M (1977) The Protein Data Bank. A computer-based archival file for macromolecular structures. J Mol Biol 112: 535–542

Bhat TN, Bentley GA, Fischmann TO, Boulot G, Poljak RJ (1990) Small rearrangements in structures of Fv and Fab fragments of antibody D1.3 on antigen binding. Nature 347:483–485

Chothia C, Lesk AM (1987) Canonical structures for the hypervariable regions of immunoglobulins. J Mol Biol 196:901–917

Chothia C, Lesk AM, Tramontano A, Levitt M, Smith-Gill SJ, Air G, Sheriff S, Padlan EA, Davies D, Tulip WR, Colman PM, Spinelli S, Alzari PM, Poljak RJ (1989) Conformations of immunoglobulin hypervariable regions. Nature 342:877–883

Co MS, Queen C (1991) Humanized antibodies for therapy. Nature 351:501–502

Co MS, Deschamps M, Whitney RJ, Queen C (1991) Humanized antibodies for antiviral therapy. Proc Natl Acad Sci USA 88:2869–2873

Colman PM, Laver WG, Varghese JN, Baker AT, Tulloch PA, Air GM, Webster RG (1987) Three-dimensional structure of a complex of antibody with influenza virus neuraminidase. Nature 326:358–363

Daugherty BL, DeMartino JA, Law M-F, Kawka DW, Singer II, Mark GE (1991) Polymerase chain reaction facilitates the cloning, CDR-grafting, and rapid expression of a murine monoclonal antibody directed against the CD18 component of leukocyte integrins. Nucleic Acid Res 19:2471–2476

Davies DR, Padlan EA (1992) Twisting into shape. Curr Biol 2:254–256

DeMartino JA, Daugherty BL, Law M-F, Cuca GC, Alves K, Silberklang M, Mark GE (1991) Rapid humanization and expression of murine monoclonal antibodies. Antib Immunoconj Radiopharm 4:829–835

Devereux J, Haeberli P, Smithies O (1984) A comprehensive set of sequence analysis programs for the VAX. Nucleic Acids Res 12:387–395

Edmundson AB, Ely KR, Girling RL, Abola EE, Schiffer M, Westholm FA, Fausch MD, Deutsch HF (1974) Binding of 2,4-dinitrophenyl compounds and other small molecules to a crystalline λ-type Bence-Jones dimer. Biochemistry 13:3816–3827

Ehrlich PH, Harfeldt KE, Justice JC, Moustafa ZA, Ostberg L (1987) Rhesus monkey responses to multiple injections of human monoclonal antibodies. Hybridoma 6:151–160

Fischmann TO, Bentley GA, Bhat TN, Boulot G, Mariuzza RA, Phillips SEV, Tello D, Poljak RJ (1991) Crystallographic refinement of the three-dimensional structure of the FabD1.3-lysozyme complex at 2.5-A resolution. J Biol Chem 266:12915–12920

Foote J, Winter G (1992) Antibody framework residues affecting the conformation of the hypervariable loops. J Mol Biol 224:487–499

George DG, Barker WC, Hunt LT (1986) The protein identification resource (PIR). Nucleic Acids Res 14:11–16

Hakimi J, Chizzonite R, Luke DR, Familletti PC, Bailon P, Kondas JA, Pilson RS, Lin P, Weber DV, Spence C, Mondini SJ, Tsien W-H, Levin JL, Gallati VH, Korn L, Waldmann TA, Queen C, Benjamin W (1991) Reduced immunogenicity and improved pharmacokinetics of humanized anti-Tac in cynomolgus monkeys. J Immunol 147:1352–1359

Herron JN, He XM, Ballard DW, Blier PR, Pace PE, Bothwell ALM, Voss EW Jr, Edmundson AB (1991) An autoantibody to single-stranded DNA: comparison of the three-dimensional structures of the unliganded Fab and a deoxynucleotide-Fab complex. Proteins 11:159–175

Jones PT, Dear PH, Foote J, Neuberger MS, Winter G (1986) Replacing the complementarity-determining regions in a human antibody with those from a mouse. Nature 321:522–525

Jonker M, Schellekens PT, Harpprecht J, Slingerland W (1991) Complications of monoclonal antibody (MAb) therapy: the importance of primate studies. Transplant Proc 23:264–265

Kabat EA, Wu TT, Bilofsky H (1977) Unusual distribution of amino acids in complementarity-determining (hypervariable) segments of heavy and light chains of immunoglobulins and their possible roles in specificity of antibody combining sites. J Biol Chem 252:6609–6616

Kabat EA, Wu TT, Perry HM, Gottesman KS, Foeller C (1991) Sequences of proteins of immunological interest, 5th edn. US Department of Health and Human Services, Public Health Service, National Institutes of Health (NIH Publ no 91-3242)

Kettleborough CA, Saldanha J, Heath VJ, Morrison CJ, Bendig MM (1991) Humanization of a mouse monoclonal antibody by CDR-grafting: the importance of framework residues on loop conformation. Protein Eng 4:773–783

Koehler G, Milstein C (1975) Continuous cultures of fused cells secreting antibodies of predefined specificity. Nature 256:495–497

Morrison SL, Oi VT (1988) Genetically engineered antibody molecules. Adv Immunol 44:65–92

Nakamaye KL, Eckstein F (1986) Inhibiton of restriction endonuclease Nci I cleavage by phosphorothioate groups and its application to oligonucleotide-directed mutagenesis. Nucleic Acid Res 14:9679–9698

Padlan EA (1977) Structural implications of sequence variability in immunoglobulins. Proc Natl Acad Sci USA 74:2551–2555

Padlan EA (1990) On the nature of antibody combining sites: unusual structural features that may confer on these sites an enhanced capacity for binding ligands. Proteins 7:112–124

Padlan EA (1991) A possible procedure for reducing the immunogenicity of antibody variable domains while preserving their ligand-binding properties. Mol Immunol 28:489–498

Padlan EA, Davies DR (1975) Variability of three-dimensional structure in immunoglobulins. Proc Natl Acad Sci USA 72:819–823

Padlan EA, Silverton EW, Sheriff S, Cohen GH, Smith-Gill SJ, Davies DR (1989) Structure of an antibody-antigen complex: crystal structure of the HyHEL-10 Fab-lysozyme complex. Proc Natl Acad Sci USA 86:5938–5942

Poljak RJ, Amzel LM, Avey HP, Chen BL, Phizackerley RP, Saul F (1973) Three-dimensional structure of the Fab' fragment of a human immunoglobulin at 2.8-A resolution. Proc Natl Acad Sci USA 70:3305–3310

Queen C, Schneider WP, Selick HE, Payne PW, Landolfi NF, Duncan JF, Avdalovic NM, Levitt M, Junghans RP, Waldmann TA (1989) A humanized antibody that binds to the interleukin 2 receptor. Proc Natl Acad Sci USA 86:10029–10033

Riechmann L, Clark M, Waldmann H, Winter G (1988) Reshaping human antibodies for therapy. Nature 332:323–327

Rini JM, Schulze-Gahmen U, Wilson IA (1992) Structural evidence for induced fit as a mechanism for antibody-antigen recognition. Science 255:959–965

Risler JL, Delorme MO, Delacroix H, Henaut A (1988) Amino acid substitutions in structurally related proteins, a pattern recognition approach: determination of a new and efficient scoring matrix. J Mol Biol 204:1019–1029

Schiffer M, Girling RL, Ely KR, Edmundson AB (1973) Structure of a λ-type Bence-Jones protein at 3.5-A resolution. Biochemistry 12:4620–4631

Schwartz RM, Dayhoff MO (1979) In:Dayhoff MO (ed) Atlas of protein sequence and structure. National Biomedical Research Foundation, Washington DC

Sheriff S, Silverton EW, Padlan EA, Cohen GH, Smith-Gill SJ, Finzel BC, Davies DR (1987) Three-dimensional structure of an antibody-antigen complex. Proc Natl Acad Sci USA 84:8075–8079

Stanfield RL, Fieser TM, Lerner RA, Wilson IA (1990) Crystal structure of an antibody to a peptide and its complex with peptide antigen at 2.8 Å. Science 248:712–719

Taylor JW, Ott J, Eckstein F (1985) The rapid generation of oligonucleotide-directed mutations at high frequency using phosphorothioate-modified DNA. Nucleic Acid Res 13:8765–8785

Tempest PR, Bremner P, Lambert M, Taylor G, Furze JM, Carr FJ, Harris WJ (1991) Reshaping a human monoclonal antibody to inhibit human respiratory syncytial virus infection in vivo. Biotechnology 9:266–271

Tramontano A, Chothia C, Lesk AM (1990) Framework residue 71 is a major determinant of the position and conformation of the second hypervariable region in the V_H domains of immunoglobulins. J Mol Biol 215:175–182

Tulip WR, Varghese JN, Laver WG, Webster RG, Colman PM (1992) Refined crystal structure of the influenza virus N9 neuraminidase-NC41 Fab complex. J Mol Biol 227:122–148

Verhoeyen M, Milstein C, Winter G (1988) Reshaping human antibodies: grafting an antilysozyme activity. Science 239:1534–1536

CHAPTER 5

Applications for *Escherichia coli*-Derived Humanized Fab' Fragments: Efficient Construction of Bispecific Antibodies

P. CARTER, M.L. RODRIGUES, and M.R. SHALABY

A. Introduction

The potential benefits of bispecific antibodies (BsAbs) for the diagnosis and therapy of human disease have long been appreciated (reviewed by CLARK et al. 1988; SONGSIVILAI and LACHMANN 1990; NOLAN and O'KENNEDY 1990; NELSON 1991). Unfortunately it has proved very difficult to generate clinically relevant amounts of highly purified BsAbs using traditional hybrid hybridoma technology (MILSTEIN and CUELLO 1983) or via directed chemical coupling of Fab' fragments derived from murine monoclonal antibodies (MAbs) (BRENNAN et al. 1985; GLENNIE et al. 1987). Nevertheless, in spite of the limited availability of BsAb a few small scale clinical studies have been undertaken (Table 1). For example, BsF(ab')$_2$ have proved useful for retargeting lymphokine-activated killer cells (NITTA et al. 1990; BOLHUIS et al. 1992), toxins (BONARDI et al. 1992) and also radionuclides (STICKNEY et al. 1989, 1991; LE DOUSSAL et al. 1992) to tumor targets in patients. This motivated us to develop an efficient and general route to the construction of BsF(ab')$_2$ fragments (SHALABY et al. 1992; RODRIGUES et al. 1992a). Our strategy relies upon separate *Escherichia coli* expression of each Fab' arm (CARTER et al. 1992a) followed by traditional directed chemical coupling to form the BsF(ab')$_2$.

In the future, BsF(ab')$_2$ fragments for therapeutic use are likely to be constructed from human antibodies obtained via one of several emerging technologies (Chaps. 1–3), the most powerful of which is antibody display phage (reviewed in Chap. 7 and also by MARKS et al. 1992). In the short term, murine antibodies which have been humanized to reduce their immunogenicity (RIECHMANN et al. 1988; HALE et al. 1988) are likely to provide the starting point for BsAb construction, as in our own studies (SHALABY et al. 1992; RODRIGUES et al. 1992). Murine antibodies are humanized by grafting the six antigen-binding complimentarity determining region (CDR) loops from their variable domains into a human antibody (see Chap. 4). The rest of the variable domains, known as framework regions as well as the entire constant domains are human in origin.

Table 1. Clinical usage of BsF(ab')$_2$ fragments

BsF(ab')$_2$		Indication	Application	Reference
Antigen specificities	Construction			
αNCAM/αCD3	In vitro coupled, SS-linked	Malignant glioma	LAK cell delivery	Nitta et al. 1990
αFBP/αCD3	Quadroma-derived, SS-linked	Ovarian carcinoma	LAK cell delivery	Bolhuis et al. 1992
2 (αCD22/αsaporin)[a]	In vitro coupled, thioether-linked	Non-Hodgkin's lymphoma	Toxin delivery	Bonardi et al. 1992
αCEA/αEOTUBE	In vitro coupled, thioether-linked	Colorectal carcinoma	RAID	Stickney et al. 1989, 1991
αCEA/αDPTA	In vitro coupled, thioether-linked	Colorectal and thyroid carcinomas	RAID	Le Doussal et al. 1992

α, anti; BEDTA IV, benzyl EDTA analog of bleomycin; CEA, carcinoembryonic antigen; DPTA, diethylenetriaminepentaacetic acid; EOTUBE, hydroxyethylthiourea-benzyl-EDTA; FBP, folate binding protein; LAK, lymphokine-activated killer; NCAM, neural cell adhesion molecule; RAID, radioimmunodetection.
[a] Two BsAb recognizing nonoverlapping epitopes on both saporin and CD22 were used.

B. Choice of Antigen Specificities for Bispecific F(ab')₂

As a driving problem for the development of an efficient route to BsAbs we chose antigen specificities that have potential clinical utility when combined in one molecule. One arm of our BsF(ab')$_2$ is a humanized version (CARTER et al. 1992a,b) of the murine MAb 4D5 (FENDLY et al. 1990), which is directed against the p185^{HER2} product of the protooncogene *HER2* (also known as c-*erbB*-2 and HER-2/*neu*). The overexpression of p185^{HER2} appears to be integrally involved in the progression of 25%–30% of primary human breast and ovarian cancers (SLAMON et al. 1987, 1989). The second arm is a humanized version (SHALABY et al. 1992; RODRIGUES et al. 1992) of the murine anti-CD3 MAb UCHT1 (BEVERLEY and CALLARD 1981). Our anti-p185^{HER2}/anti-CD3 BsF(ab')$_2$ (SHALABY et al. 1992), as well as those constructed by others (NISHIMURA et al. 1992; SUGIYAMA et al. 1992), is highly effective in retargeting the cytotoxic activity of T cells against p185^{HER2} overexpressing tumor cells. The feasibility of retargeting effector cells to specifically lyse tumor cells using BsAb has been well demonstrated in many systems in vitro, in tumor models in animals in vivo (reviewed by NELSON 1991, SEGAL et al. 1992) and in patients (Table 1).

C. Expression of Humanized Fab' Fragments in *E. coli*

Functional murine Fv (SKERRA and PLÜCKTHUN 1988) and chimeric Fab (BETTER et al. 1988) fragments have been secreted from *E. coli* by cosecretion of corresponding light and heavy chain fragments (see also Chap. 12). We extended these pioneering studies by the development of an *E. coli* expression system which secretes functional humanized Fab' fragments at gram per liter titers in the fermentor (CARTER et al. 1992a). Fab' differ from Fab fragments by the addition of a few extra residues at the COOH-terminal end of the heavy chain C$_H$1 domain, including one or more cysteines. We chose a hinge sequence containing a single cysteine (CysAlaAla) to avoid intrahinge disulfide bonding which may occur with hinges containing multiple cysteines (BRENNAN et al. 1985).

The plasmid pAK19 is designed to coexpress the light chain and heavy chain Fd' fragment of the most potent humanized anti-p185^{HER2} antibody, HuMAb4D5-8 (CARTER et al. 1992b), from a synthetic dicistronic operon (CARTER et al. 1992a; Fig. 1). *E. coli* strain 25F2 was transformed with pAK grown to high cell density (OD$_{550}$ 120–150) in the fermentor. The titer of cell-associated soluble and functional Fab' is routinely 1–2 g/l as judged by antigen binding ELISA (CARTER et al. 1992a). Only modest quantities of Fab' (≤100 mg/l) are found in the culture media under these fermentation conditions which are optimized for high titers of functional cell-associated Fab'. The plasmid pAK22 encoding the Fab' fragment of the most potent humanized anti-CD3 antibody, HuMAbUCHT1-9, was created from pAK19 by replacing the segments encoding the six CDR loops with ones corres-

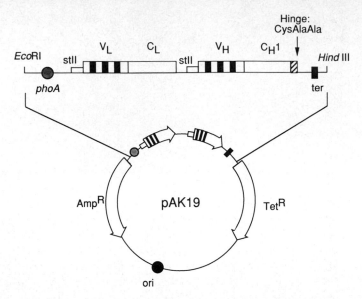

Fig. 1. Plasmid pAK19 for the expression of HuMAb4D5-8 Fab′. The discistronic operon (see Carter et al. 1992a for nucleotide sequence) is under the transcriptional control of the *E. coli* alkaline phosphatase (*phoA*) promoter (Chang et al. 1986) which is inducible by phosphate starvation. Each antibody chain is preceded by the *E. coli* heat-stable enterotoxin II (*stII*) signal sequence (Picken et al. 1993) to direct secretion to the periplasmic space of *E. coli*. The humanized variable domanis (V_L and V_H, Carter et al. 1992b) are precisely fused on their 3′ side to human $\kappa_1 C_L$ (Palm and Hilschmann 1975) and IgG1 C_H1 (Ellison et al. 1982) constant domains, respectively. The C_H1 domain is followed by the bacteriophage lambda t_0 transcriptional terminator (*ter*, Scholtissek and Grosse 1987). Fab′ fragments with alternative specificities are readily created by replacing the antigen binding CDR loops in the variable domains (*solid bars*) with those from corresponding antibodies

ponding to the anti-CD3 antibody, UCHT1 (Rodrigues et al. 1992). The anti-CD3 Fab′ fragment was expressed in *E. coli* in the fermentor at titers of up to 700 mg/l as judged by total immunoglobulin ELISA (Rodrigues et al. 1992). We have also constructed a humanized anti-CD18 Fab′ by recruiting the corresponding CDR loops into the humanized anti-p185[HER2] Fab′ and obtained titers of up to 900 mg/l (unpublished data). This suggests that the anti-p185[HER2] Fab′ may be a broadly useful template for high level expression of humanized antibody fragments created by CDR grafting.

D. Recovery of Fab′-SH Fragments

Functional anti-p185[HER2] and anti-CD3 Fab′ fragments are readily and efficiently recovered from corresponding fermentation pastes with the unpaired hinge cysteine predominantly (75%–90%) in the free thiol form (Fab′-SH, Carter et al. 1992a; Shalaby et al. 1992). The Fab′ fragments are first released from cell pastes in near quantitative yield (as judged by ELISA) by partial digestion of the bacterial cell wall (without extensive lysis) using hen egg white lysozyme. Recovery of Fab′ is performed at pH ~5 to maintain the free thiol in the less reactive protonated form and in the presence of EDTA to chelate metal ions capable of catalyzing disulfide bond formation. Next, the Fab′ fragments are affinity purified using either streptococcal protein G (CARTER et al. 1992a) or staphylococcal protein A (KELLEY et al. 1992).

We prefer protein G to protein A for purification of Fab′ fragments because it is more broadly applicable and gives more homogeneous preparations. Protein G purification has been successful for all humanized and chimeric Fab and Fab′ fragments that we have constructed to date. This might have been anticipated since the crystallographically determined binding site for protein G is on the C_H1 domain (DERRICK and WIGLEY 1992) and all of our antibody fragments share the same human IgG1 C_H1 domain. In contrast, protein A purification has been successful for only some of the humanized Fab′ fragments and for none of the corresponding chimeric molecules containing murine variable domains. Protein A binds to HuMAb4D5-8 Fab′ (CARTER et al. 1992a) as well as corresponding V_L, V_H and Fv fragments but not the C_L/C_H1 fragment (MR, PC, unpublished data). The heterogeneity of protein A purified Fab′ fragments may reflect protein A binding of subfragments generated by the action of E. coli proteases.

Fab′ fragments containing a single hinge cysteine show little propensity to form $F(ab')_2$ in the periplasmic space of E. coli in spite of the very high expression titers in the fermentor. In contrast, the disulfide bond between light chain and heavy chain is formed in virtually all molecules as shown by SDS-PAGE. This suggests that the redox potential of the periplasmic space is sufficiently oxidizing to favor formation of the disulfide bond between light and heavy chains and also the stronger intradomain disulfides but is insufficiently oxidizing to drive formation of the weaker interheavy chain disulfide bond. The high concentration of Fab′ in the periplasmic space (estimated as approximately millimolar) may itself perturb the redox potential by virtue of the free hinge cysteinyl thiol titrating the available redox components.

E. Construction of Bispecific $F(ab')_2$

The ready availability of E. coli derived Fab′-SH fragments enables the simple and efficient construction of either disulfide- or thioether-linked $BsF(ab')_2$ fragments by directed chemical coupling using the methods of BRENNAN et al. (1985) and GLENNIE et al. (1987), respectively. Thioether-linked molecules may be preferable for in vivo applications because of their greater stability (GLENNIE et al. 1987). For example, our thioether-linked anti-p185^{HER2}/anti-CD3 $BsF(ab')_2$ has an approximately threefold longer permanence time in normal mice than the corresponding single disulfide-linked BsF(ab')$_2$ (RODRIGUES et al. 1993). However the nature of the linkage between Fab' arms is apparently not critical since both disulfide- and thioether-linked BsF(ab')$_2$ have proved useful in the clinic (Table 1).

Our thioether-linked BsF(ab')$_2$ (Fig. 2) was created by reacting the anti-p185^{HER2} Fab'-SH fragment first with N,N'-1,2-phenylenedimalemide to form the maleimide derivative and then with the anti-CD3 Fab'-SH. BsF(ab')$_2$ was formed in $\geqslant 50\%$ yield and was purified away from unreacted Fab' by size exclusion chromatography (RODRIGUES et al. 1992). The level of contamination of the BsF(ab')$_2$ with monospecific (Ms) F(ab')$_2$ is likely to be very low since mock coupling reactions with either the anti-p185^{HER2}

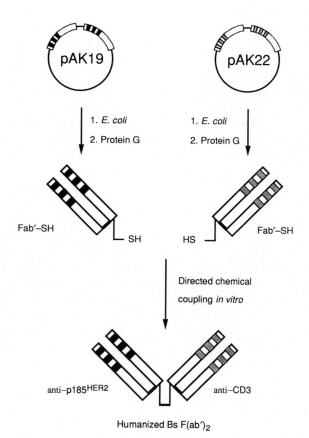

Fig. 2. Overview of the strategy used to construct the anti-p185^{HER2}/anti-CD3 humanized BsF(ab')$_2$. Humanized anti-p185^{HER2} (HuMAb4D5–8) and anti-CD3 (HuMAbUCHT1-9) Fab' fragments were expressed in *E. coli* transformed with plasmids pAK19 (CARTER et al. 1992a) and pAK22 (RODRIGUES et al. 1992a), respectively. Fab'-SH fragments were recovered from corrresponding *E. coli* fermentation pastes by affinity purification using streptococcal protein G and then efficiently coupled to form the BsF(ab')$_2$ using the method of GLENNIE et al. (1987)

Fab' maleimide derivative or anti-CD3 Fab'-SH alone did not yield detectable quantities of $F(ab')_2$. Furthermore the coupling reaction is subjected to a mild reduction step followed by alkylation to remove trace amounts of disulfide-linked $F(ab')_2$ that might be present.

Direct expression in *E. coli* allows large (gram) quantities of Fab'-SH fragments to be isolated more readily and with greater purity than by the traditional method of limited proteolysis and mild reduction of intact antibodies. The advantages of *E. coli*-derived Fab'-SH are due in part to the abolition of problems inherent in generating them from intact antibodies: differences in susceptibility to proteolysis and nonspecific cleavage resulting in heterogeneity, low yield and partial reduction that is not completely selective for the hinge disulfide bonds (PARHAM 1986). Many murine antibodies contain multiple cysteine residues in their hinge region (see KABAT et al. 1991). Engineering the hinge region to leave a single cysteine removes some sources of heterogeneity in $BsF(ab')_2$ preparations, e.g., intrahinge disulfide formation and contamination with intact parent antibody whilst greatly diminishing others, such as formation of $F(ab')_3$ fragments.

F. Uses of *E. coli*-Derived Fab' Fragments

E. coli expression offers a facile route to Fab'-SH fragments which is likely to promote their usage in a variety of ways (Fig. 3) in addition to $BsF(ab')_2$ construction. All of these applications exploit the free hinge cysteinyl thiol for site-specific coupling of Fab' to other molecules with unit stoichiometry. This has clear advantages over the commonly used strategy of coupling

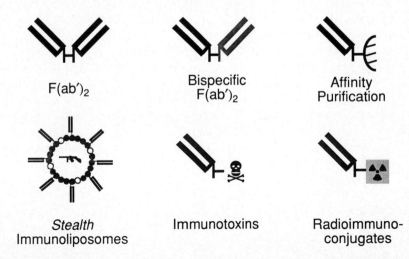

F(ab')₂

Bispecific
F(ab')₂

Affinity
Purification

Stealth
Immunoliposomes

Immunotoxins

Radioimmuno-
conjugates

Fig. 3. Potential uses of Fab'-SH fragments. So-called stealth immunoliposomes formulated for prolonged circulation time in vivo have been used for encapsulation of anti-tumor drugs and toxins (e.g., see AHMAD et al. 1993)

antibodies to other moieties through surface accessible lysine residues. The latter "random" attachment method generates heterogeneous populations of molecules with variable site and stoichiometry of attachment often resulting in differences in physicochemical and biological properties and reduced specific binding to antigen.

MsF(ab')$_2$ fragments can be constructed by simply adding half an equivalent of a coupling agent to the corresponding Fab'-SH fragment (Carter et al. 1992a). Thus MsF(ab')$_2$ fragment are even easier to construct than BsF(ab')$_2$. MsF(ab')$_2$ fragments can also be formed efficiently in the fermentor by judicious engineering of the hinge sequence. For example, HuMAb4D5-8 Fab' with a hinge containing three tandem repeats of the motif CysProPro has been recovered from fermentation pastes with up to ~75% in the bivalent form (Rodrigues et al. 1993). Furthermore this F(ab')$_2$ variant isolated directly from E. coli is functionally indistinguishable from a thioether-linked F(ab')$_2$ variant both in vitro and in vivo (Rodrigues et al. 1993). The principal advantage of obtaining F(ab')$_2$ directly from E. coli is that this obviates the need for in vitro coupling. Unfortunately, there are currently several drawbacks to F(ab')$_2$ isolated directly from E. coli. Firstly, efficient formation of F(ab')$_2$ appears to be dependent upon very high expression titers. Secondly, small quantities of unwanted F(ab')$_3$ and F(ab')$_4$ are also formed. Finally, the disulfide bonding arrangement between the three hinge cysteines is not known and is nontrivial to determine experimentally. Thus, in vitro coupling is currently the preferred route to obtain F(ab')$_2$ fragments from E. coli.

An alternative in vivo route to bivalent Ms antibody fragments was demonstrated by the use of helix bundle and also leucine zipper motifs for the dimerization of single chain Fv fragments into "miniantibodies" (Pack and Plückthun 1992). A combination of different leucine zippers (Fos and Jun) has been used as a heterodimerization motif for the construction of BsF(ab')$_2$ fragments both in vivo and in vitro (Kostelney et al. 1992). These new methods, whilst certainly elegant, do not currently offer an attractive alternative to the facile construction of bivalent monospecific and bispecific antibody fragments for human therapy using humanized Fab' fragments. Nor do antibody fragments fused to dimerization motifs seem to offer the broad utility of applications of Fab'-SH fragments (Fig. 3).

Another potential use of Fab'-SH fragments is for one step purification of proteins. Unfortunately immunoaffinity purification has only rarely been useful to date because it is arduous, time-consuming and expensive to obtain large amounts of suitable antibodies and immobilize them on appropriate matrices (reviewed by Bailon and Roy 1990). Our system for high level E. coli expression and recovery of Fab'-SH (Carter et al. 1992a) provides a solution to most of these problems. For example, we have exploited the free hinge cysteinyl thiol for site-specific immobilization of HuMAb4D5-8 Fab'-SH on an activated thiol support, enabling the p185^{HER2} ECD to be affinity purified from solution (PC, unpublished data).

E. coli-derived Fab′-SH fragments may also be useful for the construction of immunotoxins as shown recently by the coupling of an anti-CD5 Fab′ fragment to ricin A chain (BETTER et al. 1993). In some instances it is possible and preferable to construct immunotoxins using a fusion protein strategy rather than by coupling of components (reviewed in Chap. 7 and also by PASTAN and FITZGERALD 1991). However, directed chemical coupling is the only available method for attaching nonprotein toxins (e.g., calicheamicin) to antibodies.

The feasibility of site-specific attachment of phospholipid to Fab′-SH for immunospecific targeting of liposomes to cells was demonstrated over a decade ago (MARTIN et al. 1981). Much progress has subsequently been made in developing immunoliposomes for clinical use (reviewed by WEINER 1990). It seems likely that the ready availability of humanized Fab′-SH fragments will add further impetus to the construction of immunoliposomes for targeting of cytotoxic drugs or toxins to tumor cells and also for delivery of DNA for gene therapy.

G. Conclusions

Direct expression in *E. coli* offers a facile route to Fab′-SH fragments that is likely to encourage their more widespread use both in biotechnology and in the diagnosis and therapy of human disease. Currently the most promising clinical applications are in the construction of Ms and BsF(ab′)$_2$ fragments and also in the construction of immunoliposomes.

References

Ahmad I, Longenecker M, Samuel J, Allen TM (1993) Antibody-targeted delivery of doxorubicin entrapped in sterically stabilized liposomes can eradicate lung cancer in mice. Cancer Res 53:1484–1488

Bailon P, Roy SK (1990) Recovery of recombinant proteins by immunoaffinity chromatography. In: Ladisch MR, Willson RC, Painton CdC, Builder SE (eds) Protein purification: from molecular mechanisms to large-scale processes. American Chemical Society symposium series no 427, chap 11, pp 150–167

Better M, Chang CP, Robinson RR, Horwitz AH (1988) *Escherichia coli* secretion of an active chimeric antibody fragment. Science 240:1041–1043

Better M, Bernhard SL, Lei S-P, Fishwild DM, Lane JA, Carroll SF, Horwitz AH (1993) Potent anti-CD5 ricin A chain immunoconjugates from bacterially produced Fab′ and F(ab′)$_2$. Proc Natl Acad Sci 90:457–461

Beverley PCL, Callard RE (1981) Distinctive functional characteristics of human "T" lymphocytes defined by E rosetting of a monoclonal anti-T cell antibody. Eur J Immunol 11:329–334

Bolhuis RLH, Lamers CHJ, Goey SH, Eggermont AMM, Trimbos JBMZ, Stoter G, Lanzavecchia A, Di Re E, Raspagliesi F, Rivolitini L, Colnaghi MI (1992) Adoptive immunotherapy of ovarian carcinoma with BS-MAb-targeted lymphocytes: a multicenter study. Int J Cancer [Suppl] 7:78–81

Bonardi MA, Bell A, French RR, Gromo G, Hamblin T, Modena D, Tutt AL, Glennie MJ (1992) Initial experience in treating human lymphoma with a combination of bispecific antibody and saporin. Int J Cancer [Suppl] 7:73–77

Brennan M, Davison PF, Paulus H (1985) Preparation of bispecific antibodies by chemical recombination of monoclonal immunoglobulin G_1 fragments. Science 229:81–83

Carter P, Kelley RF, Rodrigues ML, Snedecor B, Covarrubias M, Velligan MD, Wong WLT, Rowland AM, Kotts CE, Carver ME, Yang M, Bourell JH, Shepard HM, Henner D (1992a) High level *Escherichia coli* expression and production of a bivalent humanized antibody fragment. Biotechnology 10:163–167

Carter P, Presta L, Gorman CM, Ridgway JBB, Henner D, Wong WLT, Rowland AM, Kotts C, Carver ME, Shepard HM (1992b) Humanization of an anti-p185^{HER2} antibody for human cancer therapy. Proc Natl Acad Sci USA 89: 4285–4289

Chang CN, Kuang W-J, Chen EY (1986) Nucleotide sequence of the alkaline phosphatase gene of *Escherichia coli*. Gene 44:121–125

Clark M, Gilliland L, Waldmann H (1988) Hybrid antibodies for therapy. In: Waldmann H (ed) Monoclonal antibody therapy. Prog Allergy 45:31–49

Derrick JP, Wigley DB (1992) Crystal structure of a streptococcal protein G domain bound to an Fab fragment. Nature 359:752–754

Ellison JW, Berson BJ, Hood LE (1982) The nucleotide sequence of a human immunoglobulin $C\gamma_1$ gene. Nucleic Acids Res 10:4071–4079

Fendly BM, Winget M, Hudziak RM, Lipari MT, Napier MA, Ullrich A (1990) Characterization of murine monoclonal antibodies reactive to either the human epidermal growth factor receptor or HER2/*neu* gene product. Cancer Res 50:1550–1558

Glennie MJ, McBride HM, Worth AT, Stevenson GT (1987) Preparation and performance of bispecific $F(ab'\gamma)_2$ antibody containing thioether-linked $Fab'\gamma$ fragments. J Immunol 139:2367–2375

Hale G, Dyer MJS, Clark MR, Phillips JM, Marcus R, Riechmann L, Winter G, Waldmann H (1988) Remission induction in non-Hodgkin lymphoma with reshaped human monoclonal antibody CAMPATH-1H. Lancet i:1394–1399

Kabat EA, Wu TT, Perry HM, Gottesman KS, Foeller C (1991) Sequences of proteins of immunological interest, 5th edn. National Institutes of Health, Bethesda MD

Kelley RF, O'Connell MP, Carter P, Presta L, Eigenbrot C, Covarrubias M, Snedecor B, Bourell JH, Vetterlein D (1992) Antigen binding thermodynamics and antiproliferative effects of chimeric and humanized anti-p185^{HER2} antibody Fab fragments. Biochemistry 31:5434–5441

Kostelny SA, Cole MS, Tso JY (1992) Formation of a bispecific antibody by the use of leucine zippers. J Immunol 148:1547–1553

Le Doussal J-M, Barbet J, Delaage M (1992) Bispecific-antibody-mediated targeting of radiolabeled bivalent haptens: theoretical, experimental and clinical results. Int J Cancer [Suppl] 7:58–62

Marks JD, Hoogenboom HR, Griffiths AD, Winter G (1992) Molecular evolution of proteins on filamentous phage. Mimicking the strategy of the immune system. J Biol Chem 267:16007–16010

Martin FJ, Hubbell WL, Papahadjopoulos D (1981) Immunospecific targeting of liposomes to cells: a novel and efficient method for covalent attachment of Fab' fragments via disulfide bonds. Biochemistry 20:4229–4238

Milstein C, Cuello AC (1983) Hybrid hybridomas and their use in immunohistochemistry. Nature 305:537–540

Nelson H (1991) Targeted cellular immunotherapy with bifunctional antibodies. Cancer Cells 5:163–172

Nishimura T, Nakamura Y, Tsukamoto H, Takeuchi Y, Tokuda Y, Iwasawa M, Yamanmoto T, Masuko T, Hashimoto Y, Habu S (1992) Human c-*erbB*-2 proto-oncogene product as a target for bispecific-antibody-directed adoptive tumor immunotherapy. Int J Cancer 50:800–804

Nitta T, Sato K, Yagita H, Okumura K, Ishii S (1990) Preliminary trial of specific targeting therapy against malignant glioma. Lancet 335:368–371

Nolan O, O'Kennedy R (1990) Bifunctional antibodies: concept, production and applications. Biochim Biophys Acta 1040:1–11

Pack P, Plückthun A (1992) Miniantibodies: use of amphipathic helices to produce functional, flexibly linked dimeric F_v fragments with high avidity in *Escherichia coli*. Biochemistry 31:1579–1584

Palm W, Hilschmann N (1975) Die primärstruktur einer kristallinen monoklonalen immunoglobulin-L-kette vom κ-typ subgruppe I (Bence-Jones-Protein Rei), isolierung und charakterisierung der tryptischen peptide; die vollständige aminosäuresequenz des proteins. Z Physiol Chem 356:167–191

Parham P (1986) Preparation and purification of active fragments from mouse monoclonal antibodies. In: Weir DM (ed) Handbook of experimental immunology, 4th edn, vol 1, chap 14. Blackwell Scientific, Oxford

Pastan I, Fitzgerald D (1991) Recombinant toxins for cancer treatment. Science 254:1173–1177

Picken RN, Mazaitis AJ, Maas WK, Rey M, Heyneker H (1983) Nucleotide sequence of the gene for heat-stable enterotoxin II of *Escherichia coli*. Infect Immun 42:269–275

Riechmann L, Clark M, Waldmann H, Winter G (1988) Reshaping human antibodies for therapy. Nature 332:323–327

Rodrigues ML, Shalaby MR, Werther W, Presta L, Carter P (1992) Engineering a humanized bispecific F(ab')$_2$ fragment for improved binding to T cells. Int J Cancer [Suppl] 7:45–50

Rodrigues ML, Snedecor B, Chen C, Wong WLT, Garg S, Blank GS, Maneval D, Carter P (1993) Engineering Fab' fragments for efficient F(ab')$_2$ formation in *Escherichia coli* and for improved in vivo stability. J Immunol 151:6954–6961

Scholtissek S, Grosse F (1987) A cloning cartridge of λ t_0 terminator. Nucleic Acids Res 15:3185

Segal DM, Qian J-H, Mezzanzanica D, Garrido MA, Titus JA, Andrew SM, George AJT, Jost CR, Perez P, Wunderlich JR (1992) Targeting of anti-tumor responses with bispecific antibodies. Immunobiology 185:390–402

Shalaby MR, Shepard HM, Presta L, Rodrigues ML, Beverley PCL, Feldmann M, Carter P (1992) Development of humanized bispecific antibodies reactive with cytotoxic lymphocytes and tumor cells overexpressing the *HER2* protooncogene. J Exp Med 175:217–225

Skerra A, Plückthun A (1988) Assembly of a functional immunoglobulin F_v fragment in *Escherichia coli*. Science 240:1038–1041

Slamon DJ, Clark GM, Wong SG, Levin WJ, Ullrich A, McGuire WL (1987) Human breast cancer: correlation of relapse and survival with amplification of HER-2/*neu* oncogene. Science 235:177–182

Slamon DJ, Godolphin W, Jones LA, Holt JA, Wong SG, Keith DE, Levin WJ, Stuart SG, Udove J, Ullrich A, Press MF (1989) Studies of the HER-2/*neu* proto-oncogene in human breast and ovarian cancer. Science 244:707–712

Songsivilai S, Lachmann PJ (1990) Bispecific antibody: a tool for diagnosis and treatment of disease. Clin Exp Immunol 79:315–321

Stickney DR, Slater JB, Kirk GA, Ahlem C, Chang C-H, Frincke JM (1989) Bifunctional antibody: ZCE/CHA [111]indium BLEDTA-IV clinical imaging in colorectal carcinoma. Antibody Immunoconj Radiopharm 2:1–13

Stickney DR, Anderson LD, Slater JB, Ahlem CN, Kirk GA, Schweighardt SA, Frincke JM (1991) Bifunctional antibody: a binary radiopharmaceutical delivery system for imaging colorectal carcinoma. Cancer Res 51:6650–6655

Sugiyama Y, Aihara M, Shibamori M, Deguchi K, Imagawa K, Kikuchi M, Momota H, Azuma T, Okada H, Alper Ö, Hitomi J, Yamaguchi K (1992) In vitro anti-tumor activity of anti-c-*erb*B-2 × anti-CD3ε bifunctional monoclonal antibody. Jpn J Cancer Res 83:563–567

Weiner AL (1990) Chemistry and biology of immunotargeted liposomes. In: Tyle P, Ram P (eds) Targeted therapeutic systems. Dekker, New York, chap 11, pp 305–336

Section III: MAb Conjugates and Fusions

CHAPTER 6
Immunotoxins

S.H. PINCUS

A. Introduction

Immunotoxins are bifunctional molecules, one portion of which is responsible for delivery to a target cell, the other consisting of the toxic moiety. Immunotoxins bind to specific target cells and cause their death. Therapeutic applications for immunotoxins include neoplastic, autoimmune, and infectious diseases (LORD 1991; OELTMANN and FRANKEL 1991; PASTAN et al. 1992; PASTAN and FITZGERALD 1991; VITETTA and THORPE 1991; WAWRZYNCZAK and DERBYSHIRE 1992). Because immunotoxins must be internalized to function, they may also be used as probes to study the cellular trafficking of the target molecule. Other, less toxic, agents may also be delivered by immunologic means and used for both diagnostic and therapeutic purposes.

Immunotoxins may be constructed by chemically combining two different molecules, or genetic engineering techniques may be applied to construct chimeric molecules with the two functions. The targeting portion of the molecule may be a monoclonal antibody, an antibody fragment, a cytokine, a ligand for cell surface receptors, a viral receptor, or other molecule that will bind specifically to the target cell. This review will concentrate on antibody-based immunotoxins, using examples of other types of immunotoxins to make specific points. The toxic moiety may consist of a plant or bacterial toxin, cytotoxic drug, high energy radioisotopes, or any agent which will cause lethal damage to the target cell without damaging neighboring cells within the tissue.

The development of immunotoxins for human therapy is proceeding rapidly. A number of potentially useful immunotoxins have been developed and shown to have in vitro efficacy. Many of these have been tested in animal models. Human clinical trials have been performed involving the direct administration of immunotoxins or their use for ex vivo purging of bone marrow and then reinfusion of the treated marrow. A U.S. Food and Drug Administration (FDA) advisory committee has recently recommended approval of an anti-T cell immunotoxin for the treatment of steroid-resistant graft vs host disease following bone marrow transplantation. Although immunotoxins hold great promise for the treatment of disease, there are significant problems associated with their use. Particular concerns include direct toxicity, immunogenicity, tissue penetration, and the selection of immunotoxin-resistant variants. Strategies are being developed to deal with these issues.

B. Considerations in Immunotoxin Development

A fairly straightforward protocol can be used in evaluating potential immunotoxins. In vitro analyses using cell lines expressing the target antigen are predictive of in vivo efficacy. This is followed by preliminary testing in animal models of disease for both efficacy and toxicity. Human clinical trials are begun only when there is adequate data indicating that the potential benefits to human subjects outweigh any risks.

I. In Vitro Testing To Identify Effective Antibodies

It is not possible to predict a priori whether a particular monoclonal antibody will function as an immunotoxin. Two absolute requirements are that the target antigen be expressed on the surface of the cell to be killed and that this antigen be internalized, or at least allow for the internalization of an immunotoxin bound to it. However, it has been shown that antibodies to different epitopes on a single molecule can differ significantly in their ability to deliver an immunotoxin and this is not necessarily a function of antibody avidity (Luo and Seon 1990; May et al. 1990; Pincus et al. 1991; Press et al. 1988). Moreover, the degree of binding of the monoclonal antibody to the surface of the target cell is not predictive of the efficacy of that antibody as an immunotoxin. Therefore, once cell surface binding has been established, it is essential to test the ability of the antibody to function as an immunotoxin. This may be performed by either directly conjugating the antibody to a toxin or by using an indirect screening assay.

Directly conjugating antibodies to be tested is more cumbersome than using an indirect screening assay, but we have found that there are circumstances in which the indirect assay fails to detect a significant proportion of useful antibodies, particularly when the antigen is expressed at low density on the target cells. Direct conjugation of monoclonal antibodies to toxins may be accomplished with commercially available materials or may be performed on a small scale by one of several companies (Pincus et al. 1991). Ricin A chain, the toxic portion of the ricin holotoxin separated away from the cell-binding B chain, may be purchased from several sources and can be used to produce highly effective and specific immunotoxins. Coupling of the toxin to the antibody requires a cross-linking reagent. We prefer to use heterobifunctional reagents which may be cleaved upon reduction; N-hydroxysuccinimidylpyridyldithiopropionate (SPDP) has been used by many groups. Details of the procedures used to couple antibodies and toxins have been published elsewhere and require only minimal skill in protein chemistry

Table 1. Decisions in immunotoxin design

Chimeric toxin vs chemical conjugate
Murine vs human antibodies
Intact antibody vs antibody fragments (Fab, Fv)
Choice of toxic moiety
Choice of cross-linking material

(CUMBER et al. 1985; FRANKEL et al. 1988). As an alternative to direct conjugation of each antibody to a toxin, indirect assays have been developed in which toxin-conjugated anti-immunoglobulin antibody is used to detect the toxic potential of a given monoclonal (TILL et al. 1988a; WELTMAN et al. 1986). This assay allows for the screening of many monoclonal antibodies, but as noted above may fail to detect useful antibodies.

A cell line expressing the target antigen is used for testing immunotoxin efficacy, either with the directly conjugated antibody or by the indirect method. Cytotoxicity is measured by incubating the cells for 48–72 h in the presence of the immunotoxin with viability determined by protein synthesis (incorporation of radiolabeled amino acids), proliferation (DNA synthesis measured as thymidine incorporation) or vital dyes (trypan blue or MTT; MOSMANN 1983). Initial studies should use the immunotoxin at relatively high concentrations ($1–10 \mu g$/ml) so that any toxicity may be detected. If toxicity is seen at high concentrations, a dose-response curve should be performed since ID_{50} in vitro is predictive of in vivo efficacy. Also important is the use of irrelevant target cells to demonstrate that the observed toxicity is specific. In the absence of a cell line expressing the target antigen, functional assays may be used to measure the desired effect. For example, the mixed lymphocyte reaction or cytotoxic T cell frequency has been used to measure the effects of immunotoxins on alloreactivity (BLAZAR et al. 1991; NAKAHARA et al. 1986) and a focal immunoassay has been used to test the effects of immunotoxins on HIV production (PINCUS et al. 1991).

Once the efficacy of an antibody as an immunotoxin has been demonstrated using in vitro test systems, several decisions need to be made before in vivo testing is begun. These are summarized in Table 1 and discussed below.

II. Immunotoxin Design

The form of the antibody used to construct the immunotoxin can influence its in vivo activity. A consensus is emerging that human or humanized antibodies have a significant advantage over murine antibodies due to decreased immunogenicity and longer serum half-life (Co and QUEEN 1991; LoBUGLIO et al. 1989; RIECHMANN et al. 1988; RYBAK et al. 1992; WINTER and MILSTEIN 1991). Using antibody fragments improves tissue penetration but decreases serum half-life (FULTON et al. 1988; GHETIE et al. 1991a;

VITETTA et al. 1987; YOKOTA et al. 1992). Because antibody valency may affect avidity of the antibody-antigen interaction, in vitro comparisons of the relative efficacy of immunotoxins made with monovalent (Fab) vs divalent $(F(ab')_2$ or intact Ig) antibodies have been performed and yielded mixed results (DEBINSKI and PASTAN 1992; GHETIE et al. 1991a; MASUHO et al. 1982). The role of antibody affinity is also debated. It is clear that high affinity antibodies have greater in vitro activity (MAY et al. 1990), but it has been suggested that in vivo lower affinity antibodies might give greater tissue penetration by not binding antigen on first encounter. However, there is little experimental evidence to support this latter assertion. Antibodies used for any in vivo experiments should be produced under the most stringent conditions (BOGARD et al. 1989; GHETIE et al. 1991b).

An important element in the design of immunoconjugates is the choice of cross-linking reagent. It is clear that the toxic moiety should be able to be released from the antibody for optimal toxicity (FRANKEL et al. 1988; MASUHO et al. 1982; VITETTA and THORPE 1991). This has led to the use of cross-linkers that will be cleaved once the immunotoxin has entered a cell, but not in the circulation (BARTON et al. 1991; FRANKEL et al. 1988). Two classes of cross-linking agents that meet this criteria are those containing disulfide bonds and those with acid-labile links, which are cleaved in endocytic vesicles (BLATTLER et al. 1985). There has been some concern that instability of disulfide bonds in the circulation may lead to premature cleavage of the toxin from the antibody (VITETTA et al. 1987; VITETTA and THORPE 1991), and this has led to the development of cross-linking reagents with hindered disulfide bonds (BARTON et al. 1991; GHETIE et al. 1991b). While the breakdown of disulfide bonds clearly does occur in the circulation, it appears that the rate of cleavage is too slow to affect the majority of site-specific localization of the immunotoxin (FULTON et al. 1988; LETVIN et al. 1986b; RAMAKRISHNAN and HOUSTON 1985).

The easy accessibility of antibody variable region genes allowed the production of chimeric immunotoxins (CHAUDHARY et al. 1989, 1990), in which single chain antibody Fv fragments are fused to a toxin (BATRA et al. 1992; BRINKMANN et al. 1991, 1992a; KREITMAN et al. 1990, 1992; RYBAK et al. 1991, 1992). The use of combinatorial phage libraries to produce antibodies in the future will serve to increase this trend. Chimeric immunotoxins have the advantages of stability and, because of their small size, enhanced tumor penetration (YOKOTA et al. 1992), The in vivo efficacy of chimeric immunotoxins has been demonstrated (BATRA et al. 1992; BRINKMANN et al. 1991).

Although in vitro testing can be used to predict the efficacy of an immunotoxin, in vivo analyses may reveal unexpected phenomena. These may be related to the preparation of the immunotoxin (GROSSBARD et al. 1992b). Unanticipated antibody cross-reactions have been detected in human immunotoxin trials (GOULD et al. 1989). The importance of careful preclinical and clinical trials cannot be overemphasized.

C. The Toxic Moiety

A wide variety of toxic compounds have been conjugated to antibodies to create immunotoxic agents. A partial listing of these agents is shown in Table 2. These include toxic proteins primarily derived from plants and bacteria, but also toxins derived from other organisms and enzymes that are toxic when directed to a novel cellular localization. Smaller molecules, such as cytotoxic antitumor drugs or radioisotopes, can also be coupled to antibodies. This review will not attempt to describe each agent. Rather, it will focus on several of the more commonly used toxins, describe modifications that have enhanced their therapeutic utility, and also describe some novel approaches.

Table 2. Some toxic agents which have been coupled to antibodies

Agent	Reference
Bacterial and fungal toxins	OLSNES et al. 1991
Pseudomonas exotoxin A	PASTAN et al. 1992; PASTAN and FITZGERALD 1991
Diphtheria toxin	CHOE et al. 1992; YOULE 1991
Clostridium perfringens phospholipase C	CHOVNICK et al. 1991
Staphylococcal enterotoxin A	DOHLSTEN et al. 1991
Aspergillus mitogillin	BETTER et al. 1992
Plant toxins	
Ricin	ROBERTUS 1991; VITETTA and THORPE 1991
Saporin	FRENCH et al. 1991; TAZZARI et al. 1992
Gelonin	REIMANN et al. 1988
Pokeweed antiviral protein	UCKUN et al. 1992; ZARLING et al. 1990
Mistletoe lectin	WIEDLOCHA et al. 1991
Nucleases	
Pancreatic ribonuclease/angiogenin	RYBAK et al. 1991, 1992
Barnase toxin	PRIOR et al. 1991
Drugs	BARTON et al. 1991
Vinca alkaloids	SCHRAPPE et al. 1992; STARLING et al. 1991
Methotrexate	AFFLECK and EMBLETON 1992
Doxorubicin	YEH et al. 1992
Daunomycin	DIENER et al. 1986
5-Fluorouracil derivatives	KRAUER et al. 1992
Maytansinoids	CHARI et al. 1992
Photosensitizing agents	GOFF et al. 1991
Drug laden liposomes	LESERMAN et al. 1981
Radioisotopes	
^{212}Bismuth	MACKLIS et al. 1988
^{90}Yttrium	ITO et al. 1992
^{131}Iodine	KHAZAELI et al. 1991; SCHWARTZ et al. 1991
Complex systems	
Biotin/avidin directed toxins	MEYER et al. 1991; SCHECHTER et al. 1992
Prodrug/enzyme	SENTER et al. 1988

I. Ricin

Ricin toxin is a ribosome-inactivating protein derived from the castor bean. It is a two chain toxin, with the B chain binding to cell surface galactose residues and the A chain functioning enzymatically to remove adenine 4324 from 28S rRNA found in the 60S subunit of eukaryotic ribosomes (PASTAN et al. 1992; ROBERTUS 1991; VITETTA and THORPE 1991). To give specificity to ricin immunotoxins, the cell-binding activity of the B chain must be removed. This led to the use of the purified ricin A chain (RTA) in immunotoxins. However, it was found that mannose and fucose residues on the native RTA mediate binding to hepatocytes, resulting in hepatotoxicity and rapid in vivo clearance of the immunotoxins. Deglycosylated forms of RTA may be produced chemically (BLAKEY et al. 1987; FULTON et al. 1988), by expressing recombinant RTA in *E. coli*, or by isolation of a naturally occurring low molecular weight form of RTA which is minimally glycosylated (TROWN et al. 1991). Immunotoxins made with these deglycosylated forms of RTA have increased plasma half-lives and enhanced tumor localization. Because it is thought that ricin B chain may play a role in the translocation of RTA from the endosome into the cytosol, immunotoxins have been constructed using a form of the holotoxin in which the galactose binding site(s) of the B chain are blocked by the covalent attachment of oligosaccharides to the toxin (GROSSBARD et al. 1992a; LAMBERT et al. 1991). The use of "blocked" ricin may enhance the activity of immunotoxins. It has been demonstrated with some antibodies that are apparently incorrectly routed during intracellular trafficking that the use of blocked ricin can produce an effective immunotoxin, whereas the same antibody coupled to RTA is ineffective (LAMBERT et al. 1991). A schematic diagram of the different of ricin that may be used in immunotoxins is shown in Fig. 1.

The nonspecific toxicity of ricin immunotoxins has been established in a number of clinical trials (BYERS and BALDWIN 1991; BYERS et al. 1990; GROSSBARD et al. 1992a; KERNAN et al. 1988; LEMAISTRE et al. 1991; SPITLER et al. 1987; VITETTA et al. 1991). The major dose-limiting toxicity seen with RTA or deglycosylated RTA was a capillary leak syndrome associated with hypoalbuminemia and edema. Constitutional symptoms (fever, malaise, fatigue, myalgia) were also seen frequently. Blocked ricin caused dose-limiting, but transient, hepatotoxicity and constitutional symptoms. Capillary leak was not seen with blocked ricin, but it was used in somewhat lower doses than RTA.

II. Pseudomonas Exotoxin A

Pseudomonas exotoxin A is a single chain toxin that ADP-ribosylates and consequently irreversibly inhibits elongation factor 2, thus resulting in the death of the cell. The elucidation of the function of the different domains of the molecule and their subsequent use in the production of immunotoxins is

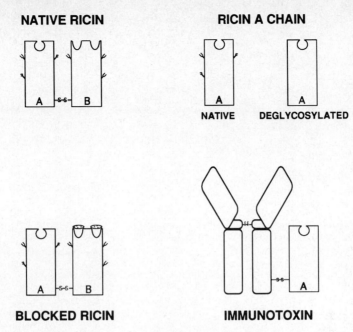

Fig. 1. Ricin derivatives for use in immunotoxins. The native ricin consists of disulfide-linked A and B chains. The A chain enzymatically degrades the 60S ribosomal subunit and has a single catalytic site. The B chain contains two galactose binding sites that allow it to act as a lectin and bind cell surface glycoproteins. Ricin itself is a glycoprotein with the A chain bearing both mannose and fucose containing carbohydrates while the B chain only has mannose side chains. Immunotoxin action requires the removal of nonspecific binding activities of ricin, either by using the A chain alone or by blocking the galactose binding sites of the B chain. An example of an immunotoxin constructed by cross-linking intact antibody with deglycosylated ricin A chain is shown

the result of the elegant protein engineering studies of Pastan and his colleagues (PASTAN et al. 1992; PASTAN and FITZGERALD 1991). The intact toxin consists of three functional domains: (1) an NH_2-terminal portion that binds the α_2-macroglobulin receptor on the target cell (WAWRZYNCZAK and DERBYSHIRE 1992), (2) a middle domain responsible for translocation across cell membranes, and (3) a COOH-terminal region which contains the ADP-ribosylating activity.

A series of modifications in the toxin have allowed for increased activity with decreased nonspecific toxicity. Removal of domain I (PE40) eliminates the nonspecific binding of the toxin to cells (KONDO et al. 1988) and enhances the therapeutic index of immunotoxins made with this construct compared to immunotoxins that utilize the intact toxin (DEBINSKI et al. 1992; PAI et al. 1992). Several alterations in PE40 have increased efficacy. By placing free amino groups toward the NH_2-terminal of PE40 and eliminating those at the

Pseudomonas Exotoxin A and Derivatives

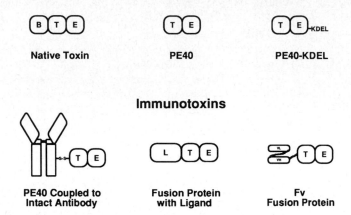

Immunotoxins

Fig. 2. Pseudomonas exotoxin A derivatives for use in immunotoxins. Pseudomonas exotoxin A consists of three functional domains responsible for: binding to the target cell (*B*), translocation to the cytosol (*T*), and enzymatic degradation of EF-2 (*E*). Constructs lacking the binding domain have been created (*PE40*). Modifications of PE40 include the placement of the amino acids KDEL at the COOH-terminal to enhance retention within the endoplasmic reticulum (*PE40-KDEL*). Immunotoxins may be created by chemically coupling the pseudomonas exotoxin A construct to intact antibodies (or their fragments), or by creating novel chimeric proteins containing a ligand for a cell surface receptor (*L*) or a single chain antibody Fv fused to the toxin

COOH-terminal, activity of an immunotoxin coupled via the amino groups on PE40 to sulfhydryl groups on an antibody fragment was significantly enhanced (DEBINSKI and PASTAN 1992). The elimination of amino acids 365–368 does not alter activity, but produces a smaller molecule (PE38) that may give better tissue penetriation (DEBINSKI and PASTAN 1992). The replacement of the COOH-terminal residues REDLK with the KDEL results in increased retention within the endoplasmic reticulum and increased toxic activity (BRINKMANN et al. 1991). Mutation of amino acid 490 (R) within a protease sensitive region resulted in a molecule without any decrease in functional activity, but resistant to proteases and with an increased serum half-life (BRINKMANN et al. 1992b). Pseudomonas exotoxin A derivatives are shown in Fig. 2.

These derivatives of PE40 have been used to create a series of chimeric toxins, in which the NH_2-terminal of the molecule consists of a ligand for a cell surface receptor or a single chain antibody fragment (PASTAN et al. 1992; PASTAN and FITZGERALD 1991). Among the nonantibody ligands that have been used are transforming growth factor-α (TGFα), interleukin-2 (IL2), IL4, IL6, acidic fibroblast growth factor, and CD4. Single chain antibody Fv fragments (V_L-EGKSSGSGSESKVD-V_H) have also been expressed as the NH_2-terminal domain of a chimeric immunotoxin using a simplified protocol of PCR amplification of antibody V genes from hybridoma mRNA and ligation into a PE40 expression vector (CHAUDHARY et al. 1989, 1990). The kinetics of the folding of these chimeric immunotoxins have been studied (BRINKMANN et al. 1992a). Antibodies made into single chain Fv-PE40 toxins include those directed against each chain of the IL2 receptor (CHAUDHARY et al. 1989; KREITMAN et al. 1990, 1992); Ley, a carbohydrate tumor associated antigen (BRINKMANN et al. 1991); erbB2, a gene that is overexpressed in tumor cells and may be involved in malignant transformation (BATRA et al. 1992); and gp160, the HIV envelope protein. Additionally, bifunctional immunotoxins have been constructed that contain two different binding domains coupled to PE40, e.g., TGFα and Fv of anti-Tac (the p55 subunit of the IL2 receptor) (BATRA et al. 1990). This toxin is active against cells expressing either receptor. Clinical trials are ongoing with several different immunotoxins made with pseudomonas exotoxin A constructs. Toxicity of the different constructs in humans has not yet been fully defined, but hepatotoxicity was seen in nonhuman primates (PAI et al. 1992).

III. Diphtheria Toxin

Diphtheria toxin is the major pathogenic factor in diphtheria. It is encoded upon a lysogenic bacteriophage of *Corynebacterium diphtheriae* and secreted as a single chain which is cleaved by extracellular proteases into an A and B chain, linked by a single disulfide bond. The B chain binds to a cell surface receptor. The toxin is endocytosed via clathrin-coated pits. Within the endocytic vesicle it is activated by acidic pH and the disulfide bond is broken. The B chain then aids in the translocation of the A chain into the cytosol, where it catalyzes the ADP-ribosylation of elongation factor 2 in a manner identical to pseudomonas exotoxin A (OLSNES et al. 1991; PASTAN et al. 1992). The 3-D structure of diphtheria toxin has recently been solved (CHOE et al. 1992). As opposed to pseudomonas toxin, in which the cell-binding domain is at the NH$_2$-terminal of the molecule, it is located at the COOH-terminal on diphtheria toxin.

Protein engineering of diphtheria toxin has allowed for the production of several mutant molecules that lack cell-binding activity, but still retain translocation and ADP-ribosylation activities (YOULE 1991). These mutant holotoxins have been used to produce extremely active immunotoxins (NEVILLE et al. 1992; YOULE 1991). Additionally, chimeric toxins have been created in which the cell-binding domain has been replaced by a targeting molecule, e.g., IL2 (WALZ et al. 1989). A major concern in the use of immunotoxins based upon diphtheria toxin is that the majority of the human population has been immunized and has antibodies against diphtheria toxin.

IV. Drug Conjugates

The major challenge in using antibodies to target cytoxic drugs is to find agents that are either active when coupled to the antibody or may be cleaved from the antibody in an active form (BARTON et al. 1991). The coupling protocols have been established for some of the more common agents (e.g., methotrexate, daunomycin and derivatives), but in general need to be individually established for each drug.

V. Novel Approaches

New methods for utilizing antibodies to target cells are continually being developed. Certain bacterial toxins, such as staphylococcal enterotoxin A, function as "superantigens" in that they activate large numbers of T cells. It has been shown that monoclonal antibodies coupled to staphylococcal enterotoxin A can direct a T cell mediated attack against tumor cells (DOHLSTEN et al. 1991). Antibodies may also be coupled to enzymes that may convert a prodrug to an active drug in the immediate vicinity of the desired target (SENTER et al. 1988). Ribonucleases (RNase) can be cytotoxic when they are delivered to the cytosol, where they inactivate ribosomes and disrupt protein synthesis. Both bovine pancreatic RNase and the human serum RNase angiogenin have been used to make immunotoxins (RYBAK et al. 1991, 1992). The latter may be more useful, since as a human protein it is less likely to be immunogenic. A chimeric toxin consisting of the extracellular RNase barnase coupled adjacent to the COOH-terminal of pseudomonas exotoxin A has been constructed (PRIOR et al. 1991). This chimeric toxin contains both activities and the membrane-translocating function of the pseudomonas toxin aids in the transport of the barnase to the correct cellular compartment to exert its toxic activity.

Several approaches have been devised that allow for the secondary delivery of toxins to cells opsonized with antibodies. Streptavidin-biotin has been used to convey toxins to cells coated with biotinylated antibodies (MEYER et al. 1991; SCHECHTER et al. 1992). By treating the intact ricin with biotin and then binding to streptavidin, the cell-binding function of the B chain is lost. This complex is still capable of binding to biotinylated antibodies since avidin is multivalent. The delivery of the ricin-biotin-avidin complex can be directed to any biotinylated antibody and the toxicity of this complex is still intact when it is internalized into the target cell. Another method of delivering ricin to antibodies utilizes a fusion protein consisting of a protein A-ricin A chain construct that is linked by a cleavable peptide derived from the diphtheria toxin (O'HARE et al. 1990). These approaches may be most useful when a cocktail of antibodies is to be used.

Bifunctional antibodies have also been used to deliver immunotoxins (FRENCH et al. 1991). In this case, one arm of the bifunctional antibody is directed against the cell-surface antigen on the target cell, while the other is directed against a toxin. The cells are first exposed to the bifunctional antibody and then the toxin. The internalized antibody-toxin complex retains toxic activity. This system has been shown to function both in vitro and in vivo. The advantages of the use of such bispecific antibodies over the direct conjugation of the toxin to the antibody remain to be demonstrated.

D. Cell Biology of Immunotoxin Action

Effective immunotoxins must bind to cell surface structures and then be internalized via one of several mechanisms: coated pits and/or uncoated pits or vesicles (OLSNES et al. 1989; WIEDLOCHA et al. 1991). Following internalization, the immunotoxin enters the endosomes. From here the immunotoxin may be routed through different pathways including: (1) translocation to the cytosol, (2) recycling back to the cell surface, (3) entry into the trans-Golgi network where it enters the biosynthetic pathway, (4) transport to the lysosome where it is degraded (OLSNES et al. 1989; RAVEL et al. 1992). That the pathways of protein secretion and toxic action cross has been shown with a hybridoma secreting anti-ricin antibody (YOULE and COLOMBATTI 1987). The IgG secreted by this cell line blocks the action of ricin by intracellular binding.

Translocation into the cytosol, the site of toxic activity for the protein toxins, may occur from different compartments for different toxins and have different processing requirements. Diphtheria toxin and pseudomonas toxin require acidification for optimal activity, whereas agents that inhibit vacuolar acidification potentiate ricin toxins. Diphtheria toxin is translocated from the endosome; pseudomonas exotoxin may enter the cytosol via the endoplasmic reticulum and ricin through the Golgi (OLSNES et al. 1989; PASTAN et al. 1992). Immunoconjugates containing radioisotopes and drugs probably release their agents in the lysosome (GEISSLER et al. 1992). These agents then enter the cell compartments where they are active (including the nucleus).

A number of factors in the processing of the immunotoxins can affect their efficacy. Cytotoxic activity of an immunotoxin is directly proportional to rate of internalization and inversely related to recycling of the antigen-immunotoxin complex back to the cell surface (PINCUS and McCLURE 1993; RAVEL et al. 1992; WARGALLA and REISFELD 1989). It has been observed that immunotoxins directed against different epitopes of the same cell surface molecule have differential efficacy, unrelated to antibody avidity (MAY et al. 1990). This effect appears to be due to differences in intracellular routing rather than rates of internalization (MAY et al. 1991).

The cellular processing of immunotoxins has also been studied by using drugs with known effects upon cells that either enhance or inhibit the action of immunotoxins (OLSNES et al. 1989). As noted above, agents that increase the pH of the endosome, e.g., NH_4Cl, inhibit the activity of diphtheria toxin but enhance efficacy of ricin immunotoxins. Monensin reduces ion gradients across cell membranes and interferes with intracellular trafficking of endosomes and secretory vesicles. In particular, monensin inhibits the transfer to lysosomes. Monension can potentiate the action of ricin immunotoxins, and this enhancement depends upon the presence of a 45–50 kDa serum protein (JANSEN et al. 1992b). Other lysosomotropic agents, such as chloroquine and β-glycylphenylnapthylamide, also potentiate immunotoxins (AKIYAMA et al.

1985; RAMAKRISHNAN and HOUSTON 1984). Calcium antagonists, such as verapamil and perhexiline, also enhance immunotoxin action, probably also by inhibiting lysosomal degradation (AKIYAMA et al. 1985; JANSEN et al. 1992b). The fungal metabolite brefeldin-A blocks the export of proteins from the Golgi. The Golgi of cells treated with brefeldin-A lose structural features and Golgi proteins are relocated to the endoplasmic reticulum. Brefeldin-A inhibits the action of intact ricin toxin, but paradoxically enhances the activity of an immunotoxin utilizing ricin A chain, suggesting that Golgi function is not required for holtoxin translocation but is tightly linked to immunotoxin translocation (HUDSON and GRILLO 1991). In addition to serving as probes of the cell biology of immunotoxin action, these drugs may serve in vivo as pharmacological enhancers of immunotoxin action.

It has been assumed that cytotoxicity of immunotoxins is a direct function of the inhibition of protein synthesis. This assumption has been challenged by a recent study in which the same monoclonal antibody was conjugated to either ricin A chain or to a binding site mutant of diphtheria toxin (SUNG et al. 1991). At equivalent concentrations, the diphtheria toxin immunotoxin inhibited protein synthesis significantly more rapidly, but achieved a substantially lower log kill than the ricin A immunotoxin. These data suggest that factors influencing the time spent by the toxin within the cytosol can influence cytotoxicity. Data indicating that cycloheximide, an inhibitor of protein synthesis itself, can protect against toxin induced cell-lysis also suggest that cytotoxicity of toxins is not a direct effect of the inhibition of protein synthesis but may involve a more active process, perhaps related to apoptosis (SANDVIG and VAN DEURS 1992).

E. Pharmacology of Immunotoxin Administration

I. Pharmacokinetics

Three important factors need to be considered in discussing immunotoxin pharmacokinetics: (1) concentration and half-life ($t_{1/2}$) of the immunotoxin within the plasma compartment, (2) delivery to the target tissue (e.g., tumor penetration), (3) in vivo stability of chemical conjugates (see Sect. B.II). In performing pharmacokinetic experiments, the presence of the target tissue is an important element that must be included to obtain accurate values, since a large amount of antigen can rapidly absorb the immunotoxin (FULTON et al. 1988; VITETTA et al. 1987). Thus, $t_{1/2}$ and peak concentrations of an immunotoxin may be considerably lower in tumor bearing animals than in normals. The presence of an immune response to the immunotoxin will inhibit both $t_{1/2}$ and peak serum concentration of the immunotoxin.

The plasma $t_{1/2}$ of an immunotoxin is shorter than $t_{1/2}$ for the equivalent antibody fragment and is, to a certain degree, dependent upon the toxic moiety. Following intravenous administration, the majority of immunotoxin is cleared from the plasma with a rapid initial phase ($at_{1/2}$) that represents equilibration with the extravascular space and hepatic uptake. Values for $at_{1/2}$ range from minutes to hours. This is followed by a slower β phase of hours to days. These values have been obtained in both experimental animals and human clinical trials (BLAKEY et al. 1987; FULTON et al. 1988; GROSSBARD et al. 1992a; LeMAISTRE et al. 1991; SIENA et al. 1988; SPITLER et al. 1987; TROWN et al. 1991; VITETTA et al. 1987, 1991). In contrast, intact chimeric (mouse V region, human C region) antibodies have an $at_{1/2}$ of 18 h and a $\beta t_{1/2}$ 4–9 days (KHAZAELI et al. 1991; LoBUGLIO et al. 1989). The $t_{1/2}$ of immunotoxins has been increased by modifications of the toxic moiety described in Sect. C, including deglycosylation of ricin A chain (BLAKEY et al. 1987; FULTON et al. 1988), use of a naturally occurring low glycosylation form of ricin A chain (TROWN et al. 1991), and the elimination of proteolytic degradation sites on pseudomonas exotoxin A (BRINKMANN et al. 1992b). In clinical trials, peak plasma levels were obtained 1–2 h following an intravenous infusion of immunotoxin and these plasma levels were in excess of the concentration needed to obtain in vitro killing of target cells (GROSSBARD et al. 1992a; LAURENT et al. 1986; LeMAISTRE et al. 1991; VITETTA et al. 1991). However, nonhuman primate studies have indicated that a number of factors may attenuate the in vivo effect of an immunotoxin, so that even though a plasma concentration in great excess of the in vitro toxic dose is obtained, in vivo killing may be limited (REIMANN et al. 1988).

The ability of the immunotoxin to penetrate to the desired target cells depends upon both the nature of the target and the form of the immunotoxin. Cells within the circulation and in the lymphoid system appear to be readily accessible to immunotoxin and coating of all of the target cells can be otained after a single intravenous injection of immunotoxin (LAURENT et al. 1986; LETVIN et al. 1986a; REIMANN et al. 1988); however, entry into solid tumors is markedly restricted. The in vivo penetration of a radioimmunotoxin into 2–10 mm tumor nodules is concentric and requires days (ITO et al. 1992). Penetration can be increased by decreasing the size of the antibody fragment. Thus, Fv and Fab fragments have been shown to penetrate tumors better than intact antibodies (FULTON et al. 1988; YOKOTA et al. 1992). Nevertheless, impressive results have been obtained by treating solid tumors in animals with immunotoxins, in some cases resulting in complete regression (BATRA et al. 1992; BRINKMANN et al. 1991; DEBINSKI et al. 1992; DEBINSKI and PASTAN 1992; GHETIE et al. 1991a; UCKUN et al. 1992).

II. Pharmacologic Enhancement of Immunotoxin Action

Several strategies exist for enhancing the activity of immunotoxins. These include: (1) the use of agents that alter the intracellular trafficking of immunotoxins as described in Sect. D. Some of these agents, especially chloroquine and calcium channel blockers, are well tolerated by patients and serum levels of the drugs that give in vitro enhancement of immunotoxin action are obtainable in vivo. (2) The use of combinations of immunotoxins. Cocktails of monoclonal antibodies, bispecific antibodies, and immunotoxins have been shown to have at least additive, and possibly synergistic, effects upon target cells (BYERS et al. 1988; CREWS et al. 1992; FRENCH et al. 1991; SUGITA et al. 1986). Combinations may be given concurrently or successively. Novel toxins that target two molecules simultaneously may be constructed using bispecific antibodies or fusion proteins (BATRA et al. 1990). (3) Specific enhancement of immunotoxin action by ligands of the target molecule. We have shown that the efficacy of anti-HIV envelope-specific immunotoxins can be enhanced 100-fold by the addition of soluble CD4, the receptor protein that is bound by the HIV envelope (PINCUS and MCCLURE 1993). This effect is caused by increased rates of internalization of the CD4-envelope complex. Other cell surface receptors are also internalized more rapidly when bound to ligand, suggesting that this ligand-specific enhancement of immunotoxin may be a general phenomenon. (4) By using agents that enhance the penetration of tumors by immunotoxins. These function by increasing the vascular permeability of the tumor tissue (PIETERSZ and MCKENZIE 1992). (5) Combining the use of immunotoxins with other therapies directed at the disease process. It is certain that in the treatment of cancer, immunotoxins will be used in conjunction with standard therapies including surgery, radiation, and the use of cytotoxic drugs. We and others have shown that anti-HIV immunotoxins act synergistically with AZT to inhibit the production of infectious virus (ASHORN et al. 1990; PINCUS and WEHRLY 1990). Optimal therapeutic responses will most likely be obtained by combining the use of immunotoxins with other treatment modalities.

III. Immunogenicity

Immune responses can develop to both the antibody and the toxic moiety. In clinical trials anti-immunotoxin responses have been seen in 20%–100% of patients (BYERS et al. 1990; FALINI et al. 1992; GROSSBARD et al. 1992a; HERTLER et al. 1987; LEMAISTRE et al. 1991; PETERSEN et al. 1991; TJANDRA et al. 1990; VITETTA et al. 1991). Antibody can arise following a single administration of immunotoxin. Giving immunotoxin in the presence of existing antibody results in enhanced clearance and lower peak levels of the immunotoxin, boosting of the antibody response (although examples where this does not occur have been reported), and infrequently manifestations of immune mediated allergic reactions. The coupling of antibody to the toxic moiety enhances the immunogenicity of the antibody (JOHNSON et al. 1991). Even humanized antibodies result in the induction of immune responses in a substantial number of cases following a single course of therapy (KHAZAELI et al. 1991; LOBUGLIO et al. 1989). In this case the immune response is directed against idiotypic determinants and neo-epitopes created in the course of chimerization.

Several different approaches have been taken to avoid such immune responses. It has been noted that patients who failed to develop an anti-immunotoxin response were receiving adjunct therapy that was immuno-suppressive (HERTLER et al. 1987). In experimental models, the concomitant administration of immunotoxins and immunosuppressive agents, such as cyclosporine, deoxyspergualin, or anti-CD4 monoclonal antibodies, can suppress the production of anti-immunotoxin antibodies (JIN et al. 1991; PAI et al. 1990; PIETERSZ and McKENZIE 1992). An alternate approach is to treat the immunotoxin in such a way as to render it either nonimmunogenic or even tolerogenic. The most effective agent for this purpose appears to be polyethylene glycol, which does not hinder the functioning of the treated molecule but tolerizes the host through the induction of active antigen-specific suppression (KITAMURA et al. 1991; TAKATA et al. 1990). Although the immunogenicity of immunotoxins has created serious concerns regarding their utility, it may be possible to surmount this problem.

F. Clinical Applications

I. Cancer

A major impetus behind the development of immunotoxins has been to produce antitumor agents. Monoclonal antibodies that recognize molecules that are either specific for tumor cells or increased in number on the surface of tumor cells have been developed. Successful treatments of human tumors with immunotoxins have been reported in immunodeficient mice. A number of clinical trials have been performed. Although less successful than proponents of immunotoxins might have hoped, the clinical trials have defined the potential problems and produced enough positive response to encourage further research. There have been a number of recent reviews that focus on the use of immunotoxins in cancer (BYERS and BALDWIN 1991; MATTHEWS et al. 1992; PASTAN and FITZGERALD 1991; PIETERSZ and McKENZIE 1992; RIETHMULLER and JOHNSON 1992; VITETTA et al. 1987; VITETTA and THORPE 1991; WAWRZYNCZAK and DERBYSHIRE 1992). Rather than go over the same details provided by these references, this section will discuss general princi-pals and conclusions.

The development of monoclonal antibodies that are absolutely specific for tumor cells has been problematic (MATTHEWS et al. 1992; RIETHMULLER and JOHNSON 1992). Many antibodies directed against tumor antigens are also expressed on normal human tissues (GOULD et al. 1989), often in a developmentally regulated manner. Many tumor-specific antigens are carbohydrates that are poorly internalized. Oncogene products and growth factor receptors make attractive targets, since their expression may be integrally related to the proliferative process (BATRA et al. 1992; DEBINSKI and PASTAN 1992; HELLSTROM and HELLSTROM 1989; KREITMAN et al. 1990,

1992; Pastan et al. 1992; Pastan and FitzGerald 1991); but they too are also expressed on normal tissues, albeit in lower numbers. Immunotoxins may either be made with monoclonal antibodies to the growth factor receptors or by attaching the toxin directly to the growth factor itself.

The current state of the art in animal testing is to establish human tumors in immunodeficient mice (nude or SCID being used most often) using either cell lines or explanted tumors. Thus, immunotoxins may be tested that are based upon antibodies that may be used in human therapy. Substantial reduction in tumor mass and even cures have been reported in these models using monoclonal antibodies directed against human gastric tumors (erbB2 gene product) (Batra et al. 1992), epidermoid carcinomas (Le^y or transferrin receptor) (Brinkmann et al. 1991; Debinski and Pastan 1992), T cell leukemia (CD7) (Jansen et al. 1992a), B cell lymphoma (CD22) (Ghetie et al. 1991a), pre-B cell leukemia (CD10 or CD-19) (Luo and Seon 1990; Uckun et al. 1992), breast carcinoma (Le^y) (Pai et al. 1992), colon carcinoma (Debinski et al. 1992; Krauer et al. 1992), glioma (Schrappe et al. 1992), and ovarian and cervical carcinomas (Starling et al. 1991; Willingham et al. 1987; Yeh et al. 1992). Among the toxic moieties that when coupled to monoclonal antibodies have yielded successful treatments in mouse models are pseudomonas exotoxin A (Batra et al. 1992; Brinkmann et al. 1991; Debinski et al. 1992; Debinski and Pastan 1992; Pai et al. 1992), ricin A chain (Ghetie et al. 1991a; Luo and Seon 1990), pokeweed antiviral protein (Uckun et al. 1992), 5-fluorouridine (Krauer et al. 1992), vinblastine (Schrappe et al. 1992; Starling et al. 1991), and doxorubicin (Yeh et al. 1992). Because most of these murine models use cell lines to represent human tumors, much of the biological diversity of human tumors is not represented. While success in treating animal models of human tumors offers encouragement to experimenters, it in no way assures success with the same antibody used in humans.

Phase 1 and 2 human clinical trials have been performed primarily with ricin immunotoxins, drug conjugates, and isotopically labeled antibodies, but also with pseudomonas exotoxin and saporin (Pietersz and McKenzie 1992). The following tumors have been treated: Cutaneous T cell lymphoma (CD5) (LeMaistre et al. 1991), Hodgkin's disease (CD30) (Falini et al. 1992), B cell lymphoma (anti-CD22 or anti-CD19) (Grossbard et al. 1992a; Vitetta et al. 1991), leukemia (Laurent et al. 1986; Waldmann 1991), breast cancer (Gould et al. 1989), and melanoma (Spitler et al. 1987). Although some impressive responses were seen, the majority of effects attributable to immunotoxins were either partial or transient. These clinical trials have also defined important problems including the host immune response and dose limiting toxicities. These results are only disappointing if one had the naive expectation that immunotoxins were the "magic bullets" that would cure cancer in one shot. More realistically, these results are the first steps in determining the role for immunotoxins in the anticancer arsenal.

Immunotoxins have also been used to purge bone marrow of tumor cells. Patients may have bone marrow removed, undergo lethal levels of irradiation or chemotherapy to treat their tumors, and then be rescued with their own bone marrow that has been purged of tumor cells. The feasibility of this approach has been shown with leukemias (MEYER et al. 1991; PREIJERSS et al. 1989) and with carcinomas (BJORN et al. 1990; GOFF et al. 1991). Several novel methodologies have been used, such as phototherapy (GOFF et al. 1991), and avidin targeting of the toxic moiety to a biotinylated antibody (MEYER et al. 1991).

Drug resistance in tumor cells may be overcome using antibody delivery systems. Because much drug resistance arises from overexpressionn of the multidrug resistance (Mdr) transporter, approaches that subvert this transporter may be useful. One approach is to use this protein as the target for an immunotoxin, since it is relatively overexpressed on the tumor cells (MICKISCH et al. 1992). Alternatively, cytotoxic drugs may be delivered via antibody, subverting the processing of the drug by the Mdr transporter (AFFLECK and EMBLETON 1992).

II. Immunosuppression

The cells of the immune system are highly accessible to immunotoxins (LETVIN et al. 1986a; REIMANN et al. 1988). A large number of monoclonal antibodies have been developed that can target cells of the immune system. Initial attempts at using immunotoxins to modulate the immune system have used broadly reactive antibodies (e.g., pan-T cell). As we understand the roles played by different lymphocyte subsets and cell surface molecules in graft rejection and autoimmune disease, we will be able to increase the specificity of targeting (PINCUS 1993).

1. Transplantation

The ability of immunotoxins to suppress the allogeneic response has been shown both in vitro and in vivo using antibodies that react with all T cells (REIMANN et al. 1988), class II MHC molecules (NAKAHARA et al. 1986), the cell surface molecule LFA-1 (BLAZAR et al. 1991), and the IL2 receptor (TAZZARI et al. 1992), and with an IL2 fusion toxin (LORBERBOUM-GALSKI et al. 1989). Because the expression of the IL2 receptor is associated with lymphocyte activation and is an integral part of that process, it is a particularly appealing target for suppressing ongoing immune responses. Immunotoxins have been made to both chains of the IL2 receptor (CHAUDHARY et al. 1989; KREITMAN et al. 1990, 1992; TAZZARI et al. 1992; WALDMANN 1991). IL2 fusion toxins, which also function by binding to the IL2 receptor, have been made with pseudomonas exotoxin A (LORBERBOUM-GALSKI et al. 1988), and diphtheria toxin (WALZ et al. 1989). It will be important to determine whether better clinical effects are seen with the antibody-based or the IL2

fusion toxins and whether combinations may produce significant enhancement of the effects.

Immunotoxins have found their greatest utility in the prevention of graft
vs host (GVH) disease following bone marrow transplantation. Two different
approaches have been taken: prophylaxis against GVH by treating the bone
marrow ex vivo, prior to transplantation, and the in vivo treatment of
patients who have established, steroid-resistant GVH. Since GVH is primarily due to the presence of mature T cells in the donor bone marrow,
treatment of bone marrow with an anti-T cell immunotoxin prior to transplantation should prevent the disease. The feasibility of this approach was
shown in 1987 (SIENA et al.) with an anti-CD5-ricin A chain conjugate. A
recent clinical trial demonstrated that depletion of T cells can be accomplished with either an anti-CD5 immunotoxin or anti-CD5 antibody plus
complement and that this reduces the occurrence of GVH (ANTIN et al.
1991). Anti-CD5 immunotoxin has also proven to be effective in the treatment of established GVH unresponsive to corticosteroid therapy (BYERS and
BALDWIN 1991; BYERS et al. 1990; KERNAN et al. 1988). Complete or partial
responses were seen in greater than 50% of patients with severe cases of this
potentially fatal condition. Treatment was associated with a rapid decrease
in peripheral blood T cells that persisted for more than a month following
cessation of therapy. Immunotoxins were given for 14 consecutive days at
doses of 0.05–0.33 mg/kg. Response was not dose-related. On the basis of
these findings the an advisory committee has recommended to the FDA that
anti-CD5-ricin A chain be approved for in vivo use for steroid-resistant
GVH, thus making this product the first officially approved immunotoxin in
the United States.

Immunotoxins may also be used in the transplantation of solid organs to
prevent the rejection of a graft by host T cells. This approach has been
shown to have merit in animal models (LORBERBOUM-GALSKI et al. 1989).
Although no clinical studies have yet been reported, the routine use of
anti-CD3 (pan-T cell) monoclonal antibody in renal transplantation
(ORTHO MULTICENTER TRANSPLANT STUDY GROUP 1985) suggests that the
use of immunotoxins would also be successful and possibly more effective
than using the unarmed monoclonal antibody.

2. Autoimmune Disease

The rationale behind using immunotoxins in autoimmune disease is to selectively eliminate those cells responsible for the deleterious immune response
without resorting to the generalized immunosuppression that is now used to
treat these conditions (BYERS and BALDWIN 1991; PINCUS 1993). Initial
studies in rheumatoid arthritis and diabetes mellitus have involved the use
of immunotoxins reacting with all T cells (anti-CD5) or with activated T
cells (anti-IL2 receptor monoclonal and IL2 fusion-diphtheria toxin fusion
protein) (ACKERMAN 1991; BYERS and BALDWIN 1991; JANSON and AREND

1992; SEWELL and TRENTHAM 1991). In phase I and phase II clinical trials, anti-CD5 immunotoxin was tested in rheumatoid arthritis patients who were unresponsive to standard antirheumatic therapies. Significant clinical improvement was seen in up to 75% of patients and the improvement outlasted the duration of suppression in numbers of peripheral T cells. The immunotoxins were well tolerated.

Although initial approaches to use immunotoxins in autoimmune disease have used rather broadly specific antibodies, the promise of the approach is to eliminate only the pathogenic subset of cells, leaving the rest of the immune response intact. A number of potential molecules may be targeted with immunotoxins (PINCUS 1993). The ultimate in therapy would be to eliminate a single clone of lymphocytes responsible for the production of autoantibody to an anti-self T cell response. The use of anti-idiotypic immunotoxins offers this possibility (BROWN and KROLICK 1988; KROLICK 1989; VERSCHUUREN et al. 1991). Immunotoxins may be the probes that allow us to define the pathogenic subsets of cells in autoimmune disease.

III. Infectious Diseases

The use of immunotoxins to treat infectious diseases is relatively new. Serious viral illnesses, especially AIDS, are the primary candidates for immunotoxin therapy. By eliminating cells that express viral antigens on their surface, and are presumably in the process of secreting live virus, an immunotoxin may eliminate the nidus of infection. Immunotoxin therapy has also been explored for diseases caused by eukaryotic parasites. Given our understanding of bacterial physiology, it is considered highly unlikely that immunotoxins would have efficacy against bacterial diseases.

A number of different immunotoxins have been designed to treat AIDS. The majority target the envelope protein(s) of HIV: gp160, gp120, and gp41. They either use monoclonal antibodies (KIM and CHANG 1992; PINCUS et al. 1989, 1991; ZARLING et al. 1990) or soluble CD4 (AULLO et al. 1992; BERGER et al. 1989; CHAUDHARY et al. 1988; TILL et al. 1988b; TSUBOTA et al. 1990) to bind to the viral envelope expressed on the cell surface. Toxins used include ricin A chain, pseudomonas exotoxin A, diphtheria toxin and pokeweed antiviral protein. The latter has also been used with lymphocyte-specific antibodies in a way so that HIV secretion is inhibited, but the cell is not killed (ZARLING et al. 1990). In addition, it has been shown that the primary source of T cells secreting HIV expresses the IL2 receptor; thus immunotoxins directed against this structure are highly effective in eliminating HIV from tissue cultures (FINBERG et al. 1991). These immunotoxins have all been tested in vitro and been shown to be highly effective in suppressing the growth and spread of multiple different HIV isolates in various model systems. The immunotoxins have been shown to act synergistically with AZT and other reverse transcriptase inhibitors (ASHORN et al. 1990; PINCUS and WEHRLY 1990). The addition of soluble CD4 has been shown to enhance the efficacy of anti-gp41 immunotoxins as much as 100-fold (PINCUS and McCLURE 1993). No in vivo studies have been reported. However, clinical trials have recently begun using CD4-pseudomonas exotoxin constructs. Important concerns exist about the use of immunotoxins to treat AIDS. Among these are: (1) If a sufficient number of cells are producing HIV, won't the treatment reproduce the worst features of the disease, i.e., elimination of large numbers of CD4+ cells? (2) Will the virus mutate rapidly enough so that immunotoxin-resistant variants will arise (PINCUS et al. 1990)? (3) Will the presence of anti-HIV antibodies in the serum of patients block the delivery of immunotoxins to target cells (PINCUS et al. 1991)? The answers to these questions will only be obtained in human clinical trials. Other viral diseases that may be candidates for immunotoxin therapy include cytomegalovirus and recurrent herpes virus infections, including varicella zoster (shingles).

Immunotoxins have been made against *Trypanosoma cruzi*, the parasite that causes Chagas' disease. This protozoon causes persistent infection and can invade the cardiac muscle. Immunotoxins inhibit the metabolism of the trypanosoma and, moreover, inhibit the release of *T. cruzi* trypomastigotes from infected cells, thus preventing infection of other cells in vitro (SANTANA and TEIXEIRA 1989; TEIXEIRA and SANTANA 1990). However, preliminary in vivo experiments in mice indicate that neither parasitemia nor survival was affected by the administration of this immunotoxin.

IV. Disordered Cellular Growth

In the process of designing immunotoxins that may be used in cancer, the receptors for a number of different growth factors have been targeted either using monoclonal antibodies against the growth receptor or making a fusion toxin out of the growth factor. Such immunotoxins may be used to treat nonmalignant disorders resulting from unregulated cellular proliferation. An example of this is the proliferation of smooth muscle cells that occurs in arterial walls following angioplasty and other vascular procedures. This proliferation is the dominant mechanism that leads to restenosis and diminution of the efficacy of the surgical procedure. A fusion toxin consisting of acidic fibroblast growth factor coupled to pseudomonas exotoxin A can inhibit the proliferation of smooth muscle cells, while sparing endothelial cells (BIRO et al. 1992). Other uses of immunotoxins in the regulation of cell growth are being explored (WAWRZYNCZAK and DERBYSHIRE 1992).

G. Conclusions

Immunotoxins hold great promise for the treatment of a number of different conditions. Both in vitro and animal studies indicate that this potential may be fulfilled. Clinical trials have been less encouraging, but certainly indicate that therapeutic effects can be obtained with immunotoxins. Only in the treatment of immune disorders have clear-cut successes been obtained in initial clinical trials, particularly in GVH disease unresponsive to standard therapies, but also to a lesser degree in some autoimmune diseases, such as rheumatoid arthritis. Ongoing clinical trials in neoplastic disease, immunologically mediated conditions, and HIV infection will help define important parameters for the successful use of imunotoxins.

A number of problems have emerged from both preclinical studies and human trials. These problems include: (1) immunogenicity (as discussed in Sect. E.III); (2) direct toxicity of the immunotoxin caused either by the toxic moiety (Sect. C) or cross-reaction of the targeting moiety with normal tissues; (3) emergence of variants that have lost expression of the target antigen (GLENNIE et al. 1987; MEEKER et al. 1985; PINCUS et al. 1990); (4) poor penetration of target tissues (Sect. E.I); (5) short serum half-life (Sect. E.I). Many of these problems will be minimized by improvements in protein engineering and by increased specificity of new monoclonal antibodies.

Another approach to the problems of immunotoxin therapy is the use of panels of immunotoxins. We have already discussed that combinations of antibodies and/or immunotoxins act in an additive, possibly synergistic, fashion (Sect. E.II). By using different immunotoxins in succession, problems due to immunogenicity and to emergence of variants may be dealt with. For example, if a patient is treated with an antibody coupled to ricin A chain and then develops an immune response to ricin and to the portions of the antibody, then treatment may be initiated with a different antibody coupled to a pseudomonas exotoxin A construct. Similarly, if escape variants are arising, the targeting antibody can be changed to one that recognizes a different structure on the target. Much as combination chemotherapy has emerged as the most effective use for antitumor drugs and for antibiotics, it is likely that combination immunotherapy will prove more efficacious than the use of a single antibody reagent.

Finally, the use of immunotoxins as probes of human disease must be considered. Because of their ability to eliminate a single set of target cells, the role of different cell subsets in the pathogenesis of disease may be ascertained. An example of this is the definition of cells expressing the IL2 receptor as the primary site of HIV replication (FINBERG et al. 1991). Similarly, the subset of lymphocytes responsible for the initiation or maintenance of autoimmune diseases may be identified. Clinical trials have already identified a CD5+ T cell expressing the IL2 receptor as improtant in sustaining disease activity in advanced rheumatoid arthritis (ACKERMAN 1991; JANSON and AREND 1992). Future trials will undoubtedly aim to narrow down the list of suspects even further.

Inherent in the study of immunotoxins is the beauty of the magic bullet concept, the ability to eliminate specifically a single subset of cells. Of course, the reality has not yet matched this ideal. But the underlying idea ensures that the use of immunotoxins as both experimental agents and as possible therapies will continue for many years.

Acknowledgements. I would like to thank Carole Smaus and Susan Smaus for secretarial assistance. I was greatly impressed by the figures in the review articles by Ira Pastan. They made clear the alterations in different molecules induced by protein engineering. Although the figures accompanying this text are original, they are clearly influenced by Pastan's.

References

Ackerman S (1991) Use of an anti-CD5 immunoconjugate in rheumatoid arthritis. In: Strand V, Scribner C, Amento E (eds) Proceedings: early decisions in DMARD development II, biologic agents in autoimmune disease. Arthritis Foundation, Atlanta

Affleck K, Embleton MJ (1992) Monoclonal antibody targeting of methotrexate (MTX) against MTX-resistant tumour cell lines. Br J Cancer 65:838–844

Akiyama S-I, Seth P, Pirker R, FitzGerald D, Gottesman MM, Pastan I (1985) Potentiation of cytotoxic activity of immunotoxins on cultured human cells. Cancer Res 45:1005–1007

Antin JH, Bierer BE, Smith BR, Ferrara J, Guinan EC, Sieff C, Golan DE, Macklis RM, Tarbell NJ, Lynch E, Reichert TA, Blythman H, Bouloux C, Rappeport JM, Burakoff SJ, Weinstein HJ (1991) Selective depletion of bone marrow T lymphocytes with anti-CD5 monoclonal antibodies: effective prophylaxis for graft-versus-host disease in patients with hematologic malignancies. Blood 78: 2139–2149

Ashorn P, Moss B, Weinstein JN, Chaudhary VK, FitzGerald DJ, Pastan I, Berger EA (1990) Elimination of infectious human immunodeficiency virus from human T-cell cultures by synergistic action of CD4-*Pseudomonas* exotoxin and reverse transcriptase inhibitors. Proc Natl Acad Sci USA 87:8889–8893

Aullo P, Alcami J, Popoff MR, Klatzmann DR, Murphy JR, Boquet P (1992) A recombinant diphtheria toxin related human CD4 fusion protein specifically kills HIV infected cells which express gp120 but selects fusion toxin resistant cells which carry HIV. EMBO J 11:575–583

Barton RL, Briggs SL, Koppel GA (1991) Monoclonal antibody drug targeting. Drug News Perspect 4:73–88

Batra J, Chaudhary VK, FitzGerald D, Pastan I (1990) TGFalpha-anti-Tac(Fv)-PE40: a bifunctional toxin cytotoxic for cells with EGF or IL2 receptors. Biochem Biophys Res Commun 171:1–6

Batra JK, Kasprzyk PG, Bird RE, Pastan I, King CR (1992) Recombinant anti-erbB2 immunotoxins containing *Pseudomonas* exotoxin. Proc Natl Acad Sci USA 89:5867–5871

Berger EA, Clouse KA, Chaudhary VK, Chakrabarti S, FitzGerald DJ, Pastan I, Moss B (1989) CD4-*Pseudomonas* exotoxin hybird protein blocks the spread of human immunodeficiency virus infection in vitro and is active against cells expressing the envelope glycoproteins from diverse primate immunodeficiency retroviruses. Proc Natl Acad Sci USA 86:9539–9543

Better M, Bernhard SL, Lei S-P, Fishwild DM, Carroll SF (1992) Activity of recombinant mitogillin and mitogillin immunoconjugates. J Biol Chem 267: 16712–16718

Biro S, Siegall CB, Fu YM, Speir E, Pastan I, Epstein SE (1992) In vitro effects of a recombinant toxin targeted to the fibroblast growth factor receptor on rat vascular smooth muscle and endothelial cells. Circ Res 71:640–645

Bjorn MJ, Manger R, Sivam G, Morgan AC Jr, Torok-Storb B (1990) Selective elimination of breast cancer cells from human bone marrow using an antibody-*Pseudomonas* exotoxin A conjugate. Cancer Res 50:5992–5996

Blakey DC, Watson GJ, Knowles PP, Thorpe PE (1987) Effect of chemical deglycosylation of ricin A chain on the in vivo fate and cytotoxic activity of an immunotoxin composed of ricin A chain and anti-Thy 1.1 antibody. Cancer Res 47:947–952

Blattler WA, Kuenzi BS, Lambert JM, Senter PD (1985) New heterobifunctional protein crosslinking reagent that forms an acid-labile link. Biochemistry 24: 1517–1524

Blazar BR, Carroll SF, Vallera DA (1991) Prevention of murine graft-versus-host disease and bone marrow alloengraftment across the major histocompatibility barrier after donor graft preincubation with anti-LFA1 immunotoxin. Blood 78:3093–3102

Bogard WC Jr, Dean RT, Deo Y, Fuchs R, Mattis JA, McLean AA, Berger HJ (1989) Practical considerations in the production, purification, and formulation of monoclonal antibodies for immunoscintigraphy and immunotherapy. Semin Nucl Med XIX:202–220

Brinkmann U, Pai LH, FitzGerald DJ, Willingham M, Pastan I (1991) B3(Fv)-PE38KDEL, a single-chain immunotoxin that causes complete regression of a human carcinoma in mice. Proc Natl Acad Sci USA 88:8616–8620

Brinkmann U, Buchner J, Pastan I (1992a) Independent domain folding of Pseudomonas exotoxin and single-chain immunotoxins: influence of interdomain connections. Proc Natl Acad Sci USA 89:3075–3079

Brinkmann U, Pai LH, FitzGerald DJ, Pastan I (1992b) Alteration of a protease-sensitive region of Pseudomonas exotoxin prolongs its survival in the circulation of mice. Proc Natl Acad Sci USA 89:3065–3069

Brown RM, Krolick KA (1988) Selective idiotype suppression of an adoptive secondary antiacetylcholine receptor antibody response by immunotoxin treatment before transfer. J Immunol 140:893–898

Byers VS, Baldwin RW (1991) Rationale for clinical use of immunotoxins in cancer and autoimmune disease. Cell Biol 2:59–70

Byers VS, Pawlucyzk I, Berry N, Durrant L, Robins RA, Garnett MC, Price MR, Baldwin RW (1988) Potentiation of anti-carcinoembryonic antigen immunotoxin cytotoxicity by monoclonal antibodies reacting with co-expressed carcinoembryonic antigen epitopes. J Immunol 140:4050–4055

Byers VS, Henslee PJ, Kernan NA, Blazar BR, Gingrich R, Phillips GL, MeMaistre CF, Gilliland G, Antin JH, Martin P, Tutscha PJ, Trown P, Ackerman SK, O'Reilly RJ, Scannon PJ (1990) Use of an anti-pan T-Iymphocyte ricin A chain immunotoxin in steroid-resistant acute graft-versus-host disease. Blood 75: 1426–1432

Chari RVJ, Martell BA, Gross JL, Cook SB, Shah SA, Blattler WA, McKenzie SJ, Goldmacher VS (1992) Immunoconjugates containing novel maytansinoids: promising anticancer drugs. Cancer Res 52:127–131

Chaudhary VK, Mizukami T, Fuerst TR, FitzGerald DJ, Moss B, Pastan I, Berger EA (1988) Selective killing of HIV-infected cells by recombinant human CD4-Pseudomonas exotoxin hybrid protein. Nature 335:369–372

Chaudhary VK, Queen C, Junghans RP, Waldmann TA, FitzGerald DJ, Pastan I (1989) A recombinant immunotoxin consisting of two antibody variable domains fused to Pseudomonas exotoxin. Nature 339:394–398

Chaudhary VK, Batra JK, Gallo MG, Willingham MC, FitzGerald DJ, Pastan I (1990) A rapid method of cloning functional variable-region antibody genes in Escherichia coli as single-chain immunotoxins. Proc Natl Acad Sci USA 87: 1066–1070

Choe S, Bennett MJ, Fujii G, Curmi PMG, Kantardjieff KA, Collier RJ, Eisenberg D (1992) The crystal structure of diptheria toxin. Nature 357:216–222

Chovnick A, Schneider WP, Tso JY, Queen C, Chang CN (1991) A recombinant, membraneacting immunotoxin. Cancer Res 51:465–467

Co MS, Queen C (1991) Humanized antibodies for therapy. Nature 351:501–502

Crews JR, Maier LA, Yu YH, Hester S, O'Briant K, Leslie DS, DeSombre D, George SL, Boyer CM, Argon Y, Bast RC Jr (1992) A combination of two immunotoxins exerts synergistic cytotoxic activity against human breast-cancer cell lines. Int J Cancer 51:772–779

Cumber AJ, Forrester JA, Foxwell BMJ, Ross WCJ, Thorpe PE (1985) Preparation of antibodytoxin conjugates. Methods Enzymol 112:207–225

Debinski W, Pastan I (1992) Monovalent immunotoxin containing truncated form of *Pseudomonas* exotoxin as potent antitumor agent. Cancer Res 52:5379–5385

Debinski W, Karlsson B, Lindholm L, Siegall CB, Willingham MC, FitzGerald D, Pastan I (1992) Monoclonal antibody C242-*Pseudomonas* exotoxin A. A specific and potent immunotoxin with antitumor activity on a human colon cancer xenograft in nude mice. J Clin Invest 90:405–411

Diener E, Diner UE, Sinha A, Xie S, Vergidis R (1986) Specific immunosuppression by immunotoxins containing daunomycin. Science 231:148–150

Dohlsten M, Hedlund G, Akerblom E, Lando PA, Kalland T (1991) Monoclonal antibody-targeted superantigens: a different class of anti-tumor agents. Proc Natl Acad Sci USA 88:9287–9291

Falini B, Bolognesi A, Flenghi L, Tazzari PL, Broe MK, Stein H, Durkop H, Aversa F, Corneli P, Pizzolo G, Barbabietola G, Sabattini E, Pileri S, Martelli MF, Stirpe F (1992) Response of refractory Hodgkin's disease to monoclonal anti-CD30 immunotoxin. Lancet 339:1195–1196

Finberg RW, Wahl SM, Allen JB, Soman G, Strom TB, Murphy JR, Nichols JC (1991) Selective elimination of HIV-1-infected cells with an interleukin-2 receptor-specific cytotoxin. Science 252:1703–1705

Frankel AE, Welsh PC, Withers DI, Schlossman DM (1988) Immunotoxin preparation and testing in vitro. In: Rodwell JD (eds) Antibody-mediated delivery systems. Dekker, New York, pp 225–244

French RR, Courtenay AE, Ingamells S, Stevenson GT, Glennie MJ (1991) Co-operative mixtures of bispecific F(ab')$_2$ antibodies for delivering saporin to lymphoma in vitro and in vivo. Cancer Res 51:2353–2361

Fulton RJ, Tucker TF, Vitetta ES, Uhr JW (1988) Pharmacokinetics of tumor-reactive immunotoxins in tumor-bearing mice: effect of antibody valency and deglycosylation of the ricin A chain on clearance and tumor localization. Cancer Res 48:2618–2625

Geissler F, Anderson SK, Venkatesan P, Press O (1992) Intracellular catabolism of radiolabeled anti-μ antibodies by malignant B-cells. Cancer Res 52:2907–2915

Ghetie M-A, Richardson J, Tucker T, Jones D, Uhr JW, Vitetta ES (1991a) Antitumor activity of Fab' and IgG-anti-CD22 immunotoxins in disseminated human B lymphoma grown in mice with severe combined immunodeficiency disease: effect on tumor cells in extranodal sites. Cancer Res 51:5876–5880

Ghetie V, Thorpe P, Ghetie MA, Knowles P, Uhr JW, Vitetta ES (1991b) The GLP large scale preparation of immunotoxins containing deglycosylated ricin A chain and a hindered disulfide bond. J Immunol Methods 1991:223–230

Glennie MJ, McBride HM, Stirpe F, Thorpe PE, Worth AT, Stevenson GT (1987) Emergence of immunoglobulin variants following treatment of a B cell leukemia with an immunotoxin composed of antiidiotypic antibody and saporin. J Exp Med 166:43–62

Goff BA, Bamberg M, Hasan T (1991) Photoimmunotherapy of human ovarian carcinoma cells ex vivo. Cancer Res 51:4762–4767

Gould BJ, Borowitz MJ, Groves ES, Carter PW, Anthony D, Weiner LM, Frankel AE (1989) Phase I study of an anti-breast cancer immunotoxin by continuous infusion: report of a targeted toxic effect not predicted by animal studies. J Natl Cancer Inst 81:775–781

Grossbard ML, Freedman AS, Ritz J, Coral F, Goldmacher VS, Eliseo L, Spector N, Dear K, Lambert JM, Blattler WA, Taylor JA, Nadler LM (1992a) Serotherapy of B-cell neoplasms with anti-B4-blocked ricin: a phase I trial of daily bolus infusion. Blood 79:576

Grossbard ML, Lambert JM, Goldmacher VS, Blattler WA, Nadler LM (1992b) Correlation between in vivo toxicity and preclinical in vitro parameters for the immunotoxin anti-B4-blocked ricin. Cancer Res 52:4200–4207

Hellstrom KE, Hellstrom I (1989) Oncogene-associated tumor antigens as targets for immunotherapy. FASEB J 3:1715–1722

Hertler AA, Spitler LE, Frankel AE (1987) Humoral immune response to a ricin A chain immunotoxin in patients with metastatic melanoma. Cancer Drug Deliv 4:245–253

Hudson TH, Grillo FG (1991) Brefeldin-A enhancement of ricin A-chain immunotoxins and blockade of intact ricin, modeccin, and abrin. J Biol Chem 266: 18586–18592

Ito T, Griffin TW, Collins JA, Brill AB (1992) Intratumoral and whole-body distributions of C110 anti-carcinoembryonic antigen radioimmunotoxin after intraperitoneal and intravenous injection: a quantitative autoradiographic study. Cancer Res 52:1961–1967

Jansen B, Vallera DA, Jaszcz WB, Nguyen D, Kersey JH (1992a) Successful treatment of human acute T-cell leukemia in SCID mice using the anti-CD7-deglycosylated ricin A-chain immunotoxin DA7. Cancer Res 52:1314–1321

Jansen FK, Jansen A, Derocq JM, Carriere D, Carayon P, Veas F, Jaffrezou JP (1992b) Golgi vacuolization and immunotoxin enhancement by monensin and perhexiline depend on a serum protein. J Biol Chem 267:12577–12582

Janson RW, Arend WP (1992) Receptor-targeted immunotherapy. Bull Rheum Dis 41:6–8

Jin F-S, Youle RJ, Johnson VG, Shiloach J, Fass R, Longo DL, Bridges SH (1991) Suppression of the immune response to immunotoxins with anti-CD4 monoclonal antibodies. J Immunol 146:1806–1811

Johnson DA, Barton RL, Fix DV, Scott WL, Gutowski MC (1991) Induction of immunogenicity of monoclonal antibodies by conjugation with drugs. Cancer Res 51:5774–5776

Kernan NA, Byers V, Scannon PJ, Mischak RP, Brochstein J, Flomenberg N, Dupont B, O'Reilly RJ (1988) Treatment of steroid-resistant acute graft-vs-host disease by in vivo administration of an anti-T-cell ricin A chain immunotoxin. JAMA 259:3154–3157

Khazaeli MB, Saleh MN, Liu TP, Meredith RF, Wheeler RH, Baker TS, King D, Secher D, Allen L, Rogers K, Colcher D, Schlom J, Shochat D, LoBuglio AF (1991) Pharmacokinetics and immune response of ^{131}I-chimeric mouse/human B72.3 (human gamma 4) monoclonal antibody in humans. Cancer Res 51: 5461–5466

Kim Y-W, Chang TW (1992) Potential use of immunoconjugates for AIDS therapy. AIDS Res Hum Retroviruses 8:1033–1038

Kitamura K, Takahashi T, Yamaguchi T, Noguchi A, Noguchi A, Takashina K-I, Tsurumi H, Inagake M, Toyokuni T, Hakomori S-I (1991) Chemical engineering of the monoclonal antibody A7 by polyethylene glycol for targeting cancer chemotherapy. Cancer Res 51:4310–4315

Kondo T, FitzGerald D, Chaudhary VK, Adhya S, Pastan I (1988) Activity of immunotoxins constructed with modified *Pseudomonas* exotoxin A lacking the cell recognition domain. J Biol Chem 263:9470–9475

Krauer KG, McKenzie IFC, Pietersz GA (1992) Antitumor effect of 2′-deoxy-5-fluorouridine conjugates against a murine thymoma and colon carcinoma xenografts. Cancer Res 52:132–137

Kreitman RJ, Chaudhary VK, Waldmann T, Willingham MC, FitzGerald DJ, Pastan I (1990) The recombinant immunotoxin anti-Tac(Fv)-*Pseudomonas* exotoxin 40 is cytotoxic toward peripheral blood malignant cells from patients with adult T-cell leukemia. Proc Natl Acad Sci USA 87:8291–8295

Kreitman RJ, Schneider WP, Queen C, Tsudo M, FitzGerald DJP, Waldmann TA, Pastan I (1992) Mik-beta1(Fv)-PE40, a recombinant immunotoxin cytotoxic

toward cells bearing the beta-chain of the IL-2 receptor. J Immunol 149:
2810–2815

Krolick KA (1989) Selective elimination of autoreactive lymphocytes with immunotoxins. Clin Immunol Immunopathol 50:273–282

Lambert JM, Goldmacher VS, Collinson AR, Nadler LM, Blattler WA (1991) An
immunotoxin prepared with blocked ricin: a natural plant toxin adapted for
therapeutic use. Cancer Res 51:6236–6242

Laurent G, Pris J, Farcet J-P, Carayon P, Blythman H, Casellas P, Poncelet P,
Jansen FK (1986) Effects of therapy with T101 ricin A-chain immunotoxin in
two leukemia patients. Blood 67:1680–1687

LeMaistre CF, Rosen S, Frankel A, Kornfeld S, Saria E, Meneghetti C, Drajesk J,
Fishwild D, Scannon P, Byers V (1991) Phase I trial of H65-RTA immunoconjugate in patients with cutaneous T-cell lymphoma. Blood 78:1173–1182

Leserman LD, Machy P, Barbet J (1981) Cell-specific drug transfer from liposomes
bearing monoclonal antibodies. Nature 293:226–228

Letvin NL, Chalifoux LV, Reimann KA, Ritz J, Schlossman SF, Lambert JM
(1986a) In vivo administration of lymphocyte-specific monoclonal antibodies in
nonhuman primates. Delivery of ribosome-inactivating proteins to spleen and
lymph node T cells. J Clin Invest 78:666–673

Letvin NL, Goldmacher VS, Ritz J, Yetz JM, Schlossman SF, Lambert JM (1986b)
In vivo administration of lymphocyte-specific monoclonal antibodies in non-
human primates. In vivo stability of disulfide-linked immunotoxin conjugates. J
Clin Invest 77:977–984

LoBuglio AF, Wheeler RH, Trang J, Haynes A, Rogers K, Harvey EB, Sun L,
Ghrayeb J, Khazaeli MB (1989) Mouse/human chimeric monoclonal antibody
in man: kinetics and immune response. Proc Natl Acad Sci USA 86:4220–
4224

Lorberboum-Galski H, Kozak RW, Waldmann TA, Bailon P, FitzGerald DJP,
Pastan I (1988) Interleukin 2 (IL2) PE40 is cytotoxic to cells displaying either
the p55 or p70 subunit of the IL2 receptor. J Biol Chem 263:18650–18656

Lorberboum-Galski H, Barrett LV, Kirkman RL, Ogata M, Willingham MC,
FitzGerald DJ, Pastan I (1989) Cardiac allograft survival in mice treated with
IL-2-PE40. Proc Natl Acad Sci USA 86:1008–1012

Lord JM (1991) Redirecting nature's toxins. Semin Cell Biol 2:1–71

Luo Y, Seon BK (1990) Marked difference in the in vivo antitumor efficacy between
two immunotoxins targeted to different epitopes of common acute lymphoblastic
leukemia antigen (CD10). J Immunol 145:1974–1982

Macklis RM, Kinsey BM, Kassis AI, Ferrara JLM, Atcher RW, Hines JJ, Coleman
CN, Adelstein SJ, Burakoff SJ (1988) Radioimmunotherapy with alpha-particle-
emitting immunoconjugates. Science 240:1024–1026

Masuho Y, Kishida K, Saito M, Umemoto N, Hara T (1982) Importance of the
antigen-binding valency and the nature of the cross-linking bond in ricin A-chain
conjugates with antibody. J Biochem 91:1583–1591

Matthews DC, Smith FO, Bernstein ID (1992) Monoclonal antibodies in the study
and therapy of hematopoietic cancers. Curr Opin Immunol 4:641–646

May RD, Finkelman FD, Wheeler HT, Uhr JW, Vitetta ES (1990) Evaluation of
ricin A chaincontaining immunotoxins directed against different epitopes on the
delta-chain of cell surface-associated igD on murine B cells. J Immunol 144:
3637–3642

May RD, Wheeler HT, Finkelman FD, Uhr JW, Vitetta ES (1991) Intracelluar
routing rather than cross-linking or rate of internalization determines the potency
of immunotoxins directed against different epitopes of sigD on murine B cells.
Cell Immunol 135:490–500

Meeker T, Lowder J, Cleary ML, Stewart S, Warnke R, Sklar J, Levy R (1985)
Emergence of idiotype variants during treatment of B-cell lymphoma with anti-
idiotype antibodies. N Engl J Med 312:1658–1665

Meyer BF, Stoner ML, Raphael CL, Davis RE, Herrmann RP (1991) Streptavidin-biotin immunotoxins: a new approach to purging bone marrow. Exp Hematol 19:710–713

Mickisch GH, Pai LH, Gottesman MM, Pastan I (1992) Monoclonal antibody MRK16 reverses the multidrug resistance of multidrug-resistant transgenic mice. Cancer Res 52:4427–4432

Mosmann T (1983) Rapid colorimetric assay for cellular growth and survival: application to proliferation and cytotoxicity assays. J Immunol Methods 65:55–63

Nakahara K, Kaplan D, Bjorn M, Fathman CG (1986) The effectiveness of anti-Ia-immunotoxins in the suppression of MLR. Transplantation 42:205–211

Neville DM Jr, Scharff J, Srinivasachar K (1992) In vivo T-cell ablation by a holoimmunotoxin directed at human CD3. Proc Natl Acad Sci USA 89: 2585–2589

O'Hare M, Brown AN, Hussain K, Gebhardt A, Watson G, Roberts LM, Vitetta ES, Thorpe PE, Lord JM (1990) Cytotoxicity of a recombinant ricin-A-chain fusion protein containing a proteolytically-cleavable spacer sequence. FEBS Lett 273:200–204

Oeltmann TN, Frankel AE (1991) Advances in immunotoxins. FASEB J 5:2334–2337

Olsnes S, Sandvig K, Peterson OW, van Deurs B (1989) Immunotoxins: entry into cells and mechanisms of action. Immunol Today 10:291–295

Olsnes S, Kozlov JV, van Deurs B, Sandvig K (1991) Bacterial protein toxins acting on intracellular targets. Semin Cell Biol 2:7–14

Ortho Multicenter Transplant Study Group (1985) A randomized clinical trial of OKT3 monoclonal antibody for acute rejection of cadaveric renal transplants. N Engl J Med 313:337–342

Pai LH, FitzGerald DJ, Tepper M, Schacter B, Spitalny G, Pastan I (1990) Inhibition of antibody response to *Pseudomonas* exotoxin and an immunotoxin containing *Pseudomonas* exotoxin by 15-deoxyspergualin in mice. Cancer Res 50:7750–7753

Pai LH, Batra JK, FitzGerald DJ, Willingham MC, Pastan I (1992) Antitumor effects of B3-PE and B3-LysPE40 in a nude mouse model of human breast cancer and the evaluation of B3-PE toxicity in monkeys. Cancer Res 52: 3189–3193

Pastan I, FitzGerald D (1991) Recombinant toxins for cancer treatment. Science 254:1173–1177

Pastan I, Chaudhary V, FitzGerald DJ (1992) Recombinant toxins as novel therapeutic agents. Annu Rev Biochem 61:331–354

Petersen BH, DeHerdt SV, Schneck DW, Bumol TF (1991) The human immune response to KS1/4-desacetylvinblastine (LY256787) and KS1/4-desacetylvinblastine hydrazide (LY203728) in single and multiple dose clinical studies. Cancer Res 51:2286–2290

Pietersz GA, McKenzie IFC (1992) Antibody conjugates for the treatment of cancer. Immunol Rev 129:57–80

Pincus SH (1993) Immunoregulation and experimental therapies. In: McCarthy DJ, Koopman WJ (eds) Arthritis and allied conditions: a textbook of rheumatology. Lea and Febiger, Philadelphia, pp 683–710

Pincus SH, McClure J (1993) Soluble CD4 enhances the efficacy of immunotoxins directed against gp41 of the human immunodeficiency virus. Proc Nat Acad Sci USA 90:332–336

Pincus SH, Wehrly K (1990) AZT demonstrates anti-HIV-1 activity in persistently infected cell lines: implications for combination chemotherapy and immunotherapy. J Infect Dis 162:1233–1238

Pincus SH, Wehrly K, Chesebro B (1989) Treatment of HIV tissue culture infection with monoclonal antibody-ricin A chain conjugates. J Immunol 142:3070–3075

Pincus SH, Wehrly K, Tschachler E, Hayes SF, Buller RS, Reitz M (1990) Variants selected by treatment of human immunodeficiency virus-infected cells with an immunotoxin. J Exp Med 172:745–757

Pincus SH, Cole RL, Hersh EM, Lake D, Masuho Y, Durda PJ, McClure J (1991) In vitro efficacy of anti-HIV immunotoxins targeted by various antibodies to the envelope protein. J Immunol 146:4315–4324

Preijers FWMB, De Witte T, Wessels JMC, De Gast GC, Van Leeuwen E, Capel PJA, Haanen C (1989) Autologous transplantation of bone marrow purged in vitro with anti-CD7-(WT1−) ricin A immunotoxin in T-cell lymphoblastic leukemia and lymphoma. Blood 74:1152–1158

Press OW, Martin PJ, Thorpe PE, Vitetta ES (1988) Ricin A-chain containing immunotoxins directed against different epitopes on the CD2 molecule differ in their ability to kill normal and malignant T cells. J Immunol 141:4410–4417

Prior TI, FitzGerald DJ, Pastan I (1991) Barnase toxin: a new chimeric toxin composed of pseudomonas exotoxin A and barnase. Cell 64:1017–1023

Ramakrishnan S, Houston LL (1984) Inhibition of human acute lymphoblastic leukemia cells by immunotoxins: potentiation by chloroquine. Science 223:58–61

Ramakrishnan S, Houston LL (1985) Immunological and biological stability of immunotoxins in vivo as studied by the clearance of disulfide-linked pokeweed antiviral protein-antibody conjugates from blood. Cancer Res 45:2031–2036

Ravel S, Colombatti M, Casellas P (1992) Internalization and intracellular fate of anti-CD5 monoclonal antibody and anti-CD5 ricin A-chain immunotoxin in human leukemia T cells. Blood 79:1511–1517

Reimann KA, Goldmacher VS, Lambert JM, Chalifoux LV, Cook SB, Schlossman SF, Letvin NL (1988) In vivo adminstration of lymphocyte-specific monoclonal antibodies in nonhuman primates. IV. Cytotoxic effect of an anti-T11-gelonin immunotoxin. J Clin Invest 82:129–138

Riechmann L, Clark M, Waldmann H, Winter G (1988) Reshaping human antibodies for therapy. Nature 332:323–327

Riethmuller G, Johnson JP (1992) Monoclonal antibodies in the detection and therapy of micrometastatic epithelial cancers. Curr Opin Immunol 4:647–655

Robertus J (1991) The structure and action of ricin, a cytotoxic N-glycosidase. Semin Cell Biol 1991:23–30

Rybak SM, Saxena SK, Ackerman EJ, Youle RJ (1991) Cytotoxic potential of ribonuclease and ribonuclease hybrid proteins. J Biol Chem 266:21202–21207

Rybak SM, Hoogenboom HR, Meade HM, Raus JCM, Schwartz D, Youle RJ (1992) Humanization of immunotoxins. Proc Natl Acad Sci USA 89:3165–3169

Sandvig K, van Deurs B (1992) Toxin-induced cell lysis: protection by 3-methyladenine and cycloheximide. Exp Cell Res 200:253–262

Santana JM, Teixeira ARL (1989) Effect of immunotoxins against Trypanosoma cruzi. Am J Trop Med Hyg 41:177–182

Schechter B, Arnon R, Wilchek M (1992) Cytotoxicity of streptavidin-blocked biotinyl-ricin is retrieved by in vitro immunotargeting via biotinyl monoclonal antibody. Cancer Res 52:4448–4452

Schrappe M, Bumol TF, Apelgren LD, Briggs SL, Koppel GA, Markowitz DD, Mueller BM, Reisfeld RA (1992) Long-term growth suppression of human glioma xenografts by chemoimmunoconjugates of 4-desacetylvinblastine-3-carboxyhydrazide and monoclonal antibody 9.2.27. Cancer Res 52:3838–3844

Schwartz MA, Lovett DR, Redner A, Gulati S, Divgi CR, Graham MC, Finn R, Gee TS, Andreeff M, Old LJ (1991) Therapeutic trial of radiolabelled monoclonal antibody M195 in relapsed or refractory myeloid leukemias. Proc Am Soc Hematol 78:54

Senter PD, Saulnier MG, Schreiber GJ, Hirschberg DL, Brown JP, Hellstrom I, Hellstrom KE (1988) Anti-tumor effects of antibody-alkaline phosphatase conjugates in combination with etoposide phosphate. Proc Natl Acad Sci USA 85:4842–4846

Sewell KL, Trentham DE (1991) Rapid improvement in refractory rheumatoid arthritis by an interleukin-2 receptor targeted immunotherapy. Clin Res 39: 314A (abstract)

Siena S, Villa S, Bonadonna G, Bregni M, Gianni AM (1987) Specific ex-vivo depletion of human bone marrow T lymphocytes by an anti-pan-T cell (CD5) ricin A-chain immunotoxin. Transplantation 43:421–426

Siena S, Lappi DA, Bregni M, Formosa A, Villa S, Soria M, Bonadonna G, Gianni AM (1988) Synthesis and characterization of an antihuman T-lymphocyte saporin immunotoxin (OKT1-SAP) with in vivo stability into nonhuman primates. Blood 72:756–765

Spitler LE, del Rio M, Khentigan A, Wedel NI, Brophy NA, Miller LL, Harkonen WS, Rosendorf LL, Lee HM, Mischak RP, Kawahata RT, Stoudemire JB, Fradkin LB, Bautista EE, Scannon PJ (1987) Therapy of patients with malignant melanoma using a monoclonal anti-melanoma antibody-ricin A chain immunotoxin. Cancer Res 47:1717–1723

Starling JJ, Maciak RS, Law KL, Hinson NA, Briggs SL, Laguzza BC, Johnson DA (1991) In vivo antitumor activity of a monoclonal antibody-*vinca* alkaloid immunoconjugate directed against a solid tumor membrane antigen characterized by heterogeneous expression and noninternalization of antibody-antigen complexes. Cancer Res 51:2965–2972

Sugita K, Majdic O, Stockinger H, Holter W, Koller U, Peschel C, Knapp W (1986) Use of a cocktail of monoclonal antibodies and human complement in selective killing of acute lymphocytic leukemia cells. Int J Cancer 37:351–357

Sung C, Wilson D, Youle RJ (1991) Comparison of protein synthesis inhibition kinetics and cell killing induced by immunotoxins. J Biol Chem 1991:14159–14162

Takata M, Maiti PK, Kubo RT, Chen Y, Holford-Strevens V, Rector ES, Sehon AH (1990) Cloned suppressor T cells derived from mice tolerized with conjugates of antigen and monomethoxypolyethylene glycol. Relationship between monoclonal T suppressor factor and the T cell receptor. J Immunol 145:2846–2853

Tazzari PL, Bolognesi A, Totero DD, Pileri S, Conte R, Wijdenes J, Herve P, Soria M, Stirpe F, Gobbi M (1992) B-B10 (Anti-CD25)-Saporin immunotoxin-A possible tool in graftversus-host disease treatment. Transplantation 54:351–356

Teixeira ARL, Santana JM (1990) Chagas' disease. Immunotoxin inhibition of *Trypanosoma cruzi* release from infected host cells in vitro. Lab Invest 63:248–252

Till M, May RD, Uhr JW, Thorpe PE, Vitetta ES (1988a) An assay that predicts the ability of monoclonal antibodies to form potent ricin A chain-containing immunotoxins. Cancer Res 48:1119–1123

Till MA, Ghetie V, Gregory T, Patzer EJ, Porter JP, Uhr JW, Capon DJ, Vitetta ES (1988b) HIV-infected cells are killed by rCD4-ricin A chain. Science 242:1166–1168

Tjandra JJ, Ramadi L, McKenzie IFC (1990) Development of human anti-murine antibody (HAMA) response in patients. Immunol Cell Biol 68:367–376

Trown PW, Reardan DT, Carroll SF, Stoudemire JB, Kawahata RT (1991) Improved pharmacokinetics and tumor localization of immunotoxins constructed with the Mr 30000 form of ricin A chain. Cancer Res 51:4219–4225

Tsubota H, Winkler G, Meade HM, Jakubowski A, Thomas DW, Letvin NL (1990) CD4-*Pseudomonas* exotoxin conjugates delay but do not fully inhibit human immunodeficiency virus replication in lymphocytes in vitro. J Clin Invest 86:1684–1689

Uckun FM, Manivel C, Arthur D, Chelstrom LM, Finnegan D, Tuel-Ahlgren L, Irvin JD, Myers DE, Gunther R (1992) In vivo efficacy of B43 (anti-CD19)-pokeweed antiviral protein immunotoxin against human pre-B cell acute lymphoblastic leukemia in mice with severe combined immunodeficiency. Blood 79:2201–2214

Verschuuren JJGM, Graus YMF, Tzartos SJ, van Breda Vriesman PJC, De Baets MH (1991) Paratope- and framework-related cross-reactive idiotopes on anti-acetylcholine receptor antibodies. J Immunol 146:941–948

Vitetta ES, Thorpe PE (1991) Immunotoxins containing ricin or its A chain. Semin Cell Biol 2:47–58
Vitetta ES, Fulton RJ, May RD, Till M, Uhr JW (1987) Redesigning nature's poisons to create anti-tumor reagents. Science 238:1098–1104
Vitetta ES, Stone M, Amlot P, Fay J, May R, Till M, Newman J, Clark P, Collins R, Cunningham D, Ghetie V, Uhr JW, Thorpe PE (1991) Phase I immunotoxin trial in patients with B-cell lymphoma. Cancer Res 51:4052–4058
Waldmann TA (1991) Monoclonal antibodies in diagnosis and therapy. Science 252:1657–1662
Walz G, Zanker B, Brand K, Waters C, Genbauffe F, Zeldis JB, Murphy JR, Strom TB (1989) Sequential effects of interleukin 2-diphtheria toxin fusion protein on T-cell activation. Proc Natl Acad Sci USA 86:9485–9488
Wargalla UC, Reisfeld RA (1989) Rate of internalization of an immunotoxin correlates with cytotoxic activity against human tumor cells. Proc Natl Acad Sci USA 86:5146–5150
Wawrzynczak EJ, Derbyshire EJ (1992) Immunotoxins: the power and the glory. Immunol Today 13:381–383
Weltman JK, Pedroso P, Johnson SA, Fast LD, Leone LA, Minna JD, Cuttitta F (1986) Indirect immunotoxin method for demonstrating antibodies against human tumor cells. Biotechniques 4:224–228
Wiedlocha A, Sandvig K, Walzel H, Radzikowsky C, Olsnes S (1991) Internalization and action of an immunotoxin containing mistletoe lectin A-chain. Cancer Res 51:916–920
Willingham MC, FitzGerald DJ, Pastan I (1987) *Pseudomonas* exotoxin coupled to monoclonal antibody against ovarian cancer inhibits the growth of human ovarian cancer cells in a mouse model. Proc Natl Acad Sci USA 84:2474–2478
Winter G, Milstein C (1991) Man-made antibodies. Nature 349:293–299
Yeh M-Y, Roffler SR, Yu M-H (1992) Doxorubicin: monoclonal antibody conjugate for therapy of human cervical carcinoma. Int J Cancer 51:274–282
Yokota T, Milenic DE, Whitlow M, Schlom J (1992) Rapid tumor penetration of a single-chain Fv and comparison with other immunoglobulin forms. Cancer Res 52:3402–3408
Youle RJ (1991) Mutations in diptheria toxin to improve immunotoxin selectivity and understand toxin entry into cells. Semin Cell Biol 2:39–45
Youle RJ, Colombatti M (1987) Hybridoma cells containing intracellular anti-ricin antibodies show ricin meets secretory antibody before entering the cytosol. J Biol Chem 262:4676–4682
Zarling JM, Moran PA, Haffar O, Sias J, Richman DD, Spina CA, Myers DE, Kuelbeck V, Ledbetter JA, Uckun FM (1990) Inhibition of HIV replication by pokeweed antiviral protein targeted to CD4+ cells by monoclonal antibodies. Nature 3477:92–95

CHAPTER 7

Antibody-Enzyme Fusion Proteins and Bispecific Antibodies

E. HABER

A. Introduction

The immunoglobulin molecule is the veritable prototype of a multidomain structure. Each of its units possesses a characteristic function, and in many instances antibody domains can be separated from one another and still retain their functions. The 150 kDa intact molecule can be cleaved to produce the 50 kDa antigen-binding fragment (Fab). A still smaller domain that actually participates in antigen binding, the 25 kDa variable region fragment (Fv), can be produced by recombinant DNA methods. For Fv, Fab, or whole antibody, the affinity for a given epitope is the same. The 50 kDa complement-binding fragment (Fc), another independent immunoglobulin unit, retains the ability to bind the first component of complement and the property of transport across the placental barrier.

Advances in recombinant DNA technology now permit the selective biosynthesis of an individual antibody domain that can be used as is or fused with some other protein of unrelated function. In a fusion protein one of the antibody's most desirable properties, the ability to recognize a specific antigen, can be imparted to another type of molecule such as an enzyme. The coupling of two different antigen-binding domains, an extension of the fusion protein idea, produces a bispecific antibody that allows for the cross-linking in a living organism of two different proteins or even two different cell types. Because of the great selectivity of the antibody's antigen recognition site, antibody fusion proteins begin to address the pharmacologist's dream of a "magic bullet," a drug that modifies only its intended target and does not affect other tissues or organs.

B. Antibody-Enzyme Fusion Proteins

I. Development of the Concept of a Bifunctional Protein

The antibody molecule is nature's original bifunctional reagent. It binds to an antigen through the Fv and then through receptors on the Fc effects a change in that antigen by a variety of means, of which mobilization of the complement system and attraction of macrophages are but two examples. It followed logically that investigators would seek to enlarge the molecule's repertoire by replacing the Fc with proteins of other functions. Indeed, the concept of antibody targeting predates a knowledge of the domain structure of the molecule (Pressman and Korngold 1953; Korngold and Pressman 1954). However, it was not until after the identification of individual antibody functional domains (Porter 1959; Inbar et al. 1972) that antibody fusion proteins could be constructed.

Antigen binding domains have been linked to toxins or cytokines to target them to cell surface antigens or cytokine receptors on the membranes of neoplastic cells (Vitetta et al. 1987; Trauth et al. 1989), virally infected cells (Till et al. 1988), and cells proliferating to form arteriosclerotic lesions (Casscells et al. 1992). The aim is either to directly destroy an undesirable cell type (with an antibody-targeted toxin) or to modulate cell behavior so that an intrinsic biologic mechanism produces cell death (with an antibody-targeted cytokine). An alternative strategy entails linking an enzyme to an antibody. Here the goal is to increase the local concentration of the enzyme at a desired site. For example, with an antibody-targeted enzyme a plasminogen activator can be concentrated at the site of a thrombus in order to enhance the local production of plasmin (Haber et al. 1989), thereby restricting the action of this powerful and nonselective protease to a location where it is required. The fusion with β-glucuronidase of an antibody to a tumor-associated antigen represents another example. In this case the locally increased concentration of the enzyme at the surface of tumor cells permits an inactive prodrug to be cleaved to produce a potent chemotherapeutic drug that can kill tumor cells while sparing normal cells (which do not possess the tumor-associated antigen) (Bosslet et al. 1992).

The Fc imparts a remarkably long physiologic half-life to the immunoglobulin G molecule. Thus it is only reasonable to apply the Fc as a fusion partner for proteins (or protein fragments) in cases in which the goal is stability and prolonged residence in plasma. The extracellular domains of plasma membrane proteins are by themselves often characterized by extremely short half-lives. CD4, for example, has been fused with the Fc to create a decoy for the human immunodeficiency virus, which has a binding protein on its surface specific for the CD4 receptor (Traunecker et al. 1989). Alone, CD4 would have an extremely short half-life in plasma. Fusion with the Fc sufficiently increases the CD4 half-life to permit a more prolonged therapeutic effect. The extracellular domain of the tumor necrosis factor receptor has also been fused with the Fc, to be used as a drug for capturing excess tumor necrosis factor circulating in the plasma of patients in septic shock (Peppel et al. 1991).

Thus when it seems that an immunoglobulin domain possesses a property that would be desirable to confer upon another protein, the cross-linking or fusion of the two proteins or their fragments often effects the needed result.

II. Chemically Cross-Linked Conjugates as Models for Fusion Proteins

Chemically cross-linking an antibody to another protein is the most rapid and straightforward method for determining whether the production of a fusion product by recombinant DNA technology would be worthwhile. The cross-linking procedure can be carried out quickly and with little difficulty if both proteins are available, and, in contrast with the production of fusion proteins, the cross-linking procedure does not require cloned genes for each of the two proteins or the working out of effective bacterial or eukaryotic expression systems. Although this is not always the case, the chemically cross-linked model conjugate also often predicts attributes of the fusion protein.

Significant disadvantages of chemically cross-linked proteins – which recombinant fusion proteins do not share – include chemical heterogeneity, low yield, lack of stability, and immunogenicity associated with the cross-linking reagent's presence in the product. Chemical cross-linking reagents are often indiscriminate about the amino acid side chains with which they interact. For example, any of the ε-amino groups on the protein surface can participate. Also, should a reactive group be near an active site (on the antibody or the enzyme), the functional properties of either molecule can be altered.

The commonly used two-stage cross-linking reactions, which are designed to produce heterodimeric complexes, are often subject to side reactions that result in multimers. These multimers make necessary size fractionation steps that often reduce the ultimate yield of the conjugate. In my laboratory and those of my collaborators, for example, it has been difficult to obtain active heterodimers in yields in excess of 10% (by methods summarized below). Disulfide conjugates also lack stability on storage and in vivo, although there has been progress in refining cross-linking reactions to produce more stable conjugates. Finally, even though there has been no clear demonstration that cross-linking reagents increase immunogenicity, one would expect these protein-linked organic groups to behave like classic haptenic antigenic determinants.

1. Cross-Linked Antibody-Plasminogen Activator Conjugates

My laboratory's work on fibrin-specific plasminogen activators provides an example of the utility of antibody-targeted conjugates as models for fusion proteins. Reasoning that it was essential that the antigen combining site be specific for a component of the clot and not cross-react with soluble serum proteins or antigens present on endothelial cells, we selected fibrin as a target because it has antigenic epitopes that differentiate it from fibrinogen, its precursor in circulating plasma. Monoclonal antibody 59D8 (HUI et al.

1983), which is specific for an epitope exposed when thrombin catalyzes the conversion of fibrinogen to fibrin, is the cornerstone of our work in this area. Another monoclonal antibody of similar specificity, 64C5, was used in some of our initial studies (Hui et al. 1983; Bode et al. 1985, 1987b).

Antibody 59D8 (and 64C5) was raised in response to immunization with a peptide of the sequence GHRPLDK(C), which represents the seven NH_2-terminal residues of the β chain of fibrin combined with a COOH-terminal cysteine for cross-linking to keyhole limpet hemocyanin. The NH_2-terminal of the β chain appears to be conformationally protected in fibrinogen, as evidenced by the fact that there is essentially no cross-reactivity between fibrin and fibrinogen when tested with this antibody. Another important consideration in selecting a target on the thrombus is whether the epitope recognized by the antibody persists during clot dissolution. We have shown that, contrary to some reservations, the epitope recognized by 59D8 is lost from the clot (during in vitro fibrinolysis) at a rate identical to the rate of clot dissolution (Chen et al. 1992). Thus, epitope availability is sufficient for antibody binding throughout the course of fibrinolysis.

Holvoet et al. (1989) have confirmed the utility of targeting fibrin in their experiments with monoclonal antibodies that have little reactivity with fibrinogen but react with fragment D of noncross-linked fibrin or fragment D dimer of cross-linked fibrin.

We first showed that a conjugate of two chain urokinase plasminogen activator (tcuPA) and anti-fibrin antibody 64C5 substantially enhanced in vitro fibrinolysis in comparison with tcuPA (Bode et al. 1985). We then showed that tcuPA conjugated to the 64C5 Fab' was equally active (Bode et al. 1987b) and that single chain urokinase plasminogen activator (scuPA) could be used in a plasminogen activator-59D8 Fab' conjugate (Bode et al. 1990). Because the antigen for 59D8 and 64C5 is a hapten, it was possible to demonstrate unequivocally that the enhancement of fibrinolytic potency was solely due to the antigen-antibody reaction: a sufficient concentration of peptide GHRPLDK reduced the fibrinolytic activity of the conjugate to that of its urokinase plasminogen activator (uPA) parent (Bode et al. 1985).

Even the activity of tissue-type plasminogen activator (tPA), by itself fibrin-selective, could be enhanced by coupling it to a fibrin-specific antibody (Runge et al. 1988). In vivo results for a tPA-59D8 conjugate in a rabbit venous thrombosis model were very encouraging (Runge et al. 1987).

2. Methods for Synthesizing and Purifying Cross-Linked Conjugates

Hayzer et al. (1991) have summarized methods used by my research group and others to chemically link plasminogen activators to fibrin-specific antibodies. In brief, the heterobifunctional cross-linking reagent N-succinimidyl 3-(2-pyridyldithio)propionate (SPDP) is used to connect the two proteins through their lysine ε-amino groups. The reagent is considered heterobifunctional because during the reaction the attachment and the cross-linking steps are separate and involve different functional groups on each protein: in the attachment step a lysine amino group on one protein is modified by SPDP, and in the cross-linking step a bond is formed between the SPDP-modified protein and a sulfhydryl residue available on the other protein. The advantage of this two-step process is that it reduces the number of unwanted homoconjugates formed by the two proteins (CARLSSON et al. 1978).

Alternative approaches are to link SPDP to an amino group on one protein and then allow it to react with a cysteine residue on the other protein, or use SPDP to introduce disulfide groups into each protein. Mild reduction with dithiothreitol releases thiopyridine moieties to give free sulfhydryl groups while leaving the disulfide bonds of each protein intact. The two protein preparations are then mixed and allowed to react at room temperature, and excess thiol groups are blocked by alkylation.

BODE et al. (1992) have recently summarized methods for purifying antibody-plasminogen activator cross-linked conjugates. The chemistry should be controlled to the extent that predominantly 1:1 conjugates of whole antibody (or Fab′) and plasminogen activator are formed. After the reaction the mixture will always contain uncoupled antibody and uncoupled plasminogen activator in addition to the desired conjugate. At this point it is possible to use gel filtration to separate the desired conjugate from reactants and higher polymers, but often the molecular size of the desired product is close enough to that of the reactants for gel filtration to prove ineffective.

The use of two sequential affinity procedures, one selecting for active enzyme and the other for functionally intact antibody, has proved to be the method of choice for purifying active conjugate. For uPA (BODE et al. 1987a,b) and tPA (RUNGE et al. 1987, 1988), which are both serine proteases, the specific inhibitor benzamidine can be used as an immobilized ligand (HOLMBERG et al. 1976). Only enzyme-antibody conjugates in which the enzymatic center remains intact during the coupling procedure (and uncoupled intact enzyme from the reaction mixture) can bind and be retained by the benzamidine-Sepharose affinity column. Uncoupled whole antibody or uncoupled antibody fragments such as Fab have no affinity for benzamidine and are thus not retained on the column. To reduce nonspecific binding on the affinity column to a minimum, a high salt washing step is performed before elution. Free antibody in the nonbinding fraction from the column can then be collected and reused for conjugation to enhance yield. The desired conjugate (and uncoupled enzyme) is eluted by a mild change in pH.

Immunoaffinity chromatography is an alternative to chromatography based on catalytic site specificity. An antibody to one component of a conjugate is immobilized on a cyanogen bromide-activated Sepharose matrix and the desired conjugate (and one of the reactants) is captured and subsequently eluted by altering buffer pH. A second method of separation is then required to differentiate the remaining reactant from the conjugate. An application of this approach is described by BODE et al. (1991). If the conjugate contains the Fc portion of immunoglobulin G, immobilized protein A can be used for the separation (SOLOMON et al. 1992).

III. Fusion Protein Construction

Building on the pioneering work of Neuberger and coworkers (NEUBERGER et al. 1984; WILLIAMS and NEUBERGER 1986) and on our experience with cross-linked antibody-enzyme model proteins, we then used recombinant technology to create a fusion protein with the activities of anti-fibrin antibody 59D8 and tPA (SCHNEE et al. 1987; LOVE et al. 1989). We assembled the fusion protein by joining the gene coding for the 59D8 immunoglobulin heavy chain to the gene coding for tPA and transfecting this chimeric construct into a hybridoma cell capable of producing only immunoglobulin light chain. The resulting cell line produced a bifunctional protein with domains for activating plasminogen and attaching to cross-linked fibrin. Using a part of the uPA gene, RUNGE et al. (1991) have described a more recent application of these methods.

1. Cloning the Rearranged Immunoglobulin Gene

During the somatic development of the B cell, germline variable (V) and joining (J) regions are juxtaposed to produce a unique rearranged VJ sequence that codes for the antigen combining site of the heavy or light chain of the immunoglobulin. Also during this process the heavy chain incorporates a diversity (D) segment at the joining region (TONEGAWA 1983). These rearrangements result in a unique VDJ or VJ exon sequence that codes for the antigen the antibody recognizes. Unfortunately, expression of a particular immunoglobulin chain can only occur after the unique rearrangement has been cloned. In constructing the expression plasmid coding for the 59D8-tPA fusion protein, we elected to clone and incorporate the VDJ segment of the rearranged heavy chain of 59D8. It was unnecessary to clone or otherwise manipulate the light chain gene because we transfected heavy chain constructs into cell lines that had lost the ability to express normal heavy chains but retained the ability to produce light chains. It is possible to obtain heavy chain VDJ sequence either by cloning complementary DNAs (cDNAs) constructed from the mRNA of the hybridoma or by cloning genomic sequences from high molecular weight DNA derived from the hybridoma; although we initially used genomic sequences, we later learned that expression plasmids could be constructed with equal facility by using cDNAs.

2. Constructing the Expression Vector

A previously described immunoglobulin expression vector (pSV2gpt) (NEAR et al. 1990) was adapted to a cassette format to facilitate expression of recombinant immunoglobulin. The pUC12 polylinker was inserted into the

EcoRI-PstI site of pSV2gpt and an XbaI fragment containing the mouse γ2b heavy chain constant region (TUCKER et al. 1979) was inserted into the polylinker. The cloned 59D8 rearrangement was inserted with a unique EcoRI site 5' of the constant region, which produced sequence coding for a complete 59D8 heavy chain.

tPA is secreted as a single chain 70 kDa protein that is subsequently cleaved by plasmin to form A and B chains attached by a single disulfide bond (HARRIS 1987; GETHING et al. 1988). Each chain contains multiple intrachain disulfide bridges. Because we were interested only in the catalytic function of the protein, we used sequence coding only for the B chain. To construct the 59D8 heavy chain-tPA fusion protein, we removed the γ2b constant region 3' of the XhoI site in the hinge region and replaced it with tPA B chain cDNA sequence. An SfaNI site in the tPA cDNA was converted to an XhoI site by the addition of a synthetic oligonucleotide adapter (which also added a glycine residue to the fusion product). The cDNA fragment was thus inserted in-frame into the XhoI site in the heavy chain hinge region, creating a new hybrid exon. No attempt was made to modify the 3' portion of the tPA cDNA, which contained a polyadenylation signal.

3. Selecting Loss Variant Cell Lines

To select for heavy chain loss variant hybridomas, we first grew the cells in soft agarose containing goat anti-mouse heavy chain antiserum. Clusters secreting heavy chain developed halos. Cells without halos were picked and subcloned.

4. Transfecting the Expression Plasmid

The hybrid heavy chain construct was transfected by electroporation into the heavy chain loss variant hybridoma line described above. Selection occurred in medium containing xanthine, hypoxanthine and mycophenolic acid.

5. Purifying and Analyzing Protein

Protein was purified from the cell supernatants and from ascites by sequential double affinity chromatography (TONEGAWA 1983). One column consisted of the GHRPLDK(C) peptide (used to generate 59D8) linked to Sepharose; the other of an anti-human tPA monoclonal antibody (TCL8) linked to Sepharose.

6. Recombinant Protein Expression Levels

Although the structure and function of the tPA-59D8 recombinant protein were as we had anticipated, the amount of protein secreted into the cell culture supernatant was only 1% the amount secreted by the original hybridoma cell line. In more recent work, we have learned that protein production can be greatly enhanced by modifying the 3′ untranslated region of the mRNA. LOVE et al. (1993) found that cell lines transfected with constructs in which the 3′ untranslated region was coded by plasminogen activator genes produced very low levels of both mRNA and protein (0.008–0.06 μg/ml) in comparison with the parental 59D8 myeloma cell line (7.6–10 μg/ml). In vitro nuclear run-off analysis indicated that these low steady state levels of mRNA did not result from a lower rate of transcription of the transfected gene (relative to the rate of transcription of the endogenous heavy chain gene in the 59D8 parent cells). In contrast, cell lines transfected with expression plasmids in which the 3′ untranslated region of the mouse γ2b heavy chain or human β-globin gene had been substituted for the 3′ untranslated region of the plasminogen activator gene showed an increase in recombinant protein secretion of 68- to 100-fold.

IV. Structural and Functional Properties of Specific Fusion Proteins

1. Antibody-Plasminogen Activator Fusion Proteins

Although the 59D8-tPA fusion protein (SCHNEE et al. 1987) described in Sect. III retained both the fibrin antigen-binding and plasminogen activator activities of its parents, we were disappointed by the fusion protein's ability to lyse plasma clots. If eventually became apparent that the tPA catalytic site had been an inappropriate choice – the loss in the construct of the NH$_2$-terminal portion of the protein prevented the requisite enhancement of catalytic activity on fibrin binding. With this experience in mind, we reasoned that the catalytic site of uPA might function better in a fusion protein because its activity is independent of fibrin binding. We chose to use the single chain form, scuPA, because it has the additional advantage of being resistant to inactivation by plasminogen activator inhibitor-1 and α$_2$-antiplasmin. As it traveled through plasma, a fusion protein containing scuPA might resist circulating inhibitors and remain incapable of activating circulating plasminogen until it reached the plasmin-rich environment of the thrombus, where it would become active through cleavage of the plasmin-susceptible Lys-158-Ile-159 peptide bond (DECLERCK et al. 1990).

    To reduce the mass of the chimeric protein to essential components, we decided to include only the Fab part of the anti-fibrin antibody. In a similar vein, we omitted the uPA kringle and growth factor regions and used the sequence of low molecular weight (32 kDa) scuPA, as described by STUMP et al. (1986), which is reported to be as active in fibrinolysis as the intact molecule. We had also initially included the CH3 domain of the antibody heavy chain as a spacer between the antibody and the plasminogen activator; later experience (Shaw, unpublished observations) indicated that this was not necessary.

    The fusion protein contained antibody 59D8 heavy chain from residues 1 to 351 and, in contiguous peptide sequence, residues 144–411 of low molecular weight scuPA (RUNGE et al. 1991). We included the 3' untranslated region from β-globin for reasons described in Sect. III.

To assemble a heterodimer that included this fusion protein and an immunoglobulin light chain, we transfected the fusion protein expression plasmid into heavy-chain loss variants (as described in Sect. III). SDS gel electrophoresis, immunoblot analysis, and DNA sequencing showed that the product, scuPA (32 kDa)-59D8, was a disulfide-linked 103 kDa heterodimer consisting of an immunoglobulin light chain linked to the fusion protein heavy chain.

The K_m of scuPA (32 kDa)-59D8 was 16.6 μM, that of tcuPA 9.1 μM. Fibrin binding was also similar between the recombinant protein and its parent. In an in vitro plasma clot assay, scuPA (32 kDa)-59D8 was six times more potent than scuPA, with considerably diminished fibrinogen degradation and α_2-antiplasmin inactivation in the supernatant. In vivo, in the rabbit jugular vein model, scuPA (32 kDa)-59D8 was 20 times more potent than scuPA (Fig. 1). It should be noted that some of this enhancement in in vivo activity must, in part, have been related to a fivefold increase in the half-life of scuPA (32 kDa)-59D8 (in comparison with scuPA) in the rabbit.

Runge (unpublished observations) has since studied the in vivo activities of scuPA (32 kDa)-59D8, scuPA, and tPA in a baboon model that allows comparison of both thrombolytic potency and inhibition of thrombus deposition in relation to the plasma concentration of each plasminogen activator. scuPA (32 kDa)-59D8 was approximately 8-10-fold more potent than

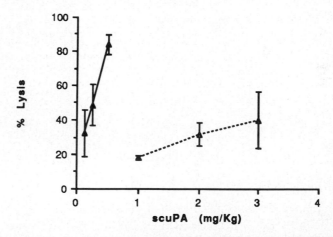

Fig. 1. Thrombolysis in vivo with scuPA (*dotted line*) and scuPA (32 kDa)-59D8 (*solid line*). Data represent the means of values from between three and eight animals at each point. The 20-fold increase in potency derived for scuPA (32 kDa)-59D8 was calculated by comparing the percent lysis curves in plasma clot and rabbit jugular vein assays, which were fit using a two-parameter exponential function (RUNGE et al. 1988). (From RUNGE et al. 1991)

tPA and 15-20-fold more potent than scuPA in the lysis of thrombi and about 11 times more potent than scuPA in the inhibition of thrombin deposition.

Of equally great interest is the observation that, at equipotent thrombolytic doses, template bleeding times for scuPA (32 kDa)-59D8 in the baboon were unchanged whereas those for tPA and scuPA were significantly prolonged. Since bleeding time prolongation seems to reflect the risk of clinical hemorrhage (GIMPLE et al. 1989), it would be of interest to determine whether scuPA (32 kDa)-59D8, in addition to being more potent, might also be safer.

HOLVOET et al. (1991) have further refined this concept by constructing an M_r 57000, single chain, chimeric plasminogen activator consisting of a 33 kDa fragment of scuPA and an Fv derived from a fibrin-specific antibody directed at a cross-linked epitope of the D dimer. This single chain molecule was expressed by baculovirus-infected cells of the insect *Spodoptera frugiperda*. The recombinant molecule showed high affinity for binding the fibrin D dimer fragment (essentially identical to that of the parent antibody molecule) and a very similar Michaelis-Menten constant for activating plasminogen. When tested in the lysis of a plasma clot in vivo, the recombinant single chain molecule was 13 times more potent than low molecular weight scuPA.

Fibrin is but one of the epitopes contained within a thrombus that may be a potential target for the development of antibody-plasminogen activator fusion proteins. Other candidates include epitopes on platelets (BODE et al. 1991; DEWERCHIN et al. 1991) and fibrin-linked α_2-antiplasmin (REED et al. 1990).

2. A Model Minimal Size Fusion Protein

As knowledge about antibody structure increased it became apparent that the 50 kDa Fab (PORTER 1959) comprised two domains, only one of which participated in antigen binding. INBAR et al. (1972) separated this domain, the 25 kDa Fv, by limited pepsin digestion of the dinitrophenol-binding myeloma protein MOPC 315. They showed that this fragment contained the variable regions of the heavy and light chains held together only by noncovalent interactions. Fv possessed the same binding affinity for dinitrophenol as did the parent immunoglobulin molecule.

3. Fv and Single Chain Fv

Using myeloma cells transfected with a plasmid coding for the variable domains of the light and heavy chains of an anti-lysozyme antibody, RIECHMANN et al. (1988) expressed a heterodimer comprising both variable domains, i.e., an Fv, that was shown to bind lysozyme. SKERRA and PLÜCKTHUN (1988) also expressed a functional Fv in *Escherichia coli*. These heterodimers were associated by noncovalent interactions: there were no stabilizing interchain disulfide bridges. Because the complementarity determining regions form about one quarter of the contacts at the interface between these domains (CHOTHIA et al. 1985), the stability of the hetero-dimer, despite considerable conservation of framework structure at the interface (NOVOTNÝ and HABER 1985), would in all likelihood vary with every unique antibody sequence. Indeed, the Fv heavy and light chain variable domains have been shown in two studies to dissociate readily (K_D of 10^{-5}–$10^{-8}M$) (GLOCKSHUBER et al. 1990; SKERRA et al. 1991), whereas in the case of a high affinity digoxin-specific Fv the two variable domains have been shown to be stably associated (ANTHONY et al. 1992). Because the concentrations of Fv necessary for pharmacologic applications would often be lower than $10^{-8}M$, it is reasonable to expect that dissociation of the light and heavy chain components would result in a loss of antigen-binding function.

The Fv has been stabilized by expressing a gene encoding the variable regions of both the heavy and light chains connected by an oligonucleotide encoding an unstructured linker peptide. The resulting single chain Fv is a contiguous polypeptide chain in which either the COOH-terminal of the heavy chain segment is connected by a peptide to the NH$_2$-terminal of the light chain (HUSTON et al. 1988) or the COOH-terminal of the light chain is connected to the NH$_2$-terminal of the heavy chain (BIRD et al. 1988).

4. Minimal Fv-Containing Fusion Protein

To explore the minimal structure that might be a prototype for a bifunctional fusion protein containing both an antigen-binding site and another func-tionality, TAI et al. (1990) combined the digoxin-binding antibody 26-10 single chain Fv with the immunoglobulin-binding fragment (FB) of staphy-lococcal protein A in a single polypeptide. By incorporating both binding sites into a 33 kDa molecule, it was possible for them to test whether ligand binding at one site interfered with the other, on the assumption that the most economically designed targetable proteins should consist of the smallest structures necessary for selective delivery and effector function.

The 26-10 Fv gene was joined with a gene encoding the FB to make a 927 base sequence. The pair of synthetic oligonucleotides used in the fusion resulted in the addition of only three extra amino acids (Ser-Asp-Pro) between the FB and the Fv sequences. The assembled gene was incor-porated into an expression vector and the plasmid was used to transform *E. coli*. Figure 2A shows a circular map of the expression plasmid, and Fig. 2B shows the linear arrangement of the protein domains.

The affinity for digoxin of the fusion protein (K_a 2.5 $\times$ $10^9 M^{-1}$) was indistinguishable from that of the parent antibody (K_d 2.4 $\times$ $10^9 M^{-1}$), and the fine specificity profiles of the two species were also similar. The affinity of the fusion protein for ^{125}I-labeled human immunoglobulin G (K_a 3.2 $\times$ $10^7 M^{-1}$) was similar to that of the recombinant 58 residue FB. These similarities show that each of the two functional sites in the fusion protein

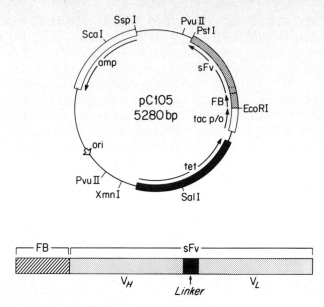

Fig. 2. *Top* Circular map of FB-anti-digoxin 26-10 single chain Fv (sFv) expression plasmid. *Bottom* Linear arrangement of protein domains and linker in the FB-sFv polypeptide chain. (From TAI et al. 1990)

was able to bind its corresponding ligand with appropriate affinity and specificity; both sites were able to act simultaneously in cross-linking the two ligands, digoxin and immunoglobulin G.

5. Prodrug Activation

Often a drug that might be useful in therapy is too toxic to be administered systemically. A means of targeting such a toxin to a lesion would accrue two advantages: the drug's localized concentration would be sufficient to render it effective, and its systemic concentration would be below toxic limits. Particularly cogent applications of targeted toxins or chemotherapeutic drugs are found in cancer chemotherapy, in which agents potent in killing tumor cells are also highly toxic to normal cells – especially in tissues characterized by high proliferative rates such as the bone marrow or gastrointestinal epithelium.

Antibodies selective for epitopes on tumor cell membranes have been used to target toxins directly to tumor cells, but an application more consistent with the thrust of this chapter is the targeting of enzymes that cleave inactive (and thereby nontoxic) prodrugs into active cytotoxic agents. Since the antibody concentrates the enzyme at the site of the tumor, the conversion of prodrug to active drug occurs preferentially at this site, thereby sparing normal tissue from a high concentration of the cytotoxic agent. A general strategy envisions the administration of an antibody-enzyme conjugate systemically. After enough time has elapsed for the conjugate to bind to antigens on the tumor cell and, most importantly, to clear from nontarget tissues and thereby avoid systemic toxicity, a prodrug is administered that is catalytically converted into an active antitumor agent by the antibody-targeted enzyme (HELLSTRÖM and SENTER 1991).

A singular advantage of this method not shared by the use of toxins directly targeted to tumor cells is that if all tumor cells in a mass do not possess the antigen for which the monoclonal antibody is selective, it should still be possible to attain a cytotoxic concentration of active drug in the vicinity of these cells. One of its disadvantages is that substances of high molecular weight, such as antibodies, do not diffuse readily into the re-latively avascular interior of a tumor. However, because the active antitumor agent released from the prodrug is of low molecular weight, it should diffuse readily from the vascular region (where it was catalyzed by the antibody-enzyme fusion protein) to the avascular interior of the tumor.

Choosing the appropriate enzyme is critical to the success of antibody-targeted prodrug activation. While it would be desirable to use human enzymes because of their lack of immunogenicity, poor selectivity may rule against their use. For example, SENTER et al. (1988, 1989) used human alkaline phosphatase linked to a cancer-selective antibody to activate etoposide phosphate at the tumor surface. However, endogenous alkaline phosphatase – in addition to the antibody-linked enzyme – also activated etoposide phosphate, thereby significantly reducing the specificity of the method. Although nonmammalian enzymes are likely to be immunogenic, they possess the potential of interacting with substrates that are resistant to all mammalian enzymes, thus affording a very high degree of selectivity to the enzyme-substrate interaction (KERR et al. 1990; SENTER et al. 1991; SPRINGER et al. 1991; BIGNAMI et al. 1992; SVENSSON et al. 1992).

Antibody-enzyme fusion proteins for prodrug targeting have been pro-duced by recombinant methods analogous to those described above. In general, it is possible to preserve the activity of both the enzyme and the antigen combining site and to humanize the nonmammalian enzyme to diminish problems of immunogenicity (BOSSLET et al. 1992).

The success of antibody-targeted prodrug activation will ultimately depend on the stability of the drug in vivo, the difference in cytotoxicity between the prodrug and the drug, the pharmacokinetics of the prodrug, the tumor selectivity of the antibody, the turnover rate of the enzyme, the retention of the conjugate at the tumor site in contrast with that at normal tissues, the presence of an endogenous enzyme that can activate the prodrug, and the potential immunogenicity of the antibody-enzyme complex (HELLSTRÖM and SENTER 1991).

C. Bispecific Antibodies

I. Development of the Concept of a Bispecific Antibody

An alternative to constructing a molecule that contains a binding (antibody) and a functional (plasminogen activator) domain is to assemble a structure that contains two antigen combining sites, one binding a target in the thrombus, such as fibrin, the other binding a plasminogen activator. Because tPA and uPA are present in low concentrations in normal plasma, we reasoned that a bispecific antibody capable of binding fibrin and either tPA or uPA might concentrate these endogenous plasminogen activators at the site of a fibrin deposit and thereby induce fibrinolysis without a need for exogenous plasminogen activator. Should the concentration of endogenous plasminogen activator be too low at the thrombus, this intrinsic concentration could be supplemented by administration of exogenous plasminogen activator. The necessary dose of administered plasminogen activator should be much lower than that ordinarily required in therapy.

II. Chemically Cross-Linked Bispecific Antibodies as Models

We first tested the feasibility of this approach with antibodies to tPA or uPA chemically cross-linked to 59D8, the fibrin-specific antibody discussed above (BODE et al. 1989; CHARPIE et al. 1990). A significant enhancement in fibrinolytic activity was demonstrated both in vitro and in vivo with the 59D8-anti-tPA conjugate (BODE et al. 1989) and in vitro with the 59D8-anti-uPA conjugate (CHARPIE et al. 1990). A conjugate of the anti-tPA Fab' and the Fab' of 59D8 was also as effective as the intact antibody conjugate (RUNGE et al. 1990). SAKHAROV et al. (1988) have demonstrated a tenfold enhancement in in vitro plasma clot lysis with cross-linked antibodies specific for fibrin and uPA.

III. Cell Fusion in the Production of Bispecific Antibodies

A more elegant method for producing bispecific antibodies, which avoids the disadvantages of chemical cross-linking, is to use somatic cell fusion to produce bivalent antibodies that possess two different antigen combining sites. This method was first elaborated by MILSTEIN and CUELLO (1983) and has been applied more recently by my laboratory in the design of bifunctional antibodies that bind both tPA and fibrin (BRANSCOMB et al. 1990).

Immunoglobulin G is a symmetrical molecule possessing two antigen combining sites of the same specificity. Since the component chains of the molecule are assembled after their individual biosynthesis, it is possible to obtain molecules of mixed specificity from cells that are synthesizing two

different antibodies (SURESH et al. 1986). In my laboratory, somatic cell
fusion between the hybridoma cell line secreting anti-fibrin antibody 59D8
(BRANSCOMB et al. 1990) and another hybridoma line secreting anti-tPA
antibody TCL8 resulted in a cell line that produced a mixture of antibodies.
In addition to the antibodies characteristic of the parental lines and inactive immunoglobulins, an antibody was secreted that was able to bind
both tPA and fibrin. This antibody was isolated from mixtures in culture
medium or ascites by two steps of affinity chromatography, using tPA as
immobilized antigen in the first and a peptide at the NH$_2$-terminal of the
fibrin β chain (GHRPLDK(C); BRANSCOMB et al. 1990) in the second. This
bispecific antibody bound simultaneously to tPA and fibrin in a solid phase
immunoassay.

IV. Functional Properties of Bispecific Antibodies

I expected that an antibody specific for both fibrin and tPA would bind to
the fibrin matrix of a thrombus but would also bind to tPA in plasma,
thereby increasing the concentration of tPA at the surface of the thrombus.
This expectation was confirmed by in vitro clot lysis experiments showing
that the 59D8-TCL8 bispecific antibody enhanced the potency of tPA 14-
fold when added to the assay system before tPA and 22-fold when mixed
with tPA to form an immunoconjugate before addition to the assay system.
When the 59D8-TCL8 bispecific antibody was tested in vivo in the rabbit
jugular vein model, a 1.6-fold enhancement in fibrinolytic activity was
observed (BRANSCOMB et al. 1990). KUROKAWA et al. (1989, 1990) have
extended this work and produced by cell fusion bispecific antibodies that
bind uPA and fibrin and tPA and fibrin.

The production of bifunctional antibodies by somatic cell fusion is
severely limited by product yield. Because the method depends on a stochastic assortment of immunoglobulin chains within the fused cell, none of
the products are secreted in large amounts. In the future it is likely that
bifunctional antibodies will be produced as fusion proteins by recombinant
DNA methods.

D. Conclusion

Advances in molecular biology allow almost limitless variation in the construction of new proteins. For a given pharmacologic application, functional
domains from molecules considered to complement one another can be
joined into a single molecule. One of the most useful of these applications is
perhaps the targeting of an enzyme to a site where its catalytic function is
essential – without exposing the entire organism to the unneeded impact of
the intervention.

References

Anthony J, Near R, Wong S-L, Iida E, Ernst E, Wittekind M, Haber E, Ng S-C
(1992) Production of stable anti-digoxin Fv in *Escherichia coli*. Mol Immunol
29:1237–1247

Bignami GS, Senter PD, Grothaus PG, Fischer KJ, Humphreys T, Wallace PM
(1992) N-(4'-hydroxyphenylacetyl)palytoxin: a palytoxin prodrug that can be
activated by a monoclonal antibody-penicillin G amidase conjugate. Cancer Res
52:5759–5764

Bird RE, Hardman KD, Jacobson JW, Johnson S, Kaufman BM, Lee S-M, Lee T,
Pope SH, Riordan GS, Whitlow M (1988) Single-chain antigen-binding proteins.
Science 242:423–426

Bode C, Matsueda GR, Hui KY, Haber E (1985) Antibody-directed urokinase: a
specific fibrinolytic agent. Science 229:765–767

Bode C, Runge MS, Newell JB, Matsueda GR, Haber E (1987a) Characterization of
an antibody-urokinase conjugate: a plasminogen activator targeted to fibrin. J
Biol Chem 262:10819–10823

Bode C, Runge MS, Newell JB, Matsueda GR, Haber E (1987b) Thrombolysis by a
fibrin-specific antibody Fab'-urokinase conjugate. J Mol Cell Cardiol 19:335–341

Bode C, Runge MS, Branscomb EE, Newell JB, Matsueda GR, Haber E (1989)
Antibody-directed fibrinolysis: an antibody specific for both fibrin and tissue
plasminogen activator. J Biol Chem 264:944–948

Bode C, Runge MS, Schönermark S, Eberle T, Newell JB, Kübler W, Haber E
(1990) Conjugation to antifibrin Fab' enhances fibrinolytic potency of single-
chain urokinase plasminogen activator. Circulation 81:1974–1980

Bode C, Meinhardt G, Runge MS, Freitag M, Nordt T, Arens M, Newell JB, Kübler
W, Haber E (1991) Platelet-targeted fibrinolysis enhances clot lysis and inhibits
platelet aggregation. Circulation 84:805–813

Bode C, Runge MS, Haber E (1992) Purifying antibody-plasminogen activator conjugates. Bioconjugate Chem 3:269–272

Bosslet K, Czech J, Lorenz P, Sedlacek HH, Schuermann M, Seemann G (1992)
Molecular and functional characterization of a fusion protein suited for tumour
specific prodrug activation. Br J Cancer 65:234–238

Branscomb EE, Runge MS, Savard CE, Adams KM, Matsueda GR, Haber E (1990)
Bispecific monoclonal antibodies produced by somatic cell fusion increase the
potency of tissue plasminogen activator. Thromb Haemost 64:260–266

Carlsson J, Drevin H, Axen R (1978) Protein thiolation and reversible protein-
protein conjugation. Biochem J 173:723–737

Casscells W, Lappi DA, Olwin BB, Wai C, Siegman M, Speir EH, Sasse J, Baird A
(1992) Elimination of smooth muscle cells in experimental restenosis: targeting
of fibroblast growth factor receptors. Proc Natl Acad Sci USA 89:7159–7163

Charpie JR, Runge MS, Matsueda GR, Haber E (1990) A bispecific antibody
enhances the fibrinolytic potency of single-chain urokinase. Biochemistry 29:
6375–6378

Chen F, Haber E, Matsueda GR (1992) Availability of the Bβ(15–21) epitope on
cross-linked human fibrin and its plasmic degradation products. Thromb
Haemost 67:335–340

Chothia C, Novotný J, Bruccoleri R, Karplus M (1985) Domain association in
immunoglobulin molecules. The packing of variable domains. J Mol Biol
186:651–663

Declerck PJ, Lijnen HR, Verstreken M, Moreau H, Collen D (1990) A monoclonal
antibody specific for two-chain urokinase-type plasminogen activator. Application to the study of the mechanism of clot lysis with single-chain urokinase-type
plasminogen activator in plasma. Blood 75:1794–1800

Dewerchin M, Lijnen HR, Stassen JM, De Cock F, Quertermous T, Ginsberg MH,
Plow FF, Collen D (1991) Effect of chemical conjugation of recombinant single-

chain urokinase-type plasminogen activator with monoclonal antiplatelet antibodies on platelet aggregation and on plasma clot lysis in vitro and in vivo. Blood 78:1005–1018

Gething M-J, Adler B, Boose J-A, Gerard RD, Madison EL, McGookey D, Meidell RS, Roman LM, Sambrook J (1988) Variants of human tissue-type plasminogen activator that lack specific structural domains of the heavy chain. EMBO J 7:2731–2740

Gimple LW, Gold HK, Leinbach RC, Coller BS, Werner W, Yasuda T, Johns JA, Ziskind AA, Finkelstein D, Collen D (1989) Correlation between template bleeding times and spontaneous bleeding during treatment of acute myocardial infarction with recombinant tissue-type plasminogen activator. Circulation 80:581–588

Glockshuber R, Malia M, Pfitzinger I, Plückthun A (1990) A comparison of strategies to stabilize immunoglobulin Fv-fragments. Biochemistry 39:1362–1367

Haber E, Quertermous T, Matsueda GR, Runge MS (1989) Innovative approaches to plasminogen activator therapy. Science 243:51–56

Harris TJ (1987) Second-generation plasminogen activators. Protein Eng 1:449–458

Hayzer DJ, Lubin IM, Runge MS (1991) Conjugation of plasminogen activators and fibrin-specific antibodies to improve thrombolytic therapeutic agents. Bioconjugate Chem 2:301–308

Hellström KE, Senter PD (1991) Activation of prodrugs by targeted enzymes. Eur J Cancer 27:1342–1343

Holmberg L, Bladh B, Astedt B (1976) Purification of urokinase by affinity chromatography. Biochim Biophys Acta 445:215–222

Holvoet P, Stassen JM, Hashimoto Y, Spriggs D, Devos P, Collen D (1989) Binding properties of monoclonal antibodies against human fragment D-dimer of cross-linked fibrin to human plasma clots in an in vivo model in rabbits. Thromb Haemost 61:307–313

Holvoet P, Laroche Y, Lijnen HR, Van Cauwenberge R, Demarsin E, Brouwers E, Matthyssens G, Collen D (1991) Characterization of a chimeric plasminogen activator consisting of a single-chain Fv fragment derived from a fibrin fragment D-dimer-specific antibody and a truncated single-chain urokinase. J Biol Chem 266:19717–19724

Hui KY, Haber E, Matsueda GR (1983) Monoclonal antibodies to a synthetic fibrin-like peptide bind to human fibrin but not fibrinogen. Science 222:1129–1132

Huston JS, Levinson D, Mudgett-Hunter M, Tai M-S, Novotný J, Margolies MN, Ridge RJ, Bruccoleri RE, Haber E, Crea R, Oppermann H (1988) Protein engineering of antibody binding sites: recovery of specific activity in an anti-digoxin single-chain Fv analogue produced in Escherichia coli. Proc Natl Acad Sci USA 85:5879–5883

Inbar D, Hochman J, Givol D (1972) Localization of antibodycombining sites within the variable portions of heavy and light chains. Proc Natl Acad Sci USA 69:2659–2662

Kerr DE, Senter PD, Burnett WV, Hirschberg DL, Hellström I, Hellström KE (1990) Antibody-penicillin-V-amidase conjugates kill antigen-positive tumor cells when combined with doxorubicin phenoxyacetamide. Cancer Immunol Immunother 31:202–206

Korngold L, Pressman D (1954) The localization of antilymphosarcoma antibodies in the Murphy lymphosarcoma of the rat. Cancer Res 14:96–99

Kurokawa T, Iwasa S, Kakinuma A (1989) Enhanced fibrinolysis by a bispecific monoclonal antibody reactive to fibrin and tissue plasminogen activator. Bio-technology 7:1163–1167

Kurokawa T, Iwasa S, Kahinuma A (1990) Enhancement of fibrinolysis by bispecific monoclonal antibodies reactive to fibrin and plasminogen activators. Thromb Res Suppl 10:83–89

Love TW, Runge MS, Haber E, Quertermous T (1989) Recombinant antibodies possessing novel effector functions. Methods Enzymol 178:515–527

Love TW, Quertermous T, Zavodny PJ, Runge MS, Chou C-C, Mullins D, Huang PL, Schnee JM, Kestin AS, Savard CE, Michelson KD, Matsueda GR, Haber E (1993) High-level expression of antibody-plasminogen activator fusion proteins in hybridoma cells. Thromb Res 69:221–229

Milstein C, Cuello AC (1983) Hybrid hybridomas and their use in immunohisto-chemistry. Nature 305:537–540

Near RI, Ng SC, Mudgett-Hunter M, Hudson NW, Margolies MN, Seidman JG, Haber E, Jacobson MA (1990) Heavy and light chain contributions to antigen binding in an antidigoxin chain recombinant antibody produced by transfection of cloned anti-digoxin antibody genes. Mol Immunol 27:901–909

Neuberger MS, Williams GT, Fox RO (1984) Recombinant antibodies possessing novel effector functions. Nature 312:604–608

Novotný J, Haber E (1985) Structural variants of antigen binding: comparison of immunoglobulin V_L-V_H and V_L-V_L domain dimers. Proc Natl Acad Sci USA 82:4592–4596

Peppel K, Crawford D, Beutler B (1991) A tumor necrosis factor (TNF) receptor-IgG heavy chain chimeric protein as a bivalent antagonist of TNF activity. J Exp Med 174:1483–1489

Porter RR (1959) The hydrolysis of rabbit γ-globulin and antibodies with crystalline papain. Biochem J 73:119–126

Pressman D, Korngold L (1953) The in vivo localization of anti-Wagner-osteogenic-sarcoma antibodies. Cancer 6:619–623

Reed GL III, Matsueda GR, Haber E (1990) Synergistic fibrinolysis: combined effects of plasminogen activators and an antibody that inhibits α_2-antiplasmin. Proc Natl Acad Sci USA 87:1114–1118

Riechmann L, Foote J, Winter G (1988) Expression of an antibody Fv fragment in myeloma cells. J Mol Biol 203:825–828

Runge MS, Bode C, Matsueda GR, Haber E (1987) Antibody-enhanced thrombolysis: targeting of tissue plasminogen activator in vivo. Proc Natl Acad Sci USA 84:7659–7662

Runge MS, Bode C, Matsueda GR, Haber E (1988) Conjugation to an antifibrin monoclonal antibody enhances the fibrinolytic potency of tissue plasminogen activator in vitro. Biochemistry 27:1153–1157

Runge MS, Bode C, Savard CE, Matsueda GR, Haber E (1990) Antibody-directed fibrinolysis: a bispecific (Fab')2 that binds to fibrin and tissue plasminogen activator. Bioconjugate Chem 1:274–277

Runge MS, Quertermous T, Zavodny PJ, Love TW, Bode C, Freitag M, Shaw S-Y, Huang PL, Chou C-C, Mullins D, Schnee JM, Savard CE, Rothenberg ME, Newell JB, Matsueda GR, Haber E (1991) A recombinant chimeric plasminogen activator with high affinity for fibrin has increased thrombolytic potency in vitro and in vivo. Proc Natl Acad Sci USA 88:10337–10341

Sakharov DV, Sinitsyn VV, Kratasjuk GA, Popov NV, Domogatsky SP (1988) Two-step targeting of urokinase to plasma clot provides efficient fibrinolysis. Thromb Res 49:481–488

Schnee JM, Runge MS, Matsueda GR, Hudson NW, Seidman JG, Haber E, Quertermous T (1987) Construction and expression of a recombinant antibody-targeted plasminogen activator. Proc Natl Acad Sci USA 84:6904–6908

Senter PD, Saulnier MG, Schreiber GJ, Hirschberg DL, Brown JP, Hellström I, Hellström KE (1988) Anti-tumor effects of antibody-alkaline phosphatase conjugates in combination with etoposide phosphate. Proc Natl Acad Sci USA 85:4842–4846

Senter PD, Schreiber GJ, Hirschberg DL, Ashe SA, Hellström KE, Hellström I (1989) Enhancement of the in vitro and in vivo antitumor activities of phosphorylated mitomycin C and etoposide derivatives by monoclonal antibody-alkaline phosphatase conjugates. Cancer Res 49:5789–5792

Senter PD, Su PC, Katsuragi T, Sakai T, Cosand WL, Hellström I, Hellström KE (1991) Generation of 5-fluorouracil from 5-fluorocytosine by monoclonal antibody-cytosine deaminase conjugates. Bioconjugate Chem 2:447–451

Skerra A, Plückthun A (1988) Assembly of a functional immunoglobulin Fv fragment in *Escherichia coli*. Science 240:1038–1041

Skerra A, Pfitzinger I, Plückthun A (1991) The functional expression of antibody Fv fragments in *Escherichia coli*. Improved vectors and a generally applicable purification technique. Biotechnology 9:273–278

Solomon B, Raviv O, Leibman E, Fleminger G (1992) Affinity purification of antibodies using immobilized FB domain of protein A. J Chromatogr 597: 257–262

Springer CJ, Bagshawe KD, Sharma SK, Searle F, Boden JA, Antoniw P, Burke PJ, Rogers GT, Sherwood RF, Melton RG (1991) Ablation of human choriocarcinoma xenografts in nude mice by antibody-directed enzyme prodrug therapy (ADEPT) with three novel compounds. Eur J Cancer 27:1361–1366

Stump DC, Lijnen HR, Collen D (1986) Purification and characterization of a novel low molecular weight form of single-chain urokinase-type plasminogen activator. J Biol Chem 261:17120–17126

Suresh MR, Cuello AC, Milstein C (1986) Bispecific monoclonal antibodies from hybrid hybridomas. Methods Enzymol 121:210–228

Svensson HP, Kadow JF, Vrudhula VM, Wallace PM, Senter PD (1992) Monoclonal antibody-beta-lactamase conjugates for the activation of a cephalosporin mustard prodrug. Bioconjugate Chem 3:176–181

Tai M-S, Mudgett-Hunter M, Levinson D, Wu G-M, Haber E, Oppermann H, Huston JS (1990) A bifunctional fusion protein containing Fc-binding fragment B of staphylococcal protein A amino-terminal to antidigoxin single-chain Fv. Biochemistry 29:8024–8030

Till MA, Ghetie V, Gregory T, Patzer EJ, Porter JP, Uhr JW, Capon DJ, Vitetta ES (1988) HIV-infected cells are killed by rCD4-ricin A chain. Science 242: 1166–1168

Tonegawa S (1983) Somatic generation of antibody diversity. Nature 302:575–581

Traunecker A, Schneider J, Kiefer H, Karjalainen K (1989) Highly efficient neutralization of HIV with recombinant CD4-immunoglobulin molecules. Nature 339:68–70

Trauth BC, Klas C, Peters AMJ, Matzku S, Möller P, Falk W, Debatin K-M, Krammer PH (1989) Monoclonal antibody-mediated tumor regression by induction of apoptosis. Science 245:301–305

Tucker PW, Marcu KB, Newell N, Richards J, Blattner FR (1979) Sequence of the cloned gene for the constant region of murine γ2b immunoglobulin heavy chain. Science 206:1303–1306

Vitetta ES, Fulton RJ, May RD, Till M, Uhr JW (1987) Redesigning nature's poisons to create anti-tumor reagents. Science 238:1098–1104

Williams GT, Neuberger MS (1986) Production of antibody-tagged enzymes by myeloma cells: application to DNA polymerase I Klenow fragment. Gene 43:319–324

CHAPTER 8

Three Generations of Recombinant CD4 Molecules as Anti-HIV Reagents

A. Traunecker and K. Karjalainen

A. Introduction

The technology to produce soluble forms of cell surface molecules has been crucial both for their structural and functional characterization and for their exploitation as therapeutic reagents. Usually, large numbers of different surface molecules are involved in the complex interactions between cells of the immune system. A study of the role of a given receptor-ligand pair can be greatly facilitated by "solubilizing" these components from their natural cellular environments, which greatly reduces the complexity of the system (Traunecker et al. 1991b).

Here we describe the different versions of soluble recombinant CD4 (sCD4) molecules which we have produced to illustrate the various principles of "solubilization" procedures of cell surface receptors. Besides its natural function on the lymphocyte surface, CD4 is also the receptor for the human immune deficiency virus (HIV) (Dalgleish et al. 1985; Klatzman et al. 1984). Therefore, different generations of sCD4 were generated to maximize their potency as anti-HIV reagents. This review emphasizes the technical aspects of our studies; thus, it does not represent an exhaustive review of HIV biology or strategies with which to combat the virus.

B. General Aspects of HIV Infection

CD4 is expressed on T helper (T_H) and, to a lesser degree, on macrophage/monocyte cell lineages. Therefore, these cell types, which are of crucial importance for the functioning of the immune system, are the prime targets of the initial HIV infection (Dalgleish et al. 1985; Klatzman et al. 1984). Eventually, the infection causes depletion of CD4-positive lymphocytes by an unknown mechanism, resulting in the total collapse of the immune system. Usually, the final stage of the infection is preceded by a latent phase which can last for several years. During this time the functional immune system, presumably, controls the infection by generating successive specific immune responses against new waves of viral variants, which emerge at high rates under the immune selection. It is at this stage that sCD4-based molecules could be used to help the immune system break the fatal vicious circle.

sCD4 can be viewed as an universal anti-HIV reagent, because all viral escape variants should maintain their ability to bind their receptor in order to propagate themselves. In addition, the affinity of viral gp120 for CD4 is high, $10^{-9} M$ (SMITH et al. 1987). Importantly, sCD4 does not seem to have any negative effects on the immune system, i.e., it does not interfere with the normal immunological functions of cell surface CD4. This was most dramatically shown by studies with transgenic mice which constitutively expressed high amounts of sCD4 (mouse CD4 in this case) in their body fluids (WEBER et al. 1993). All phenotypical and functional properties of these mice were unaffected. Taken together, these results suggest that sCD4 reagents are strong candidates for use as therapeutic agents to combat HIV. With this in mind, we have developed three generations of sCD4 molecules, with increasing potency against HIV, by using standard recombinant DNA technology.

C. Characteristics of Different Forms of sCD4

I. First Generation: Truncated Forms of sCD4

Early versions of sCD4 consisted only of extracellular domains of CD4 or different truncations of them. These molecules behaved as monomers in solution. Importantly, they retained both high affinity for gp120 and their antigenic properties. In addition, they were shown to neutralize HIV efficiently in vitro (SMITH et al. 1987; HUSSEY et al. 1988; DEEN et al. 1988; TRAUNECKER et al. 1988; FISHER et al. 1988). These early studies also showed that the first two NH_2-terminal domains of CD4 made up the smallest soluble part of the molecule, which retained all its anti-viral activity (TRAUNECKER et al. 1988; Fig. 1A). Additional mutational analyses pinpointed the first NH_2-terminal domain as the main binding site for HIV (CAPON and WARD 1991). However, these simple forms turned out to be inefficient in vivo, presumably due to their extremely short biological half-life (GERETY et al. 1989; SCHOOLEY et al. 1990; KAHN et al. 1990).

II. Second Generation: CD4-Immunoglobulins

The early version of sCD4 could be viewed as a passive molecule, only able to neutralize a free virus. Therefore, CD4 was chimerized to different effector moieties which could then, in addition, mediate destruction of infected cells. To this end, we and others have produced different forms of CD4-Ig molecules, in which the CD4 part was joined to different classes of Ig constant regions (TRAUNECKER et al. 1989; CAPON et al. 1989; ZEITTLMEISSL et al. 1990; Fig. 1B). These molecules were orders of magnitude more potent in vitro than earlier versions, presumably due to their multivalency, the most dramatic case being CD4-IgM (TRAUNECKER et al. 1989; Fig. 1C).

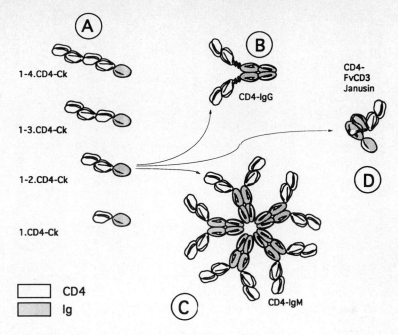

Fig. 1A–D. Three generations of sCD4 molecules. For explanation see text

In addition, CD4-Ig molecules retained the biological properties of Ig constant regions, i.e., they could activate the complement system and bind to Fc receptors of lymphocytes and macrophages responsible for antibody-mediated cellular cytotoxicity (ADCC). Importantly, Ig constant regions increased the biological half-life of sCD4 dramatically (CAPON et al. 1989) and activated their transport across the placental barrier (BYRN et al. 1990).

III. Third Generation: CD4-FvCD3 Janusins

The newest version of sCD4 exploit the feature of potent cellular cyto-toxicity as an effector arm. Those molecules, janusins, are based on the principle of bispecific antibodies. To bypass the low frequency of specific cytotoxic T lymphocytes (CTLs) within the normal immune system, bispecific antibodies were previously designed to retarget CTLs of any specificity onto desired target cells (STAERZ et al. 1985; LANZAVECCHIA and SCHEIDEGGER 1987). To induce efficient killing in this system one antibody arm is specific for the constant part of the T cell receptor (TCR), e.g., CD3 ε chain, and the other arm against the desired target epitope. Anti-HIV janusin molecules are single polypeptide chains which contain CD4 and anti-CD3 ε antibody combining site (Fv CD3) as separate functional domains in the same molecule (Fig. 1D). These molecules can efficiently retarget CTLs onto HIV-infected cells and induce killing in minute quantities in vitro (Traunecker et al. 1991a).

D. Molecular Designs and Strategies to Produce Recombinant CD4 Molecules

I. Production of sCD4 Molecules

We have selected and developed a myeloma-based expression system for our purposes. Myeloma cells are potent secretory cells that grow to high cell densities in suspension cultures in vitro. Secreted proteins thus contain all the typical protein modifications found in mammalian cells. In addition, the genetic regulatory elements that allow high level expression of antibodies in myeloma cells, such as promoter and enhancer elements, are well defined.

Our standard expression vector contains the Ig K variable region (V_K) promoter, which drives transcription in the presence of defined Ig gene enhancers. To take advantage of the stable antibody domain structure, the recombinant gene can be forced to splice, if so desired, in correct translational reading frame to a downstream Ig domain encoding sequence, i.e., Ig-Cκ for monovalent products or modified Ig heavy chain constant region for multivalent products (Fig. 2). Ig heavy chains are not usually secreted alone without Ig light chains. Therefore, to ensure secretion of chimeras containing Ig heavy chain constant regions, the first heavy chain constant region exon has to be deleted from the constructs in analogy to human heavy chain disease proteins (Traunecker et al. 1989). When non-chimeric proteins are preferred, the stop codon can be introduced behind the recombinant gene.

Several biochemical selection markers are now available, e.g., neo, gpt, hygro, histidinol and bleomycin, which make it possible to introduce successive constructs in the same cell in order to produce multisubunit molecules. Protoplast fusion, as opposed to electroporation, is the preferred method of transfection, since consistently higher production levels can be reached due to increased copy numbers of the integrated vectors.

II. CD4-Cκ Molecules

Since exons very often correspond to functional domains of the protein, we produced our first monovalent truncated forms of sCD4 (CD4-Cκ) by exploiting the genomic structure of the CD4 gene. Different numbers of exons were included in our constructs to generate 1. Cκ, 1.2. Cκ, 1.2.3. Cκ and 1.2.3.4. Cκ molecules, with 1.2.3. and 4 corresponding to different exons coding for the extracellular part of the CD4 molecule (Fig. 2A). Only 1.2. Cκ and 1.2.3.4. Cκ molecules were secreted, indicating that CD4 domains associate pairwise into two stable subunits consisting of 1.2 and 3.4 domains, respectively. This has been now partially confirmed by X-ray

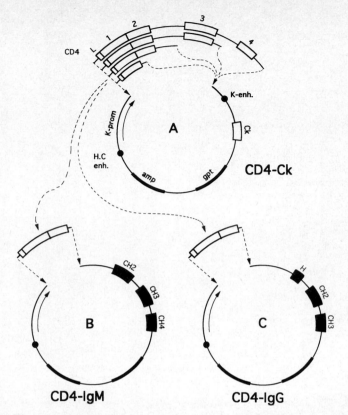

Fig. 2A–C. The expression vectors used to produce sCD4 molecules. **A** The *arrow* indicates the Vκ promoter and transcriptional orientation. *Small filled circles* represent different Ig enhancer elements: *H.C. enh*, Ig heavy chain core enhancer; *k enh*, Ig κ locus enhancer. The *open boxes* represent different exons of the CD4 gene. **B, C** The same basic vectors modified to produce multivalent forms of sCD4, i.e., CD-IgM and CD4-IgG. (From TRAUNECKER et al. 1989, 1991a)

crystallography (WANG et al. 1990; RYU et al. 1990). All four forms of CD4-Cκ, regardless of their secretory capacity, retained binding to gp120 allowing us to map the CD4-gp120 binding epitope to the first domain of CD4. Higher resolution, obtained by mutational analyses, pinpointed the loop corresponding to the second hypervariable region of Vκ as a major binding site (CAPON and WARD 1991). Since both secreted forms, 1.2. Cκ and 1.2.3.4. Cκ, neutralized HIV equal efficiency in vitro, the shorter form of CD4 was included in all subsequent constructs of sCD4 (Table 1).

III. Multivalent sCD4 Molecules: CD4 Immunoglobulins

The crucial feature of the vectors designed to produce different types of CD4-Igs is modification of the corresponding heavy chain constant region

Table 1. Summary of biological properties of different soluble CD4 derivatives

	Molecule			
	CD4-Ck	CD4-IgG	CD4-IgM	CD4-FvCD3 Janusin
CD4-mediated functions				
Secretion	+++	+++	+++	+++
Binding gp120	+++	+++	+++	+++
[IC_{50}] ng/ml[a]	~10000	150–10000	10–40	>1–10
Ig-mediated functions				
Number of valences	1	2	10	1
Binding proteinA	―――	+++	―――	―――
Binding protein G	―――	+++	―――	―――
Binding Fc γ receptor	―――	+++	―――	―――
Binding C1q	―――	+++	+++	―――
Serum half-life (in rabbit)[b]	1	60–80x	45–50x	NT
Placental transport	NT	+++	NT	NT
ADCC	NT	+++	NT	NT
Retargeting CTLs	NT	NT	NT	+++

[a] Concentrations needed for anti-viral effect.
[b] SCD4 = 1.

genes, i.e., the first constant region exon has to be deleted to ensure maximal secretion (Fig. 2B,C). Initially, we compared different types of CD4-Igs containing either the complete or modified versions of the constant regions. We noticed that this modification consistently increased secretion by 10- to 100-fold in all cases studied, which included CD4 chimeras with mouse (m) μ or $\gamma2$ and with human (h) μ, $\gamma1$ or $\gamma3$ constant regions. Our prototype CD4-Ig vectors now allow us to routinely obtain transformants secreting 10–100 μg/ml of CD4-Igs. As expected, decameric CD4-IgM was the most efficient in vitro in neutralizing HIV, followed by bivalent CD4-h IgGs, which were about two orders of magnitude more active than mono-valent sCD4 (Table 1).

IV. Bispecific Reagents: CD4-FvCD3 Janunsin

Classical bispecific antibodies that can retarget CTLs of any specificity onto target cells are usually obtained from a quadroma, which is the product of the fusion between two selected hybridomas with the desired specificities (MILSTEIN and CUELLO 1983). This procedure results in a mixture of anti-bodies, theoretically ten different Igs, due to random assortment of Ig heavy and light chains. To avoid this complexity, which usually reduces the yield of the desired product, we have developed an approach which enables us to produce bispecific reagents in a single polypeptide chain.

In our model case, one arm is the two domain CD4 and the other the anti-CD3 antibody combining site (FvCD3). FvCD3 was engineered according to standard procedures for production of a single chain antibody. No special linkers were used in the junction between CD4 and FvCD3, although the COOH-terminal of the second domain of CD4 can probably function as such. Both arms retained their functional properties, as shown by their capacity to retarget CTLs to HIV-infected cells (Table 1; TRAU-NECKER et al. 1991a).

E. Concluding Remarks

Here we have reviewed the development of different forms of sCD4 as potential anti-HIV reagents. These three generations of sCD4 can also be used to illustrate the various principles that can be exploited to solubilize cell surface receptors in biologically active forms.

The human clinical trials with simple truncated forms of sCD4 could not be considered successful (GERETY et al. 1989; SCHOOLEY et al. 1990; KAHN et al. 1990), possibly due to the extremely short biological half-lives of these molecules and their lack of effector functions. CD4-Igs and CD4-FvCD3 janusins, however, can exploit the effector functions associated with anti-bodies and cellular immunity, respectively (TRAUNECKER et al. 1989, 1991a; CAPON et al. 1989; BYRN et al. 1990). CD4-Igs have dramatically longer half-lives in vivo than do simply truncated sCD4 (CAPON et al. 1989). Interestingly, some CD4-Igs can be transferred across the placenta and are thus serious candidates for use as therapeutic agents, especially in perinatally transmitted HIV infections. As of the end of 1992, no published information was available regarding clinical trials with CD4-Ig molecules. CD4-FvCD3 janusins are still at their early experimental stage but their superior in vitro activity makes them worthy candidates for further development.

References

Byrn R, Mordenti J, Lucas C, Smith D, Marsters S, Johnson J, Cossum P, Chamow S, Wurm F, Gregory T, Groopman J, Capon D (1990) Biological properties of a CD4 immunoadhesin. Nature 344:667–670

Capon D, Ward R (1991) The CD4-gp120 interaction and AIDS pathogenesis. Annu Rev Immunol 9:649–678

Capon DJ, Chamow SM, Mordeni J, Marsters S, Gregory T, Mitsuya H, Byrn R, Lucas C, Wurm F, Groopman J, Broder S, Smith D (1989) Designing CD4 immunoadhesins for AIDS therapy. Nature 337:525–531

Dalgleish AG, Beverley PC, Clapham PR, Crawford DH, Greaves MF, Weiss RA (1985) The CD4 (T4) antigen is an essential component of the receptor for the AIDS retrovirus. Nature 312:763–767

Deen K, McDougal J, Inacker R, Folena-Wasserman G, Arthos J, Rosenberg J, Maddon P, Axel R, Sweet R (1988) A soluble form of CD4 (T4) protein inhibits AIDS virus infection. Nature 331:82–84

Fisher R, Bertonis J, Meier W, Johnson V, Costopoulos D, Liu T, Tizard R, Walker B, Hirsch M, Schooley R, Flavell R (1988) HIV infection is blocked in vitro by recombinant soluble CD4. Nature 331:76–78

Gerety R, Hanson D, Thomas D (1989) Human recombinant soluble CD4 therapy. Lancet 2:1521

Hussey R, Richardson N, Kowalski M, Brown N, Chang H-C, Silicano R, Dorfman T, Walker B, Sodroski J, Reinherz E (1988) A soluble CD4 protein selectively inhibits HIV replication and syncytium formation. Nature 331:78–81

Kahn J, Allan J, Hodges T, Kaplan L, Arri C, Fitch H, Izu A, Mordenty J, Sherwin S, Groopman J, Volberding P (1990) The safety and pharmacokinetics of recombinant soluble CD4 (rCD4) in subjects with the acquired immunodeficiency syndrom (AIDS) and AIDS related complex. A Phase I study. Ann Intern Med 112:254–261

Klatzmann D, Barre-Sinousi F, Nugeyre M, Dauguet C, Vilmer E, Griscelli C, Brun-Vezinet F, Rouzioux C, Gluckman J-C, Chermann J-C, Montagnier L (1984) Selective tropism of lymphadenopathy associated virus (LAV) for helper T lymphocythes. Science 225:59–63

Lanzavecchia A, Scheidegger D (1987) The use of hybridomas to target human cytotoxic T lymphocytes. Eur J Immunol 17:105–111

Milstein C, Cuello C (1983) Hybrid hybridomas and their unse in immunohisto-chemistry. Nature 305:537–539

Ryu S, Kwong N, Axel R, Sweet R, Hendricson W (1990) Crystal structure of an HIV-binding recombinant fragment of human CD4. Nature 348:418–426

Schooley R, Merigan T, Gaut P, Hirsch M, Holodniy M, Flynn T, Liu S, Byington B, Henochowicz S, Gubish E, Spriggs D, Kufe D, schindler J, Dawson A, Thomas D, Hanson D, Letwin B, Liu TJG, Kennedy S, Fisher R, Ho D (1990) Recombinant soluble CD4 therapy in patients with the acquired immunodeficiency syndrom (AIDS) in AIDS-related complex. Ann Intern Med 112:247–253

Smith D, Byrn R, Marsters S, Gregory T, Groopman J, Capon D (1987) Blocking of HIV-1 infectivity by a soluble, secreted form of the CD4 antigen. Science 238:1704–1707

Staerz U, Kanagawa O, Bevan M (1985) Hybrid antibodies can target sites for attack by T cells. Nature 314:628–631

Traunecker A, Lüke W, Karjalainen K (1988) Soluble CD4 molecules neutralize human immunodeficiency virus type 1. Nature 331:84–86

Traunecker A, Schneider J, Kiefer H, Karjalainen K (1989) Highly efficient neutrali-zation of HIV with recombinant CD4-immunoglobulin molecules. Nature 339:68–70

Traunecker A, Lanzavecchia A, Karjalainen K (1991a) Bispecific single chain molecules (Janusins) target cytotoxic lymphocytes on HIV infected cells. EMBO J 10:3655–3659

Traunecker A, Oliveri F, Karjalainen K (1991b) Myeloma based expression system for production of large mammalian proteins. Trends Biotech 9:109–113

Wang J, Yan Y, Garret T, Lui J, Rodgers D, Garlick R, Tarr G, Husain Y, Reinherz E, Harrisson S (1990) Atomic structure of a fragment of human CD4 containing two immunoglobulin-like domanis. Nature 348:411–418

Weber S, Traunecker A, Karjalainen K (1993) Constitutive expression of high levels of soluble mouse CD4 in transgenic mice does not interfere with their immune function. Em J Immunol 23:511–516

Zeittlmeissl G, Gregerson J, Dupont J, Mehdi S, Reiner G, Seed B (1990) Expres-sion and characterisation of homand CD4: immunoglobulin fusion proteins. DNA Cell Biol 9:347–353

Section IV: Combinatorial Libraries

CHAPTER 9
Chemical and Biological Approaches to Catalytic Antibodies

K.D. Janda and Chen Y.-C. Jack

A. Introduction

A long-standing goal of biochemists is the design and synthesis of catalysts endowed with particular activities and properties. For example, one might create an enzyme that can degrade drugs into inactive metabolites as a treatment for drug overdoses or an enzyme that acts on a prodrug to unmask its therapeutically useful form. Yet another enzyme could target the cleavage of a specific tumor or virus protein that is vital for development of that tumor or virus thus acting as a therapeutic agent. Other traits, such as thermostability, chemical stability, and oxidative stability, can also be incorporated into such catalysts.

There are two general approaches to the design of new enzymes. One method would be to improve or alter the activities of existing enzymes by site-directed mutagenesis. For example, the specificity of a particular protein might be changed or improved. Craik and Evnin (Craik et al. 1985; Evnin et al. 1990) altered the substrate specificity of trypsin by changing the amino acids in the substrate binding pocket which are responsible for the lysine/arginine specificity of the protease. Alternatively, one might improve the specificity of trypsin so that it cleaves only after a lysine in a particular array of amino acids, the protein equivalent of a DNA restriction enzyme.

Unfortunately, the attempts to create new enzymes by site-directed mutagenesis have met with limited success. One factor that weighs against the success of site-directed mutagenesis experiments is the lack of detailed information regarding the mechanism of an enzyme catalyzed reaction. This requires knowledge of the amino acids in the active site of an enzyme, the residues involved in substrate binding and substrate release, and the role the amino acids play in catalysis. These insights are best obtained from X-ray crystal structures of enzyme-substrate or enzyme-inhibitor complexes. Ideally, one would like a movie of the entire catalytic process, such as might be obtained from Laué spectroscopy (Johnson 1992). Unfortunately, high resolution X-ray crystal structures of all enzymes are not easily obtained.

Another argument against the success of such experiments is that enzymes have evolved over millions of years to their present, highly efficient forms (Knowles 1987). Any attempts to improve on "enzymatic perfection" will not be successful. There are reports in which the catalytic efficiencies of enzymes have been improved by site-directed mutagenesis (Mandecki et al. 1991). In these cases, the enzymes are not working at the diffusion limit so that room for improvement does exist.

A second tact would be the the de novo design of proteins and enzymes. By this approach we would design a completely novel protein with the particular characteristics we want. This approach has also met with limited success because we do not have a complete understanding of the rules and principles governing protein folding. It is known that the primary sequence of a protein determines its three-dimensional structure, but we have not been able to predict how a sequence of amino acids will fold. Several groups are attempting to design proteins from scratch with moderate success.

DeGrado and Eisenberg (Eisenberg et al. 1986; Ho and DeGrado 1987; Regan and DeGrado 1988) have collaborated on the design, synthesis, and characterization of a four-helix bundle protein consisting of four identical α-helices connected by three identical loops. The protein could be expressed in and isolated from *Escherichia coli* and was found to exist as a four-helix bundle packed in an antiparallel arrangement. A similar approach was used to design a polypeptide that folds into an αβ barrel motif by the concatenation of eight successive turn/β-strand/turn/α-helix motifs (Goraj et al. 1990). Refinement of these structures has led to the production of four-helix bundles that contain metal binding sites (Handel and DeGrado 1990; Regan and Clarke 1990). One designed four-helix bundle, chymohelizyme, consisting of four highly similar α-helices that incorporate the protease catalytic triad residues (serine, histidine, and aspartic acid) at the termini of three of the helices has been reported to possess a low esterolytic activity (Hahn et al. 1990). It is unclear what role, if any, the "catalytic" residues play in the reaction.

The success achieved so far with proteins designed from first principles is exciting and provides encouragement for future studies. It is possible to design proteins to fold in a desired conformation. The addition of residues to form a catalytic core adds an extra dimension of complexity. Given the problems encountered so far with designing proteins of defined structures, it is safe to say that endowing these proteins with the tools for catalysis will be an even more difficult process.

The problem of designing new enzymes or catalysts with tailored specificities can be reduced to one of designing the proper active site. Enzymes, in an oversimplified view, are merely catalytic cores embedded in a protein scaffold. It has been demonstrated that a scaffold can be made; the challenge lies in creating a core with the correct arrangement of amino acid residues or catalytic groups to effect catalysis. Indeed, chemists have been trying the complementary approach by designing synthetic receptors that will function as such catalytic cores, again with limited success. In this chapter, we describe an approach to creating new catalysts which utilizes the binding site in antibodies as these catalytic cores. In brief, synthetic molecules representing the transition states of a desired reaction can be synthesized and used to raise antibodies. Some antibodies, upon binding the homologous substrate, will utilize the energy released upon binding of the substrate to effect a chemical transformation on that substrate. To date, antibodies have been shown to catalyze a variety of chemical transformations including hydrolytic, bimolecular, carbon-carbon bond forming and redox reactions. Many of these reactions proceed with high rates and enantioselectivity. We shall summarize the underlying principles for this approach and describe various techniques for designing the required immunogens. Methods to optimize the catalytic efficiency and broaden the scope of reactions catalyzed by these antibodies will also be described. Finally, we shall discuss future roles for catalytic antibodies.

B. Background

Enzymes and antibodies are protein molecules that share some common features. Both bind particular target molecules with varying degrees of affinity; for enzymes, the K_ms for substrates range from the micromolar to the millimolar range whereas antibodies tend to bind their targets more tightly. Enzymes and antibodies can also accommodate some structural variability in the compounds to which they bind. It is known that enzymes can process a number of alternate substrates. Antibodies, in a similar manner, can cross-react with haptens bearing structural homology. There is one property that is unique to enzymes, the ability to catalyze chemical reactions. Thus, whereas antibodies function only to bind their targets, enzymes can bind *and* effect chemical transformations on their substrates. A number of questions are raised by these observations. First, why do only enzymes but not antibodies catalyze reactions, and second is it possible to induce antibodies to function as catalysts? To answer the second question, one must understand how enzymes catalyze reactions.

I. Bases of Enzymatic Catalysis

There are four major strategies used by enzymes for catalyzing reactions. The first is via approximation effects in which the enzyme functions to bring two molecules together so the chances of their interacting increases. This also orients the two molecules in the correct geometry for the reaction and reduces entropic barriers to the reaction. A second method by which enzymes catalyze reactions is to form a covalent complex with one of the reactants as in the case of the serine proteases. It is not clear what purpose this serves, but with cofactors, formation of covalent complexes leads to stabilization of otherwise high energy intermediates. Another way in which enzymes catalyze reactions is via acid-base catalysis. The amino acid residues or cofactors serve to abstract or donate protons so as to increase the nucleophilicity or the electrophilicity of functional groups involved in catalysis. The polarization of carbonyl groups by Zn^{2+} in the alcohol dehy-

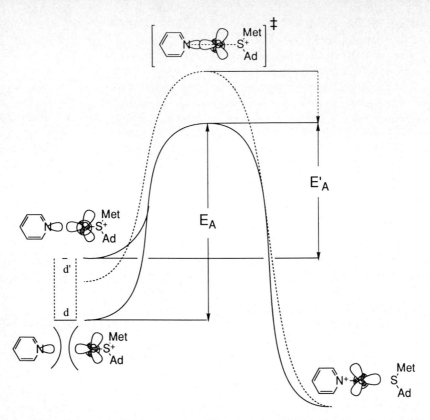

Fig. 1. Reaction profile for methyl transfer from S-adenosylmethionine to pyridine. If an enzyme were to bind the N-methylpyridinium-methionine complex more tightly than the pyridine-S-adenosylmethionine complex, the enzyme would force the two substrates together to decrease the distance from d to d'. This binding energy will be manifest as an increase in the ground state energy and result in a net decrease in the activation energy from E_A to E'_A. The net result is an increase in the reaction rate

drogenases is an example of this type of catalysis. The final strategy employed by enzymes to effect reactions is transition state stabilization. Here, the enzyme preferentially binds the reactants in a geometry that more closely represents the transition state the reaction must pass through. This is done at the expense of the ground state, thus destabilizing the ground state, and serves to lower the activation energy for the reaction. Thus, rather than passing over a high energy barrier, the reactants pass over a lower energy barrier. The difference in energies is manifest as a rate acceleration (Fig. 1).

The scientific foundation for catalytic antibodies was laid in the 1940s when Pauling (1948) suggested that an enzyme functions by selectively stabilizing the activated complex of the reaction it was catalyzing. He opined that enzymes and antibodies are similar with the difference that enzymes

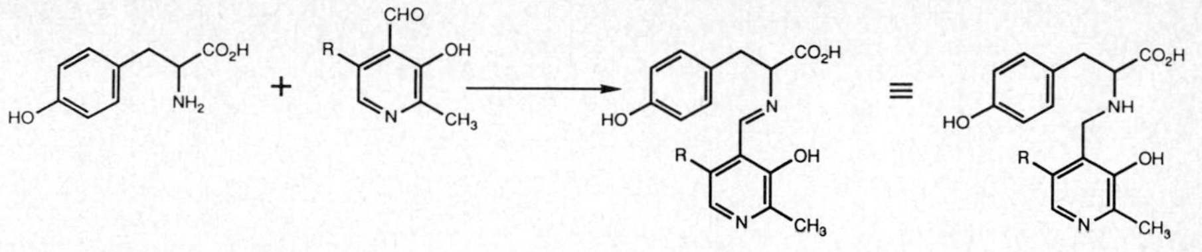

Fig. 2. Reactions attempted by Stollar and Summers (RASO and STOLLAR 1975; SUMMERS 1983). Stollar attempted the transamination using the corresponding amine *4* as the hapten. Summers used the triarylvinylacrylamide *7* as the immunogen for the tautomerization of *5* to *6*

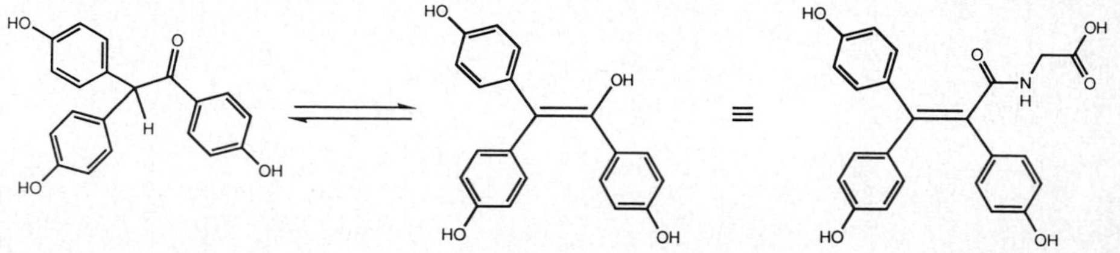

were more complementary to the transition state for the reaction than to the substrate whereas the surface of an antibody was more complementary to its antigen. This idea was echoed by JENCKS (1987) who formulated the idea more explicitly. Jencks stated that one might prepare an enzyme by raising an antibody to a molecule which is a stable representation of the transition state for the reaction of interest. The binding sites in the antibody would be complementary to the transition state for the reaction, and upon binding the homologous substrates would force the reactants to adopt a more transition state-like geometry. Thus, one way to generate new catalysts would be to use this strategy of raising antibodies to haptens that represent the transition states of reactions.

Early attempts to generate catalytic antibodies met with failure. RASO and STOLLAR (1975) used a multisubstrate analog (Fig. 2, *4*) to raise antibodies that would catalyze the condensation of pyridoxal phosphate and tyrosine (Fig. 2). SUMMERS (1983) used an amide to raise antibodies that would catalyze the isomerization of an aryl acetophenone. Both attempts were unsuccessful in raising antibodies that could perform the desired reactions.

It has been stated that the immune system consists of at least 10^{10} different antibody molecules. Given this huge diversity, it seems probable that a number of catalysts for the desired reaction might exist. Why then were these early attempts unsuccessful? The answer might lie exactly in the tremendous number of different antibodies that can be raised to a hapten.

Suppose that one is exposed to 10^5 different antigens in a lifetime. If the immunological repertoire is on the order of 10^{10} molecules, then approximately 10^5 different antibodies can be raised to one hapten. Let us assume that ten of the antibodies raised are actually catalytic. If one uses polyclonal sera, then 1 in 10000 antibodies possesses catalytic activity. Unless the catalytic efficiency is extremely high, it would be very difficult to detect that activity. An analogy might be an enzyme that is present as 0.001% of total cell protein. The catalytic activity of such an enzyme would be difficult to assay in crude cell extract.

While enzymes can be purified to homogeneity, it is not possible to purify a single catalytic antibody from a heterogeneous mixture of polyclonal antibodies. It is, however, possible to isolate the B cell that gives rise to a particular antibody by cloning it from the other B cells. This is done by fusing that B cell to a myeloma line and propagating the resulting hybrid cell in cell culture (KÖHLER and MILSTEIN 1975). Antibodies so obtained can be screened for catalytic activity. The strategy then is to design a haptenic molecule that is a stable representation of the transition state for the reaction of interest.

Since small molecules by themselves are not immunogenic, they must be coupled to a carrier protein. Typically, keyhole limpet hemocyanin (KLH) is used as the carrier protein for immunization and bovine serum albumin (BSA) is used as the carrier in ELISA assays to identify hapten-specific antibodies (ENGVALL 1980). Haptens are coupled to the ε-amino groups of the surface lysine residues on the carrier protein via amide bonds to the hapten (ERLANGER 1980; NISHIMA et al. 1974). The coupling agent consists of a spacer arm of between 5 and 10 Å, to reduce any structural interference from the carrier protein, flanked by two reactive groups. A few linkers we have designed and have found to be very successful are shown in Fig. 3. These three spacer units are termed heterobifunctional linkers (BRINKLEY

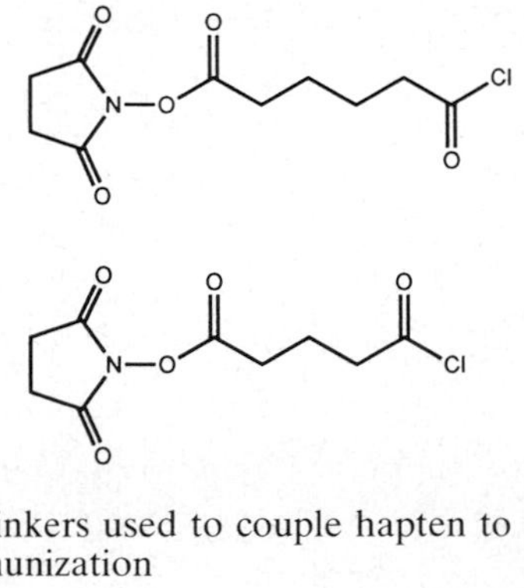

Fig. 3. Heterobifunctional linkers used to couple hapten to carrier protein (keyhole limpet hemocyanin) for immunization

1992). Unique to each is their ability to be selectively coupled to the hapten without destroying the opposite activated end, which can be coupled subsequently to the carrier protein. The carrier protein is then used in an immunization regimen to hyperimmunize mice. The mice are sacrificed, their spleen cells isolated as the source of B cells, and fused to a myeloma line. The resulting hybridomas are cloned and screened for their ability to bind the hapten used for immunization. A subpopulation of hybridomas that secrete antibodies capable of binding to the hapten are then screened for catalytic activity.

C. Hapten Design Strategies for Catalytic Antibodies

I. Transition State Stabilization

The classical method used for eliciting catalytic antibodies is to synthesize a molecule that represents the fleeting transition state for the reaction. Consider the hydrolysis of an ester which proceeds via attack of water or hydroxide on the carbonyl carbon to generate the tetrahedral complex (Fig. 4). This adduct can collapse with expulsion of water (or hydroxide) to regenerate starting material or can expel an alcohol to yield the hydrolysis product. The transition state has three notable features that should be mimicked: its charge (-1), its dimensions, and its geometry. All of these characteristics can be found in the phosphonate ester. Under physiological conditions, the phosphonate ester possesses a net -1 charge. The bond lengths of the ester are longer than those for the corresponding carbon analog but that is not a major problem. Most importantly, the phosphonate possesses the tetrahedral geometry that is characteristic of the transition state. These phosphonate esters are good mimics for the transition state for ester hydrolysis and are known to be extremely potent inhibitors for proteases (JACOBSON and BARTLETT 1981).

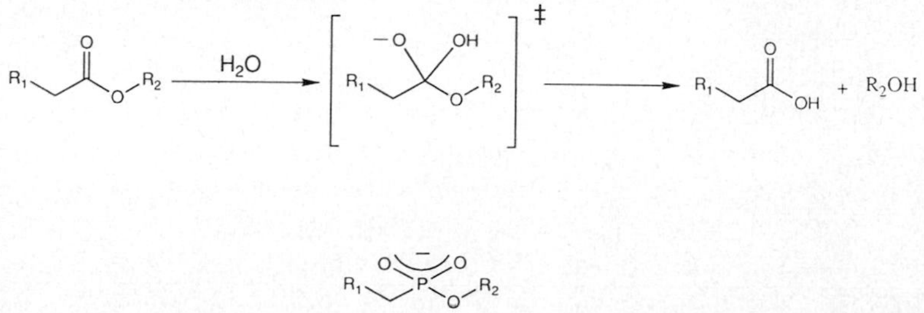

Fig. 4. Hydrolysis of an ester. The putative transition state for the reaction can be modeled by a phosphonate ester since such esters bear the requisite geometry and charge

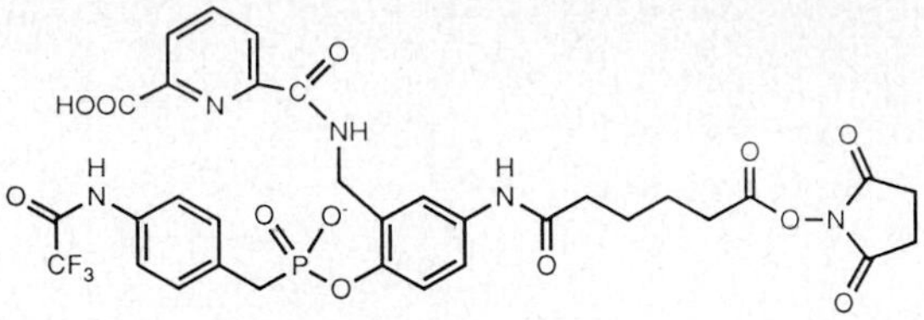

8

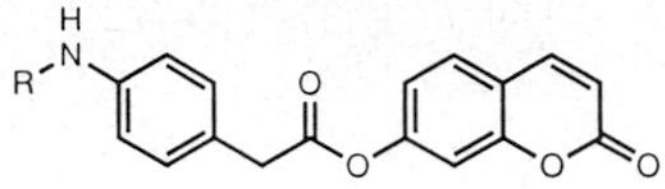

R=$COCH_3$, $COCF_3$ **8a**

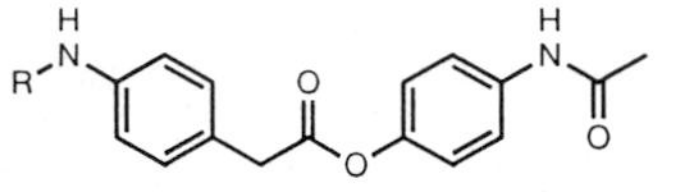

R=$COCH_3$, $COCF_3$ **8b**

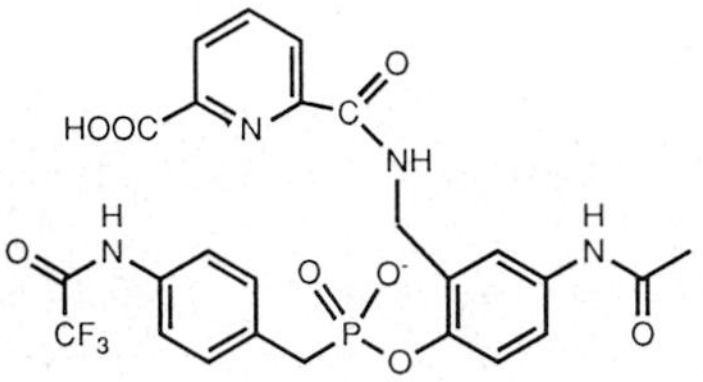

8c

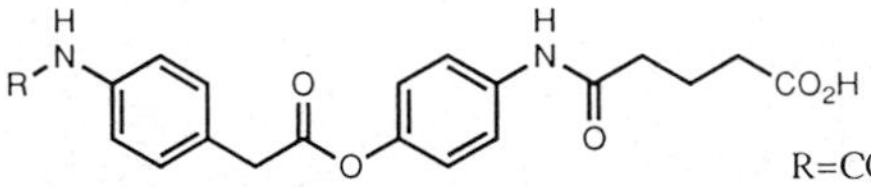

R=$COCH_3$, $COCF_3$ **8d**

Fig. 5. Phosphonate esters used to induce first catalytic antibodies

The phosphonate esters have been the compounds of choice to use in eliciting antibodies that hydrolyze aryl esters. TRAMANTANO et al. (1986a,b) used the hapten shown in Fig. 5, *8* as a transition state analog for the hydrolysis of the ester (Fig. 5, *8a*). A large number of hybridomas (50–100) capable of binding the hapten (Fig. 5, *8*) were obtained and 18 that produced the tightest binding antibodies were chosen for study. Three of these antibodies demonstrated the ability to hydrolyze the ester (Fig. 5, *8a*) and the fastest antibody (6D4) was chosen for further study. This antibody was found to effect the stoichiometric release of the coumarin ester. Addition of hydroxylamine restored activity indicating the antibody might be acylated by the ester shown in Fig. 5, *8a*. Catalytic activity was observed upon incubation of the antibody with the more congruent form of the compound (Fig. 5, *8b*). Addition of another form (Fig. 5, *8c*) inhibited the reaction with an apparent K_i of 65 nM. This would be expected since the latter resembled the hapten used to elicit the antibodies and should have bound tightly to the antibody binding pocket.

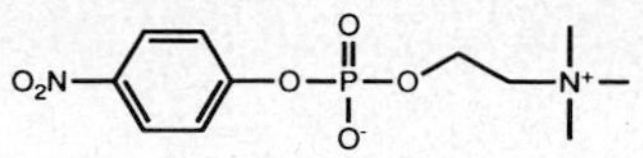

Fig. 6. Phosphorylcholine (*9a*) bound by MOPC167 and the substrate (*9*) hydrolyzed by the antibody. Phosphate ester (*10*) used to elicit antibodies hydrolyzing *10a*

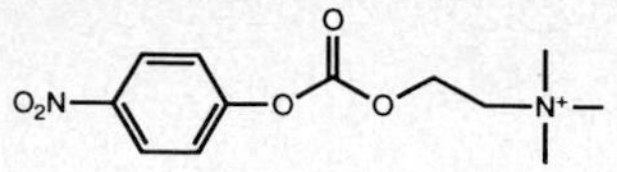

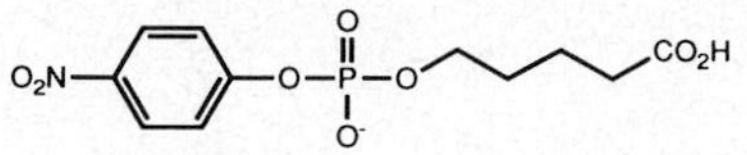

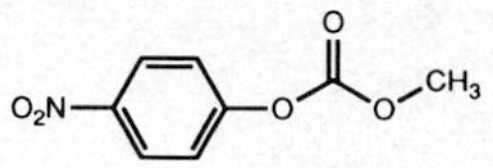

Concurrently, POLLACK et al. (1986) showed that MOPC167, an extant murine myeloma line that binds phosphorylcholine (Fig. 6, *9a*) could catalyze the hydrolysis of the carbonate form (Fig. 6, *9*). Subsequently, the Schultz group (JACOBS and SCHULTZ 1987) reported that an antibody raised against a phosphonate ester (Fig. 6, *10*) could catalyze the hydrolysis of a specific substrate (Fig. 6, *10a*). The antibody catalyzed reaction was observed to display saturation kinetics, was specific for the indicated substrate, was inhibited by addition of the hapten used for immunization (K_i = 3.3 μM), and was not catalyzed by an irrelevant antibody. Heat denaturation of the antibody destroyed activity consistent with catalysis being mediated by the antibody.

The properties exhibited by these antibodies are reminiscent of those of enzymes. Both groups of investigators reported saturation kinetics and observed that addition of the hapten used for immunization inhibited the hydrolysis in a competitive manner. Furthermore, the antibodies showed a measure of substrate specificity. The antibody 6D4 could discriminate between a trifluoroacetyl and acetyl group and the antibody raised by JACOBS and SCHULTZ (1987) could distinguish among substitutions along the aryl ring. The measured rate accelerations catalyzed by these preparations (~1000-fold) was also similar.

One advantage of using the immune system to generate catalysts is the potential for generating a variety of different antibodies with different catalytic efficiencies. TRAMANTANO et al. (1988) reported that a hapten similar to that described in Fig. 5, *8* (less the 2,6-pyridine dicarboxylic acid) gave rise to other antibodies that acted as more efficient catalysts. Five of 20 monoclonal antibodies hydrolyzed the substrate (acetyl) shown in Fig. 5, *8b*. One antibody (50D8) proved to be an extremely efficient catalyst hydrolyzing the substrate shown in Fig. 5, *8d* with a k_{cat} of 20 s^{-1}, a rate enhancement of 10^6 over that of the uncatalyzed reaction. The reaction was inhibited

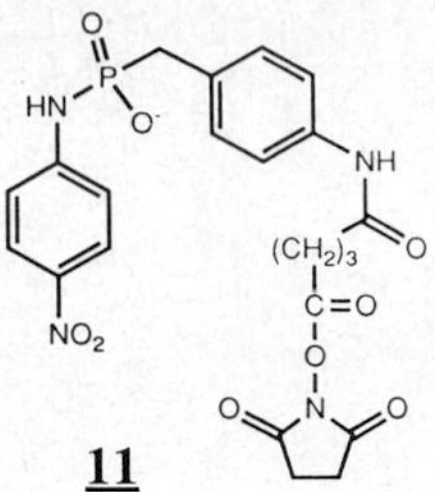

Fig. 7. Hapten (*11*) and substrate (*11a*) for amide hydrolysis. The antibody 43C9 catalyzes the hydrolysis of *11a* with a rate acceleration of 250 000-fold over background

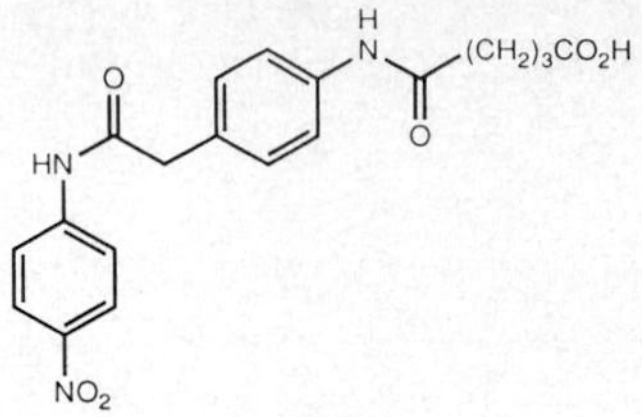

by addition of the hapten (Fig. 5, *8*) ($K_i = 50 \pm 5 \, nM$) and was observed to vary with pH with an inflection at 8.9. Treatment of the preparation with diethylpyrocarbonate had no effect but addition of tetranitromethane resulted in a time-dependent loss of activity (96%). The presence of the inhibitor reduced the effect of the chemical modifiers. Taken together, the data indicate that the antibody 50D8 behaves very much like an enzyme in terms of kinetic characteristics, sensitivity to hydrogen ion concentrations, and susceptibility to chemical modification reagents.

The use of phosphorous containing groups as the immunogens to generate catalytic antibodies underscores the interest in obtaining antibodies capable of catalyzing hydrolysis reactions, namely esterolysis or amidolysis. To this end, Janda et al. (1988a) demonstrated that an antibody raised against the *p*-nitrophenyl phosphonamidate (Fig. 7, *11*) could catalyze the hydrolysis of an activated amide. One out of 44 antibodies (43C9) catalyzed the hydrolysis of the substrate shown in Fig. 7, *11a* with a rate acceleration of 250 000-fold.

Another hallmark of enzyme-mediated reactions is the ability to discriminate between stereoisomers. The use of phosphorous has permitted the generation of antibody catalysts with the ability to differentiate between stereoisomers. Napper et al. (1987) prepared a 2,2-phenoxy-2-oxo-6-(aminomethyl)-1,2-oxaphosphorinane (Fig. 8, *12*) in enantiomerically pure form and used it as an immunogen to raise antibodies in mice. One antibody (24B11) was found to catalyze the release of phenol from the racemic carbinol shown in Fig. 8, *12a* with a rate acceleration of 167 over the uncatalyzed reaction. The oxaphosphorinane acted as a competitive inhibitor of the reaction with a $K_i = 0.25 \, \mu M$. The most notable characteristic of the reaction was that substrate consumption was observed to stop at 50% of the initial ester concentration. Addition of more substrate resulted in consumption of 50% of the added material. This suggested that the antibody was specific for only one enantiomer and was unable to catalyze the cyclization of the other. Chiral NMR analysis of the product isolated from the reaction revealed the presence of a major and a minor peak. Direct comparison of peak heights gave an enantiomeric excess of 94% after correcting for

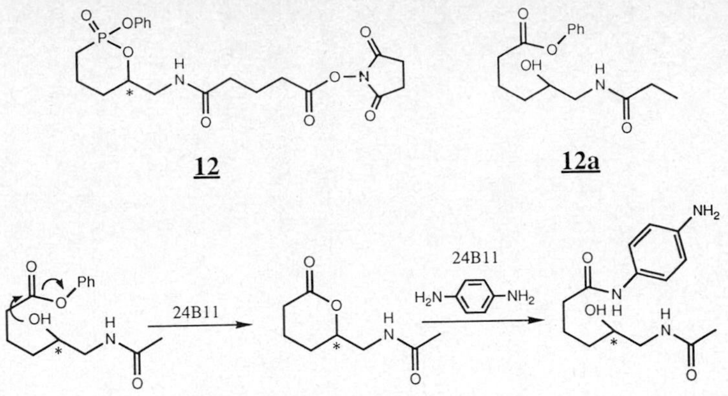

Fig. 8. Stereospecific cyclization performed by antibody raised to an enantiomerically pure immunogen. The antibody catalyzes the cyclization and amine mediated ring-opening of one enantiomer

product arising from spontaneous cyclization. This same antibody (24B11) was also able to catalyze the reverse reaction, a ring opening in the presence of phenylenediamine (BENKOVIC et al. 1988). Here, as with the cyclization reaction, only one enantiomer was processed as substrate.

JANDA et al. (1989) used a racemic antigen (Fig. 9, *13*) to raise antibodies that could catalyze the hydrolysis of the (*R,S*)-α-phenethyl acetate shown in Fig. 9, *14*. There were 18 antibodies that bound the immunogen shown in Fig. 9, *13*; these were isolated and assayed for their ability to catalyze the hydrolysis of the racemic compound shown in Fig. 9, *14*. Eleven of these antibodies were found to be catalytic. Of these, nine specifically hydrolyzed the *R*-enantiomer and two were specific for the *S*-isomer. One antibody (2H6) specific for the *R*-enantiomer hydrolyzed the substrate with a K_m of 4 m*M* and a k_{cat} of 4.6 min^{-1} and was inhibited by the racemic hapten with a K_i of 2.0 μM. An antibody specific for the other antipode (21H3) had a K_m of 400 μM and catalyzed the hydrolysis of the *S*-isomer with a k_{cat} of 0.09 min^{-1}. It was inhibited by the racemic hapten with a K_i of 0.2 μM. The rate accelerations for the antibodies 2H6 and 21H3 over the uncatalyzed reaction were 83 000 and 1700, respectively.

Antibodies can also be induced to catalyze the stereospecific hydrolysis of groups from a meso compound. The acetoxycyclopentene (Fig. 9, *15*) is a starting material for a stereospecific synthesis of prostaglandin $F_{2\alpha}$. Antibodies were raised against the hapten shown in Fig. 9, *16* (IKEDA et al. 1991). One (37E8) out of 33 antibodies that bound to the immunogen was found to catalyze the stereospecific hydrolysis of *cis*-3,5-diacetoxycyclopent-1-ene (Fig. 9, *17*) to the corresponding (1*R*, 4*S*)-(+)-4-hydroxy-2-cyclopentenyl acetate (Fig. 9, *15*).

The use of transition state mimics has been applied to eliciting antibodies capable of performing other reactions. Pericyclic rearrangements are

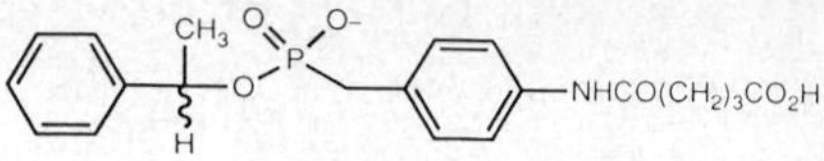

13

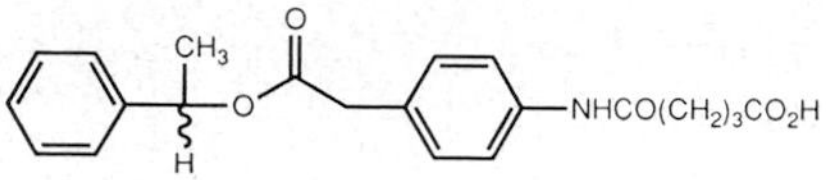

14

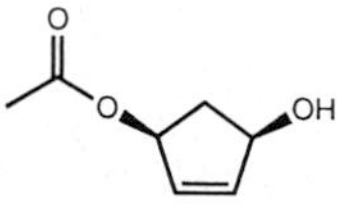

15 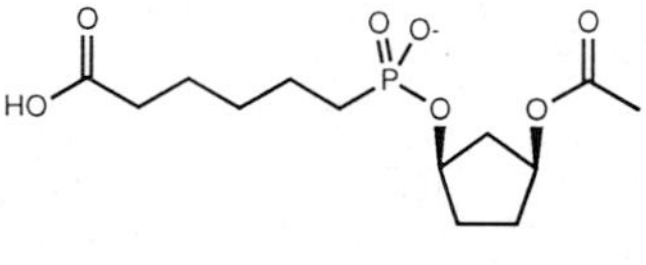   **16**

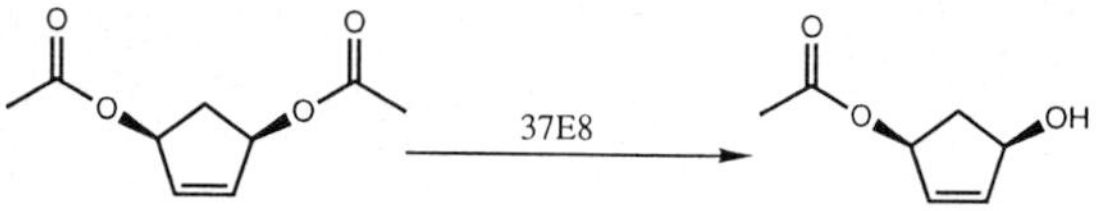

17   **15**

Fig. 9. Antibodies can catalyze stereospecific reactions. Racemic *13* was used to raise a set of antibodies that could hydrolyze one or the other isomer of *14*. The hapten *16* was used to generate antibodies capable of discriminating the acetyl groups on *cis*-3,5-diacetoxycyclopent-1-ene *17* to generate (1R, 4S)-(+)-4-hydroxy-2-cyclopentenyl acetate *15*

believed to proceed through a closed, chair-like transition state (Fig. 10). The enzyme chorismate mutase catalyzes such a pericyclic rearrangement of chorismate (Fig. 10, *19*) to prephenate (Fig. 10, *20*) in a reaction that is inhibited by the analog shown in Fig. 10, *18*, the most potent known inhibitor ($K_i = 0.15\,\mu M$) (BARTLETT and JOHNSON 1985). HILVERT et al. (1988) and JACKSON et al. (1988) independently reported antibodies that effected the transformation of chorismate to prephenate using the oxabicyclic inhibitor as the immunogen. The Schultz group reported that one out of eight antibodies obtained catalyzed the reaction with k_{cat} and K_m values of $2.7\,min^{-1}$ and $260\,\mu M$, respectively. The reaction was inhibited by addition of the hapten with a $K_i = 9\,\mu M$. The rate acceleration afforded by the antibody was reported to be 10000 over the uncatalyzed rate which compares

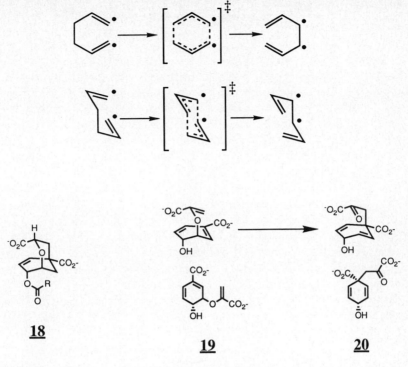

Fig. 10. Pericyclic rearrangements proceed through a chair-type transition state. The oxabicyclic compound *18* is a potent inhibitor of chorismate mutase. It has been used as the immunogen for inducing antibodies that catalyze the rearrangement of chorismic acid *19* to prephenic acid *20*

favorably with the 10^6 rate acceleration for the enzymic reaction. This is especially interesting since the transition state analog binds only about 100 times more tightly than the substrate. The lack of a deuterium isotope effect ($k_D/k_H = 1$) rules out the possibility of acid-base catalysis and the fact that the methyl ether is a substrate rules out mechanisms involving formation of cationic centers. Catalysis is believed to be effected by the antibody acting to freeze out nonproductive conformations and providing a binding pocket that mirrors the transition state. The experimentally determined values of 18.3 kcal mol^{-1} and -1.2 cal k^{-1} mol^{-1} for $\Delta H^{\ddagger}$ and $\Delta S^{\ddagger}$, respectively, support this idea.

HILVERT et al. (1988) obtained 15 antibodies that bound the immunogen and found one to be catalytically active. This antibody effected a rate acceleration of 100 over the uncatalyzed reaction. Furthermore, this antibody was enantiospecific catalyzing the rearrangement solely of $(-)$-chorismate to $(-)$-prephenate (HILVERT and NARED 1988).

II. Entropic Effects

There has been much discussion over the relation between proximity effects and the rate acceleration in biological catalysis (DAFFORN and KOSHLAND 1973; JENCKS 1975). One advantage of a reaction promoted by an enzyme is the entropy gain associated with an intramolecular vs an intermolecular reaction. By analogy with numerous experimental and theoretical measurements, a value of $10^8 M$ represents the approximate upper limit for an enzyme mediated reaction compared to its uncatalyzed bimolecular counterpart (JENCKS 1975). Similarly this advantage could be provided by sequestering two molecules in an antibody combining site in positions favorable for the reaction to take place.

The first successful demonstration of a catalytic antibody mediated bimolecular reaction is shown in Fig. 11 (JANDA et al. 1988b). Saturation kinetics were observed with both substrates (Fig. 11, *21*, *22*) as determined by measurement of initial rates. A comparison of the first order rate constant (k_{cat}) for the antibody catalyzed reaction to the second-order rate constant for the uncatalyzed reaction showed the antibody to provide an effective molarity of $10.5 M$.

Another antibody, generated against the phosphonate shown in Fig. 12, *23*, was shown to catalyze a bimolecular ester reaction (Fig. 12) (WIRSCHING

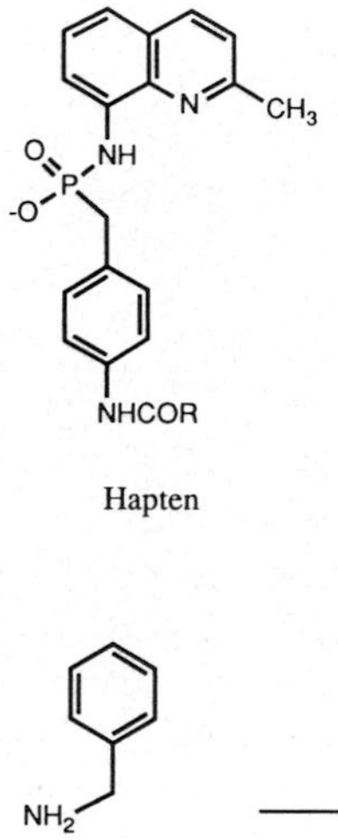

Fig. 11. First successful demonstration of a bimolecular reaction catalyzed by an antibody

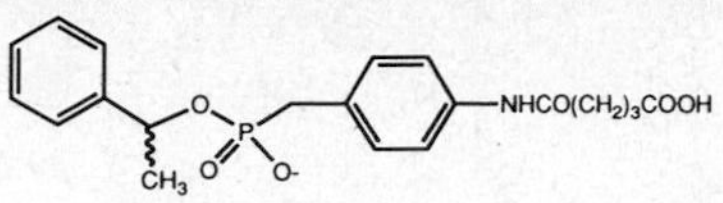

23

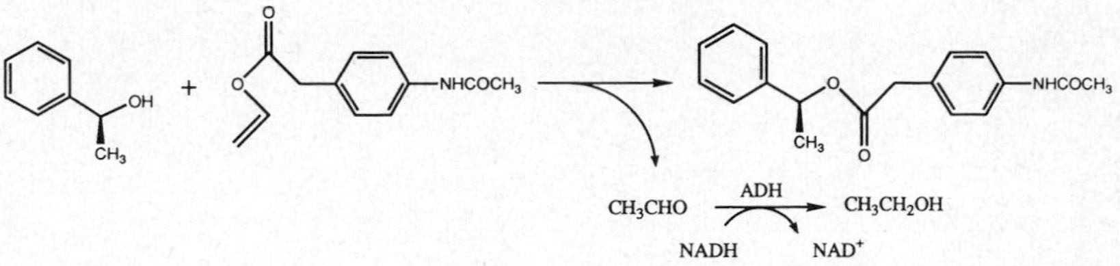

Fig. 12. Transesterification reaction catalyzed by antibody 21H3. The reaction is driven by tautomerization of the enol alcohol to acetaldehyde. The kinetics of the reaction can be followed by using alcohol dehydrogenase (*ADH*) in a coupled assay to reduce the acetaldehyde

et al. 1991). This reaction is one of the most exciting results described in the catalytic antibody field. The antibody 21H3 was able to catalyze a transesterification reaction in water between *sec*-phenethyl alcohol and an enol ester to form the corresponding chiral ester. The transesterification reaction was found to be highly efficient with an effective molarity between 10^6 and $10^8 M$ verifying previous predictions by other workers (vide infra) (JENCKS 1975). In addition, the reaction was found to be largely entropic in nature as reflected by a $\Delta S^{\ddagger}$ of 35 eu. Steady state kinetics showed ping-pong behavior in double reciprocal plots, indicating that the reaction proceeds through two half-reactions in which the initial step consists of formation of a covalent antibody complex. The most remarkable finding was that formation of the acyl-intermediate depends on an induced fit mechanism, since a close analog of *sec*-phenethyl alcohol, *sec*-phenethyl chloride, increases the rate of the acylation reaction.

The two bimolecular antibody catalyzed reactions we have shown are probably of little practical use. However, by using the strategies we have shown, it may be possible to extend this methodology to practical cases. One area would be the condensation of large peptide fragments. This is a problem in solid phase peptide synthesis because of slow reaction rates and unwanted side reactions. A catalytic antibody could be designed to perform this type of amide bond forming reaction without racemization at the α-carbon.

III. Charge Complementarity

Work by Pressman and Pauling in the late 1940s and early 1950s suggested a strategy whereby the electrostatic complementarity between haptens and antibodies could be used to introduce general acids/bases into an antibody combining site (PRESSMAN et al. 1946). It was found that positively charged haptens could elicit negatively charged amino acid residues in the antibody cleft (GROSSBERG and PRESSMAN 1960) and negatively charged haptens could induce complementary positively charged amino acid residues (MAYERS et al. 1973).

Our design methodology, which we term "bait and switch" catalysis, involves the placement of a point charge on the hapten in close proximity to, or in direct substitution for, a chemical functional group we wish to transform in the substrate. The haptenic charge will induce a complementary charge in the antibody binding site. The substrate will lack this charge, but will retain a similar overall structure. The amino acids in the binding sites of the monoclonal antibodies now have the potential to act as general acids/bases for substrates having hydrolyzable functional groups. Haptens and substrates which demonstrated our theory are shown in Fig. 13 (JANDA et al. 1990a).

We found only catalytic antibodies were obtained to N-methylpyridinium hapten (Fig. 13, 24). Antibodies raised to the uncharged pyridine hapten (Fig. 13, 25) bound the substrate but were not catalytic. A pH rate profile of one of the catalytic antibodies revealed participation by the basic form of a dissociable group whose pKa was determined to be 6.26 ± 0.05. In addition, the ratio of k_{cat}/k_o, a comparison of the pH-independent antibody catalyzed rate of the hydrolysis of the ester shown in Fig. 13, 26 to that in water, corresponded to a rate acceleration of over a million-fold. Most significantly, the pH optimum of the antibody catalyzed reaction was shifted into the neutral pH region by participation of an amino acid residue.

Catalytic group(s) have also been introduced into an antibody combining site to generate an antibody catalyzing a β-fluoride elimination reaction (Fig. 13) (SHOKAT et al. 1989). The hapten shown in Fig. 13, 27 was expected to resemble the β-fluoroketone substrate (Fig. 13, 28) with the ammonium group replacing the abstractable α-proton of the substrate. The positively charged alkyl ammonium ion was expected to induce a complementary negatively charged carboxylate residue in the antibody's binding cavity, positioned to function as a general base for β-elimination of hydrogen fluoride. As predicted, monoclonal antibodies obtained were able to catalyze the elimination reaction. Chemical modification and affinity labeling experiments confirmed the presence of an active site carboxylate (SHOKAT and SCHULTZ 1991).

Most recently, we have applied the charge complementarity approach to produce catalytic antibodies for which there are no synthetic or enzymic equivalents (JANDA et al. 1993). Shown in Fig. 14 is a cyclic ring closure reaction catalyzed by an antibody obtained to the N-oxide hapten (Fig. 14, 29). This antibody catalyzed reaction violates what are known as Baldwin's rules (BALDWIN 1976). Furthermore, the antibody catalyzed reaction is regio- and enantiospecific. This is most impressive when one considers that there was no stereochemical bias in the synthesis of the hapten. It is anticipated

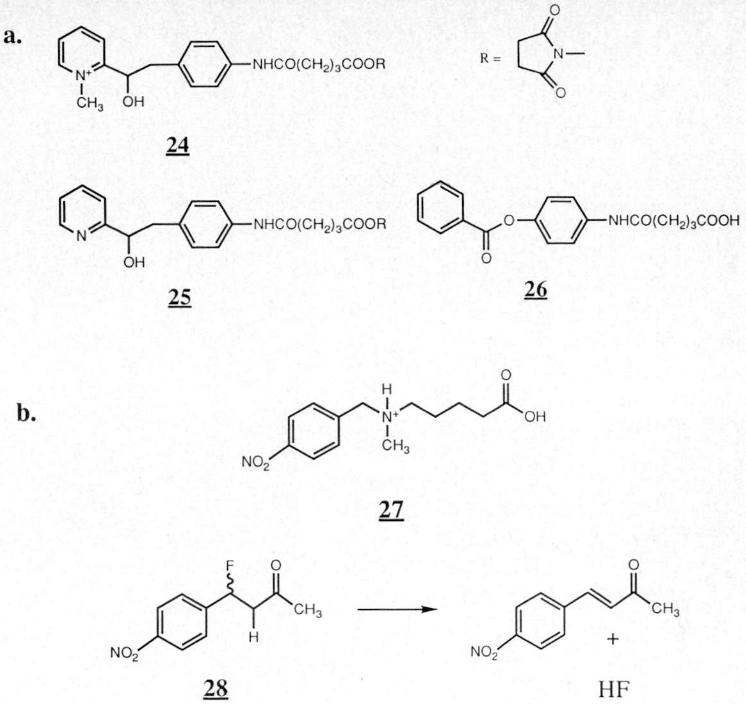

a.

24

25   26

b.

27

28   HF

Fig. 13. a Antigen *24* used in "bait and switch" strategy to elicit antibodies that hydrolyze esters. **b** System used to demonstrate the same technique can be applied toward elimination reactions

that future applications of this hapten-antibody strategy may be useful in total synthesis of natural products. Pyran ring systems like the one shown in Fig. 14, *30* are difficult to synthesize, but are found ubiquitously in the structural makeup of marine natural products (SHIMIZO et al. 1986; LIN et al. 1981) many of which have been shown to be potent antitumor agents.

IV. Solvent Effects

Antibody-hapten charge complementarity is one type of microenvironment which can be induced in an antibody combining site. Just as charged haptens are stabilized by oppositely charged entities in an antibody combining site, hydrophobic haptens are expected to be surrounded by apolar groups. A reaction which is quite susceptible to changes in the medium's dielectric constant is decarboxylation. Model studies indicate that the decarboxylation rate is strongly affected by the medium (TAYLOR 1972; CROSBY et al. 1970; KEMP and PAUL 1975). Furthermore, it is believed that local medium effects

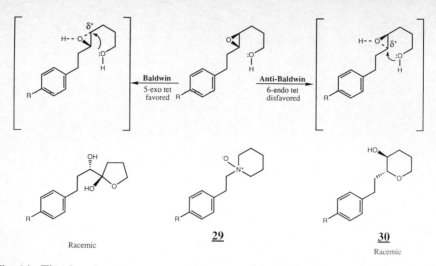

Fig. 14. The ring closure of the epoxy alcohol follows Baldwin's rules to generate the five-membered furan ring. By raising an antibody to the N-oxide *29*, the reaction was induced to proceed by the disfavored 6-endo-tet path way to generate the pyran product *30*

may be an important source of catalytic power in enzymes. Enzymatic "environmental" effects appear to have two components. One is medium polarity, while the second appears to be solvation or ion pairing. Catalytic antibodies present an excellent forum to test these notions of medium effects and catalysis.

Lewis et al. (1991) showed that an antibody raised to the antigen shown in Fig. 15, *31* could catalyze decarboxylation of the substrate shown in Fig. 15, *32*. Saturation kinetics were observed ($k_{cat} = 17 \, min^{-1}$, $K_m = 168 \, \mu M$) and tight binding inhibition was demonstrated. Interestingly, only a small percentage of the hybridomas screened (approximately 2%) were catalytic.

Recently, antibodies elicited to the pyridinium salt shown in Fig. 15, *33* were found to catalyze decarboxylation of the compound shown in Fig. 15, *34* (Ashley et al., unpublished results). Notable features in the hapten design include a cyclopentyl ring which can core out a hydrophobic pocket and the *N*-methylpyridinium moiety which provides a means to anchor the substrate carboxylate in an apolar environment. Steady state kinetics were observed ($k_{cat} = 0.03 \, min^{-1}$, $K_m = 0.2 \, M$) and the rate acceleration over background was approximately 200 000. This antibody catalyzed decarboxylation reaction may give us a better estimate on the magnitudes of "environmental" effects in enzymatic decarboxylations since pyridylacetic acid decarboxylations are thought to be excellent models for enzyme catalyzed decarboxylations (Marlier and O'Leary 1986).

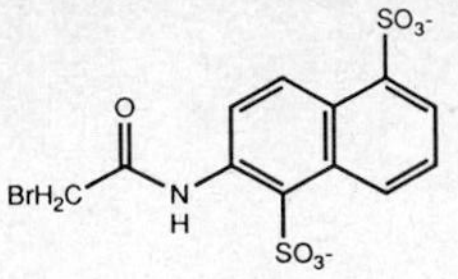

Fig. 15. Decarboxylations show that antibodies can be used to provide a hydrophobic binding pocket for compounds. Such changes in solvent can be used to catalyze reactions such as these decarboxylations

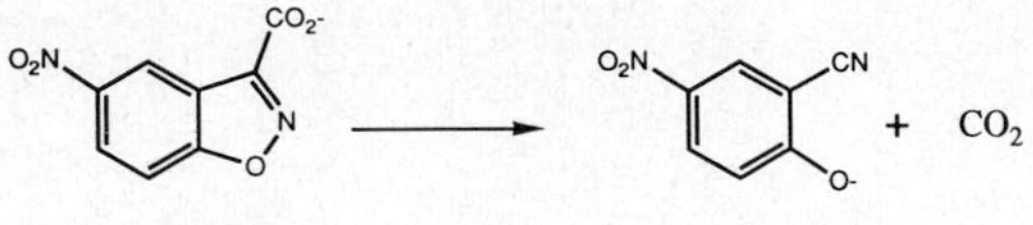

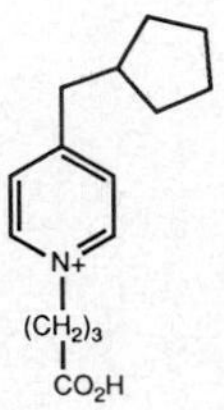

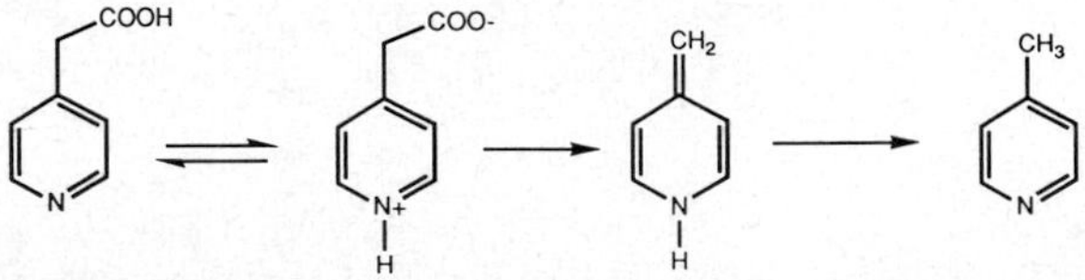

D. Catalytic Antibodies in Organic Solvents

The use of enzymes in organic synthesis is becoming more prominent and is an area that could be stimulated to a considerable degree by the availability of enzymes possessing the appropriate properties (JONES 1976). From the point of view of the synthetic organic chemist, the main attraction of enzymes as catalysts is their ability to discriminate between enantiomers and to distinguish among prochiral groups and faces of molecules. With the emergence of catalytic antibodies, the potential to create specifically tailored catalysts for synthetic transformations has become evident. Two applications of abzymes in synthesis are reactions yielding one or more defined products and the production of biologically active molecules. These goals could be realized if these reactions could be conducted in aqueous buffers as well as in water miscible/immiscible solvents.

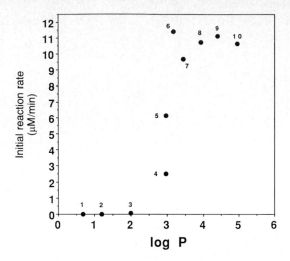

Fig. 16. Initial reaction rates of 21H3 in the transesterification reaction as a function of log P. Initial rates were all measured using $8\,mM$ alcohol and $1\,mM$ vinyl ester with 4% (v/v) buffer. The solvents used were: *1*, ethyl acetate; *2*, propyl acetate; *3*, chloroform; *4*, carbon tetrachloride; *5*, pentane; *6*, cyclohexane; *7*, hexane; *8*, heptane; *9*, octane; *10*, nonane

Antibodies have been shown to carry out reactions on water insoluble substrates in reverse micelles (DURFOR et al. 1988). We have demonstrated that catalytic antibodies (see Fig. 12) immobilized on an inorganic support retain the same activity and stereoselectivity they exhibit in free solution (JANDA et al. 1990b). In addition, the immobilization process imparts added stability in organic solvents and can yield antibodies that are catalytically active in an aqueous dipolar aprotic medium.

More recently, we have found that unimmobilized immunoglobulins can function in apolar solvent media of low water content (ASHLEY and JANDA 1992) using antibodies that catalyze the transesterification reaction (Fig. 13). We determined that one antibody (21H3) can catalyze this transesterification reaction in octane with as little as 0.12% water added. In addition, we found that though this type of media had no effect on the previous observed mechanistic pattern, a sizable increase in K_m for the vinylester was seen. Most interestingly, the reaction seemed to be very dependent on the hydrophobicity of the solvent used as the reaction medium (Fig. 16). As seen in Fig. 16, a sigmoidal correlation exists between the transesterification activity of 21H3 and log P. Activities are virtually non-existent in hydrophilic solvents having a log P <2.0, are variable in solvents having a log P between 2.0 and 3.5, and are high in hydrophobic solvents having log P >4.0. It is anticipated that these results will provide impetus for further exploration of catalytic antibodies in organic solvents.

E. Biological Aspects

I. Hybridoma Techniques

The ability to generate large amounts of *homogeneous* antibody preparations is the key step to obtaining catalytic antibodies. If only a small subset of all the antibodies raised to a hapten are catalytic, it would be extremely difficult to isolate the desired catalytic antibody. It would also be difficult to exclude the possibility the activity arises from trace amounts of a contaminating natural enzyme activity. Although it is difficult to purify one antibody from a larger set of antibodies in a polyclonal mixture, it is possible to purify an antibody by isolating the B lymphocyte that produces that particular antibody, a breakthrough accomplished in 1975 by KöHLER and MILSTEIN (1975), who demonstrated that one can fuse antibody producing cells isolated from immunized mice to immortal myeloma cells to produce hybrids (hybridomas) that carry traits from both progenitors. More importantly, these hybridomas can secrete the antibody and can be maintained in cell culture for an indefinite period of time.

In practice, the procedure involves isolating the spleen cells from a hyperimmunized mouse and fusing them to a myeloma line by treating the cells with a solution of polyethylene glycol. This fusion process gives rise to three populations of cells, unfused spleen cells, unfused myeloma cells, and fused hybridoma cells. The unfused spleen cells will die since they are not adapted for growth in cell culture. The unfused myeloma cells will also die since the culture medium contains aminopterin, a drug that blocks two points in the pathway for purine biosynthesis and the myeloma cells are deficient in genes in the salvage pathway for purine biosynthesis. Only cells containing the antibody secreting genes and genes from the myeloma line will survive. The myeloma derived genes will confer immortality to the hybridoma and the spleen cells will confer genes for resistance to aminopterin. The surviving cells are selected for their ability to bind the antigen used for immunization. Loss of functional antibody genes is manifest as a loss in antibody titer. Thus, only those cells that stably express the antibody genes will be maintained. The cell lines are cloned by limiting dilution and screened for their ability to bind the antigen used for immunization. These "binders" are then assayed for the desired catalytic activity. This usually involves growing a large batch of antibodies, either in cell culture or as ascites in mice, purifying the antibody, and using it in an assay for catalytic activity.

The process of isolating catalytic antibodies by this methodology suffers from some limitations. Of special concern is the ability to generate a large enough array of monoclonal antibodies to screen for the desired activity. The fusion process is not a very efficient one; approximately 1% of the starting cells are fused and a smaller subset form viable hybridomas. Furthermore, one must balance the number of hybridomas one can carry through the cloning process with the time, labor, and expense involved. If the total possible number of antibodies raised to a hapten is on the order of 10^5-10^6, one can only reasonably expect to be able to screen and process 10^2 hybridomas leaving a large proportion of the population unsurveyed. Nevertheless, hybridoma technology is a useful and viable method for isolating catalytic antibodies since all the catalytic antibodies isolated so far have been obtained by this procedure. There are reports in the literature of an autoantibody that can cleave a peptide bond (PAUL et al. 1989, 1990, 1991) and of a polyclonal serum preparation from sheep that catalyzes the hydrolysis of a carbonate ester with kinetic parameters that follow Michaelis-Menten kinetics (GALLACHER et al. 1991).

II. Auxotrophic Selection

Hybridoma technology is a powerful tool for the isolation of catalytic antibodies, but it leaves much room for improvement. It would be particularly helpful if one could screen all the antibodies raised to an antigen for catalytic activity rather than just the ones that survived the fusion and cloning process. Also, it would be of interest if all the binders could be screened for catalytic activity. Currently, only the cell lines that secrete the tightest binders are usually screened for activity. This arbitrary discrimination might exclude potential catalysts. Direct screening for catalytic activity would greatly improve the ease with which catalytic antibodies could be selected.

The ability to clone the immune response into *E. coli* has made it possible to screen all the antibodies raised against a particular hapten for binding, and potential catalysis (SASTRY et al. 1989; HUSE et al. 1989; KANG et al. 1991; BARBAS et al. 1991). The random combination of heavy and light chains can lead to new heavy and light chain combinations and increase the total number of antibodies and possible number of catalysts (KANG et al. 1992). With the ability to generate large libraries of antibodies (10^7-10^8) comes the problem of screening these antibodies for activity. The approach of screening for binding and then for catalysis is inefficient because of the large number of clones involved. The best way to isolate catalytic antibodies would be to assay the antibodies directly for a particular activity or trait. One suggested approach is to utilize a chromogenic assay to directly screen bacterial colonies or plaques (GONG et al. 1992). The method uses a linker connected to a substrate and a chromogenic agent (Fig. 17). Hydrolysis of the ester yields the linker attached to the indolyl group which is subsequently released by intramolecular hydrolysis. The resulting 3-hydroxyindole undergoes oxidative dimerization to form indigo which is deposited as a purple precipitate.

A general strategy would be to screen for the conferral of a particular phenotype that could only arise from the presence of a catalytic antibody. This is similar to the various methods used in molecular biology to screen

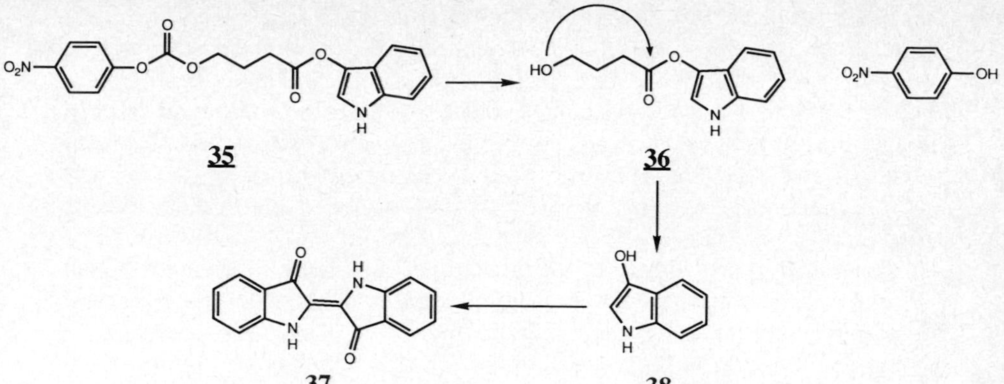

Fig. 17. A method for screening libraries of antibodies for catalytic activity by generation of a precipitatable chromophore

for the presence of plasmids by the acquisition of antibiotic resistance or the insertion of DNA fragments by the disruption of a gene function. The viability of such an approach to detect low levels of catalytic activity has been demonstrated. EVNIN et al. (1990) utilized a genetic selection to screen a library of trypsin mutations. The selection used as host cell a strain of *E. coli* (X90) which is deficient in an enzyme required for arginine biosynthesis. Mutants were assayed for their ability to complement this deficiency when cells were grown on arginine β-naphthylamide as the sole arginine source. Trypsin can hydrolyze arginine β-naphthylamide to release arginine, thus only functional mutations that hydrolyze the substrate to free arginine for use by the cells can permit the cells to grow. This selection method is extremely sensitive, being able to isolate mutations that reduce catalytic activity 10^5-fold compared to wild-type trypsin.

The validity of such a method applied to catalytic antibodies has also been demonstrated using the antibody previously described which catalyzes the chorismate to prephenate transformation (Fig. 11) (HILVERT et al. 1988; JACKSON et al. 1988). BOWDISH et al. (1991) constructed a yeast strain (YT-4Ca) deficient in chorismate mutase activity by inserting the HIS3 gene into the AR07 locus (encoding for chorismate mutase in yeast) and selecting for cells unable to grow on tyrosine. The genes encoding the light chain and Fd portion of the antibody 1F4 (HILVERT et al. 1988) were isolated and cloned and expressed in yeast. The auxotrophy of the yeast strain for tyrosine, however, was not complemented presumably due to the extremely low (0.01 min^{-1}) catalytic efficiency of the antibody. In an effort to augment the activity of the antibody, the antibody-harboring yeast cells were chemically mutagenized and screened for their ability to rescue the chorismate mutase deficiency (TANG et al. 1991). From screening 10^9 mutants, 298 colonies that grew under auxotrophic selection were obtained. Serial transformation of the plasmid into a mutant strain (351 m) derived from the chemical mutagenesis treatment conferred the ability to complement the deficiency in chorismate mutase, demonstrating the ability to complement auxotrophy was associated with the plasmid bearing the Fab genes. Transformation of an irrelevant Fab encoding for an esterolytic catalytic antibody did not suppress the deficiency in chorismate mutase. The plasmid, however, was not able to permit auxotrophic growth when transformed into the original YT-4Ca strain.

These results suggest that the growth advantage was due not to an increase in the efficiency of the catalytic antibody but to other factors, namely, an increase in basal chorismate mutase activity. One possibility is that intracellular expression or stability of the Fab has been improved. Another possibility is that the mutagenesis treatment of the whole cells has induced a revertant that expresses a slightly functional chorismate mutase; the activity of the Fab serves to push the threshold chorismate mutase activity to a level that permits complementation of the auxotrophy. A better tact for using genetic selection to select for antibodies that are more efficient catalysts using this system would be to perform chemical mutagenesis on the plasmid and transform the mutagenized DNA into the yeast strain. By using this approach, one localizes any mutagenesis to the DNA of interest and would not run the risk of inducing yeast revertants.

Genetic selection has also been used in an attempt to isolate catalytic antibodies capable of hydrolyzing peptide bonds. (The sequence-selective cleavage of peptide bonds is an attractive target for novel catalysts.) Haptens that mimic the transition state for hydrolysis of the amide bond between phenylalanine and proline or phenylalanine (Phe-Pro, Phe-Phe) via a phosphonamidate substitution for the scissile bond were synthesized and used to immunize mice. A combinatorial library of antibodies was constructed in a phagemid system and the library screened for the ability to rescue E. coli strains from auxotrophy for phenylalanine or proline. Growth of discrete colonies was observed, but analysis of the plasmids showed deletion of the genes encoding for the heavy and light chains indicating spontaneous reversion had taken place.

Antibodies capable of hydrolyzing amide bonds have also been isolated by genetic selection. Chloramphenicol is an antibiotic that acts by binding to the 50S ribosome and inhibiting protein synthesis. It is inactivated by chloramphenicol transferase which transfers an acetyl group from acetyl-CoA to one of the free hydroxyls on the propionyl group. Acetylation appears to prevent the binding of chloramphenicol to its target site in the 50S ribosome. Chloramphenicol can also be cleared by hydrolysis of the dichloroacetyl group to generate the chloramphenicol free base (Fig. 18).

We have synthesized a hapten in which the dichloroacetyl group is replaced by a dichloromethylphosphonamidate. This compound was used to hyperimmunize mice and a library of antibodies was constructed on the surface of filamentous phage. These antibodies were shown to be able

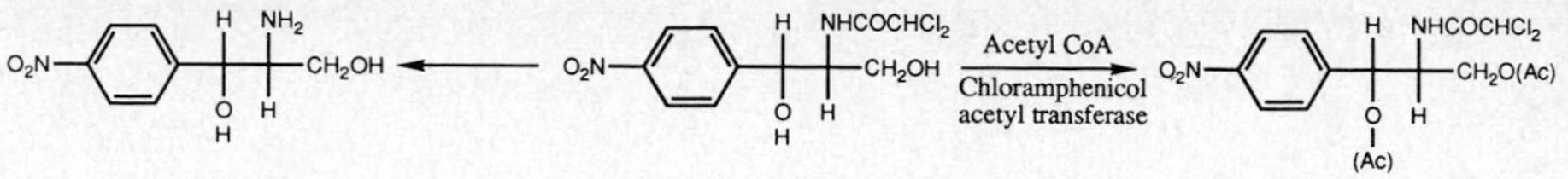

Fig. 18. Two modes for detoxification of chloramphenicol. The natural enzyme chloramphenicol acetyl transferase catalyzes the transacetylation of chloramphenicol from the acetate donor acetyl-CoA. The free base formed by hydrolysis of the dichloroacetate group is also inactive

to bind the immunogen and conferred a chloramphenicol resistance to normally sensitive *E. coli* cells. Cells infected with a control antibody previously shown to bind the Phe-Pro hapten were sensitive to chloramphenicol. Surprisingly, the cells infected with the vector without light or heavy chain genes were slightly resistant to chloramphenicol (1 μg/mL) but died upon exposure to higher levels of chloramphenicol.

It is not clear by what mechanism the antibodies are able to confer resistance to chloramphenicol. It is gratifying to see that cells infected with an irrelevant antibody are sensitive but puzzling that the wild-type vector shows some resistance. We are attempting large scale expression of one antibody in an effort to determine how it confers resistance to chloramphenicol.

III. Expression Methods

A variety of methods exist for the expression of antibodies and are covered in other chapters of this volume. Here, we focus on the problem of expression as it applies specifically to catalytic antibodies.

The choice of expression system is a complicated one and must take into account a number of issues. The ease and reliability of one method might be counterbalanced by negative factors which make it an inappropriate choice given the desired goals. In the case of catalytic antibodies, one would like a system that yields high level expression of properly folded, functional gene products, permits easy purifications, is readily accessible to all labs, and is amenable to genetic manipulations. *E. coli* would be the ideal system in which to work since it fulfills most of these criteria. If one wishes to generate a small amount of antibody for initial screens and binding studies, then the use of *E. coli* provides the fastest and easiest method to obtain antibodies. For catalytic antibodies, there is an additional issue. The turnover numbers for most of the catalytic antibodies are modest, so one necessarily needs more antibody in order to detect turnover in the assays for catalytic activity. The *E. coli* systems are powerful means for generating and screening antibodies, but they can be limited in their utility for expressing antibodies.

The ability to express antibodies in *E. coli* is well documented (CABILLY et al. 1984; BETTER et al. 1988; BIRD et al. 1988; CONDRA et al. 1990; UDAKA et al. 1990; BUCKNER and RUDOLPH 1991; DAVIS et al. 1991; CARTER et al.

1992) with overproduction levels ranging from low (CABILLY et al. 1984) to spectacular (CARTER et al. 1992). The observation that yields are variable and appear to be independent of the promoters used suggests antibody expression in *E. coli* is not limited by problems with transcription or translation, but with secretion and folding. Expression yields of functional antibody seem to depend to a large extent on the particular antibody. Thus, the expression problem must be solved for each antibody.

If an antibody can be readily expressed in *E. coli*, then that should be the method of choice. There are instances, however, in which an interesting antibody cannot be expressed to high levels in *E. coli*. One has the option of inserting the gene sequences into yeast shuttle vectors or mammalian shuttle vectors. Yeast expression appears to generate low levels of chain assembly and high levels of insoluble antibody (WOOD et al. 1985). Mammalian systems, by contrast, are reliable sources for steady production of antibodies. Nonetheless, yields can be low and problems can be encountered in generating stable transfectant lines.

An alternative method, which we and others have used with success to express antibodies (HASEMANN and CAPRA 1990, 1992; ZU PUTLITZ et al. 1990), is to use the baculovirus system originally described by Summers (SMITH et al. 1983a; SUMMERS and SMITH 1987). This system has been used to express a large variety of different proteins, some of which are not readily expressed by any other means (MIYAMOTO et al. 1985; SMITH et al. 1985; WEBB et al. 1989; DEVLIN et al. 1989; JARVIS et al. 1990; MEDIN et al. 1990). One advantage of using insect cells is that they perform posttranslational modifications such as phosphorylation and glycosylation, a function not seen with bacterial systems. This is useful for expression of full-length antibodies. A second advantage is that yields can approach 50% of total cell protein, an overproduction level not achievable by mammalian cell lines.

The primary interest lies in the polyhedrin protein since it accounts for 50%–75% of the total stainable protein in the occlusion bodies. The gene encoding for the polyhedrin protein is nonessential (SMITH et al. 1983b) and can be replaced by other genes. Indeed, it has been shown that the polyhedrin promoter can be coopted for expression of foreign gene products without any adverse affects on the virus (SUMMERS and SMITH 1987; BISHOP 1990).

The general scheme for expressing foreign genes in baculovirus is to insert the gene of interest into an *E. coli* transfer vector downstream of the polyhedrin promoter in place of or disrupting the polyhedrin gene. This vector is then cotransfected (BURAND et al. 1980) into a host cell (*Spodoptera frugiperda*) along with wild-type viral DNA (Autographa californica nuclear polyhedrosis virus, AcNPV). Both wild-type and recombinant (foreign gene containing) gene loci will be present in the population of infected cells. A certain number of recombinant viruses will be formed by homologous recombination involving sequences flanking the polyhedrin gene. These are then isolated and propagated for further study.

In practice, the isolation of recombinant viruses is a difficult process since the only phenotypic trait is the inability to form polyhedrin protein. This is screened by plating infected cells in semisolid agarose and looking for cells that are infected (plaques) but lack occlusion bodies and, therefore, do not refract light. Other methods, such as screening plaque lifts with antibodies against the gene product, coexpression of a selectable marker such as β-galactosidase, and limiting dilution can be used, though none offer a distinct advantage. The major problem appears to be the low frequency (0.01%) with which transfection and recombination takes place. The use of linearized viral DNA as the cotransfectant seems to increase the efficiency of transfection and the number of recombinants. A new procedure has been introduced that increases the efficiency of isolating recombinants. Viral DNA containing a lethal gene inserted in the loci of the polyhedrin gene is used as the cotransfectant. Insect cells containing this engineered viral DNA die whereas cells infected with recombinant viruses, in which this gene segment has been swapped, survive. The efficiency of forming recombinants is reported to be >99%.

We have isolated an antibody cloned from the immunological repertoire of a hyperimmunized mouse into λ phage. Initial studies indicated this antibody possessed catalytic activity, but the expression levels in *E. coli* were too low to obtain enough Fab for kinetic studies. In an effort to obtain enough Fab to assay, we chose to express one clone (1D) in the baculovirus expression system.

The heavy and light chain genes for this clone were individually amplified from the original plasmid via PCR to introduce the necessary restriction sites for cloning into the baculovirus transfer vector pAC360E and a consensus mouse leader sequence. A single heavy chain gene and a single light chain gene were individually cloned into the vector pAC360E, combined into a single plasmid designated pAC.1D, and transfected into Sf9 cells. Viral stocks were isolated and individual viruses were used to infect small scale cultures and the supernatants assayed for the ability to bind the hapten via ELISA. The viral line producing the greatest amount of Fab was selected and used for large scale expression. The expressed antibody was judged to be >95% pure as determined by silver staining of SDS-polyacrylamide gels and was observed to cross-react with anti-murine Fab and anti-κ antibodies on western blots. Furthermore, mock-infected cells (infected with wild-type virus) and uninfected cells did not yield protein that bound the immunogen as judged by ELISA. The antibody was observed to hydrolyze the substrate with $K_m = 115\,\mu M$ and $k_{cat} = 0.25\,min^{-1}$.

The transfer vector constructed for this study has the heavy and light chains oriented in a divergent dicistronic arrangement to ensure that both chains would end up in the virus. This is apparently not necessary. If viruses containing either the heavy or light chain are used in a double infection, some cells will express both chains. The two chains will be secreted into the media where they will combine and form functional antibody. One, therefore, has the flexibility to mix different heavy and light chains. If a stock of different heavy and light chains were available, then one can easily perform random chain shuffling experiments to generate new antibodies.

By transferring the antibody encoding genes into insect cells, we have been able to increase the expression level of a particular Fab approximately tenfold from <0.5 mg/L in *E. coli* to 5 mg/L in the baculovirus system. One should note that the expression level of foreign genes expressed in baculovirus seems to be clone-dependent, with different clones expressing with different efficiencies. Nevertheless, the increase in yield has allowed us to rigorously purify the antibody and kinetically characterize it.

The choice of an expression system must take into consideration the goals one has and how a particular system will meet or fall short of those expectations. All else being equal, the easiest system with which to work is *E. coli*. The baculovirus system serves as an alternative to *E. coli* for expression of antibodies. The antibodies expressed in insect cells are functional and are easily purified. Since the transfer vectors themselves are maintained in *E. coli*, the option of performing genetic manipulations remains. The process of isolating pure recombinants is a difficult one, though recent advances have made this less of a laborious process. Finally, the expression levels are higher than those obtained in *E. coli*.

F. Prospects

Antibodies capable of catalyzing a variety of chemical transformations including hydrolytic, bimolecular, carbon-carbon bond forming and redox reactions have been isolated. It has also been shown that antibodies can be immobilized onto solid supports and used as catalysts in small bioreactors. Such antibodies even retain their activity in organic solvents. The fact that antibodies have also been shown to perform enantioselective reactions and reactions that disobey empirical chemical rules indicates that the goal of using this method to create novel catalysts with tailored specificities and properties is achievable. The finding that these catalysts are active in non-natural settings suggests that catalytic antibodies might be useful in industrial applications.

Catalytic antibodies will have the most impact for performing reactions for which no known enzyme exists. Such targets would include sequence-specific cleavage of peptide bonds, ligation of peptide bonds, sequence-specific cleavage of RNA molecules, sequence-specific cleavage of oligosaccharides, and linkage of oligosaccharides. Antibodies might also serve a role in natural products synthesis by catalyzing difficult transformations under mild conditions. Catalytic antibodies might also serve as therapeutic agents, the "magic bullet" that combines exquisite specificity with rapid destruction of the invasive agent.

There is room for improvement of catalytic antibodies. The turnover numbers for most of the isolated catalysts are modest. Thus, it appears that these antibodies represent first-generation catalysts, such as primordial enzymes. Catalytic antibodies provide an opportunity to imitate evolution in the laboratory by increasing the catalytic efficiency of a slow catalyst or by converting an antibody that binds but has no activity into one that possesses catalytic activity. One can envision a scheme by which catalytic antibodies are subjected to multiple rounds of mutagenesis and selection to mimic the slow evolution of natural enzymes, but at a much more rapid pace. Such a project requires a facile means for generating libraries of antibodies, a sensitive means of screening for new up-mutants, and a general means for efficiently expressing large quantities of antibodies. Much progress has been made in creating libraries of antibodies; much work remains to be done in screening for and expressing catalytic antibodies.

Catalytic antibodies have had a long gestation period from the initial conception to the first demonstration. A great deal of progress has been made in a short period of time in a field in which the ideas have always far outpaced the technical capabilities. The work done so far has shown that catalytic antibodies can accomplish truly exciting deeds. The work to be done should show that catalytic antibodies will ultimately fulfill their potential.

Acknowledgements. This work was supported in part by the NIH, NSF, ALFRED P. SLOAN Foundation (KDJ) and the JANE COFFIN CHILDS Memorial Fund for Medical Research (Y.C.J.C.).

References

Ashley JA, Janda KS (1992) Antibody catalysis in low water content media. J Org Chem 57:6691–6693

Baldwin J (1976) Rules for ring closure. J Chem Soc Chem Commun, pp 734–736

Barbas CF, Kang AS, Lerner RA, Benkovic SJ (1991) Assembly of combinatorial antibody libraries on phage surfaces: the gene III site. Proc Natl Acad Sci USA 88:7978–7982

Bartlett PA, Johnson CA (1985) An inhibitor of chorismate mutase resembling the transition-state conformation. J Am Chem Soc 10:7792

Benkovic SJ, Napper AD, Lerner RA (1988) Catalysis of a stereospecific bimolecular amide synthesis by an antibody. Proc Natl Acad Sci USA 85:5355–5538

Better M, Chang CP, Robinson RR, Horwitz AH (1988) *Escherichia coli* secretion of an active chimeric antibody fragement. Science 240:1041–1043

Bird RE, Hardman KD, Jacobson JW, Johnson S, Kaufman BM, Lee S-M, Lee T, Pope S, Riordan GS, Whitlow M (1988) Single-chain antigen-binding proteins. Science 242:423–426

Bishop DHL (1990) Gene expression using insect cells and viruses. Curr Opin Biotech 1:62–67

Bowdish K, Tang Y, Hicks JB, Hilvert D (1991) Yeast expression of a catalytic antibody with chorismate mutase activity. J Biol Chem 266:11901–11908

Brinkley M (1992) A brief survey of methods for preparing protein conjugates with dyes, haptens, and cross-linking reagents. Bioconjugate Chem 3:2–13

Buckner J, Rudolph R (1991) Renaturation, purification and characterization of recombinant Fab-fragments produced in *Escherichia coli*. Bio/Technology 9: 157–162

Burand JP, Summers MD, Smith GE (1980) Transfection with baculovirus DNA. Virology 101:286–290

Cabilly S, Riggs AD, Pande H, Shively JE, Holmes WE, Rey M, Perry LJ, Wetzel R, Heyneker HL (1984) Generation of antibody activity from immunoglobulin polypeptide chains produced in *Escherichia coli*. Proc Natl Acad Sci USA 81:3273–3277

Carter P, Kelley RF, Rodrigues ML, Snedecor B, Covarrubias M, Velligan MD, Wong WLT, Rowland AM, Kotts CE, Carver ME, Yang M, Bourell JH, Shepard HM, Henner D (1992) High level *Escherichia coli* expression and production of a bivalent humanized antibody fragment. Bio/Technology 10: 163–167

Condra JH, Sardana VV, Tomassini JE, Schlaback AJ, Davies M-E, Lineberger DW, Graham DJ, Gotlib L, Colonno RJ (1990) Bacterial expression of antibody fragments that block human rhinovirus infection of cultured cells. J Biol Chem 265:2292–2295

Craik CS, Largman C, Fletcher T, Roczniak S, Barr PJ, Fletterick R, Rutter WJ (1985) Redesigning trypsin: alteration of substrate specificity. Science 228:291–297

Crosby J, Stone R, Lienhard GE (1970) Mechanisms of thiazmine-catalyzed reactions. Decarboxylations. of 2-(1-carboxy-1-hydroxyethyl)-3,4-dimethylthiazolium chloride. J Am Chem Soc 92:2891–2900

Dafforn A, Koshland DE (1973) Proximity, entropy and orbital steering. Biochem Biophys Res Commun 52:779–783

Davis GT, Bedzyk WD, Voss EW, Jacobs TW (1991) Single chain antibody (SCA) encoding gene: one-step construction and expression in eukaryotic cells. Bio/Technology 9:165–169

Devlin JJ, Devlin PE, Clark R, O'Rourke EC, Levenson C, Mark DF (1989) Novel expression of chimeric plasminogen activators in insect cells. Bio/Technology 7:286–292

Durfor CN, Bolin RJ, Sugasawara RJ, Massey RJ, Jacobs JW, Schultz PG (1988) Antibody catalysis in reverse micelles. J Am Chem Soc 110:8713–8714

Eisenberg D, Wilcox W, Eshita SM, Pryciak PM, Ho SP, DeGrado WF (1986) The design, synthesis and crystallization of an alpha-helical peptide. Proteins 1:16–22

Engvall E (1980) Enzyme immunoassay ELISA and EMIT. Methods Enzymol 70:419–438

Erlanger B (1980) The preparation of antigenic hapten-carrier conjugates: a survey. Methods Enzymol 70:85–103

Evnin LB, Vásquez JR, Craik CS (1990) Substrate specificity of trypsin investigated by using a genetic selection. Proc Natl Acad Sci USA 87:6659–6663

Gallacher G, Jackson CS, Searcey M, Badman GT, Goel R, Topham CM, Mellor GW, Brocklehurst K (1991) A polyclonal antibody preparation with Michaelian catalytic properties. Biochem J 279:871–881

Gong B, Lesley SA, Schultz PG (1992) A chromogenic assay for screening large antibody libraries. J Am Chem Soc 114:1486–1487

Goraj K, Renard A, Martial JA (1990) Purification and initial structural characterization of octarellin, a de novo polypeptide modelled on the α/β barrel proteins. Protein Eng 3:259–266

Grossberg AL, Pressman D (1960) Nature of the combining site of antibody against a hapten bearing a positive charge. J Am Chem Soc 82:5478–5482

Hahn KW, Wieslaw AK, Stewart JM (1990) Design and synthesis of a peptide having chymotrypsin-like esterase activity. Science 248:1544–1547

Handel TM, DeGrado WF (1990) De novo design of a Zn^{2+} binding protein. J Am Chem Soc 112:6710–6711

Hasemann CA, Capra JD (1990) High-level production of a functional immunoglobulin heterodimer in a baculovirus expression system. Proc Natl Acad Sci USA 87:3942–3946

Hasemann CA, Capra JD (1992) Baculovirus expression of antibodies: a method for the expression of complete immunoglobulins in a eukaryotic host. Methods 2:146–158

Hilvert D, Nared KD (1988) Stereospecific Claisen rearrangement catalyzed by an antibody. J Am Chem Soc 110:5593–5594

Hilvert D, Carpenter SH, Nared KD, Auditor M-TM (1988) Catalysis of concerted reactions by antibodies: the Claisen rearrangement. Proc Natl Acad Sci USA 85:4953–4955

Ho SP, DeGrado WF (1987) Design of a 4-helix bundle protein: synthesis of peptides which self-associate into a helical protein. J Am Chem Soc 109:6751–6758

Huse WD, Sastry L, Iverson SA, Kang AS, Alting-Mees M, Burton DR, Benkovic SJ, Lerner RA (1989) Generation of a large combinatorial library of the immunoglobulin repertoire in phage lambda. Science 246:1275–1281

Ikeda S, Weinhouse MI, Janda KD, Lerner RA (1991) Asymmetric induction via a catalytic antibody. J Am Chem Soc 113:7763–7764

Jackson DY, Jacobs JW, Sugasawara R, Reich SH, Bartlett PA, Schultz PG (1988) An antibody-catalyzed Claisen rearrangement. J Am Chem Soc 110:4841–4842

Jacobs J, Schultz PG (1987) Catalytic antibodies. J Am Chem Soc 109:2174–2176

Jacobson NE, Bartlett PA (1981) A phosphonamidate dipeptide analogue as an inhibitor of carboxypeptidase A. J Am Chem Soc 103:654–657

Janda KD, Schloeder D, Benkovic SJ, Lerner RA (1988a) Induction of an antibody that catalyzes the hydrolysis of an amide bond. Science 241:1188–1191

Janda KD, Lerner RA, Tramontano A (1988b) Antibody catalysis of bimolecular amide formation. J Am Chem Soc 110:4835–4837

Janda KD, Benkovic SB, Lerner RA (1989) Catalytic antibodies with lipase activity and R or S substrate selectivity. Science 244:437–440

Janda KD, Weinhouse MI, Schloeder DM, Lerner RA, Benkovic SJ (1990a) Bait and switch strategy for obtaining catalytic antibodies with acyl-transfer capabilities. J Am Chem Soc 112:1274–1275

Janda KD, Ashley JA, Jones TM, Mcleod DA, Schloeder DM, Weinhouse MI (1990b) Immobilized catalytic antibodies in aqueous and organic solvents. J Am Chem Soc 112:8886–8888

Janda KD, Shevlin CG, Lerner RA (1993) Antibody catalysis of a disfavored chemical transformation. Science 259:490–493

Jarvis DL, Fleming J-AGW, Kovacs GR, Summers MD, Guarino LA (1990) Use of early baculovirus promoters for continuous expression and efficient processing of foreign gene products in stably transformed lepidopteran cells. Bio/Technology 8:950–955

Jencks WP (1975) Binding energy, specificity, and enzymic catalysis: the Circe effect. Adv Enzymol 43:219–410

Jencks WP (1987) Catalysis in chemistry and enzymology. Dover Mineola

Johnson LN (1992) Time-resolved protein crystallography. Protein Sci 1:1237–1243

Jones JB (1976) On the potential of soluble and immobilized enzymes in synthetic organic chemistry. Methods Enzymol 44:831–843

Kang AS, Barbas CF, Janda KD, Benkovic SJ, Lerner RA (1991) Linkage of recognition and replication functions by assembling combinatorial antibody Fab libraries along phage surfaces. Proc Natl Acad Sci USA 88:4363–4366

Kang AS, Jones TM, Burton DR (1992) Antibody redesign by chain shuffling from random combinatorial immunoglobulin libraries. Proc Natl Acad Sci USA 88:11120–11123

Kemp DS, Paul KG (1975) The physical organic chemistry of benzisoxazoles. III. The mechanism and the effects of solvents on rates of decarboxylation of benzisoxazole-3-carboxylic acids. J Am Chem Soc 97:7305–7317

Knowles JR (1987) Tinkering with enzymes: what are we learning? Science 236:1252–1258

Köhler G, Milstein C (1975) Continuous cultures of fused cells secreting antibodies of predefined specificities. Nature 256:495–497

Lewis C, Kramer T, Robinson S, Hilvert D (1991) Medium effects in antibody-catalyzed reactions. Science 253:1019–1022

Lin YY, Risk M, Ray SM, Van Engen D, Clardy JC, Golick J, James JC, Nakanishi K (1981) Isolation and structure of brevetoxin B from the "red tide" dinoflagellate *Ptychodicus brevis (Gymnodiniumbreve)*. J Am Chem Soc 107:6773–6775

Mandecki W, Shallcross MA, Sowadski J, Tomazic-Allen S (1991) Mutagenesis of conserved residues within the active site of *Escherichia coli* alkaline phosphatase yields enzymes with increased k_{cat}. Protein Eng 4:801–804

Marlier JF, O'Leary MH (1986) Solvent dependence of the carbon kinetic isotope effect on the decarboxylation of 4-pyridylacetic acid. A model for enzymatic decarboxylations. J Am Chem Soc 108:4896–4899

Mayers GL, Grossberg AL, Pressman D (1973) Arginine and lysine in binding sites of anti-4-azophthalate antibodies. Immunol Chem 10:37–41

Medin JA, Hunt L, Gathy K, Evans RK, Coleman MS (1990) Efficient, low-cost protein factories: expression of human adenosine deaminase in baculovirus-infected insect larvae. Proc Natl Acad Sci USA 87:2760–2764

Miyamoto C, Smith GE, Farrell-Towt J, Chizzonite R, Summers MD, Ju G (1985) Production of human c-*myc* protein in insect cells infected with baculovirus expression vector. Mol Cell Biol 5:2860–2865

Napper AD, Benkovic SJ, Tramantano A, Lerner RA (1987) A stereospecific cyclization catalyzed by an antibody. Science 237:1041–1043

Nishima T, Tsuji A, Fukushima DK (1974) Site of conjugation of bovine serum albumin to corticosteroid hormones and specificity of antibodies. Steroids 24:861–874

Paul S, Volle DJ, Beach CM, Johnson DR, Powell MJ, Massey RJ (1989) Catalytic hydrolysis of vasoactive intestinal peptide by human autoantibody. Science 244:1158–1162

Paul S, Volle DJ, Powell MJ, Massey RJ (1990) Site-specificity of a catalytic vasoactive intestinal peptide antibody: an inhibitory VIP subsequence distant from the scissile peptide bond. J Biol Chem 265:11910–11913

Paul S, Johnson DJ, Massey R (1991) Binding and multiple hydrolytic sites in epitopes recognized by catalytic anti-peptide antibodies. Ciba Found Symp 159:156–173

Pauling L (1948) Chemical achievement and hope for the future. Am Sci 36:51–58

Pollack SJ, Jacobs JW, Schultz PG (1986) Selective chemical catalysis by an antibody. Science 234:1570–1573

Pressman D, Grossberg AC, Pence LH, Pauling L (1946) The reactions of antiserum homologous to the p-azophenyltrimethylammonium group. J Am Chem Soc 68:250–255

Raso V, Stollar D (1975) The antibody-enzyme analogy. Comparison of enzymes and antibodies specific for phosphopyridoxyltyrosine. Biochemistry 14:584–599

Regan L, Clarke ND (1990) A tetrahedral zinc(II)-binding site introduced into a designed protein. Biochemistry 29:10879–10883

Regan L, DeGrado WF (1988) Characterization of a helical protein designed from first principles. Science 241:976–978

Sastry L, Alting-Mees M, Huse WD, Short JM, Sorge JA, Hay BN, Janda KD, Benkovic SJ, Lerner RA (1989) Cloning of the immunological repertoire in *Eschericia coli* for generation of monoclonal catalytic antibodies: construction of a heavy chain variable region-specific cDNA library. Proc Natl Acad Sci USA 86:5728–5732

Shimizo Y, Chou H-N, Bandu H, Van Dvyne G, Varky JC (1986) Structure of brevetoxin A (GB-1 toxin), the most potent toxin in the Florida red tide organism, *Gymnodinium breve (Ptychodiscus brevis)*. J Am Chem Soc 108:514–515

Shokat KM, Schultz PG (1991) The generation of antibody combining sites containing catalytic residues. Ciba Found Symp 159:118–135

Shokat KM, Leumann CL, Sugasawara R, Schultz PG (1989) A new strategy for the generation of catalytic antibodies. Nature 338:269–271

Smith GE, Summers MD, Fraser MJ (1983a) Production of human beta interferon in insect cells infected with a baculovirus expression vector. Mol Cell Biol 3:2156–2165

Smith GE, Fraser MJ, Summers MD (1983b) Molecular engineering of the *Autographa californica* nuclear polyhedrosis virus genome: deletion mutations within the polyhedron gene. J Virol 46:584–593

Smith GE, Ju G, Ericson BL, Moschera J, Lahm H-W, Chizzonite R, Summers MD (1985) Modification and secretion of human interleukin 2 produced in insect cells by a baculovirus expression vector. Proc Natl Acad Sci USA 82:8404–8408

Summers JB (1983) Catalytic principles of enzyme chemistry. PhD thesis, Harvard University, New Haven

Summers MD, Smith GE (1987) A manual of methods for baculovirus vectors and insect cell culture procedures. Texas Agricultural Experiment Station, College Station, Texas

Tang Y, Hicks JB, Hilvert D (1991) In vivo catalysis of a metabolically essential reaction by an antibody. Proc Natl Acad Sci USA 88:8784–8786

Taylor PJ (1972) The decarboxylation of some heterocyclic acetic acids. J Chem Soc, pp 1077–1086

Tramantano A, Janda KD, Lerner RA (1986a) Chemical reactivity at an antibody binding site elicited by mechanistic design of a synthetic antigen. Proc Natl Acad Sci USA 83:6736–6740

Tramantano A, Janda KD, Lerner RA (1986b) Catalytic antibodies. Science 234:1566–1570

Tramantano A, Ammann AA, Lerner RA (1988) Antibody catalysis approaching the activity of enzymes. J Am Chem Soc 110:2282–2286

Udaka K, Chua M-M, Tong L-H, Karush F, Goodgal SH (1990) Bacterial expression of immunoglobulin V_H proteins. Mol Immunol 27:25–35

Webb NR, Madoulet C, Tosi P-F, Broussard DR, Sneed L, Nicolau C, Summers MD (1989) Cell-surface expression and purification of human CD4 produced in baculovirus-infected insect cells. Proc Natl Acad Sci USA 86:7731–7735

Wirsching PJ, Ashley JA, Benkovic SJ, Janda KD, Lerner RA (1991) An unexpectedly efficient catalytic antibody operating by ping-pong and induced fit mechanisms. Science 252:680–685

Wood CR, Boss MA, Kenten JH, Calvert JE, Roberts NA, Emtage JS (1985) The synthesis and in vivo assembly of functional antibodies in yeast. Nature 314:446–449

Zu Putlitz J, Kubasek WL, Duchêne M, Marget M, von Specht B-U, Domdey H (1990) Antibody production in baculovirus-infected insect cells. Bio/Technology 8:651–654

CHAPTER 10
The Combinatorial Approach to Human Antibodies

C.F. BARBAS III

A. Introduction

The development of hybridoma technology 18 years ago appeared to promise antibodies for every application. Unfortunately, extension of the technique used for the production of mouse monoclonal antibodies (mAbs) to the production of human mAbs has proven to be difficult. Indeed, it is only within the past few years that advances in the field of molecular biology, coupled with the "combinatorial approach," have allowed for the routine production of human mAbs. This chapter is meant to serve as an introduction to this new methodology, to review its success and to highlight future directions for the production of antibodies.

B. The Combinatorial Approach

The combinatorial approach, as it exists today and continues to develop, seeks to mimic nature's strategies for the evolution of antibodies. This strategy is at every level (i.e., gene, protein and peptide) combinatorial. In its most basic form, the combinatorial approach involves construction of heavy and light chain libraries which are randomly combined and screened or selected for clones which bind a target antigen. This method became possible due to two developments. Firstly, it was demonstrated in 1988 that the antigen binding fragments Fv and, Fab could be expressed and functionally assembled in *Escherichnia coli* (SKERRA and PLÜCKTHUN 1988; BETTER et al. 1988). These experiments were successful because the antibody fragments were secreted form the cytoplasm to the oxidizing environment of the periplasm in *E. coli* under the guidance of bacterial leader sequences. The oxidizing environment and possibly the secretory event were necessary for disulfide bond formation and proper folding of the antibody domains. Secondly, development of the polymerase chain reaction (PCR) technique allowed for rapid cloning of antibody genes from hybridomas (LARRICK et al. 1989) and mixed populations of antibody producing cells (ORLANDI et al. 1989; SASTRY et al. 1989). Within the year, these two developments were utilized to produce the first combinatorial antibody library (HUSE et al. 1989). At the same time a more simplistic approach, consisting of production of single, heavy chain variable domains

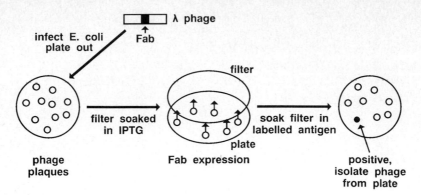

Fig. 1. Screening of combinatorial libraries in lambda phage

was reported by Ward et al. (1989). This approach, however, has not proven to be generally applicable.

The first combinatorial library experiment, reported in 1989, involved antigen screening of antibody Fab fragments from libraries expressed in lambda phage (Huse et al. 1989). In these experiments a mouse was immunized with the hapten designed to elicit catalytic antibodies and RNA was prepared from the spleen. After reverse transcription, the cDNAs of antibody heavy chains (Fd part of IgG1) and light chains were amplified by the PCR reaction and ligated into modified lambda phage vectors to give libraries of heavy and light chains. The two libraries were then combined by digestion of opposite arms of the vectors and religation to generate a random combinatorial library containing the genetic information for production of Fab fragments. The library was screened by transfer of Fabs produced by lambda phage-lysed *E. coli* onto nitrocellulose filter. The filters were then probed with ^{125}I-labeled hapten-conjugate and revealed a high frequency of positives (about 1 in 5000), which allowed identification of 200 monoclonal Fab fragments following an examination of 10^6 Fabs (Fig. 1). The examination of this library of a million clones, which as we will see later is a relatively modest library size, required 20 filter lifts. Analysis of 22 of the positive Fabs showed sequence diversity and apparent binding affinities of the order of $10^7 M^{-1}$.

With the demonstration of the applicability of the system for the cloning of mouse antibodies, the next step required a model system for cloning human antibodies. This was achieved utilizing an individual recently boosted with tetanus toxoid (Persson et al. 1991). Peripheral blood lymphocytes (PBLs) were utilized to construct an IgG1/κ library. The antibodies generated in this case showed considerable sequence diversity and apparent affinities in the range of 10^8–$10^9 M^{-1}$. Similar studies have also been reported elsewhere (Mullinax et al. 1990). Recent boosting was very important in that it was not possible to isolate antigen binding Fabs from an individual with a high anti-tetanus toxoid titer but who had not been boosted. This probably reflects the presence of antigen-specific plasma cells with their high concentration of specific mRNA in the peripheral blood of boosted subjects as compared to the low resting level of plasma cells in PBLs (Lum et al. 1990; Schibler et al. 1978). The method has also been applied the to the generation of anti-thyroglobulin antibodies from the thyroid tissue of an autoimmune individual (Hexam et al. 1991).

C. From Screening to Selection

Selective procedures are generally more efficient than screening procedures. Indeed, this was the most limiting feature of the lambda system. The screening procedure limits the size of the library which may be examined. For example, the screening of a library of 10^7 antibodies would require examination of a minimum of 200 filter lifts. Furthermore, the screening procedure places restrictions on the antigens which are being examined, in that the antigen must be available in significant quantities in purified form, be amenable to labeling with ^{125}I or enzymes and should not stick significantly to filters in the absence of antibody. This is very restrictive if your interest is in isolating antibodies against proteins which have yet to be identified or characterized but which would be found on the surface of a cell or in a crude protein extract, for example, from a viral lysate.

A logical next step was to attempt to mimic the immune system's linkage of recognition and replication, or phenotype and genotype, and to express antibody fragments on the surface of bacteria or phage (Smith 1985). This would allow selection of specific antibodies based on their ability to bind to immobilized antigen largely circumventing the problems above. Expression on the surface of bacteria met with limited success (Fuchs et al. 1991) but expression on phage was achieved (McCafferty et al. 1990; Kang et al. 1991a; Barbas et al. 1991; Clackson et al. 1991). I shall concentrate here on one particular system we have developed, pComb3 (Barbas et al. 1991).

In this system, the heavy (Fd region) and light chains are cloned sequentially into the phagemid vector pComb3 (see also Barbas and Lerner 1991) using the same restriction sites described for the lambda phage system (Fig. 2). The vector is constructed to fuse Fd with the COOH-terminal domain of the M13 minor coat protein III (cpIII). The COOH-terminal domain of cpIII was chosen for the site of fusion because previous reports had shown this domain to be essential in capping the tail of the phage as it is extruded through the membranes of *E. coli*. It was important not to utilize the entire cpIII protein as a fusion partner since its NH_2-terminal is involved in infectivity and host cell immunity to superinfection by additional phage. Utilization of the entire cpIII necessitates additional steps in library expression (Marks et al. 1991). The Fd-cpIII fusion protein is targeted to the

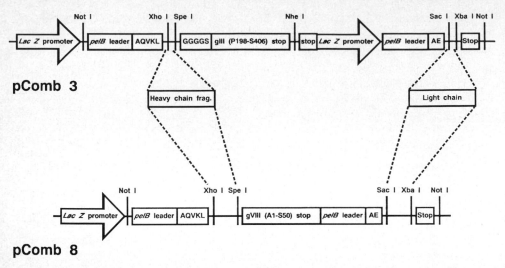

Fig. 2. Surface display vectors for the selection of combinatorial libraries

periplasmic space of *E. coli* where it is anchored in the inner membrane by the cpIII domain. The light chain then assembles on the heavy chain template to give an Fab fragment. Since the volume of the periplasmic space of *E. coli* is on the order of 10^{-16} liters, a hundred molecules produces a micromolar concentration and a driving force for chain assembly and subsequent interchain disulfide bond formation. Inclusion of the F1 intergenic region in the vector and subsequent superinfection of the host with M13 helper phage leads to packaging of the phagemid, which carries antibody heavy and light chain genes, in single-stranded form by the helper phage-produced proteins. Normal phage morphogenesis leads to incorporation of the Fab-cpIII fusion and the native helper phage-produced cpIII into the virion. Native cpIII is necessary for infection as the infectivity domain is not present in the Fab-cpIII fusion. The fusion, generally one copy per phage particle, is displayed in functional form on the surface of the phage and is available for antigen selection (Fig. 3).

The display of the antibody on the surface of the phage allows for the selection of clones by panning against antigen in ELISA wells. This is analogous to an affinity chromatography step. Antigen immobilized on beads or whole cells may also be utilized. After vigorous washing, the bound phage, which is now enriched for those bearing antigen-specific Fabs, is eluted with acid or antigen. This phage is then amplified and reselected by further rounds of panning (Fig. 4). Each step selects for antigen-specific clones and for clones of the highest affinity. In this way one can rapidly generate a panel of antigen-specific Fabs. For example (Barbas et al. 1991), the first experiments began with a human combinatorial anti-tetanus toxoid library in which the frequency of positives was about 1 in 5000. This number

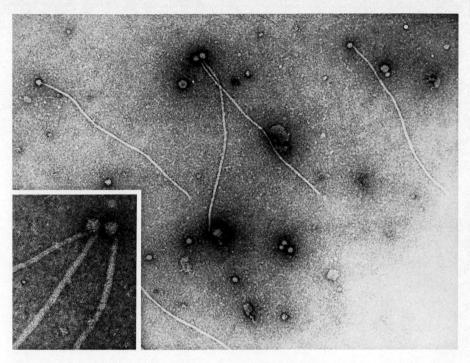

Fig. 3. Electron micrograph showing the specific binding of filamentous phage to hepatitis B surface antigen via a human antibody Fab fragment displayed at the tail of the phage. The phage encapsulates pComb3 phagemid DNA, which encodes the Fab fragment displayed on its surface. (Kindly provided by Suzanne Zebedee; see ZEBEDEE et al. 1992)

was 13 in 57 after one round of panning against the antigen, seven out of ten following two rounds and nine out of ten following three rounds. Experiments which involved the doping of phage displaying one specificity into an excess of phage displaying a Fab of another specificity or no Fab showed that a single panning step can enrich for specific phage by 10^3- to 10^5-fold. In another study, a library was prepared which included known anti-tetanus toxoid clones at a frequency of about 1 in 170 000. Three rounds of panning against toxoid were found to give enrichment such that 20/20 clones were antigen-specific, indicating the method could access clones of low abundance. Subsequent investigations of more diverse libraries suggest the technique may be utilized to isolate clones present at 1 part per 10 million if not less. The efficiency of the selection process is due in part to over-sampling of the library. This is a distinct advantage of phage since 10^{12} phage in a volume of $50\,\mu l$ may be applied to a single ELISA well. Thus, for a library of a million clones, each type will be present on a million phage. The method also allows sorting of clones based on affinity, as shown by a 250-fold enrichment of a tight binding tetanus toxoid clone (apparent K_a

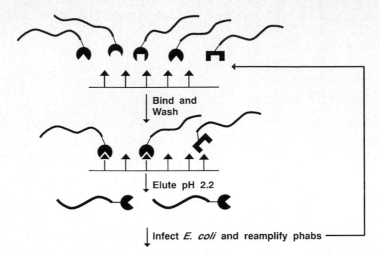

Fig. 4. The panning technique for selection of antigen-specific phage displaying antibody combining sites (*phabs*)

approximately $10^9 M^{-1}$) relative to a weaker binding clone ($10^7 M^{-1}$) in a mixture of the two following one round of panning. This is a distinct advantage of the monovalent display of the Fab on the phage surface with the pComb3 system. A monovalent display avoids the avidity effects which accompany the display of multiple copies on the phage surface. This was clearly demonstrated by the lack of considerable enrichment with the pComb8 system, which produces phage which bear multiple copies of Fab, in a similar experiment and has likely resulted in the isolation of low affinity clones in other systems (CLACKSON et al. 1991).

The pComb3 system was designed such that, once a surface display phagemid has been selected, the cpIII gene can be excised in a *Nhe*I/*Spe*I digestion and the vector religated via the compatible cohesive ends. The phagemid can the be used to express soluble Fab fragment in the conventional manner (Fig. 2). Subsequently it has been determined that excision of the cpIII gene is not absolutely necessary and that considerable quantities of functional soluble Fab will be produced by proteolysis of the fusion within *E. coli*.

D. Features of the Combinatorial Approach

The experiments just described and others which will be discussed later were successful despite the scrambling of heavy and light chains inherent in the construction of random combinatorial libraries. One might have expected that this scrambling would render the chances of a productive heavy-light chain combination extremely improbable. Indeed, arguments have been

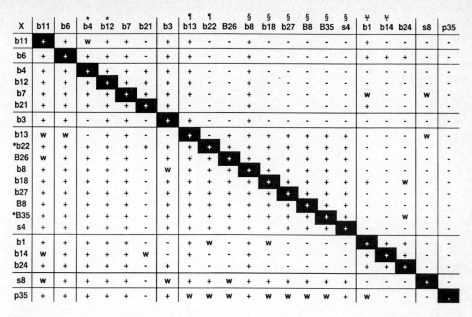

X	b11	b6	*b4	*b12	b7	b21	b3	¶b13	¶b22	B26	§b8	§b18	b27	§B8	§B35	§s4	¥b1	¥b14	b24	s8	p35
b11	■	+	w	+	+	-	+	+	-	-	+	-	-	-	-	-	+	-	-	-	-
b6	+	■	+	+	+	-	+	+	-	-	+	-	-	-	-	-	+	+	+	-	-
b4	+	+	■	+	+	+	+	+	-	-	+	-	-	-	-	-	-	-	-	-	-
b12	+	+	+	■	+	+	+	+	-	-	+	-	-	-	-	-	-	-	-	-	-
b7	+	+	+	+	■	+	+	+	-	-	+	-	-	-	-	-	w	-	-	w	-
b21	+	+	+	+	+	■	+	-	-	-	+	-	-	-	-	-	+	-	-	-	-
b3	+	+	-	+	+	-	■	+	-	-	+	-	-	-	-	-	-	-	-	-	-
b13	w	w	-	+	+	-	-	■	-	+	+	+	+	+	+	+	-	-	-	w	-
*b22	+	+	+	+	+	+	+	+	■	+	+	+	+	+	+	+	-	-	-	-	-
B26	w	+	+	+	+	-	+	+	+	■	+	+	+	+	+	+	-	-	-	-	-
b8	+	+	+	+	+	-	w	+	+	+	■	+	+	+	+	+	-	-	-	-	-
b18	+	+	+	+	+	-	+	+	+	+	+	■	+	+	+	+	+	-	w	-	-
b27	+	+	+	+	+	-	+	+	+	+	+	+	■	+	+	+	-	-	-	-	-
B8	+	+	+	+	+	-	+	+	+	+	+	+	+	■	+	+	-	-	-	-	-
*B35	+	+	+	+	+	-	+	+	+	+	+	+	+	+	■	+	-	-	w	-	-
s4	+	+	+	+	+	-	+	+	+	+	+	+	+	+	+	■	-	-	-	-	-
b1	+	+	+	+	+	-	-	+	w	-	+	w	-	-	-	-	■	+	+	-	-
b14	w	+	+	+	+	w	-	+	-	-	+	-	-	-	-	-	+	■	+	-	-
b24	+	+	+	+	+	-	+	-	-	-	+	-	-	-	-	-	+	+	■	-	-
s8	w	+	+	+	+	-	w	+	+	w	+	+	+	+	+	+	-	-	-	■	-
p35	+	+	+	+	+	-	+	w	w	w	+	w	w	w	w	+	w	-	-	-	■

Fig. 5. A reconstitution matrix showing the results of a number of directed crosses between heavy and light chains isolated from a combinatorial library. Heavy chains are listed *horizontally* and light chains are listed *vertically*. Different groups are seperated by *horizontal* and *vertical lines*. All clones except p35 bind to gp120 IIIb. p35 binds to an MN peptide. ELISA results: −, negative; +, positive (comparable to the signal of the original combination); w, intermediate value. Identical chains carry the same identifier. (From COLLET et al. 1992)

made by others that this approach is not applicable to the preparation of specific high-affinity antibodies (GHERADI and MILSTEIN 1992). However two factors seem to work to generate a reasonable frequency of binders in the library (BURTON 1991; BURTON and BARBAS 1992). One is the immune mRNA source, which leads to a high representation in the library of chains arising from in vivo binders. The second is chain promiscuity, i.e., the ability, particularly of heavy chains, to accept more than one partner in productive antigen binding (Figs. 5, 6).

Questions then arise as to whether the heavy-light chain combinations isolated in the combinatorial approach resemble in vivo pairs or antibody produced by the hybridoma approach. In one instance where this has been examined functional heavy chains were found at a frequency of 1 in 50 and functional light chains at a frequency of 1 in 250, giving a frequency of functional pairs of 1 in 13750. This examination utilized complementarity determining region 3 (CDR3)-specific nucleotide probes which were constructed using information derived from hybridoma derived sequences, so differences may exist in other regions of the antibody. Nevertheless, this experiment indicates that the diversity of IgG libraries of immunized animals

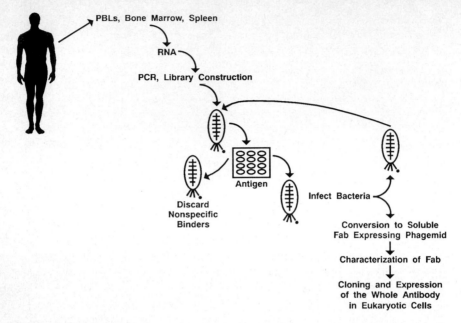

Fig. 6. The steps involved in cloning and selecting human antibodies using the combinatorial approach

can be more restricted than theoretical arguments allow. An exact match of antibodies prepared from the combinatorial and hybridoma approaches should not be expected, since the combinatorial approach uses mRNA as a starting material and should reflect antibodies from plasma cells (a plasma cell produces orders of magnitude more antibody encoding mRNA than a resting B cell) whereas the hybridoma approach is thought to reflect antibodies from activated but not terminally differentiated B cells.

The role of chain promiscuity in combinatorial antibody libraries is most vividly demonstrated in Fig. 5, where the results of a number of directed crosses using replicon-compatible plasmids are shown (COLLET et al. 1992). Within the panel a number of features should be noted. For example, heavy chain b6 can pair with every other light chain and still bind antigen. This is an extreme case of heavy chain promiscuity but it should be noted that binding was not observed when light chain was not present. This can be compared with p35 which pairs only with its light chain. Blocks of clones such as b4,12,7 and 21 or b1,14 and 24 have heavy chain CDR3 sequences which are identical or very similar to others within the same block but differ elsewhere in the gene. All of these family blocks can exchange light chains freely. These results do not imply that the resulting pairs have the same affinity for the antigen as the originally isolated pair, only that binding activity is maintained. The system was further used to examine combinations of chains from Fabs binding to gp120 with those binding to tetanus toxoid.

The results clearly supported the notion that specificity for antigen is dominated by the heavy chain. Heavy chains from some gp120 binders could retain affinity for gp120 with a light chain from a tetanus toxoid binder. A heavy chain from a tetanus toxoid binder could retain affinity for toxoid with light chains from gp120 binders. However none of the light chains in the experiment could dictate a new specificity to heavy chain partners (see also KANG et al. 1991b). Defined crosses within libraries using the pComb3, system, which selects on the basis of affinity, indicate that heavy and light chains sort to find similar partners at least within the CDR3 sequences (BARBAS et al. 1993). This approach can be utilized to produce a spectrum of related clones, of which some may show enhanced affinity and fine specificity differences.

An important consequence of this extensive promiscuity is that it may be difficult, if not impossible, to know what the exact original chain pairings were in vivo (BURTON and BARBAS 1992). But, is it important to know the exact original pairing and to use this pairing in therapeutic applications? In the study of antibody responses the knowledge of pairings would be of use in studying the evolution of binding activity across both chains. In the absence of this information the fact that the heavy chain appears to play a dominant role in binding, as demonstrated by chain shuffling experiments, would at least allow examination of the evolution of the heavy chain. For therapeutic use it is unlikely to be of importance unless an unnatural combination led to an immunogenic protein. Chain promiscuity may be utilized advantageously for the generation of families of functionally related clones, which could be of use in therapy if anti-Id responses prove to be important in the repeated long-term administration of therapeutic antibodies.

E. Human Antiviral Antibodies

I. Introduction

The utility of human mAbs in antiviral therapy remains for the most part untested. Although there has been considerable success in the development of pharmaceuticals to combat the more complex bacterial invaders, analogous antiviral strategies are lacking. There appears to be some resistance in the field of immunology to the importance of the development of antibodies in postinfection antiviral strategies. This is due, in large part, to the current dogma that T cells are needed to clear virus from infected cells. There are, however, two recent reports which emphasize the role of antibodies in just such tasks. The first is a report by the group of Griffin which demonstrated antibody-mediated clearance of α-virus infection from neurons using mAbs (LEVINE et al. 1991). In this case transfer of sensitized T lymphocytes had no effect on viral replication. The second case was reported by the group of Walter Gerhard, who suggested that antibodies, not virus-specific T cells, are required for the recovery of mice from pulmonary influenza virus infection (Scherle et al. 1992).

Antibodies may have a distinct advantage over T cells. Antibodies can be produced which recognize linear amino acid sequences or conformational epitopes (distinct shapes or surfaces). T cells are limited to the recognition of linear sequences presented in the context of MHC. The ability of antibodies to recognize shape may be an important feature in the development of antiretroviral agents since viruses utilize receptors (shapes) for entry into cells. Thus, even for viruses which readily vary their linear sequences, the maintenance of shape is important for receptor-mediated entry of the virus and may be utilized in the development of antiretroviral agents and strategies. I would now like to summarize the results of the application of the combinatorial approach to antiviral antibodies.

II. Antibodies to HIV-1

1. Rationale

The driving forces for this work include investigation of the use of human mAbs for passive immunization and prophylaxis in humans and the study of human immune responses. Studies of passive immunization in HIV-1 infected humans have been reported but are controversial (Jackson et al. 1988; Karpas et al. 1990). A correlation has been described between the presence of maternal antibodies against the V3 loop and the failure of infected mothers to transmit the virus to their offspring (Goedert et al. 1989; Devash et al. 1990; Rossi et al. 1989). Interpretation is difficult and firm conclusions must await studies in which the maternal antibodies are investigated in the context of the strain of virus transmitted. Nevertheless approximately 10%–20% of HIV-infected mothers transmit infection to their offspring, and in this situation it would seem worthwhile to evaluate the ability of antibodies given to the mother to prevent transmission to the fetus or newborn. Furthermore, recent studies (Goedert et al. 1991) on identical twins of HIV-positive mothers have implicated viral contact in the birth canal as an important factor in infection, a risk which could be reduced by systemic and topical administration of antibody to the mother.

The incredible variability of the HIV-1 virus makes it desirable to have a cocktail of human mAbs rather than a single antibody for any clinical intervention. A cocktail of antibodies against a variety of distinct epitopes on several proteins should prevent the emergence of escape mutations. This approach requires the ability to generate relatively large numbers of human mAbs. As discussed, the combinatorial approach offers this capability. The cocktail is likely to find application in the treatment of individuals accidentally infected with HIV-1, e.g., due to needle sticks, and in prophylaxis of infected mothers to hinder transmission of virus to the fetus or newborn. Given the complexity of AIDS, the value of antibodies in therapy remains uncertain until this approach is tried. However it seems reasonable to suppose that reducing the viral load is always going to be a desirable goal and, even if antibodies alone do not reverse the course of the disease, they may be valuable in conjunction with new generations of antiretroviral drugs.

2. Source of RNA

The individual involved in this study was a 31-year-old homosexual male who had been HIV-positive for 6 years but had no symptoms of disease. Serological studies showed the presence of a significant ELISA titer (1:3000) against the HIV-1 surface glycoprotein gp120 (IIIB strain). The tissue source in this case was bone marrow. Bone marrow has been shown, in humans, to be a major repository for differentiated B cells that spontaneously produce antibodies to maintain circulating antibody titers (LUM et al. 1990) and therefore is probably the most convenient source of mRNA encoding a diverse set of antibodies. For PBLs to be a viable source of specific mRNA, recent boosting (within a few days) is probably required, as discussed above. We have also successfully used spleen as a source of RNA and isolated Fab fragments specific for measles virus from the corresponding library. However, this and other organs such as lymph nodes are not so readily obtained. The overall strategy is shown in Fig. 6.

For this HIV-1 seropositive study (BURTON et al. 1991), the amplified antibody genes were cloned into pComb3 to give a library of 10^7 members. This phage surface expression library was panned against recombinant gp120 (strain IIIB) coated on ELISA wells. Four rounds of panning produced an amplification in eluted phage of a factor of about 100, indicating enrichment for specific antigen binding clones. A total of 40 reconstructed clones secreting soluble Fab fragments were grown up and the supernates screened in an ELISA assay for reactivity with recombinant gp120. The supernates from more than three quarters of the clones showed clear reactivity. The supernates did not react with BSA-coated wells and anti-tetanus toxoid Fab supernates did not react with gp120-coated wells.

3. Characterization of Antibodies

The diversity of antibodies selected is most easily assessed by sequencing, which is greatly facilitated by the existence of the genes in a phagemid vector. This approach revealed that the heavy chains could be grouped into six families by CDR3 sequence (Fig. 7) with evidence of somatic mutation within a given family (compare clones b3 and b5). The light chains showed even greater diversity. Chain promiscuity was observed in the sense that a very similar or identical heavy chain was found paired with a different light chain, e.g., clones b3 and b5 (BARBAS et al. 1993).

To search for additional anti-HIV antibodies, the library was also panned against a number of related antigens. These were gp160 (strain IIIB), gp120 (strain SF2) and a constrained peptide having the central part of the gp120

V_H Sequences

Clone	FR1	CDR1	FR2	CDR2	FR3	CDR3	FR4
b1	LEESGTEFKPPGSSVKVSCKASGGTFG	DYASNYAIS	WVRQAPGQGLEYIG	GITPTSGSADYAQKFQG	RVTISADRFTPILYMELRSLRIEDTAIYYCAR	ERRERGWNPRALRGALDF	WGSQGTRVFVSP
b2	....AAVQK....R...Q......D	NF.....V.	......WM.	......T.T.S....	......APL..I.......DD..V....	.........V....V	......T.
b14	....	....	....	....	....	....	....
b24	....A.V.K........I.S	.F......	......M.	....	.....AA..RV.....S....VF.	........EV	.....T.I.
b3	LEESGGRLVKPGGSLRLSCEGSGFTFT	NAWMT	WVRQSPGKGLEWVA	SIKSKFDGGSPHYAAPVEG	RFSISRNDLEDKMFLEMSGLKAEDTGVYYCAT	KYPRYSDMVTGVRNHFYMDV	WGKGTTVIVSS
b5	.Q...G......	.....	....	....	..T.......L....	....F..MA.......	.T....
B20	.Q...G......	.....	....	....	..T.......L....	....MA....L...	......
s2	............A......	.S...	....	......R....	....	.........M....	......
b4	LEQSGAEVKKPGASVKVSCQASGYRFS	NFVIH	WVRQAPGQRFEWMG	WINPYNGNKEFSAKFQD	RVIFTADTSANTAYMELRSLRSADTAVYYCAR	VGPYSWDDSPQDNYYMDV	WGKGTTVIVSS
b7	....	.....	....	....	.....D......	....T......	......K....
b12	....	.....	....	....	....I......	....T......	....
b21	....	.....	....	....	.....D.....T..I....	....T......	......K....
b6	LEESGGGLVKPGGSLRLSCVGSGFTFS	SAWVA	WVRQAPGRGLEWVG	LIKSKADGETIDYATPVKG	RFSISRNNLEDITVYLQMDSLRADDTAVYYCAT	QKPRYFDLLSGQYRRVAGAFDV	WGHGTTVTVSP
b20	....A......	.....	.G......	....	....	..S.YN....	....
s6	.....I......	....T	....K..I.	....	..T...N....	....	....
b8	LEESGEAVVQPGRSLRLSCAASGFIFR	NYAMH	WVRQAPGKGLEWVA	LIKYDGRNKYADSVKG	RETISRDNSKNTLYLQMNSLRAEDTAVYYCAR	DIGLKGEHYDLLTAYGPDY	WGQGTLVTVSS
b13	.Q......T......	.....	....	....	..S.......E....	....	......A....
b18	.Q......T......	.....	....	....	....	....	....
b22	......T...T....	.....	....	....	..S.......E....	....	......A....
b11	LEQSGGGVVKPGGSLRLSCEGSGFTFP	NAWMT	WVRQSPGKGLEWVA	SIKSKFDGGSPHYAAPVEG	RFTISRNDLEDKVFLQMNGLKAEDTGVYYCAT	RYPRYSEMVGGVRKHFYMDV	WGKGTTVSVSS
b29	..E......	.....	....	....	....	....	....

V_K Sequences

Clone	FR1	CDR1	FR2	CDR2	FR3	CDR3	FR4
b1	ELTQSPSSLSASVGDRVTITC	RASQGISNYLA	WYQQKPGKVPRLLIY	AASTLQP	GVPSRFSGSGSGTDFTLTISSLQPEDVATYYC	QKYNSAPRT	FGQGTKVEIKRT
b2	.........I......	....N....	.......R....	.......S	....T...	....V...	..G....
b14				.......S			
b24	....GT..L.P.E.A.LS.	....SVISNYLA	......QA....	GV.NRAT	....R.E...F.V.S.	.Q.GTS.W.	....
b3	ELTQSPGTLSLSPGERATLSC	RASHRVNNNFLA	WYQQKPGQAPRLLIS	GASTRAT	GIPDRFSGSGSGTDFTLTISRLEPDDFAVYYC	QQYGDSPLYS	FGQGTKLEIKRT
b5	ELTQSPASVSASVGDTVTITC	RASQDIHNWLA	WYQQPGKAPKLLIY	AASSLQS	GVPSRFSGRGSGTDFTLTISSLQPEDFATYYC	QQGNSFPK	FGPGTVVDIKR
b20	ELTQSPGTLSLSPGERATLSC	RASQSLSNNYLA	WYQQKPGQAPRLLIY	GSSTRGT	GIPDRFSGSGSGTDFTLTISRLEPEDFAVYYC	QHYGNSVVT	FGQGTKLEIKR
s2	QSPDTLSLNPGERATLSC	RASHRISSKRLA	WYQHKRGQAPRLLIY	VCPNRAG	GVPDRFSGSGSGTDFTLTYSRLEPEDFAMYYC	QYYGGSSYT	FGQGTKVEITR
b4	ELTQSPGTLSLSPGERATFSC	RSSHSIRSRRVA	WYQHKPGQAPRLVIH	GVSNRAS	GISDRFSGSGSGTDFTLTITRVEPEDFALYYC	QVYGASSYT	FGQGTKLERKRT
b7	...T......T......L.	.T......L.	..V.G......L.Y		..P.......S.L....V...	.Q..S.R.	.........I...
b12	..A......	.........R					...DF....
b21	..A...D....	....N....			......L.......		
b6	ELTQSPGTLSLSPGERATLSC	RAGQSISSNYLA	WYQQKPGQAPRLLIY	GASNRAT	GIPDRFSGSGSGTDFTLSISRLEPEDFAVYYC	QQYGTSPYT	FGQGTQLDIKRT
s6	..........A...	.S..L.N...		.S.T...	....G......T.......	..N.V...	....K.E...
b20	....SS..A.V.D.V.IT.	.TS.G..NYLA	......KV.K....	...TLQS	..G.S......T.NS.Q..V.T.S.	.N.DSA.W.	....KV....
b8	ELTQSPSSLSASVGDRVTITC	RASQSISNYIN	WYQQKPGKAPKLLIY	AASSLQR	GVPSRFSGSGSGTDFTLISSLQPEDFATYYC	QQSYSIPPLI	FGGGTKVEIKRT
b13		Q...D.R....	....	D..NSET	.........R..FT......V.....	..HQNV.LT	....
b18		.....N.N...	......E.....H	T.FN..S	....TA..E...T.R....	...T.YT	...Q....
b22		....S....	.........S	....S	........T........	...T.YT	...Q..L...
b11	ELTQSPGTLSLSPGERATLSC	RASQRVNSNYLA	WYQQKPGQTPRVVIY	STSRRAT	GVPDRFSGSGSGTDFTLTISRLEPEDFAVYYC	QQFGDAQYT	FGQGTKLEIKRT
b29							

Fig. 7. A selection of variable domain sequences cloned from an HIV seropositive individual. For a more complete listing see BARBAS et al. (1993)

V3 loop sequence (strain MN/SF2). The sequences of the Fabs obtained by panning against gp160 (IIIB) or gp120 (SF2) were, with one exception, closely related to those described above from panning against gp120 (IIIB), showing some robustness in the system. Several Fabs were obtained by panning against the constrained peptide but only one reacted with gp120. In fact this Fab reacted with gp120 (SF2) but not gp120 (IIIB) which is fully consistent with the nature of the peptide used in panning.

The affinities of the Fab fragments for gp120 were determined by inhibition ELISAs using soluble gp120. The examination of 21 clones showed that most inhibition constants were less than $10^{-8} M$ implying monomer Fab-gp120 apparent binding constants of the order of or greater than $10^8 M^{-1}$.

Characterization of the interaction of Fabs, gp120 and soluble CD4 was investigated by competition ELISAs (BARBAS et al. 1992a,c). All of the Fabs obtained by panning against gp120/160 were found to be competed by CD4 for binding to gp120. Fab binding to the V3 loop was not competitive. Therefore the predominant Fabs isolated from this donor by the library approach appear to be strain cross-reactive and CD4-inhibited. This is consistent with the observation that more than 50% of the reactivity of the donor serum with gp120 (IIIB) is inhibitable by CD4. Furthermore, the Fabs appear to be directed to major epitopes on gp120, in that a cocktail of three of the Fabs was able to inhibit >50% of the serum reactivity with gp120 (IIIB) in more than 90% of a selection of seropositive donors (R. Burioni, private communication). Hence from this study we conclude that the specificities of the antibodies arising from a library approach do have relevance to the in vivo response.

The ability of these Fabs to neutralize virus in vitro has been examined (BARBAS et al. 1992c). Neutralization was defined as the ability of Fabs to inhibit infection as measured in both p24 ELISA and syncytia assays. One group of closely sequence related Fabs was found to neutralize virus in both assays with a titer (50% neutralization) at approximately $1 \mu g/ml$ (Fig. 8). Another Fab neutralized in the p24 ELISA but not the syncytial assay. The majority of Fabs showed weak or no neutralizing ability. The results imply that virion aggregation or cross-linking of gp120 molecules on the virion surface is not an absolute requirement for HIV-1 neutralization. Furthermore, the observation that all of the Fabs were competitive with soluble CD4 for binding to gp120 and yet few neutralized effectively implied that the mechanism of neutralization in this case may not have involved receptor blocking. Subsequent investigations have demonstrated the efficacy of these Fabs in neutralization of fresh isolates from patients.

III. Antibodies to Respiratory Syncytial Virus

Respiratory syncytial virus (RSV) is the major pediatric viral respiratory tract pathogen, outranking all others as a cause of pneumonia and bronchiolitis in infants and young children (reviewed in Mc INTOSH and CHANOCK

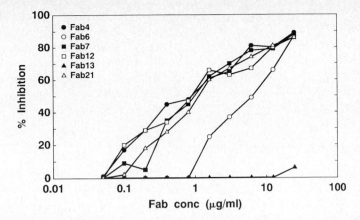

Fig. 8. Neutralization of HIV-1 by purified recombinant Fabs. The results shown are derived from a quantitative infectivity assay using the MN strain of HIV-1 and show inhibition of infection as a function of Fab concentration. (From BARBAS et al. 1992c)

1990). Very recently, it was concluded that RSV infection in bone marrow transplant patients is a serious and life-threatening infection with a high mortality rate once pneumonia develops (HARRINGTON et al. 1992). In one case, 14 of 31 patients in a bone marrow transplant center died as a result of infection during the immunosuppressed state. Several lines of evidence indicate that antibodies mediate resistance to RSV infection and illness. A clinical study of pooled human IgG containing a high titer of RSV neutralizing antibodies provided preliminary indications that these antibodies can exert a therapeutic effect in serious RSV infection in infants and young children (HEMMING et al. 1987). Given this evidence, there is considerable interest in having available human neutralizing mAbs to RSV for prophylaxis in protecting infants and immunocompromised adults at high risk of serious disease and for therapy in cases of serious RSV lower respiratory tract disease.

Since ELISA analysis of the serum of the donor used in the HIV study above indicated a high titer to RSV FG glycoprotein (a recombinant fusion of the two major surface proteins), the same library was panned against the RSV antigen. A number of Fab fragments interacting with the F glycoprotein were identified (BARBAS et al. 1992d). One of these neutralized a wide range of virus isolates, ten subgroup A and nine subgroup B isolates, with a titer (60% neutralization) of approximately $0.1-1.0\,\mu g/ml$. These viruses represent a temporal collection spanning 30 years and geographic isolates from Australia to Sweden to West Virginia. Another Fab neutralized diverse isolates at a concentration somewhat higher.

Evaluation of these Fabs will proceed in experimental animals as soon as the immunoglobulin fragments are incorporated into full length molecules. These will be required for passive immunoprophylaxis because of the need to use immunoglobulins that circulate in the host as long as possible. Such molecules may also prove to be most efficient for use in therapy. The availability of such potent monoclonals may allow intramuscular administration, potentially avoiding hospital stays for those at risk. However, consideration will also be given to the use of Fabs or F(ab')2s administered by aerosol because of the effectiveness of the latter in a model system (PRINCE et al. 1987).

IV. Antibodies to Hepatitis B Virus

This study is analogous to the tetanus toxoid study in that a vaccination protocol was available. In this case (ZEBEDEE et al. 1992), two individuals were vaccinated with recombinant hepatitis B surface antigen and PBLs were extracted 7–14 days later (although 4–5 days is probably optimal). Libraries (IgG1/κ and IgG1/λ) were constructed on the surface of phage and panned against the hepatitis antigen. Specific Fab were identified and, due to the particulate nature of the antigen, binding of phage expressing specific Fabs to virus antigen particles could be clearly visualized in the electron microscope (Fig. 3). All but one phage were found to be associated with a single surface antigen particle. One phage was found bound to two particles. These results support the view that most phage display a single antibody fragment on their surface. Sequencing of positive clones showed a limited diversity with a remarkable example of promiscuity, in that a given heavy chain was shown to bind antigen with either a κ or λ light chain. In this case the light chain partner appeared to affect fine specificity, as measured by the ability of excess Fab to compete with mouse mAbs for virus antigen. This example illustrates the way in which chain promiscuity can be utilized to modify the characteristics of existing Fabs.

F. Alternatives to the Use of Seropositive Humans

There are a number alternative sources for antibody libraries which may prove complementary to the use of seropositive humans. These are naive libraries, synthetic or semisynthetic libraries, chimpanzees and severe combined immunodeficiency (SCID) mice populated with human cells. Taken together these sources allow every compartment of the immune response to be examined and even allow for synthetic extension of the repertoire (Fig. 9).

I. Naive Libraries

The term "naive" in this sense refers to a library of antibodies which has not been "educated" or biased by the immune system towards the recognition of any particular antigen or antigens. Preparation of naive libraries involves the

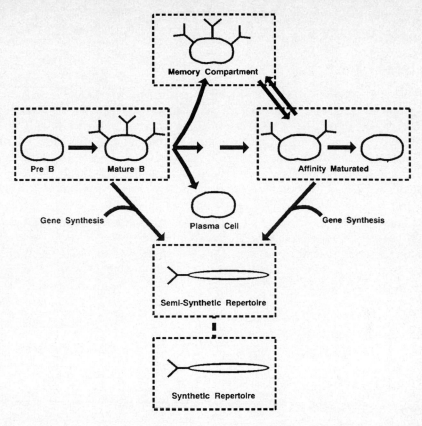

Fig. 9. Compartments of the immune system which are accessible to the combinatorial approach. The memory compartment is accessed utilizing SCID mice. Additional artificial components of the immune system are shown and involve chemical synthesis of antibody genes in combination with natural genes

use of RNA from nonimmune sources and amplification of μ or ∂ heavy chains which are the starting point in the natural response. Naive libraries have been constructed from both humans and mice and have been selected from with some success for the production of low or medium affinity antibodies (MARKS et al. 1991; GRAM et al. 1992). There are two possible problems associated with antibodies selected from naive repertoires, relatively low affinity for the antigen and possible deleterious cross-reactivity. Two strategies have been utilized to increase the affinity of antibodies isolated from naive libraries. The first has been the use of error-prone PCR to mimic random somatic mutation (GRAM et al. 1992). This resulted in a 30-fold increase in affinity from a very low affinity starting clone. Second, chain shuffling has been utilized to improve the affinity of a human anti-hapten antibody. By performing two sequential shuffling and selection steps an almost 300-fold increase in affinity was reported (MARKS et al. 1992). In this case the final clone had only one of its original CDRs.

II. Synthetic and Semisynthetic Antibodies

Synthetic or semisynthetic antibody libraries are the second alternative to an immune source. Initial explorations of semisynthetic antibodies utilized a single clone selected from a library from an immune patient (BARBAS et al. 1992b). A 16 amino acid random sequence was then introduced over the CDR3 region of the heavy chain to generate a vast library of antibodies. Selection of the library against a variety of antigens allowed for the cloning of new specifici ties. The complete randomization of 16 amino acids would require generation of a library of greater than 10^{20} clones, far in excess of the number which is obtainable by transformation of *E. coli*. However, libraries can be constructed which match or exceed the diversity of clones examined by an animal at a given moment, approximately 10^7. Selection of this library, derived from a human anti-tetanus clone against a new antigen, fluorescein, resulted in the isolation of clones which bound flourescein with affinities approaching that obtained by secondary boost of a mouse. This strategy is proving to be useful in the generation of antibodies against a variety of antigens. Extension of this strategy to the synthesis of all the CDRs or the use of natural libraries of FR1-FR3 fragments in combination with synthetic CDR3s should yield libraries from which almost any given specificity is retrievable. This approach may be limited by the same pitfalls of the naive approach; however, it has one distinct advantage. The diversity of these libraries is controlled at the level of nucleic acid synthesis, whereas the diversity of a naive library is limited by the source of RNA, which is susceptible to bias by RNA derived from plasma cells or activated B cells. Recently, synthetic antibodies have been isolated which neutralize HIV. There is certainly reason to believe that synthetic antibodies may, in many cases, alleviate the need for the use of animals in the production of antibodies (LERNER et al. 1992).

III. Human Antibodies from Severe Combined Immunodeficiency Mice

Mice with SCID and populated with human cells offer the possibility of antigen stimulation of human responses outside the human body. The strategy involves transplantation of human immune components into mice. We have shown that SCID mice can be used in conjunction with the combinatorial library approach (DUCHOSAL et al. 1992). SCID mice were populated with PBLs from a donor who had not had contact with tetanus toxoid for more than 17 years. These cells generated a secondary response in the mouse, and antigen-specific, high affinity, human Fab fragments were cloned. For seropositive donors this sort of approach might be useful in stimulating antibodies essentially relegated to the memory compartment. One also has the advantage of being able to boost with defined components of the original immunogen, which may be a virus. Thus, for instance, a response to a particular peptide could be stimulated and human monoclonal Fab fragments rescued. Future development may allow for the creation of both primary and secondary responses from seronegative donors.

IV. Antibodies from Chimpanzees

The differences between chimpanzees and humans is constant domain sequences have recently been characterized and are similar to those between allotypes of humans (EHRLICH et al. 1990, 1991), suggesting that immunized chimpanzees may serve as an alternative source to seropositive humans. Furthermore, variation within the variable regions might prove to be negligible. We have indeed found that the human PCR primers will successfully amplify γ 1 heavy and κ and λ light chains. It would thus appear that the library approach should be feasible. A distinct advantage of utilizing an immune source, human or chimpanzee, is that the antibodies have been both highly positively and negatively selected. Those antibodies which are tried and tested in the animal are likely to be the best source of therapeutic antibodies, although refinement by way of in vitro methods may prove to be important.

G. Production of Whole Antibodies and Gene Rescue from Cell Lines

For most therapeutic applications of antibodies, the whole antibody molecule is necessary, at least in part because of the glycosylation requirement of the Fc portion, of the molecule. This condition therefore necessitates the use of eukaryotic cell lines. The favorites in this regard have been myeloma (WRIGHT and SHIN 1991) or Chinese hamster ovary (CHO) cells (BEBBINGTON 1991), although baculovirus has also been used (HASEMANN and CAPRA 1991). Modification of existing vectors enables facile expression of whole antibodies utilizing Fab cassettes from the phage system. In principle any Fc can be linked to heavy chain Fd from the phage system. This allows the experimenter to choose a suitable isotype for a given task. The reader is referred to Chap. 14 for a more complete discussion. Epstein-Barr virus
    (EBV) transformation and fusion have been successfully used to generate human cell lines secreting mAbs against a number of antigens (JAMES and BELL 1987). Such lines can be unstable or low secretors of antibody. Alternatively, the antibody may not be of the isotype desired. In any of these cases, it may be appropriate to rescue heavy and light chain genes from mRNA from the cell line and express the Fab in bacteria. This process has been described for the cloning and expression of a human anti-rhesus D

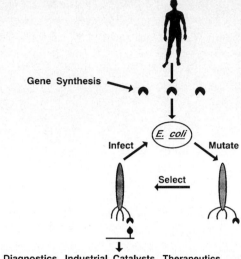

Fig. 10. The future of antibodies will include vast antibody libraries derived from natural and synthetic sources. These will be coupled with genetic systems which will allow for experimenter controlled evolution of antibodies in computer controlled machines

antibody (WILLIAMSON et al. 1991). In this case five out of five clones examined had the correct heavy and light chains. This is often not the case for hybridomas, which may contain mRNA from other chains that has been PCR amplified and cloned. For instance, for one mouse hybridoma only 1 in 1000 recombinants had antigen binding activity. Therefore the most prudent general strategy is to clone from the cell line into a phage display vector such as pComb3 and then pan against antigen to select positive clones.

H. The Future of Antibodies

The successful cloning of an antigen-specific antibody need not be an end in itself. The cloning of genes in a phage display system will likely lead to antibodies which can be further evolved to exceed even nature's best antibodies. One strategy which we have mentioned is chain shuffling, which is an extension of the basic combinatorial concept. Alternative strategies for the refinement of clones involve mutagenesis and reselection. This, in effect, attempts to carry evolution of the antibody farther than the body has. Random mutagenesis of the variable regions is possible by a number of different approaches such as chemical mutagenesis (MYERS et al. 1985), polymerase induced mutagenesis (LEUNG et al. 1989) and in vivo mutagenesis using mutator strains of *E. coli* (SCHAAPER 1988). Focused mutagenesis, in which several residues are targeted, does not mimic the supposed random mutation and selection of the immune system but allows all possible mutations in a defined region to be explored. This strategy has been successful in the generation of high affinity variants of human growth hormone (Lowman et al. 1991). For antibodies, the strategy involves targeting specific CDR regions for mutagenesis since this is most likely to improve affinity and least likely to create problems of immunogenicity.

Although positive selection fro variants of increased affinity is an obvious aim, mutagenesis could also be employed to increase or decrease cross-reactivity. For example, it may be desirable to produce antibodies that are highly cross-reactive in their binding to gp120 from a number of different strains of HIV-1. The selection strategy would then involve panning against alternate types of gp120 or a mixture. Specificity can conversely be increased by including the antigen with which the antibody is cross-reactive in the wash solution during selection or by preselection of the phage with the antigen. This effectively allows negative selection to be performed.

The future of antibodies is likely to involve application of all the aforementioned strategies and techniques (Fig. 10). If the current pace of developments continues we may expect that machines will produce antibodies to suit the experimenter's every requirement by the turn of the century.

Acknowledgements. I would like to acknowledge the considerable support and enthusiasm of Richard A. Lerner and discussions and assistance with the manuscript from Dennis R. Burton. CFB is a Scholar of the American Foundation for AIDS Research and the recipient of an Investigator Award from the Cancer Research Institute.

References

Barbas CF III, Kang A, Lerner RA et al. (1991) Assembly of combinatorial antibody libraries on phage surfaces: the gene III site. Proc Natl Acad Sci USA 88:7978–7982
Barbas CF III, Lerner RA (1991) Combinatorial immunoglobulin libraries on the surface of phage (Phabs): rapid selection of antigen-specific Fabs. In: Lerner RA, Burton DR (eds) Methods: a companion to methods in enzymology, vol 2. Academic, Orlando, pp 119–124
Barbas CF III, Persson MAA, Koenig S et al. (1992a) A large array of human monoclonal antibodies to HIV-1 from combinatorial libraries of an asymptomatic seropositive individual. In: Brown F (ed) Vaccines '92. Modern approaches to new vaccines including prevention of AIDS. Cold Spring Harbor Laboratory, Cold Spring Harbor, pp 9–12
Barbas CF III, Bain JB, Hoekstra DM et al. (1992b) Semisynthetic combinatorial antibody libraries: a chemical solution to the diversity problem. Proc Natl Acad Sci USA 89:4457–4461
Barbas CF III, Björling E, Chiodi F, Dunlop N, Cababa D, Jones TM, Zebedee SL, Persson MAA et al. (1992c) Recombinant human Fab fragments neutralize human type 1 immunodeficiency virus in vitro. Proc Natl Acad Sci USA 89: 9339–9343

Barbas CF III, Crowe JE Jr, Cababa D et al. (1992d) Human monoclonal Fab
    fragments derived from a combinatorial library bind to respiratory syncytial
    virus F glycoprotein and neutralize infectivity. Proc Natl Acad Sci USA 89:
    10164–10168
Barbas CF III, Collet TA, Roben P et al. (1993) Molecular profile of an antibody
    response to HIV-I as probed by combinatorial libraries. J Mol Biol 230:812–
    823
Bebbington CR (1991) Expression of antibody genes in nonlymphoid mammalian
    cells. In: Lerner RA, Burton DR (eds) Methods: a companion to methods in
    enzymology, vol 2. Academic, Orlando, pp 136–145
Better M, Chang CP, Robinson PR et al. (1988) *Escherichia coli* secretion of an
    active chimeric antibody fragment. Science 240:1041–1043
Burton DR (1991) Human and mouse monoclonal antibodies by repertoire cloning.
    Trends Biotechnol 9:169–175
Burton DR, Barbas CF III (1992) Antibodies from libraries. Nature 359:782–783
Burton DR, Barbas CF III, Persson MAA et al. (1991) A large array of human
    monoclonal antibodies to type 1 human immunodeficiency virus from com-
    binatorial libraries of asymptomatic seropositive individuals. Proc Natl Acad Sci
    USA 88:10134–10137
Clackson T, Hoogenboom HR, Griffiths AD, Winter G (1991) Making antibody
    fragments using phage display libraries. Nature 352:624–628
Collet TA, Roben P, O'Kenedy R, Barbas CF III et al. (1992) A binary plasmid
    system for shuffling combinatorial antibody libraries. Proc Natl Acad Sci USA
    89:10026–10030
Devash Y, Calbelli TA, Wood DG et al. (1990) Vertical transmission of human
    immunodeficiency virus is correlated with the absence of high-affinity/avidity
    maternal antibodies to the gp120 principal neutralizing domain. Proc Natl Acad
    Sci USA 87:3445–3449
Duchosal MA, Eming S, Fischer P et al. (1992) Immunization of hu-PBL-SCID mice
    and the rescue of human monoclonal Fab fragments through combinatorial
    libraries. Nature 355:258–262
Ehrlich PH, Moustafa ZA, Harfeldt KE et al. (1990) Potential of primate mono-
    clonal antibodies to substitute for human antibodies: nucleotide sequence of
    chimpanzee Fab fragments. Hum Antibod Hybridomas 1:23–26
Ehrlich PH, Moustafa ZA, Osterberg L (1991) Nucleotide sequence of chimpanzee
    Fc and hinge regions. Mol Immunol 28:319–322
Fuchs P, Breitling F, Dubel S et al. (1991) Targeting recombinant antibodies to the
    surface of *Escherichia coli*: fusion to a peptidoglycan associated lipoprotein.
    Biotechnology 9:1369–1372
Gheradi E, Milstein C (1992) Original and artificial antibodies. Nature 357:201–202
Goedert JJ, Drummand JE, Minhoff HL et al. (1989) Mother-to-infant transmission
    of human immunodeficiency virus type 1: association with prematurity or low
    anti-gp120. Lancet 2:1351–1354
Goedert JJ, Duliege AM, Amos CI et al. (1991) High risk of HIV-1 infection for
    first-born twins. Lancet 338:1471–1475
Goudsmit J, Back NKT, Nara PL et al. (1991) Genomic diversity and antigenic
    variation of HIV-1: links between pathogenesis, epidemiology and vaccine de-
    velopment. Faseb J 5:2427–2436
Gram H, Marconi L-A, Barbas CF III et al. (1992) In vitro selection and affinity
    maturation of antibodies from a naive combinatorial immunoglobulin library.
    Proc Natl Acad Sci USA 89:3576–3580
Harrington RD, Hooton TM, Hackman RC et al. (1992) An outbreak of respiratory
    syncytial virus in a bone marrow transplant center. J Infect Dis 165:987–993
Hasemann CA, Capra JD (1991) Baculovirus expression of antibodies: a method for
    the expression of complete immunoglobulins in a eukaryotic host. In: Lerner
    RA, Burton DR (eds) Methods: a companion to methods in enzymology, vol 2.
    Academic, Orlando, pp 146–158

Hemming VG, Rodriguez W, Kini HW et al. (1987) Intravenous immunoglobulin treatment of respiratory syncytial virus infections in infants and young children. Antimicrob Agents Chemother 31:1882–1886

Hexham JM, Furmaniak J, Persson MAA et al. (1991) Cloning and expression of a human thyroglobulin autoantibody. Autoimmunity 11:69–70

Huse WD, Sastry L, Iverson SA et al. (1989) Generation of a large combinatorial library of the immunoglobulin repertoire in phage lambda. Science 246: 1275–1281

Jackson GG, Rubenis M, Knigge M et al. (1988) Passive immunoneutralisation of human immunodeficiency virus in patients with advanced AIDS. Lancet 2: 647–652

James K, Bell GT (1987) Human monoclonal antibody production. Current status and future prospects. J Immunol Methods 100:15–40

Kang AS, Barbas CF III, Janda KD et al. (1991a) Linkage of recognition and replication functions by assemblying combinatorial antibody Fab libraries along phage surfaces. Proc Natl Acad Sci USA 88:4363–4366

Kang AS, Jones TM, Burton DR (1991b) Antibody redesign by chain shuffling from random combintorial immunoglobulin libraries. Proc Natl Acad Sci USA 88: 11120–11123

Karpas A, Hewlett IK, Hill F et al. (1990) Polymerase chain reaction evidence for human immunodeficiency virus 1 neutralization by passive immunization in patients with AIDS and AIDS-related complex. Proc Natl Acad Sci USA 87: 7613–7617

Larrick JW, Danielsson L, Brenner CA et al. (1989) Rapid cloning of rearranged immunoglobulin genes from human hybridoma cells using mixed primers and the polymerase chain reaction. Biochem Biophys Res Commun 160:1250–1256

Lerner RA, Kang AS, Bain JD et al. (1992) Antibodies without immunization. Science 258:1313–1314

Leung DW, Chen E, Goeddel DV (1989) A method for random mutagenesis of a defined DNA segment using a modified polymerase chain reaction. Technique. J Methods Cell Mol Biol 1:11–15

Levine B, Hardwick JM, Trapp BD et al. (1991) Antibody mediated clearance of alphavirus infection from neurons. Science 254:856–860

Lowman HB, Bass SH, Simpson N et al. (1991) Selecting high-affinity binding proteins by monovalent phage display. Biochemistry 30:10832–10838

Lum LG, Burns E, Janson MM et al. (1990) IgG anti-tetanus toxoid antibody synthesis by human bone marrow. I. Two distinct populations of marrow B cells and functional differences between marrow and peripheral blood B cells. J Clin Immunol 10:255–264

Marks JD, Griffiths AD, Malmgvist M, Clackson TP, Bye JM, Winter G (1992) By-passing immunization: building high affinity human antibodies by chain shuffling. Biotechnology 10:779–783

Marks JD, Hoogenboom HR, Bonnert TP et al. (1991) By-passing immunization human antibodies from V-gene libraries displayed on phage. J Mol Biol 222: 581–597

McCafferty J, Griffiths AD, Winter G et al. (1990) Phage antibodies: filamentous phage displaying antibody variable domains. Nature 348:552–554

McIntosh K, Chanock RM (1990) Respiratory syncytial virus. In: Fields BN, Knipe DM (eds) Virology, 2nd edn. Raven, New York, pp 1045–1072

Mullinax RL, Gross EA, Amberg JF et al. (1990) Identification of human antibody fragment clones specific for tetanus toxoid in a bacteriophage lambda immuno-expression library. Proc Natl Acad Sci USA 87:8095–8099

Myers RM, Lerman LS, Maniatis T (1985) A general method for saturation mutagenesis of cloned DNA fragments. Science 229:242–247

Orlandi R, Gussow DH, Jones PT et al. (1989) Cloning immunoglobulin variable domains for expression by the polymerase chain reaction. Proc Natl Acad Sci USA 86:3833–3837

Persson MAA, Caotluen RH, Burton DR (1991) Generation of diverse high-affinity human monoclonal antibodies by repertoire cloning. Proc Natl Acad Sci USA 88:2432–2436

Prince GA, Hemming VG, Horswood RL et al. (1987) Effectiveness of topically administered neutralizing antibodies in experimental immunotherapy of respiratory syncytial virus infection in cotton rats. J Virol 61:1851–1854

Rossi P, Moschese V, Broliden PA et al. (1989) Presence of maternal antibodies to human immunodeficiency virus 1 envelope glycoprotein gp120 epitopes correlates with the uninfected status of children born to seropositive mothers. Proc Natl Acad Sci USA 86:8055–8058

Sastry L, Alting-Mees M, Huse WD et al. (1989) Cloning of the immunological repertoire in Escherichia coli for generation of monoclonal catalytic antibodies: construction of a heavy chain variable region-specific cDNA library. Proc Natl Acad Sci USA 86:5728–5732

Schaaper RM (1988) Mechanisms of mutagenesis in the Escherichia coli mutator mutD5: role of DNA mismatch repair. Proc Natl Acad Sci USA 85:8126–8130

Scherle PA, Palladino G, Gerhard W (1992) Mice can recover from pulmonary influenza virus infection in the absence of class-I restricted cytotoxic T cells. J Immunol 148:212–217

Schibler U, Marcu KB, Perry RP et al. (1978) The synthesis and processing of the messenger RNAs specifying heavy and light chain immunoglobulins in MPC-ll cells. Cell 15:1495–1509

Skerra A, Pluckthun A (1988) Assembly of a functional immunoglobulin Fv fragment in Escherichia coli. Science 240:1038–1041

Smith GP (1985) Filamentous fusion phage: novel expression vectors that display cloned antigens on the virion surface. Science 228:1315–1317

Ward ES, Güssow D, Griffiths AD, Jones PT, Winter G (1989) Binding activities of a repertoire of single immunoglobulin variable domains secreted from Echerichia coli. Nature 341:544–546

Williamson RA, Persson MAA, Burton DR (1991) Expression of a human monoclonal anti(rhesus D) Fab fragment in Escherichia coli with the use of bacteriophage lambda vectors. Biochem J 277:561–563

Wright A, Shin S-U (1991) Production of genetically engineered antibodies in myeloma cells: design, expression and applications. In: Lerner RA, Burton DR (eds) Methods: a companion to methods in enzymology, vol 2. Academic, Orlando, pp 125–135

Zebedee SL, Barbas CF III, Hom Y-L et al. (1992) Human combinatorial antibody libraries to hepatitis B surface antigen. Proc Natl Acad Sci USA 89:3175–3179

Section V:
Expression of MAbs/MAB Fragments

CHAPTER 11
Antibodies from *Escherichia coli*

A. PLÜCKTHUN

A. Introduction

Antibody engineering is still at its very beginning. No matter what the goal of the study, it is likely that a number of different ideas and variants of the recombinant antibody will have to be tested. It is therefore useful to be able to make new variants of the antibody protein easily, reliably and fast. Mostly for historic reasons, the molecular biology of *Escherichia coli* is more advanced than that of any other species; in fact essentially all molecular cloning is done today with this bacterial strain. It is therefore attractive to carry out antibody expression with this bacterium as well.

This chapter will describe the molecular biology of antibody expression in *Escherichia coli*. Some of the problems encountered will be common to all protein expression in *E. coli*, and they will not be discussed in detail. As will become apparent, the unique problem of making antibodies in *E. coli* is not so much one of ribosomal biosynthesis of the polypeptide (and certainly not one of efficient transcription, as this problem is more or less solved), but one of efficient folding of the polypeptide in the cell or, for some applications, in vitro. For this reason, it will be necessary to discuss the peculiarities of the antibody protein structure: its domain-like organization, its conserved disulfide bonds and its distinct sequence variability, which all have an effect on expression.

Due to their glycosylation and their large size, the production of functional whole antibodies in *E. coli* is still impractical, and the relation of the particular antibody fragment chosen, its biophysical characteristics and the effects on expression (that is, folding efficiency) must also be analyzed. In this context, solutions of the bivalency problem uniquely suited for bacterial expression will be discussed.

The bacterial system offers the possibility to produce fragments of the antibody directly, which can be useful for applications in which only binding is required. Examples might be in vitro diagnostics, affinity chromatography, antigen stabilization or catalysis. In other instances, the desired final product may be a whole antibody, humanized or human, to be used, for example, in therapy. In such cases, it can still be advantageous to carry out all testing of variants of the antigen binding site with *E. coli* fragments and then transplant the finished version to the eukaryotic expression system of the whole antibody. Finally, the easy availability of the recombinant product invites construction of many new molecules: hybrid proteins and antibodies with engineered metal binding sites, or with peptide tags for easy detection, or molecules with substantial alterations in the framework and quaternary structure.

Besides these "rational" alterations, *E. coli* permits a unique access to random approaches. Because of the unparalleled efficiency with which transformation with DNA or transfection by phages can be achieved in *E. coli*, many types of libraries of antibody variants can be assembled and screened by methods discussed below. With this background as a rationale, the emphasis of this article on the process of in vivo folding and assembly as the basis of all engineering and screening may become apparent.

B. Expression of Functional Antibody Fragments in *E. coli* by Secretion

I. General Overview

There are two basic strategies for obtaining recombinant antibodies and their fragments from *E. coli* (Fig. 1). The first is to produce inclusion bodies

---

Fig. 1a–c. The different expression strategies found to be useful for antibody production in *E. coli*. **a** Functional expression: In this case, the two chains making up the antibody combining site must be transported to the oxidizing milieu of the periplasm. In this compartment, there is a disulfide forming enzyme (DsbA) which allows the formation of the crucial intramolecular disulfide bonds, found in each domain and required for stability. *A* The pathway for two independent chains (as in a Fv or Fab fragment) is shown. Both are made as precursors containing a signal sequence and are then transported to the periplasm, where the signal sequence is cleaved off. There, folding, disulfide formation and assembly takes place. *B* Alternatively, both variable domains can be linked to form a continuous polypeptide chain (scFv fragment) which can then also be secreted.

For reasons not understood in mechanistic detail, secretion of these proteins presents a stress to *E. coli* and results in leakiness of the outer membrane after some time. This appears to occur to some degree with all types of antibody fragments and signal sequences tested, but its magnitude depends on the primary sequence of the antibody. The phenomenon is observed more readily in shake flasks than in controlled fermentation conditions, where it can be minimized. To prevent leakage and periplasmic inclusion body formation, it is crucial to grow cells at low temperatures (e.g., 25°C).

b Periplasmic inclusion body formation. This is observed for many fragments, but has been exploited preparatively mostly for scFv fragments. This phenomenon is temperature-dependent and is most easily induced at a temperature of 37°C or higher. The protein is apparently transported, processed and then precipitates. The protein must be refolded in vitro as in **c**.

c Cytoplasmic inclusion body formation. In this case, the protein is expressed without a signal sequence under as strong a promoter and translation signal as possible. Inclusion body formation appears to be more successful at temperatures of 37°C and higher. The protein must be refolded in vitro

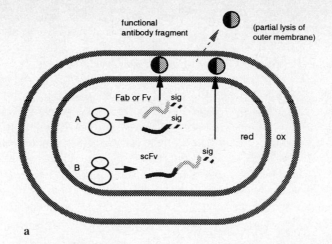

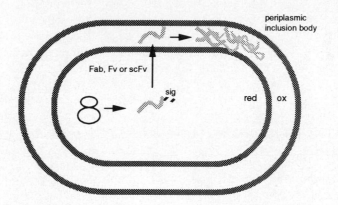

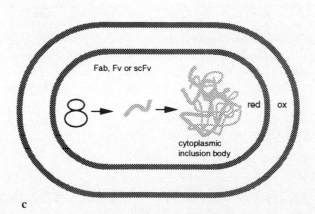

and refold the protein in vitro. The second is to imitate the situation in the eukaryotic cell and secrete the protein. With this method, completely functional antibody protein can be obtained (SKERRA and PLÜCKTHUN 1988; BETTER et al. 1988), and this strategy will be discussed first.

In *E. coli* a secretion machinery exists which leads to the transport of those proteins to the periplasmic space which carry a signal sequence (for a recent review, see PUGSLEY 1993). These are usually proteins which need to be outside of the "main" cell, for example, to degrade some biomolecules for easier uptake (e.g., peptidases, phosphatases). This secretion allows the protein to escape from the reducing environment of the cytoplasm, which in general, while perhaps not always (CABILLY 1989), appears to prevent normal disulfide formation (GILBERT 1990). Antibodies, as proteins which are normally secreted and equipped with disulfide bonds which are generally, but perhaps not for every primary sequence, necessary for stability (GLOCKSHUBER et al. 1992), ought to be secreted in bacteria, too, in order to reach a folded state identical to that formed in eukaryotes.

The *E. coli* secretion machinery directs the protein to the periplasm. There are specialized systems for transport out of the cell (PUGSLEY 1993), but they have not yet been used for high level production of foreign proteins (HOLLAND et al. 1990). It should be noted that, under conditions of high temperature and depending on the particular primary sequence and fragment of the antibody, a leakiness of the outer membrane is frequently observed with a variety of signal sequences (PLÜCKTHUN and SKERRA 1989). In these circumstances, periplasmic markers such as β-lactamase and alkaline phosphatase are also found in the medium, just as is the recombinant antibody protein. Nevertheless, some authors still refer to this phenomenon loosely as 'secretion to the medium" (see, e.g., TAKKINEN et al. 1991). This leakiness, whose molecular cause is still unknown, is possibly related to the formation of insoluble periplasmic protein from an aggregation process that occurs after the protein has been transported through the membrane (Fig. 1), but it may indicate the interference of the antibody with the transport of some crucial outer membrane component. After some time, the cells can also lyse completely, and this limits the production phase.

Periplasmic secretion has permitted the functional expression of a wide variety of antibody fragments with many antigen binding specificities. Because of the general importance and widespread application of this technology, it may be useful to discuss both the physiological limitations of the process in this chapter and the potential solutions to these problems. However, using this technology, it is already possible to conveniently obtain recombinant antibody fragments in sufficient amounts for essentially all studies.

II. Relation of Functional Secretion to Phage Libraries

An interesting and important consequence of the successful periplasmic folding of antibody fragments is the compatibility of the antibody folding

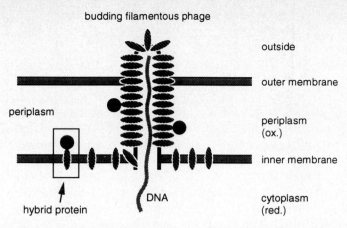

Fig. 2. The coating of a filamentous phage with hybrid proteins, present as intermediates in the inner membrane of *E. coli*

with display on the surface of a filamentous phage, such as M13 or fd. Since the NH_2-terminus of both the minor coat protein g3p (the product of gene III) and the major coat protein g8p (product of gene VIII) are probably exposed to the aqueous surrounding in the phage coat, NH_2-terminal extensions are still compatible with phage assembly (Fig. 2). This was first demonstrated for peptides displayed on the phage coat (SMITH 1985; PARMLEY and SMITH 1988; SCOTT and SMITH 1990; CWIRLA et al. 1990; DEVLIN et al. 1990; GREENWOOD et al. 1991; FELICI et al. 1991). During phage assembly, these coat proteins are present as intermediates in the inner membrane of *E. coli* (reviewed in RUSSEL 1991), with their NH_2-termini exposed to the periplasmic space. The budding phage then coats its DNA with these proteins waiting in the membrane. Consequently, any protein folding of fusion proteins consisting of NH_2-terminal antibody fragments and COOH-terminal phage coat proteins would take place in the periplasm, but while anchored to the inner membrane via the COOH-terminal membrane domain. Therefore, the same kind of fragments which correctly assemble in the periplasm can also ultimately be displayed on filamentous phage, with important applications in affinity screening of libraries. This has the important consequence of coupling genotype (the antibody gene on packaged phagemid) and phenotype (the displayed antibody fragment). Such experiments have been carried out with both Fab and single chain Fv fragments (McCAFFERTY et al. 1990; BARBAS et al. 1991; CLACKSON et al. 1991; BREITLING et al. 1991; MARKS et al. 1991; KANG et al. 1991; CHANG et al. 1991; HOOGENBOOM et al. 1991; GARRARD et al. 1991; GRAM et al. 1992; BARBAS et al. 1992a) and will be discussed in more detail elsewhere in this volume. It is occasionally surmised that this process will automatically select for efficient folding and thus expression. However, while extremely poor expression will undoubtedly be selected against, it is unclear to what degree, at the extremely low expression levels of the minor phage coat proteins, small differences in aggregation tendency of the displayed antibody fragment would be apparent.

The use of filamentous phage in antibody libraries must be contrasted with the use of phage λ (HUSE et al. 1989). In this case, the phage only delivers the DNA to the *E. coli* cell. In the λ phage genome, a plasmid constructed according to the principles laid out by SKERRA and PLÜCKTHUN (1988) and BETTER et al. (1988), is incorporated. Since production of viable phage would severely interfere with protein production (which of course requires healthy cells), the plasmid must be rescued after library formation before useful production is possible.

The display of proteins on the surface of *E. coli* has been described as well (FRANCISCO et al. 1992, 1993; KLAUSER et al. 1992; FUCHS et al. 1991), but as of yet no work with libraries has been reported. It remains to be seen how such screening methods would compare to the screening of filamentous phage libraries.

III. Description of the Secretion Process

At room temperature, a major portion of the Fv fragments and a significant fraction of the Fab fragments investigated in detail (SKERRA and PLÜCKTHUN 1991; KNAPPIK et al. 1993, unpublished) go to the native state (Fig. 3), but it is now clear that this portion crucially depends on the primary sequence of the variable domains. It was shown that the antibody fragments are processed correctly, contain their disulfide bonds, assemble to heterodimers and bind antigen with the same affinity as the normal antibody (SKERRA and PLÜCKTHUN 1988). The two chains of the Fv or Fab fragment therefore find

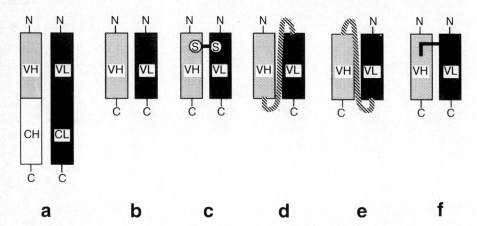

Fig. 3a–f. Monovalent fragments of antibodies functionally expressed in *E. coli.* **a** Fab fragment; **b** Fv fragment; **c** disulfide-linked Fv fragment; **d** single chain Fv fragment with the orientation V_H-linker-V_L; **e** single chain Fv fragment with the orientation V_L-linker-V_H; **f** Fv fragment which has been stabilized by chemical cross-linking after purification. (See GLOCKSHUBER et al. 1990a)

each other even if they are not covalently linked, as they do in eukaryotes. This has now been demonstrated for a variety of Fv fragments (see, e.g., SKERRA and PLÜCKTHUN 1988; WARD et al. 1989; GLOCKSHUBER et al. 1991; McMANUS and RIECHMANN 1991; ANTHONY et al. 1992; STEMMER et al. 1993a) and Fab fragments (see, e.g., BETTER et al. 1988; PLÜCKTHUN and SKERRA 1989; ANAND et al. 1991a; CARTER et al. 1992; BARBAS et al. 1992b; BETTER et al. 1993), demonstrating the generality of the method. In conclusion, the antibody binding site of the fragments produced in bacteria is functionally identical to the natural antibody (see also Sect. 1). Low bacterial growth temperature is the most efficient method to minimize periplasmic aggregation and maximize the yield of folded antibody protein (SKERRA and PLÜCKTHUN 1991; KNAPPIK et al. 1993). This may have to do with the stability of folding intermediates and the rate of protein biosynthesis or secretion, both of which are functions of temperature, but the relative importance of these phenomena is not yet clear.

From an investigation of the limiting step in the process of secreting antibody fragments, it was concluded (SKERRA and PLÜCKTHUN 1991) that, if a vector with strong transcription and translation initiation signals is used, it is periplasmic folding and/or assembly which limits the level of functional expression. During antibody folding, the insoluble periplasmic protein is formed as a by-product, presumably via the aggregation of an intermediate. This conclusion was arrived at as follows: With increasing promoter strength, the amount of correctly folded antibody does not increase significantly. The insoluble, correctly processed antibody protein increases, and only at the highest promoter strength is there a significant sign of precursor. This insoluble protein was shown, at least in one case (GLOCKSHUBER et al. 1992), to be accessible to externally added proteases after producing spheroblasts under conditions in which soluble cytoplasmic precursor was not degraded. One may conclude that the signal sequence is cleaved off and that part of the protein folds correctly to produce fully functional antibody fragments, although some part of it does not achieve the native state and instead aggregates and precipitates (Fig. 1).

In the antibody McPC603 (PERLMUTTER et al. 1984; PLÜCKTHUN 1993a), the Fv and Fab fragments are produced at about similar amounts on the ribosome from the same vector, and the total amount of protein produced is comparable (SKERRA and PLÜCKTHUN 1991; KNAPPIK et al. 1993). However, the amount of functional protein is greater for the Fv fragment. This finding has suggested that particular folding problems in the C_H1 or C_L domains may be responsible for the less efficient folding of the Fab fragment. Alternatively, the presence of the constant domains may just potentiate folding problems in the variable domains. There is evidence supporting the latter view, as different variable domains in the same Fab vector fold in different proportions (KNAPPIK and PLÜCKTHUN, unpublished). This phenomenon may be masked if an Fv fragment is particularly prone to proteolysis (SCHWEDER and PLÜCKTHUN, unpublished).

While protein folding is an exergonic and spontaneous reaction, it is now clear that proteins exist which guide this process to prevent the side reaction of aggregation (Jaenicke 1993). No general periplasmic molecular chaperone has been unambiguously identified at this time (see, e.g., Pugsley 1993). There is no evidence, nor is it likely, that cytoplasmic molecular chaperones such as GroEL or DnaK (Jaenicke 1993) have any direct influence on periplasmic protein folding processes. They may, however, help the assembly of phage particles displaying the antibody under certain conditions (Söderlind et al. 1993), and phages are known to require chaperone assistance in their assembly (Zeilstra-Ryalls et al. 1991), but there is no evidence to suggest that the chaperone influences antibody folding. Cytoplasmic chaperones may have an indirect effect on cell stability.

IV. The Role of Periplasmic Protein Folding

There is great variation in the literature about the reported efficiency of the secretory expression method for antibodies. This fact has to do with variations among vectors, procedures and quantification and the use of different fragments of antibodies varying in sequence. While the problem is far from being understood and still further from being solved, enough experiments have now been carried out to at least attempt some kind of correlation.

From a variety of experiments (Carter et al. 1992; Knappik and Plückthun, unpublished), evidence is accumulating that the primary se-

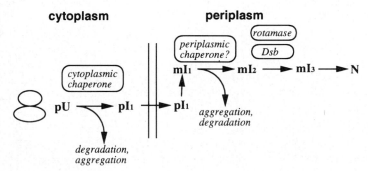

Fig. 4. Hypothetical folding pathway of a secreted protein. The protein is made on the ribosome as an unfolded precursor (pU) and probably kept in a transport-competent state (pI_1) by association with an as yet unidentified cytoplasmic factor (*chaperone*). This factor presumably prevents premature folding. It is unknown whether a fraction of the protein is degraded on its way to the membrane. After transport, the signal sequence is cleaved to give the mature folding intermediate mI_1, which must then fold via other hypothetical intermediates (mI_2, mI_3) to the native state N. During periplasmic folding, as yet unidentified periplasmic chaperones may act on the protein, and the disulfide forming activity DsbA acts on the antibody. It is unknown whether the resident proline *cis-trans* isomerase (*rotamase*) acts on the antibody. Overexpression of rotamase and DsbA do not seem to change the amount going to N, and thus the diversion to aggregates appears to happen before these steps, or at least to be independent of their extent

quence of the antibody plays a decisive role. The primary sequence of the antibody determines the critical partitioning of the protein intermediates (presumably after transport) between folding to the native structure, aggregation and degradation (Fig. 4). This deduction requires constructs to be compared with each other which differ in nothing else but the primary sequence, i.e., the same type of fragments (Fab, Fv, single chain Fv) in exactly the same vector. As a case in point, an Fab fragment of a humanized antibody gave much more favorable (10- to 50-fold) partitioning between folding and aggregation than the chimerized version of the same molecule containing the mouse variable domains (CARTER et al. 1992).

Other antibody fragments may conceivably be prone to proteolysis and thus not reach the expression limit possible by folding. In poorly designed vectors, it may of course also be possible that other steps, such as transcription or translation, are limiting and expression is so low that insoluble protein is never seen. In low expression systems, it is thus possible that an improvement of these processes will increase the overall yield of folded protein (STEMMER et al. 1993a). It should be generally feasible, however, to reach the limit set by protein folding with suitable vectors (see below and, e.g., SKERRA and PLÜCKTHUN 1991), and it was shown that with such vector systems the antibody protein can be one of the most prominent soluble proteins in the cell, if it has a sequence and structure which allows efficient folding (KNAPPIK and PLÜCKTHUN, unpublished; CARTER et al. 1992).

V. Catalysis of Periplasmic Protein Folding

1. Disulfide Bond Formation

The observed difference in folding between the Fv and the Fab fragment of the same antibody (SKERRA and PLÜCKTHUN 1991) led to the question of whether particular structural features in the constant domains can be delineated which might be responsible for this difference, perhaps by leading to particularly slow folding steps of the Fab fragment. Two types of slow processes in protein folding have been discovered which can be pinpointed to particular chemical events (FISCHER and SCHMID 1990): proline *cis-trans* isomerization and disulfide formation and isomerization. It is now clear, however, that these are not the only slow events and that noncovalent rearrangements of the protein during folding, once the protein has already attained a native-like structure, can be slower still and thus can be rate determining. Unfortunately, the rate determining step for most antibody fragments is not known and, again, may well depend on the primary sequence.

The importance of proline *cis-trans* isomerization and disulfide isomerization can be tested by mutagenesis experiments and by catalysis with specific proteins in vitro and in vivo. Such experiments have been carried out with different fragments of the antibody McPC603 (SKERRA and PLÜCK-

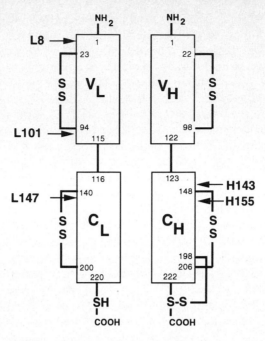

Fig. 5. Cysteine residues and *cis*-peptide bonds adjacent to proline residues in the Fab fragment of the antibody McPC603, a mouse IgA

THUN 1991; Knappik et al. 1993) in *E. coli* and a single chain Fv-toxin fusion protein in vitro (Buchner et al. 1992a).

The Fab fragment of McPC603, a mouse IgA, carries five disulfide bonds (Satow et al. 1986). One consensus S-S bond is present in each domain linking the two β-sheets, and there is an extra one in C_H1, which is a feature of mouse IgA (Satow et al. 1986; Cockle and Young 1985) (Fig. 5). Removing this additional disulfide creates an Fab fragment which is fully functional but is not obtained at significantly higher yield (Skerra and Plückthun 1991), although the removal of two cysteines might at first be expected to diminish the possibilities for incorrect disulfide linkages. The light chain of this Fab ends in a free cysteine probably linked to the other Fab in the mouse IgA (Abel and Grey 1968), but similarly, its removal does not change the partitioning of Fab between folded and aggregated protein. If the C_H1 domain of an IgG1 is introduced instead of the C_H1 of the mouse IgA, a covalent link is obtained between H and L, yet again at about the same level of correctly assembled protein. These data are consistent with the idea that disulfide formation does not limit the periplasmic folding process.

The formation of disulfide bonds is known to occur in the bacterial periplasm (Pollitt and Zalkin 1983) (Fig. 4), and recently an enzymatic system has been discovered which is responsible for it (Bardwell et al.

1991, 1993; KAMITANI et al. 1992). Briefly, the protein DsbA, itself containing a reversibly opening disulfide bond, is involved in the formation and perhaps rearrangement of disulfide bonds of the proteins to be folded after they have reached the periplasmic space. There has been some debate about whether disulfide rearrangement involving stable intermediates (as opposed to the serial formation of only correct disulfide bonds) is a physiologically relevant reaction (CREIGHTON 1978; WEISSMAN and KIM 1991; GOLDENBERG 1992). In the best investigated case in vitro, that of the bovine pancreatic trypsin inhibitor BPTI, disulfide formation occurs in a very late state of folding, after the tertiary fold is essentially complete (WEISSMAN and KIM 1991).

In the case of the antibody domains, the mechanistic details of disulfide formation have remained unclear. In their natural environment (the B cell or plasma cell) antibodies develop disulfide links in the endoplasmic reticulum (BERGMANN and KUEHL 1979), catalyzed by the eukaryotic disulfide isomerase (FREEDMAN et al. 1989). The conserved intradomain disulfide, however, is completely buried, and it seems that its formation must occur at an early step in folding.

Using a $dsbA^-$ deletion mutant, no formation of Fv fragment could be observed in the bacterial periplasm (KNAPPIK et al. 1993). Fv formation could, however, be restored by plasmid-encoded DsbA. These experiments show that bacterial DsbA takes part in the formation of active Fv fragment, and is *required* for this assembly.

An obvious question was therefore whether the overexpression of DsbA would increase the amount of active Fv, single chain Fv (scFv) or Fab fragment. For the antibody McPC603, for which such an experiment was carried out, the answer was negative (KNAPPIK et al. 1993). This is consistent with the results from the experiments on removing disulfides by mutagenesis (SKERRA and PLÜCKTHUN 1991), and it suggests that aggregation of the periplasmic protein occurs either before disulfide formation or is at least independent of its extent. This experiment does not, however, automatically lead to the generalization that disulfide formation may never be limiting for any fragments or fusion protein. There might conceivably be cases in which overexpression of a disulfide formation catalyst and/or the inclusion of redox couples (i.e., reduced and oxidized glutathione at various proportions) might make a difference. However, in the investigated cases, the aggregation phenomena seemed to be independent of the extent of disulfide formation and perhaps to precede it (Fig. 4).

In vitro, a strong dependence of folding rates and yields on disulfide formation has been seen for antibody folding (HABER 1964; ROWE and TANFORD 1973; ROWE 1976; GOTO and HAMAGUCHI 1979, 1982, 1986; GOTO et al. 1988; HUSTON et al. 1991; BUCHNER et al. 1992a,b; BUCHNER and RUDOLPH 1991). Nevertheless, different scFv fragments seemed to show rather different requirements for the type of oxidative folding: the method most frequently used for disulfide containing proteins, simultaneous oxidation and refolding (by dilution from denaturant into a buffer containing a redox couple) with or without prior formation of mixed disulfides, was not always successful (summarized in Huston et al. 1991). Some antibody Fv and scFv fragments require oxidative formation of disulfide bonds already in the completely denatured state, suggesting problems with accessibility of the cysteines (Hochman et al. 1976; Glockshuber et al. 1992).

In conclusion, the conserved disulfides of the variable domains, which are important for structural integrity (Glockshuber et al. 1992), can be formed in vitro and in vivo (in eukaryotes and bacteria) and they do not seem to constitute the reason for the amount of aggregation which may accompany antibody expression in bacteria.

2. Proline *cis-trans* Isomerization

Another reaction which can be the slow step in protein folding is proline *cis-trans* isomerization (Brandts et al. 1975). The peptide bond contains partial double-bond character and thus presents a significant barrier to rotation. Two stable configurations exist, but for all amino acids except proline, the *trans* configuration (labeled with respect to the two C_α atoms) avoids steric crowding, whereas for proline, the difference is marginal (summarized in Stewart et al. 1990). Proline *cis-trans* isomerization can be a slow step of protein folding in vivo, even of antibody domains (Goto and Hamaguchi 1982; Lang and Schmid 1988; Buchner et al. 1992a). However, even for very slow folding proteins containing *cis* prolines, there may be packing rearrangements that occur still more slowly than proline *cis-trans* isomerization and thus constitute the rate-determining steps.

There are two unrelated, ubiquitous classes of proteins with proline-*cis-trans* isomerase activity (Fischer and Schmid 1990; Trandinh et al. 1992). So far, actual demonstration of their involvement in folding in vivo is still lacking, but their acceleration of folding in vitro has been demonstrated for numerous substrate proteins. *E. coli* has two such enzymes, one in the cytoplasm, and one in the periplasm (Liu and Walsh 1990; Hayano et al. 1991).

Antibodies contain *cis* prolines both in their constant and in their variable domains (Fig. 5). In the particular case of the Fab fragment of the mouse IgA McPC603, there are five of them (Satow et al. 1986) (two in V_L at L8 and L101, one in C_L at L147 and two in C_H1 at H143 and H155). Thus, the Fv fragment would contain two *cis* bonds and the Fab fragment all five. The observation that the Fab fragment and the Fv fragment of the same antibody fold to various efficiencies in the periplasm, whereas they are produced in about the same amount on the ribosome, made it worthwhile to test proline *cis-trans* isomerization as a possible cause for the different behavior of Fv and Fab fragments.

For this purpose, a loop in the C_H1 domain at the opposite end of the molecule from the binding site was altered from its wild-type sequence (containing two *trans*-prolines and one *cis*-proline) to a loop containing only 1 *trans*-proline, but without dramatic effect on periplasmic folding (Skerra and Plückthun 1991). Since the putative cause of aggregation might be in any of the *cis*-prolines, the *E. coli* periplasmic proline *cis-trans* isomerase was also overexpressed together with the Fv fragment, or both versions of the scFv fragment (V_H-linker-V_L and V_L-linker-V_H) or the Fab fragment (Knappik et al. 1993). There was no effect on the yield of folded protein, with the possible exception of one of the scFv fragments. In this particular case, V_L was preceded by the linker and perhaps isomerization of the peptide bond at L8 was sterically more hindered than in any other fragment. Nevertheless, there is no evidence that proline *cis-trans* isomerization has a limiting role on the folding of these antibody fragments in *E. coli*. This conclusion is also not changed by overexpressing E. coli disulfide isomerase DsbA together with proline *cis-trans* isomerase (Knappik et al. 1993).

It appears therefore that neither proline *cis-trans* isomerization nor disulfide isomerization can be held responsible for the absence of quantitative folding of antibody fragments in the bacterial periplasm. Rather, aggregation events which occur before these reactions or are at least independent of their extent appear to be the cause (Fig. 4). Nevertheless, periplasmic functional expression is possible and provides a fast, convenient and versatile method to directly obtain folded antibody in quantities sufficient for essentially any experiments, up to grams per liter in fermentation (Carter et al. 1992).

VI. Design of Secretion Vectors

The previous discussion has emphasized that it is the periplasmic folding process which appears to limit the expression level of functional antibody protein. Therefore, it is not necessarily useful to choose as high an expression level as possible. Instead, it is crucial that expression is under the control of a repressible promoter, since secretion (or the concomitant production of some insoluble periplasmic protein) appears to stressful for the cells, which may respond with poor growth, plasmid loss, and, in extreme cases (observed for a single chain T cell receptor; Wülfing and Plückthun, unpublished), complete rearrangement of plasmids. Many different antibody secretion vectors have been designed by now, but they all essentially follow the principles laid out by Skerra and Plückthun (1988) and Better et al. (1988). Phages use the *E. coli* machinery to make proteins in just the same way. Therefore, they must contain similar expression cassettes. This is true for phage λ, in which the phage is merely used for transfecting the genetic information efficiently, and for filamentous phages (see above), in which a fusion protein is made from the antibody and a coat protein, using periplasmic secretion as an intermediate.

In a walk around the vector (Fig. 6), the considerations leading to the choice of promoter of the recombinant immunoglobulin will be discussed

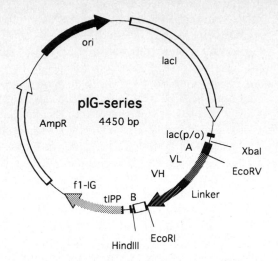

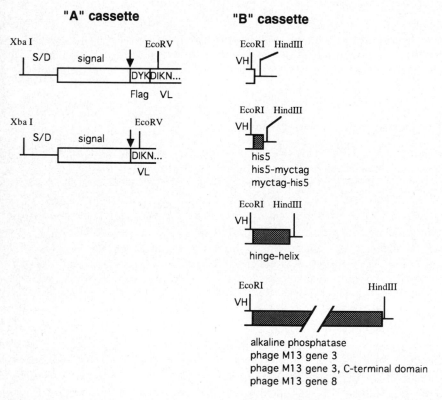

Fig. 6A

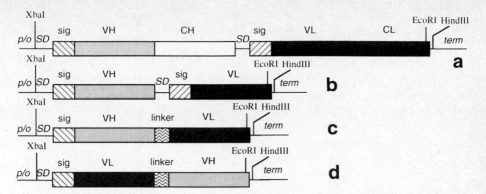

Fig. 6A,B. Secretion vectors for antibody expression in *E. coli*. As an example, a series of improved vectors suitable for expressing single chain Fv (scFv) Fv and Fab fragments is shown (L. GE, A. KNAPPIK and A. PLÜCKTHUN, unpublished). These vectors incorporate restriction sites within the antibody gene for convenient PCR cloning and cassettes for detection, purification and phage display.

A The design is very modular, allowing easy exchange of antibiotic resistance, the fragment of the antibody and the *A* and *B* cassettes.

The *A cassette* contains a bacterial signal sequence, directly fused to the mature antibody part or preceded by only three additional *Flag* amino acids, which can be detected with extremely high specificity and sensitivity using a Ca^{2+}-dependent antibody (KNAPPIK et al., unpublished) using a much shorter epitope than originally presumed to be necessary (PRICKETT et al. 1989).

The *B cassette* may be used to introduce a COOH-terminal purification or detection tag, to fuse a hinge and helix for dimerization, to fuse a phage gene for surface display or an enzyme for easy detection of the antibody.

PCR of antibody genes can be carried out with primers carrying extensions using *Eco*RI and *Eco*RV sites, which are both rare in antibody sequences.

B The schematic arrangement of genes for functional expression. Since in this case both chains of the antibody must be secreted to the same periplasmic space to assemble, they must either be produced as two different secreted protein chains in the same cell (*a,b*) or they must be linked via a peptide linker (*c,d*). It is advantageous to express the independent chains of the Fab fragment (*a*) or the Fv fragment (*b*) in a dicistronic operon (*a,b*), as discussed in the text. Two different orientation of the scFv fragment are shown which have both been shown to function (*c,d*). *p/o* denotes a promoter/operator structure; *SD*, a Shine-Dalgarno sequence; and *term*, a transcription terminator

first. While it should be strong, it needs to be, first and foremost, very tightly regulated. Two natural promoters have proven to be particularly useful in this respect, the first being the *lac* promoter, which is regulated not only by the *lac* inducer IPTG, but also by glucose (BECKWITH and ZIPSER 1970). The second is the *pho*A promoter (WANNER 1987), which is turned off in the presence of phosphate. The latter is more useful in reproducible fermentations, but perhaps less so on a laboratory scale, since the initial phosphate must be precisely calculated to run out at a particular point. A third system, which is completely tight and has been shown to be useful for T cell receptor expression, is that of invertible promotors (WÜLFING and

Plückthun, 1993; Podhajska et al. 1985), using phage λ integrase for inverting a cassette in which any promotor can be placed.

Less suitable are promoters such as *pL* when coupled with the thermolabile cI857 repressor, since they have to be induced at high temperatures. This is counterproductive with regard to correct folding (Colcher et al. 1990; Gibbs et al. 1991), which occurs more efficiently at low temperature (Takagi et al. 1988; Glockshuber et al. 1990a; Skerra and Plückthun 1991), and it requires that the temperature is shifted back to room temperature in the expression phase.

The translation initiation region is usually taken from a well expressed protein and it might be taken from the same bacterial gene as the signal sequence. The principle of a mini-cistron preceding the actual antibody gene or genes, themselves arranged in two cistrons in the case of an Fab or Fv fragment, has been found useful (Skerra et al. 1991; Schoner et al. 1990). Despite intensive research (summarized in McCarthy and Gualerzi 1990), there is still only incomplete rational understanding of efficient translation initiation, especially the positive or negative effect played by secondary structure of the mRNA, which on the one hand prevents translation and on the other hand protects the mRNA from degradation (see, e.g., Ehretsmann et al. 1992). Therefore, pragmatic approaches using well expressed *E. coli* genes as a framework still prevail. At low expression levels transcription and translation may be limiting (Stemmer et al. 1993a).

The signal sequence directs the antibody protein to the periplasmic space. There is no evidence that there are particular signal sequences which can direct the antibody protein to the medium; it appears rather that the outer membrane becomes leaky due to the mature antibody protein. Eukaryotic signal sequences often work successfully in bacteria, but the original heavy chain signal sequence of the antibody T15 failed in *E. coli* (Skerra and Plückthun 1991), probably because it cannot be cleaved properly due to its cysteine residue at the -1 position. Among bacterial signal sequences used successfully are those of the bacterial outer membrane protein A (*ompA*), alkaline phosphatase (*phoA*) and pectate lyase of *Erwinia carotovora* (*pelB*). The latter tolerates a sequence change, in which a very rare restriction site can be introduced, useful for PCR cloning (see below).

Beckwith and coworkers made the remarkable observation that a small amount of secretion to the periplasm of some proteins is detectable even if no signal sequence at all is used, especially in a *prlA* strain of *E. coli* (Derman et al. 1993). It remains to be investigated whether the observation of some amount of functional Fab, expressed without signal sequence in *E. coli* (Cabilly 1989), is related to this phenomenon.

For efficient secretion, the antibody gene must surely be fused precisely to a bacterial signal sequence. In the case of a scFv fragment, both domains making up the binding site are connected in a single protein (Figs. 3, 6, 7). In the case of Fv fragments and Fab fragments, however, both chains are unconnected. The most efficient way to guarantee coexpression and cose-

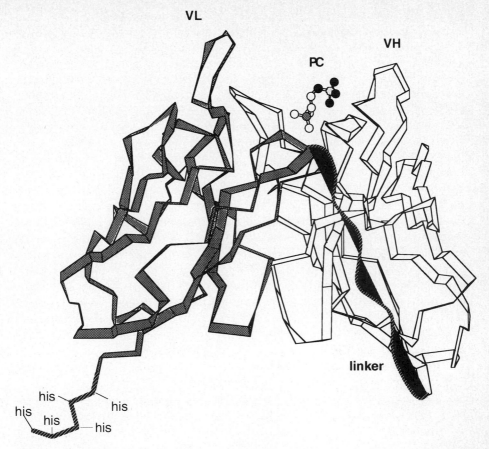

Fig. 7. A single chain Fv (scFv) fragment containing a histidine tail for purification with immobilized metal affinity chromatography. The linker is drawn very schematically, as it is now known to have no defined structure. (For details, see FREUND et al. 1993; see text)

cretion of both chains has been to design an aritificial operon, in which both genes, each fused to its signal sequence, are encoded in tandem on a single mRNA (SKERRA and PLÜCKTHUN 1988; BETTER et al. 1988) (Fig. 6). This way, simultaneous folding of heavy and light chains can occur in the periplasm. There is no indication that there is a higher kinetic barrier to association of V_H and V_L in an unlinked Fv fragment than in a scFv fragment, as evidenced by the similar expression yields (GLOCKSHUBER et al. 1990a). However, at equilibrium, some Fv fragments show significant dissociation into V_H and V_L which is of course concentration dependent (see below).

For protein production, the original dicistronic approach (Fig. 6) (SKERRA and PLÜCKTHUN 1988; BETTER et al. 1988) still seems to be the most advantageous, as only one promoter needs to be tightly regulated and only one plasmid needs to be maintained at high copy number. Nevertheless, other approaches have now been used as well. A two-promoter system has been used in a phage vector (BARBAS et al. 1991), as have been two-plasmid based systems (COLLET et al. 1992). These vector systems may add versatility to combinatorial libraries, but they will also add complications in stable production systems.

The mRNA ends at a transcription terminator, and there is evidence that efficient termination also protects the mRNA against exonucleolytic degradation (BELASCO and HIGGINS 1988 and references therein). Other elements of the plasmid include an antibiotic resistance, which, despite being a standard procedure, is worth a comment. The periplasmic leakiness induced by the antibody can lead to massive amounts of β-lactamase in the medium, degrading the antibiotic and making it possible for plasmid-free cells to grow (PLÜCKTHUN and SKERRA 1989). It is therefore better to resort to other antibiotics in prolonged growth experiments and fermentation, such as kanamycin or tetracycline. The origin of the expression plasmids based on the pUC series will give a high copy number (YANISCH-PERRON et al. 1985), but only at high temperature, and at room temperature, it falls below pBR322 (LIN-CHAO et al. 1992). Both for mutagenesis purposes and phage display (see below) an origin for a filamentous phage on the plasmid is useful (Fig. 6).

VII. Fermentation

Two reasons make it worthwhile to discuss the fermentation of bacteria producing antibodies in this context. First, the most efficient method to increase the amount of folded antibody per volume is by producing more cells, if the amount produced per cell is limited. Second, the fermenter allows for more careful control of the growth conditions. This is not only crucial for reproducibility, but also helps in understanding the physiology of the process.

An unexpected observation was the lack of periplasmic leakiness under fed-batch fermentation conditions (PACK et al. 1993). One possible explanation might be the artificially slowed growth in the fed-batch process, which might lead to a higher degree of cross-linking of the peptidoglycan cell wall (PARK 1987) and thus to a greater stability of the cell. Alternatively, the defined medium may be lacking a leakiness-inducing component. It was found to be crucial to use tightly regulated promoters (CARTER et al. 1992; BETTER et al. 1993; PACK et al. 1993), since otherwise plasmid loss is observed during cultivation. Before induction, nitrogen levels stay constant, indicating a metabolic balance of the repressed cells. After induction, this balance is lost, demonstrating the stress that antibody secretion constitutes for the bacterial cell. This stress appears to be dependent on the particular antibody sequence (KNAPPIK and PLÜCKTHUN, unpublished).

Fermentation now allows the production of functional antibody fragments (depending on the fragment and its sequence) with yields between 100 mg and 1 g per liter *E. coli* culture (CARTER et al. 1992; BETTER et al. 1993; PACK et al. 1993), demonstrating the general utility of the bacterial secretion technology.

VIII. Cloning Antibodies by Polymerase Chain Reaction

The availability of a large number of antibody sequences obtained by conventional cloning and the advent of PCR made it possible to define consensus primers to amplify antibody sequences (reviewed in LARRICK et al. 1992). PCR amplification of mRNA after reverse transcription and the amplification of rearranged V-D-J heavy chain or V-J light chain genes have been described (reviewed in LARRICK et al. 1992; HOOGENBOOM et al. 1992).

The "COOH-terminal" primers (reading toward the 5′ end of the RNA)[1] are not a problem: all constant domain sequences of mice and humans are known, and precisely matching primers can be made to amplify mRNA. Together with appropriate "NH_2-terminal" primers (reading towards the 3′ end of the mRNA)[1] sequences encoding the Fab fragment can be amplified. Alternatively, primers located in the J region (also completely known) can be used as the COOH-terminal primers. To amplify genomic V genes, different COOH-terminal primers have to be used for hybridizing at the end of the V genes (TOMLINSON et al. 1992). The NH_2-terminal primer can be located at the very beginning of the mature region of the V gene (ORLANDI et al. 1989; SASTRY et al. 1989). Alternatively, primers hybridizing to the signal sequence have been used successfully (JONES and BENDIG 1991) for amplifying mRNA. While the latter strategy conserves the NH_2-terminus of the mature gene, it requires an additional PCR step to introduce the gene into the bacterial expression vector and has been used mostly with eukaryotic expression vectors, and is less suitable for library construction.

The former method of PCR amplification may change the identity of the amino acids at the very beginning of the gene, since one of the primer-encoded sequences will be obtained by necessity. One may, however, convert these residues to any consensus sequence or to an experimentally determined protein sequence in a second round of PCR amplification or site-directed mutagenesis.

Most conveniently, the PCR product is cloned directly into the expression vector or phage display vector. This may be achieved by directly extending the PCR primers to include an "overhang" encoding a restriction site plus some extra bases to ensure cutting of the restriction enzyme close

---

[1] Winter and coworkers (see, e.g., HOOGENBOOM et al. 1992) have used a different nomenclature, in that the primer reading toward the 5′ end of the mRNA is called "forward," whereas the one priming synthesis toward the 3′ end is called "back."

to the end of the PCR fragment. Alternatively, a short precise primer can be used first to amplify the eukaryotic DNA or cDNA to eliminate any false priming by the overhang; a second round of amplification is then used to introduce the restriction site.

A variety of methods for introducing the PCR products into the vector, e.g. a bacterial secretion vector (Fig. 6), have been used. Many variations on this theme are possible, depending on whether bacterial secretion vectors, vectors for inclusion body formation, eukaryotic cDNA based vectors or eukaryotic genomic based vectors are to be used for receiving the PCR products.

IX. Purification

Purification of whole antibodies has usually relied on classical chromatography, antigen affinity chromatography or affinity chromatography using bacterial immunoglobulin-binding proteins such as staphylococcal proteins A, B, G or L (for summaries see BOYLE 1990; BOYLE and REIS 1987; FAULMANN et al. 1991; NILSON et al. 1992). However, the usefulness of this strategy for Fv or scFv fragments is fairly limited, as the bacterial proteins bind mostly to constant domains and only a few subgroups of V domains are recognized (INGANÄS et al. 1980; NILSON et al. 1992).

However, using affinity tails, any fragment can now be purified by rather convenient and reproducible procedures, and this technology can be carried out on a very large scale. The most convenient strategy is probably the use of a stretch of histidines at the COOH-terminus (SKERRA et al. 1991; LINDNER et al. 1992) (Fig. 7). This has been successfully tested with a scFv fragment of the form V_H-linker-V_L-His-5, with a V_L domain (V_L-His-5) and also with an Fab fragment, in which the His-5 tail was fused to C_H1. Since the heavy chain is practicably insoluble if not paired with a light chain, this amounts to a purification of assembled Fab fragments (KNAPPIK et al., unpublished) even though the two chains are not covalently linked. In the mouse κ V_L domain, the last two amino acids, Arg-Ala (numbers 108 and 109 according to Kabat), were replaced by histidines and only three additional His residues had to be added to the end. X-ray crystallography showed that this had no influence on the structure of the V_L domain (LINDNER et al. 1992).

Recent developments with immobilized metal ion affinity chromatography (IMAC) now also include a convenient detection system: by combining the metal ligand nitrilotriacetic acid (NTA) with biotin, the His-5 or His-6 containing protein can be detected in Western blots using phosphatase-labeled avidin (HOCHULI and PIESECKI 1992). Furthermore, the tail can also be used for obtaining a pseudocovalent binding: using Co^{2+} as the metal bound to NTA, which is attached to a solid support, the protein can be adsorbed. A later oxidation of Co^{2+} to Co^{3+} makes it exchange-inert, thereby effectively "covalently" binding the protein to a solid surface in a predetermined orientation with the binding site still intact (SMITH et al. 1992).

A whole number of different affinity tags have been suggested for detection and purification, such as an epitope of the myc protein (EVAN et al. 1985; MUNRO and PELHAM 1986; WARD et al. 1989), a Ca^{2+}-dependent epitope useful both at the NH_2-terminal and the COOH-terminus (WELS et al. 1992a; PRICKETT et al. 1989; KNAPPIK and PLÜCKTHUN, unpublished), a peptide biotinylated in vivo in *E. coli* (MÜLLER and PLÜCKTHUN, unpublished; WEISS, personal communication), peptides binding to streptavidin (DEVLIN et al. 1990; SCHMIDT and SKERRA 1993), or epitopes encoded in the linker of a scFv fragment (BREITLING et al. 1991), but scale-up would be more costly than with the metal affinity procedure. Nevertheless, for detection purposes, these tags can be useful. For example, only three additional amino acids, Asp-Tyr-Lys, fused to the NH_2-terminal Asp residue of the V_L domain are sufficient to specifically detect the protein (KNAPPIK and PLÜCKTHUN, unpublished) with a specific antibody (Fig. 6). In the case of novel types of constructs or new antibody-like domains, it may also be useful to test the integrity of the protein with the simultaneous use of an NH_2-terminal and a COOH-terminal tag sequence, which has been shown to be possible with scFv fragments (GE et al., unpublished).

C. Expression of Antibody Fragments as Inclusion Bodies

The production of antibody proteins as cytoplasmic inclusion bodies in *E. coli* is also possible, and it does not differ greatly from the production of other recombinant proteins by this method. This was the strategy used in the first reports about expressing antibodies in *E. coli* (Boss et al. 1984; CABILLY et al. 1984). All types of antibody fragments (Fab, Fv, scFv and even the chains for the whole antibody) have since been produced this way (see, e.g., Boss et al. 1984; CABILLY et al. 1984; WOOD et al. 1984; BIRD et al. 1988; HUSTON et al. 1988; FIELD et al. 1989; PANTOLIANO et al. 1991; CHEADLE et al. 1992; FREUND et al. 1993) and a variety of strains, plasmids and promoters have been used. There are no apparent requirements for the strain or expression system which would be specific for antibodies; any established production strain for *E. coli* inclusion bodies and most inducible strong promoters should be suitable. The use of the T7 system, as a particularly strong, but regulatable system, was found useful (see, e.g., HUSTON et al. 1991; FREUND et al. 1993). Fermentation of *E. coli* can also be carried out according to established principles (summarized by RIESENBERG 1991).

Too little is currently known to predict the ability of different sequences of form inclusion bodies (desired in this strategy) and their susceptibility to proteases. Recombinant proteins occasionally show signs of some degra-

dation, even when produced as inclusion bodies. The problem with antibodies is that there is considerable variability in the sequence and not all
behavior observed for the antibodies tested so far may be generally valid.
It is, for example, not yet known whether protease deficient strains are
generally useful for the yield of inclusion bodies, and this may depend
somewhat on the particular fragment and the particular sequence of the
antibody of interest.

Usually, the inclusion body approach is carried out using genes not
encoding signal sequences. Therefore, the antibody fragments stay in the
cytoplasm and largely precipitate. Since precipitation is desired, it is useful
to do exactly what needs to be avoided when secreting the antibody, namely,
to grow the cells at higher temperature, e.g., 37°C. At lower temperature,
some soluble Fab fragment has been observed which can be isolated from
the cytoplasm (CABILLY 1989), but it has not been completely characterized
in terms of the extent of its disulfide formation and stability.

Using secretion vectors, one may also isolate that portion of the secreted
protein which precipitates after transport to the periplasm. This has been
described for scFv fragments (COLCHER et al. 1990; GIBBS et al. 1991;
WHITLOW and FILPULA 1991) and Fab fragments (SHIBUI et al. 1993). At
higher temperatures (37°C), the protein still can be transported, but folding
in the periplasm is often severely impaired, although apparently not for all
antibodies (CARTER et al. 1992). Therefore, a heat inducible promoter is
usually problematic for soluble expression and secretion, but ideal for
inclusion body formation. The attraction of this at first paradoxical approach
of refolding from periplasmic inclusion bodies comes from the fact that the
periplasmic location protects the protein better from proteases. Therefore,
some smaller antibody fragments may not reach the critical concentration
required for precipitation because of competition from proteolysis in the
cytoplasm. In the oxidizing milieu of the periplasm, some of the precipitated
protein has disulfide linkages (PANTOLIANO et al. 1991), but it is not known
what percentage of molecules has them and how many are correct. It is
likely that a direct comparison of the yield from refolding periplasmic and
cytoplasmic inclusion bodies depends on the exact vector constructions.

Cell growth, vector construction and inclusion body enrichment are
straightforward (as there are no obvious specific differences from other
recombinant proteins), but it should be noted that, if the protein is produced
without a signal sequence, the 5'-coding region is derived from the mature
eukaryotic protein, and not the prokaryotic signal sequence, and its mRNA
secondary structure then plays a more important role. Consequently, the
nucleotide sequence may have to be modified to avoid hairpin structures
(WOOD et al. 1984). In one case, even additional amino acids had to be
fused to the heavy chain of the Fab fragment to obtain good inclusion body
formation (BUCHNER and RUDOLPH 1991). However, in other cases the
inclusion body formation of a scFv fragment was found to be very efficient
and straightforward with a T7 based expression system (see, e.g., HUSTON et

al. 1991; FREUND et al. 1993) but several other promoters have been used as well.

Several research groups have established refolding protocols. Fab fragments have been refolded at 10%–40% yield (see, e.g., BUCHNER and RUDOLPH 1991; SHIBUI et al. 1993), and scFv fragments have usually been refolded at 10%–20% yield (see, e.g., HUSTON et al. 1991; BUCHNER et al. 1992a,b; FREUND et al. 1993), in both cases with distinct variations due to the primary sequence. After refolding, the protein must be purified again and, especially, separated from incorrectly folded but perhaps soluble contaminating antibody protein (BUCHNER et al. 1992b). This is straightforward if an antigen affinity column is available, but it may require several steps of conventional chromatography if this is not available. It is thus not uncommon to obtain a yield of only a few percent of purified refolded protein (BUCHNER et al. 1992b) relative to the protein initially present in the inclusion body. In comparing the productiving of different *E. coli* strategies, it is crucial to keep this in mind.

What are the factors influencing the yield of in vitro refolding? Again, the refolding of antibodies is not principally different from that of other disulfide containing proteins (RUDOLPH 1990). First and foremost, the disulfide formation must be kinetically catalyzed and thermodynamically allowed. Using redox couples of reduced and oxidized glutathiones, concentrations of $1-2\,mM$ reduced and $0.1-0.2\,mM$ oxidized glutathione have been found useful (see, e.g., RUDOLPH 1990; BUCHNER and RUDOLPH 1991; HUSTON et al. 1991; BUCHNER et al. 1992a,b; FREUND et al. 1993), but the optimum may depend somewhat on the particular antibody (HUSTON et al. 1991). These conditions thermodynamically allow for formation of disulfide bonds, even if they appear at first sight to be reducing conditions, since the redox equilibrium of the protein disulfide bonds depends on the free energy of the folded protein, which stabilizes the oxidized form with respect to free cystine. Because of the importance of disulfide formation, it is useful to carry out refolding at high pH in order to speed up the disulfide reactions, since the reactive species is the thiolate anion. The aggregation of folding intermediates is a severe problem and probably the single most important side reaction lowering the yield in vitro and in vivo. Thus, rather low protein concentrations have to be used, but the unfolded protein may be added to the refolding mix in small portions since the folded protein has a much higher solubility. Additionally, additives such as $1\,M$ arginine are often found useful, as they appear to increase the solubility of intermediates (RUDOLPH 1990). Too low a protein concentration may lead to gigantic volumes and prevents chain association in heterodimeric Fab fragments. Usually, refolding concentrations of $0.1-5\,mg/ml$ are found useful (see, e.g., BUCHNER and RUDOLPH 1991; HUSTON et al. 1991, 1993; BUCHNER et al. 1992a,b; FREUND et al. 1993).

The addition of molecular chaperones in vitro has been investigated (BUCHNER et al. 1992a), yet without dramatic effects, just as in vivo (KNAPPIK et al. 1993). Only very slight improvements in yield are seen, but the effort of providing stoichiometric amounts of such proteins makes this approach daunting on a technical scale.

All in all, in vitro refolding is a feasible strategy for a variety of antibody fragments. Yet, it is more laborious than production by secretion. Secretion is only now beginning to be optimized and, depending on the antibody primary sequence, already often levels of 1 g antibody per liter *E. coli* (Carter et al. 1992) can be achieved. Crucial sequence determinants are beginning to be defined (Knappik et al., unpublished). Therefore, if quantities of folded protein similar to those found in inclusion bodies can be obtained by secretion directly, it will always be the method of choice. If one particular antibody fragment needs to be produced routinely, however, optimizing a refolding/purification scheme can be an attractive option. Additionally, special applications such as the production of isotope-labeled proteins, as in NMR studies (Freund et al. 1993), may make use of in vitro refolding because of the considerable expense of the label.

D. Antibody Fragments

Much of the preceding discussion on expression focused on protein folding in vivo and in vitro. A recurring theme was to make the molecule smaller, in the hope of increasing the yield of folding both in vitro and in vivo. It is necessary, however, to clearly understand the implications of working with small fragments of the antibody. Generally, working with smaller fragments can be very advantageous (see below) for many applications, and certain stability issues can now be overcome.

One may first ask: why make the antibody smaller at all. In most applications, binding the antigen is the central goal, and making the protein smaller simply removes much competing protein surface leading to non-specific reactions. In clinical applications, pharmacokinetics are an important issue. For instance, in tumor imaging experiments it is useful if the background clears rapidly, and clearance is dependent on the molecular size (Colcher et al. 1990; Yokoda et al. 1992). Furthermore, smaller molecules penetrate tumor tissue much more efficiently (Yokoda et al. 1992). It is possible that smaller antibody fragments are per se less immunogenic, but as of yet there have been no quantitative investigations. The issue of antibody antigenicity has been summarized by Adair (1992). Perhaps most attractive, however, is the convenient accessibility of these small molecules by bacterial expression technology.

All domains of the antibody have a function, of course, but in applying antibodies in research, technology and medicine, binding the antigen is frequently the only function used. All antigen contacting regions are within V_H and V_L, and we must therefore also discuss which properties of the antibody might be lost if fragments consisting just of V_H and V_L are used. For instance, the constant domains of the Fab fragment contribute to stability in preventing the dissociation of V_H and V_L. However, V_H and V_L

can be linked covalently by a variety of methods (GLOCKSHUBER et al. 1990a), (see below). Many detection systems have relied on the constant domains, using either antibodies directed against the C regions or bacterial proteins which have an affinity for them (see Sect. B.IX). This problem can also be circumvented by using a "tag" sequence at the NH$_2$-terminus (WELS et al. 1992a), at the COOH-terminus (WARD et al. 1989) or in the linker of an scFv fragment (BREITLING et al. 1991). Alternatively, in ELISA applications, the detection enzymes (e.g., alkaline phosphatase) can be fused directly (WELS et al. 1992a), and in RIA applications, a metal binding domain can be fused to the antibody fragment (SAWYER et al. 1992).

The sugars of the C$_H$2 domains have often been used as a means of covalently linking the antibody to other chemicals, proteins or solid supports, since modifications there do not disturb the antigen binding activity. However, the same specific binding can be achieved with fragments not carrying C$_H$2 domains or any sugars, for instance, by encoding an additional cysteine at the end of the antibody fragment which can be selectively derivatized with maleimide or iodoacetyl derivatives (see, e.g., BERRY et al. 1991; BERRY and DAVIES 1992; CARTER et al. 1992; CUMBER et al. 1992; BERRY and PIERCE 1993; McCARTNEY et al. 1993). This would be the only free cysteine in the molecule, all others being normally involved in disulfide bonds.

An additional feature of whole antibodies, which is lost in making fragment, is bivalency; however, even this feature can be restored with small fragments. Bivalent mini-antibodies (PACK and PLÜCKTHUN 1992; PACK et al. 1993) have been designed (see below) which self-assemble in E. coli and show the same avidity as a bivalent whole antibody, but have the size of only one Fab fragment.

There are nonetheless antibody functions which are lost when not using whole antibodies and for which no bacterial solutions have been reported yet. The most important ones are the binding of the F$_C$ part to the F$_C$ receptor (in the case of IgG, the receptors FcγRI, RII, and RIII), causing antibody-dependent cellular cytotoxicity (ADCC), and the ability to bind the complement factor C1q, the crucial step for complement activation (reviewed in MORGAN and WEIGLE 1987; SEGAL 1990; SCHUMAKER and POON 1990; SHIN et al. 1992; MORRISON 1992). Both of these functions need antibody glycosylation, presumably for the structural integrity of the C$_H$2 region of the molecule, even if the sugars themselves are not directly involved in the contact to the Fc receptor or C1q (DUNCAN and WINTER 1988; TAO and MORRISON 1989; GILLIES and WESOLOWSKI 1990; LUND et al. 1990). It remains to be seen whether functional analogs, perhaps involving variable domains binding to the effector molecule, can be found for bacteria.

I. Fv Fragments

From the available 3-D structures of antibodies, it is now obvious that all contacts with the antigen are within the V$_H$ and V$_L$ domains. While the proteolytic digestion of an antibody to make an Fab fragment is usually straightforward (WEIR 1986), the proteolytic preparation of Fv fragments is less so. GIVOL and coworkers (INBAR et al. 1972; HOCHMAN et al. 1973, 1976; GIVOL 1991) could show that the proteolytically obtained Fv fragments they investigated were functional. However, the cleavage sites are not particularly preferred, and only certain antibodies give rise to good Fv preparations (SHARON and GIVOL 1976; TAKAHASHI et al. 1991), while in other cases complete functionality was not obtainable (SEN and BEYCHOK 1986) or only qualitatively ascertained (KAKIMOTO and ONOUE 1974; LIN and PUTNAM 1978; RETH et al. 1979).

These preparative problems have been overcome by using recombinant technology, and Fv fragments (Fig. 3) have been functionally expressed in *E. coli* (SKERRA and PLÜCKTHUN 1988) and in myeloma cells (RIECHMANN et al. 1988). Since a number of recombinant Fv fragments have now been made (see, e.g., WARD et al. 1989; FIELD et al. 1990; GLOCKSHUBER et al. 1991; McMANUS and RIECHMANN 1991; TAKAHASHI et al. 1991; CHEADLE et al. 1992; ANTHONY et al. 1992), more conclusions can be drawn about their properties.

Fv fragments appear to have a lower interaction energy of V_H and V_L than Fab fragments, which are held together by the constant domains C_H1 and C_L as well (BIGELOW et al. 1974; AZUMA et al. 1974, 1978; HOCHMAN et al. 1976; MAEDA et al. 1976; KLEIN et al. 1979; STEVENS et al. 1980; HORNE et al. 1982; GLOCKSHUBER et al. 1990a). Nevertheless, many Fv fragments are stable and a few have been studied in detail. A certain range of interaction energies between V_H and V_L would be expected, since the interface also includes complementarity determining region 3 (CDR3) and part of CDR1 (CHOTHIA et al. 1985). It is possible that some very unstable Fv fragments have not been reported in the literature.

In the case of the phosphorylcholine binding antibody McPC603, the V_H-V_L association constant was measured and found to be about $10^6 M^{-1}$ (GLOCKSHUBER et al. 1990a). Since the antigen makes contact to both V_H and V_L, it stabilizes this interaction. In binding studies, a low *apparent* antigen binding constant may result because of chain dissociation. If it is deconvoluted into the V_H-V_L association constant and the antigen binding constant (identical to that of the Fab fragment or the whole antibody), the experimental data can be reproduced (GLOCKSHUBER et al. 1990a). Other Fv fragments, e.g., of the lysozyme binding antibody D1.3 (WARD et al. 1989) or of the anti-digoxin antibody 26-10 (ANTHONY et al. 1992) may have a higher V_H-V_L association constant, although exact quantitative data are not available. Taken together, these results suggest that the V_H-V_L heterodimer faithfully reproduces the binding site of the whole antibody.

Nevertheless, to make this a more general approach, methods are available (GLOCKSHUBER et al. 1990a) to stabilize the V_H-V_L interaction in antibodies: (a) by chemical crosslinking, (b) disulfide bond engineering and (c) genetic linking by a peptide linker (Fig. 7), to create a so-called single chain

Fv fragment (BIRD et al. 1988; HUSTON et al. 1988). It may be pointed out again that the reason for linking the domains lies in thermodynamic stability, but not in facilitating assembly in vivo, as the two unlinked domains do associate.

II. Single Chain Fv Fragments

Both orientations, V_H-linker-V_L and V_L-linker-V_H, have been realized (summarized in HUSTON et al. 1993). In the antibody McPC603, both were compared and no significant difference in the free energy of folding was found. Also, both were expressed at about the same level in *E. coli*, indicating a similar partitioning between folding and aggregation (KNAPPIK et al. 1993). However, different expression levels between both orientations have been noted in another case yet without molecular cause the molecular cause being pinpointed (ANAND et al. 1991b). It is conceivable that the presence of the linker might interfere with binding of some antigens, if, e.g., the NH_2-terminal residues make crucial contacts. In this case, it may be useful to switch the genetic order of the domains to free the particular NH_2-terminus.

Interestingly, the equilibrium denaturation curve of the scFv fragment is consistent with a two-state system (PANTOLIANO et al. 1991; KNAPPIK et al. 1993). This requires either that there is some coupling energy between V_H and V_L, i.e., that a state in which only one of the two domains is unfolded is not a stable intermediate (although it almost certainly is a kinetic intermediate, see below). Alternatively, the free energy of folding of V_H and V_L may be accidentally similar. However, the free energy of folding of the scFv of McPC603 is about 4.7 kcal/mol, whereas that of the V_L domain of the same antibody is only about 3 kcal/mol (LUPAS et al., unpublished). This indicates that more probably the folding of both domains is coupled.

A wide variety of linkers for connecting V_L and V_H have been tested (summarized in HUSTON et al. 1993), and it appears that there is great tolerance, as the linker seems to be a very passive entity contributing rather little to thermodynamic stability. It only appears to be critical that the linker has a length of around 15 residues, and of course it is important in this context how the end of the variable domain is defined. It is crucial for stability that the complete domain is present, as defined by the 3-D structure. One of the most frequenty used linkers has the sequence $(Gly_4\text{-}Ser)_3$ (HUSTON et al. 1988, 1991, 1993).

Recently, NMR experiments have made it possible to define the structural properties of this linker in a scFv fragment (FREUND et al. 1993). This was possible by comparing the NMR spectrum of the Fv fragment to the scFv fragment of the same antibody. The spectra were essentially superimposable, except for the linker region. This shows that the linker has essentially no influence on the structure of the variable domains at all (Fig. 7). Since the Fv fragment was obtained by periplasmic secretion and

that the same structure is obtained in both cases. To confirm that the additional peaks were indeed due to the linker, the scFv was selectively labeled with ^{15}N-glycine and ^{15}N-glycine/^{15}N-serine, and the spectrum correctly integrated to the expected number of amino acids in the linker.

The identification of the linker residues thus allowed the following conclusions to be drawn: the chemical shift of the glycine and serine residues shows almost no spread and is very similar to these amino acids in water, indicating that the linker must be largely exposed to solvent. Furthermore, the T_2 relaxation times are significantly longer than comparable residues elsewhere in the protein, indicating that the linker must be very flexible (FREUND et al. 1993). The linker can therefore adopt to a wide variety of structures. There are very few crosspeaks with the rest of the protein, indicating that there are few if any persistent contacts of the linker with the rest of the protein.

A number of laboratories have noticed a tendency of the VH and VL domains of some scFv fragments to not only associate intramolecularly but also intermolecularly (GRIFFITHS et al. 1993; WHITLOW et al. 1994; DESPLANCQ et al. 1994; MEZES, personal communication). This appears to depend on the primary sequence and on the length of the linker: The multimerization decreases with increasing linker length (DESPLANCQ et al. 1994). Whether the multimers reequilibrate rapidly or slowly also appears to depend on the system under study. HOLLIGER et al. (1993) made use of this phenomenon to force dimer formation by decreasing linker length to zero, and to also force two different scFv to come together to form bispecific scFvs.

Because of the genetic simplicity of the molecular system, a large number of fusion proteins have now been made with scFv fragments, including fusions with domains of protein A (GANDECHA et al. 1992; TAI et al. 1990), toxins (CHAUDARY et al. 1989, 1990; BATRA et al. 1990, 1991; KREITMAN et al. 1990; SEETHARAM et al. 1991; BRINKMANN et al. 1991, 1992; BUCHNER et al. 1992a,b; WELS et al. 1992b; NICHOLLS et al. 1993), alkaline phosphatase (WELS et al. 1992; KOHL et al. 1991), maltose binding protein (BRÉGÉGÈRE and BEDOUELLE 1992), interleukin-2 (SAVAGE et al. 1993), DNAse, RNAse (SPOONER and EPENETOS, personal communication), the COOH-terminal domain of the E. coli carboxyl carrier protein (BCCP) which is biotinylated in vivo in E. coli (MÜLLER and PLÜCKTHUN, unpublished; WEISS, personal communication) and avidin (SPOONER and EPENETOS, personal communication). In the production of these fusion proteins, both native secretion and in vitro refolding have been used. Fusion proteins do not have to be limited to single chain antibodies, however, and with bacterial systems Fab fragments made in bacteria have been fusion partners as well (see, e.g., SAWYER et al. 1992), following the work of NEUBERGER et al. (1984). Due to their genetic simplicity (requiring only a single gene), scFv fragments have been the antibody fragments of choice to establish expression systems in other hosts such as Bacillus subtilis (WU et al. 1993), the yeast Schizosaccharomyces pombe (DAVIS et al. 1991) and plants (OWEN et al. 1992, discussed elsewhere in this volume). Nevertheless, because of its convenience, *E. coli* has remained the strain of choice for producing scFv fragments. A complete list of the reported scFv fragments has been compiled by HUSTON et al. (1993).

III. Disulfide-Linked Fv Fragments

Another strategy for linking V_H and V_L has been to design an intermolecular disulfide bond (GLOCKSHUBER et al. 1990a). Initial experiments of this type concentrated on the antibody McPC603 as a model system. Using a purely geometric approach (PABO and SUCHANEK 1986), all positions of V_H and V_L were searched for the best root mean square fit of the main chain atoms of any candidate pair of amino acids with any disulfide bond taken from the database. Using the best root mean square fits and excluding pairs involving proline residues or residues involved in antigen binding, several candidate disulfide bonds were found for McPC603. Two of these were tested experimentally, L56-H106 and L55-H108 (sequential numbering). The Fv fragment can be obtained in a disulfide-linked form directly from the periplasm, and it shows an almost indistinguishable antigen binding constant. The periplasmic protein can be purified directly by antigen affinity chromatography, and the protein obtained by this procedure is covalently linked (GLOCKSHUBER et al. 1990a).

The disulfide-linked Fv fragment appears to be much more resistant to irreversible denaturation than the unlinked Fv fragment. The stabilization is also much more dramatic than for the scFv fragment and greater than found in a chemically cross-linked Fv fragment. This may be the most important argument for pursuing this strategy. It thus appears that covalent linking is necessary, but not sufficient, for stabilizing the protein against irreversible denaturation. Rather, the type of covalent linking is important.

Least effective is the single chain strategy. Apparently, a rather loose link is created which does not prevent the Fv fragment from aggregation and precipitation after heating, consistent with the structural results from NMR (see above). Somewhat more efficient is chemical cross-linking (GLOCK-SHUBER et al. 1990a), but this is not as easily reproducible for different antibodies as the other methods because of the different surface residues which can be cross-linked, and it is probably not as suitable as a general method or on a large scale. Most effective is the disulfide bond strategy. Probably, the location of the disulfide bonds does not allow much reversible opening and closing of the Fv fragment and may thus prevent aggregation of the Fv fragment at higher temperatures.

While very effective and demonstrating the principles, the particular disulfide bonds initially investigated are not necessarily of general utility. They connect CDR3 of V_H and CDR2 of V_L yet without interfering with binding of the antigen phosphorylcholine (PC). Because of the enormous structural variability, modeling of CDR3 is fairly difficult (CHOTHIA et al. 1989), and for most antibodies a similar juxtaposition of residues would not

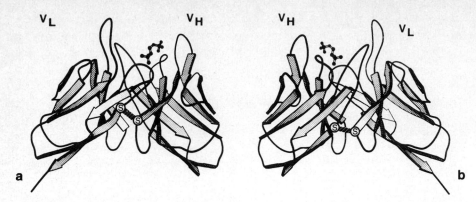

Fig. 8a,b. a Monoview of the Fv fragment of the antibody McPC603 with the positions labeled which were found suitable for linking V_H and V_L in the framework in a variety of Fv fragments. **b** Same fragment as in **a** rotated 180° about the y-axis. The similarity of the two possible positions, due to the molecular pseudo-twofold axis, is apparent. In each case, framework region 2 is connected to framework region 4. (For nomenclature see KABAT et al. 1991)

necessarily be expected. Therefore, a general solution to this problem was searched for (PLÜCKTHUN 1993; BRINKMANN et al. 1993; CARTER, personal communication). The Fv portions of ten different antibodies of known structure were superimposed, and all suitable positions for interchain disulfide bonds were calculated for all of them. Possible cross-links were superimposed, and two types of possible framework cross-links were identified for a majority of fragments (PLÜCKTHUN 1993) (Fig. 8). Because of the pseudo-twofold axis of the Fv fragment (due to the similarity of V_H and V_L), the two positions are structurally related by a rotation about this axis, and in both cases framework region 2 is linked to framework region 4 of the opposite chain (as defined in KABAT et al. 1991). None of the proposed positions will work well for all antibodies, and there is always some uncertainty about the exact geometry of the V_H-V_L interaction of a new antibody. Nevertheless, either of these positions appears to be reasonably promising (BRINKMANN et al. 1993) and need now to be tested on a sufficiently large number of different antibodies to evaluate their generality as a means of linking the component chains of an Fv fragment.

IV. Mini-antibodies

Nature has equipped antibodies with at least two binding sites. This way, they can bind to a surface (e.g., a bacterial surface or a virus particle) with higher functional affinity (sometimes called avidity) (CROTHERS and METZGER 1972; KARUSH 1976, 1978) (Fig. 9). Furthermore, a collection of different bivalent antibodies recognizing different epitopes on the same antigen can aggregate the antigen. This phenomenon will also be noticeable in solid phase binding assays such as ELISA.

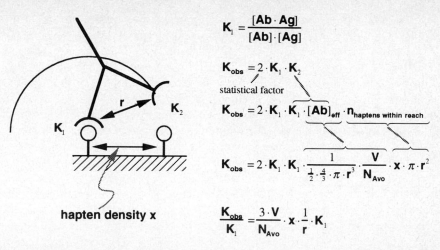

$$K_1 = \frac{[Ab \cdot Ag]}{[Ab] \cdot [Ag]}$$

$$K_{obs} = 2 \cdot K_1 \cdot K_2$$

statistical factor

$$K_{obs} = 2 \cdot K_1 \cdot K_1 \cdot [Ab]_{eff} \cdot n_{haptens\ within\ reach}$$

$$K_{obs} = 2 \cdot K_1 \cdot K_1 \cdot \frac{1}{\frac{1}{2} \cdot \frac{4}{3} \cdot \pi \cdot r^3} \cdot \frac{V}{N_{Avo}} \cdot x \cdot \pi \cdot r^2$$

$$\frac{K_{obs}}{K_1} = \frac{3 \cdot V}{N_{Avo}} \cdot x \cdot \frac{1}{r} \cdot K_1$$

hapten density x

Fig. 9. Enhancement of apparent equilibrium constant due to bivalency effects, according to the treatment of CROTHERS and METZGER (1972). K_1 is the observed binding constant for a monovalent fragment; K_{obs}, for a bivalent fragment. *[Ab], [Ag] and [Ab·Ag]* are the molar concentrations of antibody, antigen and complex, respectively (as if the number of molecules in the reaction well were evenly distributed). K_2 is the binding constant of the second binding site, made unitless by the constant effective antibody concentration $[Ab]_{eff}$, which results from constraining it within the hemisphere of radius r. The hapten density x is a two-dimensional concentration (molecules per area). N_{Avo} is Avogadro's number, used to convert molecular into molar concentrations

Why do dimeric or multimeric antibody molecules bind better and by how much? While the first part of the question is intuitively obvious, the second part is nontrivial. A number of quantitative approaches have been developed (summarized in KARUSH 1976, 1978), of which perhaps the most intuitive is that of CROTHERS and METZGER (1972). This says that the gain contributed by the second binding site is only observed if the antigen is on a surface or is polymeric. In this case, the gain is the product of the two binding constants of the two sites, (Fig. 9) the first being that which an Fab fragment would show, the second is the (dimensionless) constant of the second site, once the first site is bound. This is the binding at the average molar concentration of the second site constrained in the neighborhood of its epitope. The essence of the derivation is that the gain of having a second binding site should be proportional to the true intrinsic association constant K_1 and inversely proportional to the distance r of the two binding sites. Furthermore, it is proportional to the epitope density x on the surface. Therefore this gain is not a constant but dependent on many variables of a particular molecular system. This derivation neglects any energy needed to "bend" either antibody or antigen, complications from already occupied sites, surface layer effects (different ion concentrations, inaccessibility) and uneven microscopic distributions of the antigen on the surface. Nevertheless, it makes clear that a minimum distance is needed to have a limited chance of binding two different antigen molecules simultaneously.

While it is very desirable to access functional bivalent antibodies by bacterial technology, so far, no successful attempts to make functional whole antibodies in *E. coli* have been reported. Part of the problem is that, at least in IgG, for which a crystal structure is known (Marquart et al. 1980; Harris et al. 1992), the two C_H2 domains make no protein-protein contact but their contact is entirely mediated by glycosylation (Sutton and Phillips 1983), which of course does not take place in *E. coli*. Therefore, other means of dimerization are required.

One may chemically link recombinant Fab fragments to $(Fab)_2$ fragments via the free cysteines. The most efficient and stable dimer is probably not obtained by forming a disulfide, but by using a *bis*-maleimide (Brennan et al. 1985). This can be done with recombinant Fab fragments produced in *E. coli* (see, e.g., Carter et al. 1992). Similarly, scFv fragments can be linked this way (Cumber et al. 1992).

It may be asked why disulfide-linked dimers do not form efficiently from proteins containing hinge peptides with cysteines in the periplasm in *E. coli* (Carter et al. 1992; De Sutter et al. 1992) when formation of the intramolecular disulfide bond is possible. One reason may be that disulfide bonds do not cause dimer formation, they merely make covalent existing dimers which have previously formed by noncovalent forces. Intermolecular disulfide formation has been successfully obtained in *E. coli* between V_H and V_L (Glockshuber et al. 1990a), between C_H1 and C_L (Better et al. 1988; Skerra and Plückthun 1991) or between two coiled-coil helices (Pack and Plückthun 1992) (see below). While equilibrium dimer formation has been seen in vitro with a model peptide of the hinge region (Wünsch et al. 1988), there may not be equilibrium conditions in the periplasm, and other proteins or peptides may instead be cross-linked to the hinge (De Sutter et al. 1992), or the structure may be trapped in nonnative intradomain disulfide bonds. Therefore, it appears crucial to provide a specific, noncovalent dimerization interface, in addition to any cysteine. Such examples will now be discussed.

1. Mini-antibodies Based on Coiled-Coil Helices

Methods have now been devised by which scFv fragments dimerize by themselves in vivo. These have been based on the tendency of amphipathic helices to dimerize or tetramerize. Two different principles have been exploited, that of antiparallel four-helix bundles and that of parallel coiled-coils (Figs. 10, 11).

Most useful is probably the attachment of these dimerization handles to scFv fragments (Pack and Plückthun 1992), although in principle they can be added to Fab fragments (Kostelny et al. 1992) or Fv fragments. In the scFv fragment, unique heterodimers can be made, because wrong V_H-V_L pairing (which might occur during simultaneous in vivo expression or during

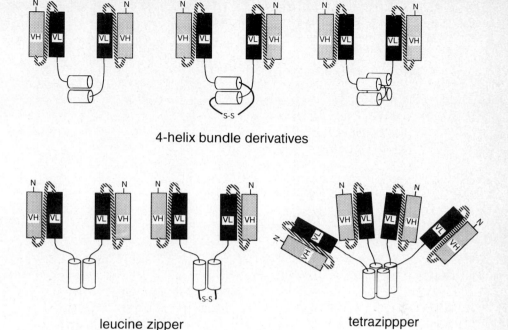

4-helix bundle derivatives

leucine zipper   tetrazippper

coiled coil derivatives

Fig. 10. Bivalent fragments which have been shown to assemble in *E. coli*. In each case, a single chain Fv (scFv) fragment is connected to a hinge region followed by an amphipathic helix. *Top row*, the helix comes from a 4-helix bundle design by deGrado and coworkers. *Top left*, only one helix is fused, but the predominant molecular species are dimers. *Top middle*, the helices are connected by a peptide which ends in a cysteine. *Top right*, two helices are fused in tandem and a 4-helix bundle is probably obtained, as very stable dimers are formed in vivo. *Bottom row*, a parallel coiled-coil helix from a leucine zipper is used. This design is suitable for making heterodimers. *Bottom right*, the sequence of the zipper is changed, as described in the text. (For details see PACK and PLÜCKTHUN 1992, PACK et al. 1993)

in vitro refolding) is not an issue. In these cases, the dimerization handle was not added to the scFv fragment directly, but rather separated by a hinge (the upper hinge from IgG3, known to be very flexible, as summarized by BURTON 1990). This way, an orientation and a distance between the two binding sites, similar to those in a whole antibody, are possible; this arrangement is also known to be very flexible (HARRIS et al. 1992).

Coiled-coil helices occur, for example, as dimerization devices in eukaryotic transcription factors (LANDSCHULZ et al. 1988). Because of their preference for leucine in every seventh position, they have been termed "leucine zippers." Leucine zippers have also been used as dimerization devices in other proteins (HU et al. 1990; BLONDEL and BEDOUELLE 1991). In a preliminary series of experiments, the zipper from the yeast transcrip-

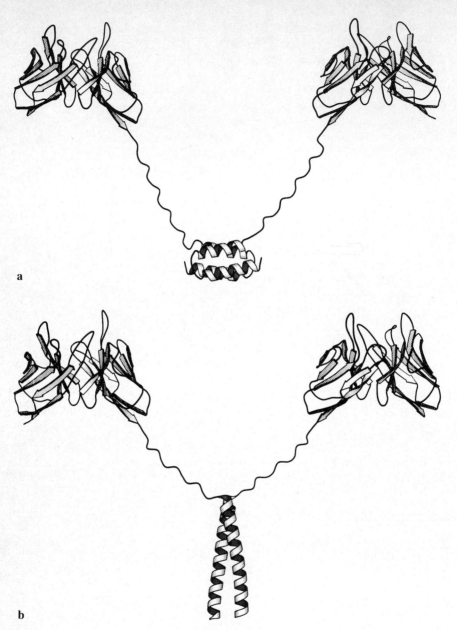

Fig. 11a,b. Molecular models of the dimeric mini-antibodies derived from the single chain Fv fragment of the mouse antibody McPC603 (SATOW et al. 1986). The hinge region was modeled according to a polyproline-II helix with $\phi = -78°$ and $\Psi = 149°$. **a** The 4-helix bundle design is shown, modeled on the *E. coli* protein Rop for packing. **b** The zipper design is shown, taken from the leucine zipper of the yeast transcription factor *GCN4*

tion factor GCN4 was used and shown to be suitable as a dimerization device, although it does not perform as well as four-helix bundles (see below). The reason for this difference is not quite clear, but it might have to do with the parallel arrangement of the helices, further constraining the two hinge regions. In the four-helix bundle, which is antiparallel, the length of the helix itself contributes to the distance the mini-antibody can span between two binding sites, and it may therefore increase the likelihood of two binding sites binding simultaneously to the surface.

Coiled-coil helices offer an approach to self-assembling heterodimers from *E. coli* eukaryotes (KOSTELNY et al. 1992). A number of different strategies have previously been employed to obtain heterodimeric antibodies or antibody fragments: (a) covalent cross-linking via disulfides or via hetero-bifunctional cross-linkers of whole antibodies (STAERZ et al. 1985; PEREZ et al. 1985) or monovalent fragments (NISONOFF and MANDY 1962; RASO and GRIFFIN 1981; BRENNAN et al. 1985), (b) forming two hybridomas to make a so-called heterohybridoma or quadroma (MILSTEIN and CUELLO 1983) or (c) cotransfection of a plasmid encoding the second antibody in a hybridoma producing the first (LENZ and WEIDLE 1990). All coexpression suffers from the statistical H-L pairing, in which the desired pairs may form only in small proportion, although preferential assumbly of the original H-L pairs is sometimes observed. Using scFv fragments and two complementary coiled-coil helices, such as from the transcription factors fos and jun, heterodimers can be made in vivo and the chain scrambling problem is greatly simplified. However, further alterations to the wild-type zipper sequences will be necessary to improve the yield of heterodimers (PACK and PLÜCKTHUN, unpublished).

It is now also possible to make tetrameric mini-antibodies. In this case, advantage was taken of the analysis of point mutants of coiled-coil helices (ALBER and KIM, personal communication). In the coiled-coil helix, repetitive heptads are found, with distinct preferences for hydrophobic β-branched amino acids in position a and d of the heptad (lettered from a to g) (COHEN and PARRY 1990; O'SHEA et al. 1989, 1991). If the naturally occurring Val (position a) and Leu (position d) are replaced with Leu and Ile, the molecule responds by tetramerizing (KIM and ALBER, personal communication). Again, this "tetrazipper" was linked to the scFv fragment by a hinge region, and gel filtration chromatography showed the tetrameric nature of these molecules, and an increase in binding avidity over dimers (PACK and PLÜCKTHUN, unpublished). Since the gain from multivalence is dependent on many molecular variables, it is very probable that there are cases in which this gain is large.

2. Mini-antibodies Based on Four-Helix Bundles

Four-helix bundles are compact folding motifs of natural proteins. Eisenberg, deGrado and coworkers described a synthetic four helix bundle (EISENBERG et al. 1986) made from either four single helices, two helix-turn-helix peptides or one continuous chain (Ho and DeGrado 1987; Regan and DeGrado 1988; Hill et al. 1990). In this design, all four-helices have the same sequence, and a surprisingly high thermodynamic stability has been measured (Ho and DeGrado 1987).

Using this association principle, three different versions have been tested with scFv fragments. In the first, only one helix was fused to the scFv fragment, linked by the upper hinge region from mouse IgG3 (Figs. 10, 11). No tetramers are formed, however, but rather a distribution between monomers and dimers is obtained, presumably because the association energy of the helices is too weak. This can be improved by extending the helix with a hydrophilic peptide ending in a cysteine (Pack and Plückthun 1992). The covalently linked scFv mini-antibody is stable and remains in the dimer state. In this case, the peptide linker may be in the way of tetramer association.

Most stable, however, is the construct carrying a helix-turn-helix motif (Figs. 10, 11). In this case, very little degradation is observed and avidities are obtained identical to whole IgA (Pack et al. 1993). It appears that these amphipathic helices are compatible with transport through the bacterial membrane and cause no problems in folding of the scFv fragments.

E. Conclusions

Producing and characterizing an engineered antibody fragment is the prerequisite for improving its performance, no matter what the application. Since all cloning and mutagenesis is carried out in *E. coli*, it is convenient to use these bacteria for expression as well. It is possible to produce surprisingly complex multisubunit structures in *E. coli* with correct folding and assembly, provided the expression strategy is adapted to the physiology of the growing cell. In this case not only can small amounts of the antibody constructs be obtained rapidly for laboratory testing, but amounts useful for clinical and industrial applications can now be produced by fermentation. Combined with library selection and current developments in imitating affinity maturation, bacterial antibody technology will become an integral part of any research involving immunoglobulins.

References

Abel CA, Grey HM (1968) Studies on the structure of mouse γA-myeloma proteins. Biochemistry 7:2682–2688

Adair JR (1992) Engineering antibodies for therapy. Immunol Rev 130:5–40

Anand NN, Dubuc G, Phipps J, MacKenzie CR, Sadowska J, Young NM, Bundle DR, Narang SA (1991a) Synthesis and expression in *Escherichia coli* of cistronic DNA encoding an antibody fragment specific for a *Salmonella* serotype B O-antigen. Gene 100:39–44

Anand NN, Mandal S, MacKenzie CR, Sadowska J, Sigurskjold B, Young NM, Bundle DR, Narang SA (1991b) Bacterial expression and secretion of various single-chain Fv genes encoding proteins specific for a *Salmonella* serotype B O-antigen. J Biol Chem 266:21874–21879

Anthony J, Near R, Wong SL, Iida E, Ernst E, Wittekind M, Haber E, Ng SC (1992) Production of stable anti-digoxin-Fv in *Escherichia coli*. Mol Immunol 29:1237–1247

Azuma T, Hamaguchi K, Migita S (1974) Interactions between immunoglobulin polypeptide chains. J Biochem 76:685–693

Azuma T, Kobayashi O, Goto Y, Hamaguchi K (1978) Monomer-dimer equilibria of a Bence Jones protein and its variable fragment. J Biochem 83:1485–1492

Barbas CF III, Kang AS, Lerner RA, Benkovic SJ (1991) Assembly of combinatorial antibody libraries on phage surfaces: the gene III site. Proc Natl Acad Sci USA 88:7978–7982

Barbas CF III, Bain JD, Hoekstra DM, Lerner RA (1992a) Semisynthetic combinatorial antibody libraries: A chemical solution to the diversity problem. Proc Natl Acad Sci USA 89:4457–4461

Barbas CF III, Björling E, Chiodi F, Dunlop N, Cababa D, Jones TM, Zebedee SL, Persson MAA, Nara PL, Norrby E, Burton DR (1992b) Recombinant human Fab fragments neutralize human type 1 immunodeficiency virus in vitro. Proc Natl Acad Sci USA 89:9339–9343

Bardwell JCA, McGovern K, Beckwith J (1991) Identification of a protein required for disulfide bond formation in vivo. Cell 67:581–589

Bardwell JCA, Lee JO, Jander G, Martin N, Belin D, Beckwith J (1993) A pathway for disulfide bond formation in vivo. Proc Natl Acad Sci USA 90:1038–1042

Batra JK, FitzGerald D, Gately M, Chaudhary VK, Pastan I (1990) Anti-Tac(Fv)-PE40, a single chain antibody *Pseudomonas* fusion protein directed at interleukin 2 receptor bearing cells. J Biol Chem 265:15198–15202

Batra JK, FitzGerald DJ, Chaudhary VK, Pastan I (1991) Single-chain immunotoxins directed at the human transferrin receptor containing *Pseudomonas* exotoxin A or diphtheria toxin: anti-TFR(Fv)-PE40 and DT388-anti-TFR(Fv). Mol Cell Biol 11:2200–2205

Beckwith JR, Zipser D (eds) (1970) The lactose operon. Cold Spring Harbor Laboratory, Cold Spring Harbor

Belasco JG, Higgins CF (1988) Mechanisms of mRNA decay in bacteria: a perspective. Gene 72:15–23

Bergmann LW, Kuehl WM (1979) Formation of an intrachain disulfide bond on nascent immunoglobulin light chains. J Biol Chem 254:8869–8876

Berry MJ, Davies J (1992) Use of antibody fragments in immunoaffinity chromatography: comparison of Fv fragments, VH fragments and paralog peptides. J Chromatogr 597:239–245

Berry MJ, Pierce JJ (1993) Stability of immunoadsorbents comprising antibody fragments – comparison of Fv fragments and single-Chain Fv fragments. J Chromatogr 629:161–168

Berry MJ, Davies J, Smith CG, Smith I (1991) Immobilization of Fv antibody fragments on porous silica and their utility in affinity chromatography. J Chromatogr 587:161–169

Better M, Chang CP, Robinson RR, Horwitz AH (1988) *Escherichia coli* secretion of an active chimeric antibody fragment. Science 240:1041–1043

Better M, Bernhard SL, Lei SP, Fishwild DM, Lane JA, Carroll SF, Horwitz AH (1993) Potent anti-Cd5 ricin-A chain immunoconjugates from bacterially produced Fab' and Fab$_2'$. Proc Natl Acad Sci USA 90:457–461

Bigelow CC, Smith BR, Dorrington KJ (1974) Equilibrium and kinetic aspects of subunit association in immunoglobulin G. Biochemistry 13:4602–4608

Bird RE, Hardman KD, Jacobson JW, Johnson S, Kaufman BM, Lee S-M, Lee T, Pope SH, Riordan GS, Whitlow M (1988) Single-chain antigen-binding proteins. Science 242:423–426

Blondel A, Bedouelle H (1991) Engineering the quaternary structure of an exported protein with a leucine zipper. Protein Eng 4:457–461

Boss MA, Kenten JH, Wood CR, Emtage JS (1984) Assembly of functional anti-
    bodies from immunoglobulin heavy and light chains synthesised in *E. coli*.
    Nucleic Acids Res 12:3791–3806
Boyle MDP (ed) (1990) Immunoglobulin binding proteins, vol 1. Academic Press,
    San Diego
Boyle MDP, Reis KJ (1987) Bacterial Fc receptors. Biotechnology 5:697–703
Brandts JF, Halvorson HR, Brennan M (1975) Consideration of the possibility that
    the slow step in protein denaturation reactions is due to cis-trans isomerism of
    proline residues. Biochemistry 14:4953–4963
Brégégère F, Bedouelle H (1992) Expression, exportation et purification des frag-
    ments d'anticorps fusionnés à la protéine affine du maltose d'*Escherichia coli*. C
    R Acad Sci [III] 314:527–532
Breitling F, Dübel S, Seehaus T, Klewinghaus I, Little M (1991) A surface expres-
    sion vector for antibody screening. Gene 104:147–153
Brennan M, Davison PF, Paulus H (1985) Preparation of bispecific antibodies by
    chemical recombination of monoclonal immunoglobulin G_1 fragments. Science
    229:81–83
Brinkmann U, Pai LH, FitzGerald DJ, Willingham M, Pastan I (1991) B3(Fv)-
    PE38KDEL, a single-chain immunotoxin that causes complete regression of a
    human carcinoma in mice. Proc Natl Acad Sci USA 88:8616–8620
Brinkmann U, Buchner J, Pastan I (1992) Independent domain folding of *Pseudo-
    monas* exotoxin and single-chain immunotoxins: influence of interdomain con-
    nections. Proc Natl Acad Sci USA 89:3075–3079
Brinkmann U, Reiter Y, Jung SH, Lee B, Pastan I (1993) A recombinant immuno-
    toxin containing a disulfide-stabilized Fv fragment. Proc Natl Acad Sci USA
    90:7538–7542
Buchner J, Rudolph R (1991) Renaturation, purification and characterization of
    recombinant Fab-fragments produced in *Escherichia coli*. Biotechnology 9:
    157–162
Buchner J, Brinkmann U, Pastan I (1992a) Renaturation of a single-chain immuno-
    toxin facilitated by chaperones and protein disulfide isomerase. Biotechnology
    10:682–685
Buchner J, Pastan I, Brinkmann U (1992b) A method for increasing the yield of
    properly folded recombinant fusion proteins: single-chain immunotoxins from
    renaturation of bacterial inclusion bodies. Anal Biochem 205:263–270
Burton DR (1990) In: Metzger H (ed) Fc receptors and the action of antibodies.
    American Society of Microbiology, Washington, pp 31–54
Cabilly S (1989) Growth at sub-optimal temperatures allows the production of
    functional, antigen-binding Fab fragments in *Escherichia coli*. Gene 85:553–557
Cabilly S, Riggs AD, Pande H, Shively JE, Holmes WE, Rey M, Perry LJ, Wetzel
    R, Heyneker HL (1984) Generation of antibody activity from immunoglobulin
    polypeptide chains produced in *Escherichia coli*. Proc Natl Acad Sci USA
    81:3273–3277
Carter P, Kelley RF, Rodrigues ML, Snedecor B, Covarrubias M, Velligan MD,
    Wong WLT, Rowland AM, Kotts CE, Carver ME, Yang M, Bourell JH,
    Shepard HM, Henner D (1992) High level *Escherichia coli* expression and
    production of a bivalent humanized antibody fragment. Biotechnology 10:
    163–167
Chang CN, Landolfi NF, Queen C (1991) Expression of antibody Fab domains on
    bacteriophage surfaces. Potential use for antibody selection. J Immunol 147:
    3610–3614
Chaudhary VK, Queen C, Junghans RP, Waldmann TA, FitzGerald DJ, Pastan I
    (1989) A recombinant immunotoxin consisting of two antibody variable domains
    fused to *Pseudomonas exotoxin*. Nature 339:394–397
Chaudhary VK, Gallo MG, FitzGerald DJ, Pastan I (1990) A recombinant single-
    chain immunotoxin composed of anti-Tac variable regions and a truncated
    diphtheria toxin. Proc Natl Acad Sci USA 87:9491–9494

Cheadle C, Hook LE, Givol D, Ricca GA (1992) Cloning and expression of the variable regions of mouse myeloma protein MOPC315 in E. coli: recovery of active Fv fragments. Mol Immunol 29:21–30

Chothia C, Novotny J, Bruccoleri R, Karplus M (1985) Domain association in immunoglobulin molecules. The packing of variable domains. J Mol Biol 186: 651–663

Chothia C, Lesk AM, Tramontano A, Levitt M, Smith-Gill SJ, Air G, Sheriff S, Padlan EA, Davies D, Tulip WR, Colman PM, Spinelli S, Alzari PM, Poljak RJ (1989) Conformations of immunoglobulin hypervariable regions. Nature 342: 877–883

Clackson T, Hoogenboom HR, Griffiths AD, Winter G (1991) Making antibody fragments using phage display libraries. Nature 352:624–628

Cockle SA, Young NM (1985) The thiol groups of mouse immunoglobulin A. Incomplete formation of the Cα1-domain disulphide bridge. Biochem J 225: 113–125

Cohen C, Parry DAD (1990) α-Helical coiled coils and bundles: how to design an α-helical protein. Proteins 7:1–15

Colcher D, Bird R, Roselli M, Hardman KD, Johnson S, Pope S, Dodd SW, Pantoliano MW, Milenic DE, Schlom J (1990) In vivo tumor targeting of a recombinant single-chain antigen-binding protein. J Natl Cancer Inst 82: 1191–1197

Collet TA, Roben P, O'Kennedy R, Barbas CF, Burton DR, Lerner RA (1992) A binary plasmid system for shuffling combinatorial antibody libraries. Proc Natl Acad Sci USA 89:10026–10030

Creighton TE (1978) Experimental studies of protein folding and unfolding. Prog Biophys Mol Biol 33:231–297

Crothers DM, Metzger H (1972) The influence of polyvalency on the binding properties of antibodies. Immunochemistry 9:341–357

Cumber AJ, Ward ES, Winter G, Parnell GD, Wawrzynczak EJ (1992) Comparative stabilities in vitro and in vivo of a recombinant mouse antibody FvCys fragment and a bisFvCys conjugate. J Immunol 149:120–126

Cwirla SE, Peters EA, Barrett RW, Dower WJ (1990) Peptides on phage: a vast library of peptides for identifying ligands. Proc Natl Acad Sci USA 87:6378–6382

Davis GT, Bedzyk WD, Voss EW, Jacobs TW (1991) Single-chain antibody (SCA) encoding genes: one-step construction and expression in eukaryotic cells. Biotechnology 9:165–169

De Sutter K, Remaut E, Fiers W (1992) Disulphide bridge formation in the periplasm of Escherichia coli: beta-lactamase:: human IgG3 hinge fusions as a model system. Mol Microbiol 6:2201–2208

Derman AI, Puziss JW, Bassford PJ, Beckwith J (1993) A signal sequence is not required for protein export in prlA mutants of Escherichia coli. EMBO J 12: 879–888

Desplancq D, King DJ, Lawson ADG, Mountain A (1994) Multimerisation behaviour of single-chain Fv variants for the tumour-binding antibody B72.3. Protein Eng, in press

Devlin JJ, Panganiban LC, Devlin PE (1990) Random peptide libraries: a source of specific protein binding molecules. Science 249:404–406

Duncan AR, Winter G (1988) The binding site for C1q on IgG. Nature 332:738–740

Ehretsmann CP, Carpousis AJ, Krisch HM (1992) mRNA degradation in prokaryotes. FASEB J 6:3186–3192

Eisenberg D, Wilcox W, Eshita SM, Pryciak PM, Ho SP, DeGrado WF (1986) The design, synthesis, and crystallization of an alpha-helical peptide. Proteins 1:16–22

Evan GI, Lewis GK, Ramsay G, Bishop JM (1985) Isolation of monoclonal antibodies specific for human c-myc proto-oncogene product. Mol Cell Biol 5: 3610–3616

Faulmann EL, Duvall JL, Boyle MDP (1991) Protein B: a versatile bacterial Fc-binding protein selective for human IgA. Biotechniques 10:748–755

Felici F, Castagnoli L, Musacchio A, Jappelli R, Cesareni G (1991) Selection of antibody ligands from a large library of oligopeptides expressed on a multivalent exposition vector. J Mol Biol 222:301–310

Field H, Yarranton GT, Rees AR (1990) Expression of mouse immunoglobulin light and heavy chain variable regions in *Escherichia coli* and reconstitution of antigen-binding activity. Protein Eng 3:641–647

Fischer G, Schmid FX (1990) The mechanism of protein folding. Implications of in vitro refolding models for de novo protein folding and translocation in the cell. Biochemistry 29:2205–2212

Francisco JA, Earhart CF, Georgiou G (1992) Transport and anchoring of β-lactamase to the external surface of *Escherichia coli*. Proc Natl Acad Sci 89:2713–2717

Francisco JA, Stathopoulos C, Warren RAJ, Kilburn DG, Georgiou G (1993) Specific adhesion and hydrolysis of cellulose by intact *Escherichia coli* expressing surface anchored cellulase or cellulose binding domains. Biotechnology 11:491–495

Freedman RB, Bulleid NJ, Hawkins HC, Paver JL (1989) Role of protein disulphide-isomerase in the expression of native proteins. Biochem Soc Symp 55:167–192

Freund C, Ross A, Guth B, Plückthun A, Holak T (1993) Characterization of the linker peptide of the single-chain Fv fragment of an antibody of NMR spectroscopy. FEBS Lett 320:97–100

Fuchs P, Breitling F, Dübel S, Seehaus T, Little M (1991) Targeting recombinant antibodies to the surface of *Escherichia coli*: fusion to a peptidoglycan associated lipoprotein. Biotechnology 9:1369–1372

Gandecha AR, Owen MRL, Cockburn B, Whitelam GC (1992) Production and secretion of a bifunctional staphylococcal protein A:: antiphytochrome single-chain Fv fusion protein in *Escherichia coli*. Gene 122:361–365

Garrard LJ, Yang M, O'Connell MP, Kelley RF, Henner DJ (1991) Fab assembly and enrichment in a monovalent phage display system. Biotechnology 9:1373–1377

Gibbs RA, Posner BA, Filpula DR, Dodd SW, Finkelman MAJ, Lee TK, Wroble M, Whitlow M, Benkovic SJ (1991) Construction and characterization of a single-chain catalytic antibody. Proc Natl Acad Sci USA 88:4001–4004

Gilbert HF (1990) Molecular and cellular aspects of thiol-disulfide exchange. Adv Enzymol 63:69–172

Gillies SD, Wesolowski JS (1990) Antigen binding and biological activities of engineered mutant chimeric antibodies with human tumor specificities. Hum Antibod Hybridomas 1:47–54

Givol D (1991) The minimal antigen-binding fragment of antibodies – Fv fragment. Mol Immunol 28:1379–1386

Glockshuber R, Malia M, Pfitzinger I, Plückthun A (1990a) A comparison of strategies to stabilize immunoglobulin Fv-fragments. Biochemistry 29:1362–1367

Glockshuber R, Steipe B, Huber R, Plückthun A (1990b) Crystallization and pre-liminary X-ray studies of the V$_L$ domain of the antibody McPC603 produced in *Escherichia coli*. J Mol Biol 213:613–615

Glockshuber R, Stadlmüller J, Plückthun A (1991) Mapping and modification of an antibody hapten binding site: a site-directed mutagenesis study of McPC603. Biochemistry 30:3049–3054

Glockshuber R, Schmidt T, Plückthun A (1992) The disulfide bonds in antibody variable domains: effects on stability, folding in vitro, and functional expression in *Escherichia coli*. Biochemistry 31:1270–1279

Goldenberg DP (1992) Native and non-native intermediates in the BPTI folding pathway. Trends Biochem 17:257–261

Goto Y, Hamaguchi K (1979) The role of the intrachain disulfide bond in the conformation and stability of the constant fragment of the immunoglobulin light chain. J Biochem 86:1433–1441

Goto Y, Hamaguchi K (1982) Unfolding and refolding of the constant fragment of the immunoglobulin light chain. J Mol Biol 156:891–910

Goto Y, Hamaguchi K (1986) Conformation and stability of the constant fragment of the immunoglobulin light chain containing an intramolecular mercury bridge. Biochemistry 25:2821–2828

Goto Y, Ichimura N, Hamaguchi K (1988) Effects of ammonium sulfate on the unfolding and refolding of the variable and constant fragments of an immunoglobulin light chain. Biochemistry 27:1670–1677

Gram H, Marconi L-A, Barbas III CF, Collet TA, Lerner RA, Kang AS (1992) In vitro selection and affinity maturation of antibodies from a naive combinatorial immunoglobulin library. Proc Natl Acad Sci USA 89:3576–3580

Greenwood J, Willis AE, Perham RN (1991) Multiple display of foreign peptides on a filamentous bacteriophage. Peptides from *Plasmodium falciparum* circumsporozoite protein as antigens. J Mol Biol 220:821–827

Griffiths AD, Malmquist M, Marks JD, Bye JD, Embleton MJ, McCafferty J, Baier M, Holliger KP, Gorick BD, Hughes-Jones NC, Hoogenboom HR, Winter G (1993) Human anti-self antibodies with high specificity from phage display libraries. EMBO J 12:725–734

Haber E (1964) Recovery of antigenic specificity after denaturation and complete reduction of disulfides in a papain fragment of antibody. Proc Natl Acad Sci USA 52:1099–1106

Harris LJ, Larson SB, Hasel KW, Day J, Greenwood A, McPherson A (1992) The three-dimensional structure of an intact monoclonal antibody for canine lymphoma. Nature 360:369–372

Hayano T, Takahashi N, Kato S, Maki N, Suzuki M (1991) Two distinct forms of peptidylprolyl-*cis-trans*-isomerase are expressed separately in periplasmic and cytoplasmic compartments of *Escherichia coli* cells. Biochemistry 30:3041–3048

Hill CP, Anderson DH, Wesson L, DeGrado WF, Eisenberg D (1990) Crystal structure of α_1: implications for protein design. Science 249:543–546

Ho SP, DeGrado WF (1987) Design of a 4-helix bundle protein: synthesis of peptides which self-associate into a helical protein. J Am Chem Soc 109:6751–6758

Hochman J, Inbar D, Givol D (1973) An active antibody fragment (Fv) composed of the variable portions of heavy and light chains. Biochemistry 12:1130–1135

Hochman J, Gavish M, Inbar D, Givol D (1976) Folding and interaction of subunits at the antibody combining site. Biochemistry 15:2706–2710

Hochuli E, Piesecki S (1992) Interaction of hexahistidine fusion proteins with nitrilotriacetic acid-chelated Ni^{2+} ions. Methods: A Companion to Methods Enzymol 4:68–72

Holland IB, Kenny B, Steipe B, Plückthun A (1990) Secretion of heterologous proteins in *Escherichia coli*. Methods Enzymol 182:132–143

Holliger P, Prospero T, Winter G (1993) "Diabodies": small bivalent and bispecific antibody fragments. Proc Natl Acad Sci USA 90:6444–6448

Hoogenboom HR, Griffiths AD, Johnson KS, Chiswell DJ, Hudson P, Winter G (1991) Multi-subunit proteins on the surface of filamentous phage: methodologies for displaying antibody (Fab) heavy and light chains. Nucleic Acids Res 19:4133–4137

Hoogenboom HR, Marks JD, Griffiths AD, Winter G (1992) Building antibodies from their genes. Immunol Rev 130:41–68

Horne C, Klein M, Polidoulis I, Dorrington KJ (1982) Noncovalent association of heavy and light chains of human immunoglobulins. III. Specific interactions between V_H and V_L. J Immunol 129:660–664

Hu JC, O'Shea EK, Kim PS, Sauer RT (1990) Sequence requirements for coiled-coils: analysis with λ repressor-GCN4 leucine zipper fusions. Science 250:1400–1403

Huse WD, Sastry L, Iverson SA, Kang AS, Alting-Mees M, Burton DR, Benkovic SJ, Lerner RA (1989) Generation of a large combinatorial library of the immunoglobulin repertoire in phage lambda. Science 246:1275–1281

Huston JS, Levinson D, Mudgett-Hunter M, Tai M-S, Novotny J, Margolies MN, Ridge RJ, Bruccoleri RE, Haber E, Crea R, Oppermann H (1988) Protein engineering of antibody binding sites: recovery of specific activity in an anti-digoxin single-chain Fv analogue produced in Escherichia coli. Proc Natl Acad Sci USA 85:5879–5883

Huston JS, Mudgett-Hunter M, Tai M-S, McCartney J, Warren F, Haber E, Oppermann H (1991) Protein engineering of single-chain Fv analogs and fusion proteins. Methods Enzymol 203:46–88

Huston JS, McCartney J, Tai MS, Mottola-Hartshorn C, Jin D, Warren F, Keck P, Oppermann H (1993) Medical applications of single-chain antibodies. Int Rev Immunol (in press)

Inbar D, Hochman J, Givol D (1972) Localization of antibody-combining sites within the variable portions of heavy and light chains. Proc Natl Acad Sci USA 69:2659–2662

Inganäs M, Johansson SGO, Bennich HH (1980) Interaction of human polyclonal IgE and IgG from different species with protein A from Staphylococcus aureus: demonstration of protein-A-reactive sites located in the Fab$_2$ fragment of human IgG. Scand J Immunol 12:23–31

Jaenicke R (1993) Role of accessory proteins in protein folding. Curr Opin Struct Biol 3:104–112

Jones ST, Bendig MM (1991) Rapid PCR-cloning of full-length mouse immunoglobulin variable regions. Biotechnology 9:88–89

Kabat EA, Wu TT, Perry HM, Gottesman KS, Foeller C (1991) Sequences of proteins of immunological interest, 5th edn. National Institutes of Health, Bethesda

Kakimoto K, Onoue K (1974) Characterization of the Fv fragment isolated from a human immunoglobulin M. J Immunol 112:1373–1382

Kamitani S, Akiyama Y, Ito K (1992) Identification and characterization of an Escherichia coli gene required for the formation of correctly folded alkaline phosphatase, a periplasmic enzyme. EMBO J 11:57–62

Kang AS, Barbas CF, Janda KD, Benkovic SJ, Lerner RA (1991) Linkage of recognition and replication functions by assembling combinatorial antibody Fab libraries along phage surfaces. Proc Natl Acad Sci USA 88:4363–4366

Karush F (1976) Multivalent binding and functional affinity. Contemp Top Mol Immunol 5:217–228

Karush F (1978) The affinity of antibody: range, variability and the role of multivalence. In: Litman GW, Good RA (eds) Immunoglobulins. Plenum Publishing, New York, pp 85–116

Klauser T, Pohlner J, Meyer TF (1992) Selective extracellular release of cholera toxin B subunit by Escherichia coli: dissection of Neisseria Igaβ-mediated outer membrane tranport. EMBO J 11:2327–2335

Klein M, Kortan C, Kells DIC, Dorrington KJ (1979) Equilibrium and kinetic aspects of the interaction of isolated variable and constant domains of light chain with the Fd' fragment of immunoglobulin G. Biochemistry 18:1473–1481

Knappik A, Krebber C, Plückthun A (1993) The effect of folding catalysts on the in vivo folding process of different antibody fragments expressed in Escherichia coli. Biotechnology 11:77–83

Kohl J, Rüker F, Himmler G, Razazzi E, Katinger H (1991) Cloning and expression of an HIV-1 specific single-chain Fv region fused to Escherichia coli alkaline phosphatase. Ann NY Acad Sci 646:106–114

Kostelny SA, Cole MS, Tso JY (1992) Formation of a bispecific antibody by the use of leucine zippers. J Immunol 148:1547–1553

Kreitman RJ, Chaudhary VK, Waldmann T, Willingham MC, FitzGerald DJ, Pastan I (1990) The recombinant immunotoxin anti-Tac(Fv)-*Pseudomonas* exotoxin 40 is cytotoxic toward peripheral blood malignant cells from patients with adult T-cell leukemia. Proc Natl Acad Sci USA 87:8291–8295

Landschulz WH, Johnson PF, McKnight SL (1988) The leucine zipper: a hypothetical structure common to a new class of DNA binding proteins. Science 240:1759–1764

Lang K, Schmid FX (1988) Protein-disulphide isomerase and prolyl isomerase act differently and independently as catalysts of protein folding. Nature 331:453–455

Larrick JW, Wallace EF, Coloma MJ, Bruderer U, Lang AB, Fry KE (1992) Therapeutic human antibodies derived from PCR amplification of B-cell variable regions. Immunol Rev 130:69–85

Lenz H, Weidle UH (1990) Expression of heterobispecific antibodies by genes transfected into producer hybridoma cells. Gene 87:213–218

Lin-Chao S, Chen WT, Wong TT (1992) High copy number of the pUC plasmids results from a Rom/Rop-suppressible point mutation in RNAII. Mol Microbiol 6:3385–3390

Lin L-C, Putnam FW (1978) Cold pepsin digestion: a novel method to produce the Fv fragment from human immunoglobulin M. Proc Natl Acad Sci USA 75:2649–2653

Lindner P, Guth B, Wülfing C, Krebber C, Steipe B, Müller F, Plückthun A (1992) Purification of native proteins from the cytoplasm and periplasm of *Escherichia coli* using IMAC and histidine tails: a comparison of proteins and protocols. Methods: a companion to Methods Enzymol 4:41–56

Liu J, Walsh CT (1990) Peptidyl-prolyl *cis-trans*-isomerase from *Escherichia coli*: a periplasmic homolog of cyclophilin that is not inhibited by cyclosporin A. Proc Natl Acad Sci USA 87:4028–4032

Lund J, Tanaka T, Takahashi N, Sarmay G, Arata Y, Jefferis R (1990) A protein structural change in aglycosylated IgG3 correlates with loss of huFcγRI and huFcγRIII binding and/or activation. Mol Immunol 27:1145–1153

Maeda H, Engel J, Schramm HJ (1976) Kinetics of dimerization of the variable fragment of the Bence-Jones protein Au. Eur J Biochem 69:133–139

Marks JD, Hoogenboom HR, Bonnert TP, McCafferty J, Griffiths AD, Winter G (1991) By-passing immunization. Human antibodies from V-gene libraries displayed on phage. J Mol Biol 222:581–597

Marquart M, Deisenhofer J, Huber R (1980) Crystallographic refinement and atomic models of the intact immunoglobulin molecule Kol and its antigen-binding fragment at 3.0 Å and 1.9 Å resolution. J Mol Biol 141:369–391

McCafferty J, Griffiths AD, Winter G, Chiswell DJ (1990) Phage antibodies: filamentous phage displaying antibody variable domains. Nature 348:552–554

McCarthy JEG, Gualerzi C (1990) Translational control of prokaryotic gene expression. Trends Genet 6:78–85

McCartney JE, Tai M-S, Opperman H, Jin D, Warren FD, Weiner LM, Bookman MA, Stafford WF III, Houston LL, Huston JS (1993) Refolding of single-chain Fv with C-terminal cysteine (sFv'): formation of disulfide-bonded homodimers of anti-*c-erb*B-2 and anti-digoxin sFv'. ICSU Short Reports 13. ICSU, Stockholm

McManus S, Riechmann L (1991) Use of 2D NMR, protein engineering, and molecular modeling to study the hapten-binding site of an antibody Fv fragment against 2-phenyloxazolone. Biochemistry 30:5851–5857

Milstein C, Cuello AC (1983) Hybrid hybridomas and their use in immunohistochemistry. Nature 305:537–540

Morgan EL, Weigle WO (1987) Biological activities residing in the Fc region of immunoglobulin. Adv Immunol 40:61–134

Morrison SL (1992) In vitro antibodies: strategies for production and application. Annu Rev Immunol 10:239–265

Munro S, Pelham HRB (1986) An Hsp70-like protein in the ER: identity with the 78 kd glucose-regulated protein and immunoglobulin heavy chain binding protein. Cell 46:291–300

Neuberger MS, Williams GT, Fox RO (1984) Recombinant antibodies possessing novel effector functions. Nature 312:604–608

Nicholls PJ, Johnson VG, Andrew SM, Hoogenboom HR, Raus JCM, Youle RJ (1993) Characterization of single-chain antibody (sFv)-toxin fusion proteins produced in vitro in rabbit reticulocyte lysate. J Biol Chem 268:5302–5308

Nilson BHK, Solomon A, Björck L, Åkerström B (1992) Protein L from *Peptostreptococcus magnus* binds to the *K* light chain variable domain. J Biol Chem 267:2234–2239

Nisonoff A, Mandy WJ (1962) Quantitative estimation of the hybridization of rabbit antibodies. Nature 194:355–359

Orlandi R, Güssow DH, Jones PT, Winter G (1989) Cloning immunoglobulin variable domains for expression by the polymerase chain reaction. Proc Natl Acad Sci USA 86:3833–3837

O'Shea EK, Rutkowski R, Kim PS (1989) Evidence that the leucine zipper is a coiled coil. Science 243:538–542

O'Shea EK, Klemm JD, Kim PS, Alber T (1991) X-ray structure of the GCN4 leucine zipper, a two-stranded, parallel coiled coil. Science 254:543–544

Owen M, Gandecha A, Cockburn B, Whitelam G (1992) Synthesis of a functional anti-phytochrome single-chain Fv protein in transgenic tobacco. Biotechnology 10:790–794

Pabo CO, Suchanek EG (1986) Computer-aided model-building strategies for protein design. Biochemistry 25:5987–5991

Pack P, Plückthun A (1992) Miniantibodies: use of amphipathic helices to produce functional, flexibly linked dimeric Fv fragments with high avidity in *Escherichia coli*. Biochemistry 31:1579–1584

Pack P, Kujau M, Schroeckh V, Knüpfer U, Wenderoth R, Riesenberg D, Plückthun A (1993) Improved bivalent miniantibodies with identical avidity as whole antibodies, produced by high cell density fermentation of Escherichia coli. Biotechnology 11:1271–1277

Pantoliano MW, Bird RE, Johnson S, Asel ED, Dodd SW, Wood JF, Hardman KD (1991) Conformational stability, folding, and ligand-binding affinity of single-chain Fv immunoglobulin fragments expressed in *Escherichia coli*. Biochemistry 30:10117–10125

Park JT (1987) The murein sacculus. In: Neidhardt FC, Ingraham JL, Low KB, Magasanik B, Schaechter M, Umbarger HE (eds) *Escherichia coli* and *Salmonella typhimurium*. American Society for Microbiology, Washington, pp 23–30

Parmley SF, Smith GP (1988) Antibody-selectable filamentous fd phage vectors: affinity purification of target genes. Gene 73:305–318

Perez P, Hoffman RW, Shaw S, Bluestone JA, Segal DM (1985) Specific targeting of cytotoxic T cells by anti-T3 linked to anti-target cell antibody. Nature 316:354–356

Perlmutter RM, Crews ST, Douglas R, Sorensen G, Johnson N, Nivera N, Gearhart PJ, Hood L (1984) The generation of diversity in phosphorylcholine-binding antibodies. Adv Immunol 35:1–37

Plückthun A (1993a) Antibody engineering to study protein-ligand interactions and catalysis: the phosphorylcholine binding antibodies. Front Bioorg Chem 3:26–65

Plückthun A (1993b) Stability of engineered antibody fragments in: *Stability and Stabilization of Enzymes* (van den Tweel WJJ, Harder A and Buitelaar RM, eds.) Elsevier Science Publishers, pp 81–90

Plückthun A, Skerra A (1989) Expression of functional antibody Fv and Fab fragments in *Escherichia coli*. Methods Enzymol 178:497–515

Podhajska AJ, Hasan N, Szybalski W (1985) Control of cloned gene expression by promoter inversion in vivo: construction of the heat-pulse-activated att-nutL-p-att-N module. Gene 40:163–168

Pollitt S, Zalkin H (1983) Role of primary structure and disulfide bond formation in β-lactamase secretion. J Bacteriol 153:27–32

Prickett KS, Amberg DC, Hopp TP (1989) A calcium-dependent antibody for identification and purification of recombinant proteins. Biotechniques 7:580–589

Pugsley AP (1993) The complete general secretory pathway in gram-negative bacteria. Microbiol Rev 57:50–108

Raso V, Griffin T (1981) Hybrid antibodies with dual specificity for the delivery of ricin to immunoglobulin-bearing target cells. Cancer Res 41:2073–2078

Regan L, DeGrado WF (1988) Characterization of a helical protein designed from first principles. Science 241:976–978

Reth M, Imanishi-Kari T, Rajewsky K (1979) Analysis of the repertoire of anti-(4-hydroxy-3-nitrophenyl)acetyl (NP) antibodies in C 57 BL/6 mice by cell fusion. II. Characterization of idiotopes by monoclonal anti-idiotope antibodies. Eur J Immunol 9:1004–1013

Riechmann L, Foote J, Winter G (1988) Expression of an antibody Fv fragment in myeloma cells. J Mol Biol 203:825–828

Riesenberg D (1991) High-cell-density cultivation of *Escherichia coli*. Curr Opin Biotechnol 2:380–384

Rowe ES (1976) Dissociation and denaturation equilibria and kinetics of a homogeneous human immunoglobulin Fab fragment. Biochemistry 15:905–916

Rowe ES, Tanford C (1973) Equilibrium and kinetics of the denaturation of a homogeneous human immunoglobulin light chain. Biochemistry 12:4822–4827

Rudolph R (1990) Renaturation of recombinant, disulfide-bonded proteins from "inclusion bodies". In: Tschesche H (ed) Modern methods in protein- and nucleic acid research. de Gruyter, Berlin, pp 149–171

Russel M (1991) Filamentous phage assembly. Mol Microbiol 5:1607–1613

Sastry L, Alting-Mees M, Huse WD, Short JM, Sorge JA, Hay BN, Janda KD, Benkovic SJ, Lerner RA (1989) Cloning of the immunological repertoire in *Escherichia coli* for generation of monoclonal catalytic antibodies: construction of a heavy chain variable region-specific cDNA library. Proc Natl Acad Sci USA 86:5728–5732

Satow Y, Cohen GH, Padlan EA, Davies DR (1986) Phosphocholine binding immunoglobulin Fab McPC603. An X-ray diffraction study at 2.7 Å. J Mol Biol 190:593–604

Savage P, So A, Spooner RA, Epenetos AA (1993) A recombinant single-chain antibody interleukin-2 fusion protein. Br J Cancer 67:304–310

Sawyer JR, Tucker PW, Blattner FR (1992) Metal-binding chimeric antibodies expressed in *Escherichia coli*. Proc Natl Acad Sci USA 89:9754–9758

Schmidt TGM, Skerra A (1993) The random peptide library-assisted engineering of a C-terminal affinity peptide, useful for the detection and purification of a functional Ig Fv fragment. Protein Eng 6:109–122

Schoner BE, Belagaje RM, Schoner RG (1990) Enhanced translational efficiency with two-cistron expression system. Methods Enzymol 185:94–103

Schumaker VN, Poon PH (1990) Activation of the classical and alternative pathways of complement by immune complexes. In: Metzger H (ed) Fc receptors and the action of antibodies. American Society for Microbiology, Washington, pp 181–207

Scott JK, Smith GP (1990) Searching for peptide ligands with an epitope library. Science 249:386–390

Seetharam S, Chaudhary VK, FitzGerald DJ, Pastan I (1991) Increased cytotoxic activity of *Pseudomonas* exotoxin and two chimeric toxins ending in KDEL. J Biol Chem 266:17376–17381

Segal DM (1990) Antibody-mediated killing by leukocytes. In: Metzger H (ed) Fc receptors and the action of antibodes. American Society for Microbiology, Washington, pp 291–301

Sen J, Beychok S (1986) Proteolytic dissection of a hapten binding site. Proteins
    1:256–262
Sharon J, Givol D (1976) Preparation of the Fv fragment from the mouse myeloma
    XRPC-25 immunoglobulin possessing anti-dinitrophenyl activity. Biochemistry
    15:1591–1594
Shibui T, Munakata K, Matsumoto R, Ohta K, Matsushima R, Morimoto Y,
    Negahari K (1993) High-level production and secretion of a mouse-human
    chimeric Fab fragment with specificity to human carcino embryonic antigen in
    Escherichia coli. Appl Microbiol Biotechnol 38:770–775
Shin S-U, Wright A, Bonagura V, Morrison SL (1992) Genetically-engineered
    antibodies: tools for the study of diverse properties of the antibody molecule.
    Immunol Rev 130:87–107
Skerra A, Plückthun A (1988) Assembly of a functional immunoglobulin Fv frag-
    ment in Escherichia coli. Science 240:1038–1041
Skerra A, Plückthun A (1991) Secretion and in vivo folding of the Fab fragment of
    the antibody McPC603 in Escherichia coli: influence of disulphides and cis-
    prolines. Protein Eng 4:971–979
Skerra A, Pfitzinger I, Plückthun A (1991) The functional expression of antibody Fv
    fragments in Escherichia coli: improved vectors and a generally applicable purifi-
    cation technique. Biotechnology 9:273–278
Smith GP (1985) Filamentous fusion phage: novel expression vectors that display
    cloned antigens on the virion surface. Science 228:1315–1317
Smith MC, Cook JA, Smanik PA, Wakulchik M, Kasher MS (1992) Kinetically inert
    Co(III) linkage through an engineered metal binding site: specific orientation of
    recombinant human papillomovirus type 16 E7 protein on a solid support.
    Methods: A companion to Methods Enzymol 4:73–78
Söderlind E, Simonsson Lagerkvist AC, Dueñas M, Malmborg AC, Ayala M,
    Danielsson L, Borrebaeck CAK (1993) Chaperonin assisted phage display of
    antibody fragments on filamentous bacteriophages. Biotechnology 11:503–507
Staerz UD, Kanagawa O, Bevan MJ (1985) Hybrid antibodies can target sites for
    attack by T cells. Nature 314:628–631
Stemmer WPC, Morris-SK, Kautzer CR, Wilson BS (1993a) Increased antibody
    expression from Escherichia coli through wobble-base library mutagenesis by
    enzymatic inverse PCR. Gene 123:1–7
Stemmer WPC, Morris SK, Wilson BS (1993b) Selection of an active single chain Fv
    antibody from a protein linker library prepared by enzymatic inverse PCR.
    Biotechniques 14:256–265
Stevens FJ, Westholm FA, Solomon A, Schiffer M (1980) Self-association of human
    immunoglobulin κI light chains: role of the third hypervariable region. Proc Natl
    Acad Sci USA 77:1144–1148
Stewart DE, Sarkar A, Wampler JE (1990) Occurrence and role of cis peptide bonds
    in protein structures. J Mol Biol 214:253–260
Sutton BJ, Phillips DC (1983) The three-dimensional structure of the carbohydrate
    within the Fc fragment of immunoglobulin G. Biochem Soc Trans 11:130–132
Tai M-S, Mudgett-Hunter M, Levinson D, Wu G-M, Haber E, Oppermann H,
    Huston JS (1990) A bifunctional fusion protein containing Fc-binding fragment
    B of staphylococcal protein A amino terminal to antidigoxin single-chain Fv.
    Biochemistry 29:8024–8030
Takagi H, Morinaga Y, Tsuchiya M, Ikemura H, Inouye M (1988) Control of folding
    of proteins secreted by a high expression secretion vector, pIN-III-ompA: 16-
    fold increase in production of active subtilisin E in Escherichia coli. Biotech-
    nology 6:948–950
Takahashi H, Igarashi T, Shimada I, Arata Y (1991) Preparation of the Fv fragment
    from a short-chain mouse IgG2a anti-dansyl monoclonal antibody and use of
    selectively deuterated Fv analogues for two-dimensional ^{1}H NMR analyses of
    the antigen-antibody interactions. Biochemistry 30:2840–2847

Takkinen K, Laukkanen ML, Sizmann D, Alfthan K, Immonen T, Vanne L, Kaartinen M, Knowles JKC, Teeri TT (1991) An active single-chain antibody containing a cellulase linker domain is secreted by *Escherichia coli*. Protein Eng 4:837–841

Tao M-H, Morrison SL (1989) Studies of aglycosylated chimeric mouse-human IgG. Role of carbohydrate in the structure and effector functions mediated by the human IgG constant region. J Immunol 143:2595–2601

Tomlinson IM, Walter G, Marks JD, Llewelyn MB, Winter G (1992) The repertoire of human germline V_H sequences reveals about 50 groups of V_H segments with different hypervariable loops. J Mol Biol 227:776–798

Trandinh CC, Pao GM, Saier MH (1992) Structural and evolutionary relationships among the immunophilins: two ubiquitous families of peptidyl-prolyl cis-trans isomerases. FASEB J 6:3410–3420

Wanner BL (1987) Phosphate regulation of gene expression in *Escherichia coli*. In: Neidhardt FC, Ingraham JL, Low KB, Magasanik B, Schaechter M, Umbarger HE (eds) *Escherichia coli* and *Salmolella typhimurium*. American Society for Microbiology, Washington, pp 1326–1333

Ward ES, Güssow D, Griffiths AD, Jones PT, Winter G (1989) Binding activities of a repertoire of single immunoglobulin variable domains secreted from *Escherichia coli*. Nature 341:544–546

Weir DM (ed) (1986) Immunochemistry. Blackwell, Oxford (Handbook of experimental immunology, vol 1, 4th edn)

Weissman JS, Kim PS (1991) Reexamination of the folding of BPTI: predominance of native intermediates. Science 253:1386–1393

Wels W, Harwerth I-M, Zwickl M, Hardman N, Groner B, Hynes NE (1992a) Construction, bacterial expression and characterization of a bifunctional single-chain antibody-phosphatase fusion protein targeted to the human erbB-2 receptor. Biotechnology 10:1128–1132

Wels W, Harwerth I-M, Mueller M, Groner B, Hynes NE (1992b) Selective inhibition of tumor cell growth by a recombinant single-chain antibody-toxin specific for the erbB-2 receptor. Cancer Res 52:6310–6317

Whitlow M, Filpula D (1991) Single-chain Fv proteins and their fusion proteins. Methods: a companion to Methods Enzymol 2:97–105

Whitlow M, Filpula D, Rollence ML, Feng SL, Wood JF (1994) Multivalent Fvs: characterization of single-chain Fv oligomers and preparation of a bispecific Fv. Protein Eng, in press

Wood CR, Boss MA, Patel TP, Emtage JS (1984) The influence of messenger RNA secondary structure on expression of an immunoglobulin heavy chain in *Escherichia coli*. Nucleic Acids Res 12:3937–3950

Wu XC, Ng SC, Near RI, Wong SL (1993) Efficient production of a functional single-chain antidigoxin antibody via an engineered *Bacillus subtilis* expression secretion system. Biotechnology 11:71–76

Wülfing C, Plückthun A (1993) A versatile and highly repressible E. coli expression system based on invertible promoters: expression of a gene encoding a toxic product. Gene 136:199–203

Wünsch E, Moroder L, Göhring-Romani S, Musiol H-J, Göhring W, Bovermann G (1988) Synthesis of the bis-cystinyl-fragment 225–232/225'–232' of the human IgG1 hinge region. Int J Pept Protein Res 32:368–383

Yanisch-Perron C, Vieira J, Messing J (1985) Improved M13 phage cloning vectors and host strains: nucleotide sequences of the M13mp18 and pUC19 vectors. Gene 33:103–119

Yokota T, Milenic DE, Whitlow M, Schlom J (1992) Rapid tumor penetration of a single-chain Fv and comparison with other immunoglobulin forms. Cancer Res 52:3402–3408

Zeilstra-Ryalls J, Fayet O, Georgopoulos C (1991) The universally conserved GroE (Hsp60) Chaperonins. Annu Rev Microbiol 45:301–325

CHAPTER 12
Structure, Function and Uses of Antibodies from Transgenic Plants and Animals

A. Hiatt and M. Hein

A. Introduction

The natural production of soluble antibodies is restricted to a specialized population of lymphoid cells in vertebrates. Development of immortalized murine cell lines secreting monoclonal antibodies (Kohler and Milstein 1976) revolutionized the development of antibody reagents and provided a critical experimental component for the investigation of the structure and function of immunoglobulins and the genes which encode them. Since the development of murine hybridomas, a variety of prokaryotic and eukaryotic cell types, as demonstrated in this volume, have been shown to synthesize immunoglobulin peptides from mammalian transgenes. However, few of these cell types are capable of assembling, processing and subsequently secreting functional antibodies with the fidelity or efficiency of B cells or hybridomas. A surprising exception to this observation is the capacity of plant cells to synthesize and assemble functional antibodies with fidelity and in relatively high concentration (Hiatt et al. 1989). Plants do not contain B cells or other lymphoid cells and the specific plant cell types capable of antibody production have yet to be identified.

The production of antibodies in various cell cultures has been important for the investigation of a wide range of biological questions and for the production of reagent antibodies for research and medicine. The more recent development of techniques for recovery of whole transgenic organisms has allowed new experimental approaches for investigating in vivo processes which are integrated and controlled at the organismal level. By far, most studies of multicellular eukaryotic organisms (as opposed to those of cell cultures) bearing immunoglobulin transgenes have utilized transgenic mice. However, recent progress has resulted in the generation of both transgenic large animals expressing immunoglobulin genes (Lo et al. 1991) and the production of antibodies in plants (Hiatt et al. 1989). The focus of this review will be on the potential uses of antibodies derived from transgenic animals and plants and the structure and function of various antibodies derived from different transgenic systems.

B. Transgenic Antibodies from Mice

Transgenic mice can be produced by several methods, the most frequently used being microinjection of the cloned genes into one of the pronuclei of a fertilized egg (Brinster et al. 1985). The microinjected embryos are developed to term in the uterus of a pseudopregnant mouse. About 25% of the resulting pups contain the injected gene and they generally contain foreign DNA stably integrated in all cells which is propagated in the germ line.

In general, immunoglobulin transgenes with their own control regions are correctly expressed in the B lymphocytes of transgenic mice. These mice therefore provide an ideal tool to investigate many parameters of the control of immunoglobulin synthesis. High levels of immunoglobulin transgene expression can occur and this does not appear to be at the expense of the health or longevity of the animal, suggesting that immunity against a variety of pathogens is not impaired. However, the expression of an immunoglobulin transgene can result in significant inhibition of gene rearrangement and expression of endogenous immunoglobulin genes (Ritchie et al. 1984; Rusconi and Kohler 1985; Storb 1990; Nussenzweig et al. 1987; Stall et al. 1988). Therefore, it is a matter of ongoing concern to determine whether the transgene approach could potentially have a deleterious effect on the host immune system or whether the host immune system could in fact be augmented by expression of a novel antibody.

C. Potential Uses of Antibody Expression in Transgenic Animals

I. Investigation of Immune System Regulation

Transgenic animals expressing immune system loci have provided an especially valuable tool for analysis of the control of expression of immunoglobulin genes in animals. Because of the extreme heterogeneity of the B cell pool with respect to the specific immunoglobulin genes expressed, the ability to introduce a well characterized immunoglobulin transgene and to follow its fate in the majority of B cells has provided new insights into regulation of the immune system. Phenomenon such as the control of allelic and isotypic exclusion (Kitamura and Rajewsky 1992), rearrangement and somatic hypermutation can probably best be solved using transgenic animals harboring an antibody trangene of known sequence. Other emerging areas in which these transgenic animal systems are finding applications are the investigation of tolerance to autoantibodies (Offen et al. 1992), the activation and control of immunoglobulin recombinases in pre-B cells (Storb 1990), and the regulation of immunoglobulin isotypes involved in allergic diseases (Adamczewski et al. 1991).

II. Human Monoclonal Antibodies in Animals

The potential of transgenic animal technology is not limited to basic biological questions since the technology can be developed to include commercial and clinical applications. One potential therapeutic application is the production in animals of monoclonal antibodies of human origin. Most immortalized human cell lines or hybridomas do not stably express large amounts of antibody; in vivo immunization of humans is not feasible for many antigens and in vitro priming is usually inefficient. The introduction of an unrearranged repertoire of human immunoglobulin gene segments into transgenic mice could potentially result in the production of a corresponding repertoire of human antibodies. The test this possibility, mice have been created that carry a human heavy chain minilocus comprising unrearranged immunoglobulin variable, diversity, and joining elements linked to a human μ chain gene (BRUGGEMANN et al. 1989). The gene segments of the minilocus were found to be rearranged in a large proportion of cells in thymus and spleen but not in nonlymphoid tissue. Approximately 4% of the B lymphocytes synthesized human μ chains resulting in a serum titer of about $50 \mu g$ transgenic IgM per ml. In addition, hybridomas could be established from transgenic mice that stably secreted mg/ml quantities of antibodies containing human μ chains.

A similar approach has sought to create transgenic mouse strains that express human immunoglobulin genes from 100 kb of cosmid DNA (BRUGGEMANN et al. 1989). Rearrangements of the human derived sequences in the transgenic mice, similar to those seen in human DNA, were found only in spleen but not in thymus. Random hybridomas made from these transgenic mice show heterogeneous rearrangement of human transgenes. The results suggest that utilization of human immunoglobulin genome segments does occur in transgenic mice making possible the derivation of mouse strains that produce authentic human antibodies from inserted heavy and light chain genetic loci.

III. Pathogen Protection in Agricultural Animals

Expression of a transgenic immunoglobulin specific for a common pathogen could provide an animal with congenital immunity for that pathogen. If successful, this approach would be of tremendous use in large agricultural animals. Since the production of antibody to various polysaccharide antigens can be protective against pathogenic bacteria, transgenic mice, pigs and sheep carrying genes encoding mouse antibodies against phosphorylcholine have been investigated to determine whether the transgene antibody might be used to influence susceptibility to common diseases.

It was found, in one set of experiments involving only mice (PINKERT et al. 1989), that the transgenic offspring produced elevated levels of anti-phosphorylcholine antibodies constitutively, at 16 days of age, when normal nontransgenic mice were not fully immunocompetent. A triggering antigenic stimulus was not necessary to evoke anti-phosphorylcholine immunoglobulin production. Additionally, the frequency of phosphorylcholine B cells in these transgenic mice was further increased upon specific immunization.

In a more extensive study involving mice, pigs and sheep carrying genes encoding the mouse α and κ chains for antibodies against phosphorylcholine (Lo et al. 1991), it was found that high serum levels of the transgenic IgA could be detected in transgenic mice and pigs, but not in transgenic sheep. In one mouse line, expression of the transgene resulted in less than 10% of the spleen B cells expressing endogenous IgM. This is consistant with previous observations that immunoglobulin transgenes can suppress both immunoglobulin gene rearrangement and endogenous immunoglobulin expression. Despite this effect, significant levels of endogenous IgM were secreted into the serum. Suppression of endogenous IgM expression was not observed in all mouse lines nor did this effect occur in transgenic pigs.

In the transgenic pigs, the mouse IgA was detected in the serum despite the inability to detect the transgenic κ chain gene. Presumably, the secreted mouse IgA was assembled with pig light chains and this was reflected in the absence of a phosphorylcholine binding specificity of the transgenic mouse IgA.

In the sheep, mouse IgA was detected in peripheral lymphocytes but not in serum. Mouse κ expression was not detected in transgenic sheep harboring an intact κ transgene.

These results illustrate both the potential and some of the problems inherent in introducing beneficial traits such as germ line-encoded immunity into large mammalian species. The complexities of introducing an immunoglobulin transgene into a population of endogenous immunoglobulin genes necessitates a broader understanding of the factors which contribute to suppression of both endogenous and transgenic immunoglobulin expression.

D. Transgenic Antibodies from Plants

DNA-mediated transformation of plant cells can be accomplished in a variety of ways (Saunders 1989; Knight 1991; Weaver and Powell 1989; Boynton et al. 1988; Nester and Kosuge 1981; Bevan and Chilton 1982; Schell 1983). The most commonly used method employs *Agrobacterium tumefaciens* as the agent for introduction of recombinant DNA to the plant cell nucleus (Nester and Kosuge 1981; Bevan and Chilton 1982; Schell 1983). This method of plant cell transformation is accomplished simply by cocultivation of leaf segments with the bacterium containing vector DNA derived from a Ti plasmid, which is the causal agent in neoplastic transformation of many types of plant cells (Nester and Kosuge 1981; Bevan and Chilton 1982; Schell 1983). Cells on the margin of the leaf segment are infected by *A. tumefaciens*, a process in which a portion of the Ti plasmid DNA is transferred to the plant nucleus, and in some cells the transferred DNA is stably integrated into the nuclear genome.

Whole plants can be regenerated from this population of transformed cells in some species of plants by manipulation of phytohormones and nutrients in the medium on which the cells are cultured. Although the process of regeneration required to obtain transgenic plants can take months, the basic manipulations and reagents required to induce regeneration are very simple. This process and the development of antibiotic markers for the selection of transformed cells has provided a general and facile method for plant transformation.

Vectors have been developed which contain plant selectible markers, promoters upstream from polylinkers, and *Escherichia coli* and *Agrobacterium* origins of replication (ROGERS et al. 1987). Since the process of transformation is mediated by a recombinant bacterium, a spectrum of transformation events is possible. The *Agrobacterium* can contain a single gene or cDNA of interest, a defined population of cDNAs or a library of DNA fragments (DITTA et al. 1980). Not all plants are amenable to the manipulations required for the stable introduction of foreign DNA. Tobacco is the most commonly used plant since it is easily transformed and regenerated and was the first plant shown to accumulate assembled mammalian antibodies (HIATT et al. 1989).

There are no hard and fast rules for achieving high level expression of foreign gene products in a new plant environment; techniques which may work for one gene may not result in the accumulation of another gene product. In the case of immunoglobulin genes, it is unlikely that native mammalian control regions would promote expression of antibodies in differentiated plant cells. However, there are a range of options available for expression of foreign proteins in plants in which temporally controlled (developmental or inducible), tissue-specific or cell-specific expression of genes can lead to accumulation of a protein product.

In the first report of antibody expression in plants, a catalytic IgG_1 antibody (6D4) was chosen for expression in tobacco (HIATT et al. 1989). This antibody recognizes a synthetic phosphonate ester, P3, and can catalyze the hydrolysis of certain carboxylic esters. The γ and κ cDNAs derived from the 6D4 hybridoma were cloned into the plant expression vector pMON530 (ROGERS et al. 1987). This vector contains the control region from the 35S RNA transcript of the cauliflower mosaic virus and promotes expression of foreign genes in a number of plant tissues (ODELL et al. 1985; BARNES 1990; BENFEY and CHUA 1990). Transgenic plants containing γ or κ chains were crossed to produce progeny expressing both chains. The F_1 plants contained assembled functional antibody as determined by the following criteria: (1) western blots of plant extracts under reducing conditions contained equimolar amounts of γ and κ chains which migrated at 50 kDa and 25 kDa, respectively. Under nonreducing conditions both γ and κ bands migrated at about 160 kDa. (2) ELISA assays in which extracts were added to microtiter plates coated with goat anti-mouse-γ then detected with goat anti-mouse κ-HRPO indicated the presence of equimolar amounts of γ bound to κ with no detectible free γ or κ chains. (3) ELISA assays with a P3-BSA conjugate as antigen gave similar results: the affinity of γ-κ complexes for P3 was identical to that of the hybridoma-derived antibody. The specificity for P3 was indicated by inhibition of P3-BSA binding by free P3 in which half-maximal inhibition was about $10\,\text{m}M$ for plant-derived or hybridoma-derived antibody.

A surprisingly high level of accumulation of functional antibody was observed. In the case of 6D4, greater than 1% of total extractable protein was found to be functional antibody. Other antibodies, which have subsequently been expressed in tobacco using the same strategy, have resulted in similar levels of accumulation. During et al. (1990) used an expression strategy employing a single vector carrying both γ and κ chain cDNAs of an antibody in the R_0 generation, obviating the need for crossing to achieve expression of both chains simultaneously. Different promoters were used to drive expression of the κ and γ chains which may be partially responsible for the comparatively low level of accumulation of these immunoglobulins.

Plants are capable of accumulating antibody-derived binding proteins in conformations other than assembled heterotetramers. Other investigators have demonstrated the ability of plant tissues to accumulate single heavy chain domain peptides at nearly 1% of extractable leaf protein (Benvenuto et al. 1991), and more recently a single chain Fv antibody recognizing the plant photoreceptor phytochrome has been expressed in tobacco (Owen et al. 1992). With the potential for manipulation of immunoglobulin gene transcription rates, the accumulation of an antibody either in whole plants or in specific tissues could be increased significantly.

E. Structure and Function of Antibodies from Plants

One of the more surprising results from the initial expression of antibodies in plants was the quantitative assembly of heterotetrameric IgG in plants which expressed both γ and κ chains (Hiatt et al. 1989; Hein et al. 1991). When both γ and κ chain genes were introduced with identical transcriptional control regions no individual γ or κ chains were detectible. Furthermore, the level of assembled antibody which accumulated in F_1 plants expressing both chains was generally higher than that found for the individual chains in the respective parent plants. These observations were the first indications that plant cells possessed cognate systems for assembly of mammalian heteromultimers and of the inherent stability of the assembled antibody in the cells or extracellular water (apoplast) of plants.

Further analysis of the plant antibody was performed after purification on Sephacryl-FPLC and protein-A Sepharose (Hein et al. 1991). After this simple size exclusion and affinity purification the antibody was evaluated by

Coomassie blue staining and western blotting and was found to be virtually pure. The quantitative retention of the antibody on protein A Sepharose indicates that the interdomain conformation of Fc recognized by protein A (between CH2 and CH3) is intact. In addition to the substrate binding and general characteristics of the protein, the catalytic activity of purified protein was measured. The catalytic activity of plant-derived antibody, in the absence or presence of a competitive inhibitor, were indistinguishable from the ascites-produced antibody (HEIN et al. 1991). The results demonstrate that the antibody maintains its catalytic capability.

I. Glycosylation of Antibodies Produced in Plants

Glycosylation of heavy chain was investigated by lectin binding analysis of western blotted antibody. Several biotinylated lectins were used to decorate the blots after which lectin binding was visualized with streptavidin-alkaline phosphatase and bromo-chloro-indoyl-phosphate (KIJIMOTO-OCHIAI et al. 1989; HEIN et al. 1991). Only concanavalin A (specific for mannose and glucose) bound to the plant-derived γ chain whereas γ chain derived from ascites fluid was recognized by both concanavalin A and the lectins from Ricinus communis (specific for terminal galactose and N-acetylgalactosamine) and wheat germ agglutinin (specific for N-acetylglucosamine dimers, terminal sialic acid) (GOLDSTEIN and HAYES 1978). A variety of other lectins did not bind to either plant- or ascites-derived γ chain. These results indicate that the transgenic plant antibody is processed in a similar fashion to complex mammalian glycoproteins but has a different secondary glycosyl elaboration and lacks a terminal sialic acid residue.

N-Linked glycosylation of proteins in plants is similar to that in mammals (JONES and ROBINSON 1989; STURM et al. 1987). A core high mannose oligosaccharide is attached to the asparagines contained within the canonical Asn-X-Ser/Thr sequence. Glycosylation occurs in the endoplasmic reticulum (ER) and can be modified in the Golgi apparatus where α-mannosidase removes some mannose residues and terminal sugars are attached. In mammals, the predominant terminal residue is N-acetyl neuraminic acid (NANA); this carbohydrate has not been identified in plants. Terminal residues in plants have been found to consist of xylose, fucose, N-acetylglucosamine, mannose or galactose (STURM et al. 1987). This suggests a distinct composition of terminal residues on the plant glycan and is consistent with the absence of NANA in plants.

II. Antibody Processing and Assembly

Evidence that the antibody has been proteolytically processed in the lumen of the ER is provided by the NH_2-terminal amino acid sequence of the κ chain from assembled antibodies isolated from plants. The light chain NH_2-terminal sequence was Asp-Val-Val-Leu for both plant and mouse antibody

(Hein et al. 1991). This demonstrates appropriate proteolytic processing of the mouse signal sequence by the plant ER. Heavy chain from either plant antibody or hybridoma was intractable to sequencing indicating a blocked NH$_2$-terminal.

Plants must contain an assembly and processing apparatus which can recognize mouse immunoglobulins. The efficiency of assembly in plants was surprising since there was generally an increase in the levels of expression of assembled antibody over the levels of individual γ or κ chains in the parental plants. Typically a heavy chain parent produces 20- to 50-fold more immunoglobulin than the light chain parent (Hiatt et al. 1989). Progeny of crosses between parents with these widely differing levels of γ and κ chain typically produced fivefold or higher levels of γ chain in assembled antibody than the level of the unassembled chain produced by the γ chain parent. This suggests that the assembled antibody is more stable in plants than the individual heavy or light chains. It is apparent that cognate mechanisms function in plants to coordinate the assembly of oligomers, direct N-glycosylation and effect processing in the Golgi such that immunoglobulins are recognized and efficiently processed.

Assembly of immunoglobulin chains in mammalian cells is thought to involve a native component of the ER, immunoglobulin heavy chain binding protein (BiP), which has been shown to be involved in the posttranslational processing of heavy chains (Rothman 1989). BiP has been found associated with unassembled immunoglobulin heavy chains prior to assembly with light chains. Unassembled heavy chains remain associated with BiP and are not transported to the Golgi apparatus (Rothman 1989). A Bip-like protein in plants has recently been characterized (Fontes et al. 1991; Boston et al. 1991) as has a protein disulfide isomerase (Shorrosh and Dixon 1991, 1992).

III. Mutagenesis to Remove N-Linked Glycosylation

Glycosylation of the heavy chain constant region of the 6D4 antibody occurs at an Asn-Ser-Thr site in the CH2 region beginning at nucleotide 953. This is the site that is glycosylated in antibodies derived from mammalian cells. We changed the Asn codon to an His codon to inactivate the glycosylation target site. To assess the efficiency of synthesis and assembly, the mutagenized heavy chain and complementary light chain genes were electroporated into tobacco protoplasts. Protein isolated from the electroporated cells or the protoplast culture medium was subjected to ELISA assays to test for assembly into a γ-κ complex and to assess functionality by binding to the 6D4 antigen (P3). The results showed that the aglycosylated heavy chain was completely assembled with κ chains and was capable of binding antigen. Both the assembly into γ-κ complexes and the antigen binding of the aglycosylated antibody was indistinguishable from the glycosylated counterpart.

IV. Deletion of Heavy Chain Constant Regions

Fab fragments are ordinarily derived from the proteolytic digestion of full length antibodies, resulting in the removal of the CH2 and CH3 region of the heavy chain constant region. These fragments are fully functional with respect to antigen binding but do not contain specific sites (such as the glycosylation sequence) which mediate effector functions in the host organism. To investigate the assembly of Fabs in plants, we first made by PCR a truncated 6D4 heavy chain containing a new stop codon at nucleotide position 745. This construct eliminates the three intermolecular disulfide bridges which ordinarily form between the heavy chains in the full length antibody. In addition, we made a slightly longer construct (stop codon at nucleotide position 763) which includes the three cysteine codons. Expression in protoplasts revealed that in both cases functional Fab fragments were secreted into the medium. The construct containing the additional cysteines, however, did not form disulfide bridges between the heavy chains.

Other truncated antibody constructs have also been shown to accumulate in plant tissues. BENVENUTO et al. (1991) demonstrated the stability of heavy chain antigen binding domains in plants which accumulated in large amounts in stably transformed leaf tissues. OWEN et al. (1992) more recently demonstrated expression of a small, single chain, antigen binding protein containing the variable regions of both heavy and light chains. Collectively these examples illustrate the general capacity of plant cells to synthesize and assemble a wide variety of antigen binding proteins derived from antibodies.

F. Potential Uses of Antibodies Expressed in Transgenic Plants

I. Scale and Economics of Plantibody Production

The reported levels of antibody accumulation in plant tissues is greater than 1% of extractable plant protein for whole assembled IgG (HIATT et al. 1989) and for a heavy chain protein alone (BENVENUTO et al. 1991). Little effort has been expended in optimizing the accumulation of antibodies in plants. A number of strategies for increasing the antibody content of plant tissues are currently possible. These include accumulation of multiple copies of antibody genes by sexual crossing or by multiple gene transformations, stabilization of immunoglobulin mRNA by inclusion of message-stabilizing nontranslated sequences and manipulation of promoter or enhancer sequences to increase transcription rates. With transgenic protein expressed at the 1% level, the cost of production for a transgenic antibody in green tissue or in a seed grain would be less than $1.00/g. This low cost estimate indicates that plants may provide a relatively cheap source of antibodies which could be used in applications for which current sources of monoclonal antibodies are not economical.

II. Potential Medical Uses of Plant Produced Antibodies

Since plants do not possess the cellular or humoral immune systems of vertebrates, the presence of immunoglobulin transgenes provides no insight into the regulatory mechanisms of the immune system as outlined for animal systems. It does however provide a eukaryotic system with the potential for large scale production of reagent antibodies.

Plant produced antibodies could potentially be used in the same biomedical applications as antibodies produced in cell culture or fermentation systems since function and assembly have been shown to be generally conserved for antibodies produced in plants. A number of specific areas warrant further investigation to evaluate the potential of plant produced antibodies for these purposes. For instance, while the Fc portion of assembled antibodies from plants appears to be intact, there has been no specific evaluation of the ability of antibodies produced in plants to participate in antibody-dependent cell-mediated cytotoxicity. The occurrence of plant-specific glycosylation patterns could pose unique challenges to therapeutic, imaging or other parenteral applications of plant-produced antibodies. Whether the plant-specific glycosylation patterns will result in specific or deleterious host-anti-plant antibody (HAPA) responses has yet to be investigated. Modification of the coding sequence of immunoglobulins to remove glycosylation sites is one possible means of obviating this type of HAPA response. But the effectiveness of the resulting aglycosylated antibodies will likely have to be determined for each individual antibody.

Other potential biomedical applications for antibodies produced in plants include extracorporeal shunts for plasma purification, diagnostics or passive immunization of mucosal surfaces. Extracorporeal filtration and other affinity purification processes would benefit from an inexpensive bulk source for specific monoclonal antibodies. Diagnostic antibodies represent a small component of cost in diagnostic kits and are required in much smaller quantities for each use than for these larger scale applications. Therefore there would be little benefit from the potential production scale and low cost of plant-produced antibodies.

One potential application for bulk antibodies produced in plants is passive immunization of mucosal surfaces. The potential for employing plant-derived antibodies for protection of mucosal surfaces was recently described by HIATT and MOSTOV (1992). Most infectious agents enter the mammalian body through exposed mucosal surfaces, such as the gastrointestinal, respiratory, urinary, reproductive and ocular systems. Mammals have a specialized mucosal or secretory immune system (CHILDERS et al. 1989). The primary antibody component of this system is secretory IgA and in some cases secretory IgM. These antibodies are unusually complex and consist of four different polypeptides normally synthesized by two cells: plasma cells and epithelial cells. IgA and IgM, especially when found in secretions, are usually polymers and for this reason they are known as polymeric immunoglobulins (pIg) (UNDERDOWN and SCHIFF 1986). Secretory immunoglobulins prevent infection by several mechanisms (CHILDERS et al. 1989) including: agglutination, direct killing, inhibition of attachment and invasion of epithelial surfaces, opsonization, and inhibition of bacterial toxins and enzymes. The antibodies also inhibit entry of other antigens thereby modulating the immune response.

Polymeric IgA is classically a dimer (i.e., four heavy or α chains, and four light chains), although there are also higher polymers and IgM is usually a pentamer. These polymeric immunoglobulins usually contain an extra polypeptide known as the J or joining chain, a 15 600 MW peptide. Typically one molecule of dimeric IgA contains one J chain and IgM contains one or more J chains. In mammals the J chain-containing polymeric immunoglobulins are synthesized by plasma cells. These plasma cells tend to underlie exposed mucosal surfaces. The epithelial cells that line these surfaces express a polymeric immunoglobulin receptor that functions to transport the polymeric immunoglobulin across the epithelial cells and release it at the mucosal or luminal surface (MOSTOV and SIMISTER 1985). This delivers the polymeric immunoglobulin to the optimal location where it can form the first line of defense against invading orgnisms. During this process the receptor-ligand complex is endocytosed and carried by vesicles to the apical or mucosal surface in a process called transcytosis. After reaching the apical cell surface the extracellular ligand-binding domain of the polymeric immunoglobulin receptor is endoproteolytically cleaved from the membrane-spanning portion of the receptor. This cleaved fragment is called secretory component (SC) and is released together with the polymeric immunoglobulin into the external secretions. The complex of SC and polymeric immunoglobulin is generically called secretory immunoglobulin; secretory IgA (sIgA) and secretory IgM (sIgM) refer to SC complexed to IgA or IgM, respectively. It may be possible to express all four chains required for formation of secretory antibodies in single plant cells by a number of strategies (HIATT and MOSTOV 1992), including transfecting genes for the individual polypeptides into separate plants and then creating a strain expressing all four by sexual crosses and selections. Since plants assemble two immunoglobulin chains into assembled anitbodies with a high degree of fidelity, they may also assemble all four secretory chains into intact, functional secretory antibodies.

These plants would provide passive oral immunity to numerous enteric pathogensand could be used to protect both animal herds and humans. They would be especially useful in newborn humans and animals, which are particularly susceptible to enteric illnesses (CHILDERS et al. 1989).

III. Pathogen Protection in Agricultural Plants

Expression of antibodies in plants potentially offers many of the same agricultural benefits as expression in large animals. These might include resistance to certain intractible plant pathogens and a relatively inexpensive production system. Since plants lack the general immune system mechanisms for challenging disease organisms, the inherent antigen binding function of the immunoglobulin must be sufficient for neutralization of plant pests. Therefore the most likely targets for antibody-mediated pathogen protection in plants will be those in which the molecular mechanisms of disease progression or pathogen attack and replication are well characterized. In these cases molecular targets for antibody-based neutralization can be addressed. To date no reports of successful plant protection against pathogens using an antibody-mediated mechanism have been reported, though a number of laboratories worldwide are currently attempting to generate plants which express antibodies specific to plant viral antigens.

References

Adamczewski M, Kohler G, Lamers MC (1991) Expression and biological effects of high levels of serum IgE in epsilon heavy chain transgenic mice. Eur J Immunol 21(3):617–626

Barnes WM (1990) Variable patterns of expression of luciferase in transgenic tobacco leaves. Proc Natl Acad Sci USA 87:9183–9187

Benfey PN, Chua NH (1990) The cauliflower mosaic virus 35S promoter: combinatorial regulation of transcription in plants. Science 250:959–966

Benvenuto E, Ordas RJ, Tavazza R, Ancora G, Biocca S, Cattaneo A, Galeffi P (1991) Phytoantibodies: a general vector for the expression of immunoglobulin domains in transgenic plants. Plant Mol Biol 17:865–874

Bevan MW, Chilton MD (1982) T-DNA of the *Agrobacterium* Ti and Ri plasmids. Annu Rev Microbiol 35:531–565

Boston RS, Fontes EBP, Shank BB, Wrobel RL (1991) Increased expression of the maize immunoglobulin binding protein homolog b-70 in three zein regulatory mutants. Plant Cell 3(5):497–505

Boynton JE, Gillham N, Harris EH, Hosler JP, Johnson AM, Jones AR Randolph-Anderson BL, Robertson D, Klein TM, Shark KB, Sanford J (1988) Chloroplast transformation in Chlamydomonas with high velocity microprojectiles. Science 240(4858):1534–1538

Brinster RL, Chen HY, Trumbauer M, Yagle MK, Palmiter RD (1985) Factors affecting the efficiency of introducing foreign DNA into mice by microinjecting eggs. Proc Natl Acad Sci USA 82:4438

Bruggemann M, Caskey HM, Teale C, Waldmann H, Williams GT, Surani MA, Neuberger MS (1989) A repertoire of monoclonal antibodies with human heavy chains from transgenic mice. Proc Natl Acad Sci USA 86(17):6709–6713

Bruggemann M, Spicer C, Buluwela L, Rosewell I, Barton S, Surani MA, Rabbitts TH (1991) Human antibody production in transgenic mice: expression from 100 kb of the human IgH locus. Eur J Immunol 21(5):1323–1326

Childers NK, Bruce MG, McGhee JR (1989) Molecular mechanisms of immunoglobulin A defense. Annu Rev Microbiol 43:503–536

Ditta M, Stanfield S, Corbin D, Helinski D (1980) Broad host range DNA cloning system for gram-negative bacteria: construction of a gene bank of *Rhizobium meliloti*. Proc Natl Acad Sci USA 77:7347–7351

During K, Hippe S, Kreuzaler E, Schell J (1990) Synthesis and self-assembly of a functional antibody in transgenic *Nicotiana tabacum*. Plant Mol Biol 15:281–293

Fontes EB, Shank BB, Wrobel RL, Moose SP, O'Brien GR, Wurtzel ET, Boston RS (1991) Characterization of an immunoglobulin binding protein homolog in the maize floury-2 endosperm mutant. Plant Cell 3(5):483–496

Goldstein IJ, Hayes C (1978) The lectins: carbohydrate-binding proteins of plants and animals. Adv Carbohydr Chem Biochem 35:127–340

Hein MB, Tang Y, McLeod DA, Janda KD, Hiatt AC (1991) Evaluation of immunoglobulins from plant cells. Biotech Prog 7:455–461

Hiatt AC, Mostov K (1992) Transgenic plants fundamentals and applications: assembly of multimeric proteins in plant cell: characteristics and uses of plant-derived antibodies. Dekker, New York, pp 231–238

Hiatt AC, Cafferkey R, Bowdish K (1989) Production of antibodies in transgenic plants. Nature 342(6245):76–78

Jones RL, Robinson DG (1989) Protein secretion in plants. Tansley Rev 17:567–588

Kijimoto-Ochiai S, Katagiri YU, Hatae T, Okuyama H (1989) Type analysis of the oligosaccharide chains on microheterogeneous components of bovine pancreatic DNAase by the lectin-nitrocellulose sheet method. Biochem J 257(1):43–49

Kitamura D, Rajewsky K (1992) Targeted disruption of mu chain membrane exon causes loss of heavy-chain allelic exclusion. Nature 356(6365):154–156

Knight DE (1991) Rendering cells permeable by exposure to electric fields. In: Baker PF (ed) Techniques in cellular physiology, vol 113. Elsevier/North-Holland, Amsterdam, pp 1–20

Kohler G, Milstein C (1976) Continuous cultures of fused cells secreting antibody of predefined specificity. Nature 256:495–497

Lo D, Pursel V, Linton PJ, Sandgren E, Behringer R, Rexroad C, Palmiter RD, Brinster RL (1991) Expression of mouse IgA by transgenic mice, pigs and sheep. Eur J Immunol 21:1001–1006

Mostov KE, Simister NE (1985) Transcytosis. Cell 43:389–390

Nester EW, Kosuge T (1981) Plasmids specifying plant hyperplasias. Annu Rev Microbiol 35:531–565

Nussenzweig MC, Shaw AC, Sinn E, Danner DB, Holmes KL, Morse HC, Leder P (1987) Allelic exclusion in transgenic mice that express the membrane form of immunoglobulin mu. Science 236(4803):816–819

Odell JT, Nagy F, Chua NH (1985) Identification of DNA sequences required for activity of the cauliflower mosaic virus 35S promoter. Nature 313:810–812

Offen D, Spatz L, Escowitz H, Factor S, Diamond B (1992) Induction of tolerance to an IgG autoantibody. Proc Natl Acad Sci USA 89(17):8332–8336

Owen M, Gandecha A, Cockburn B, Whitelam G (1992) Synthesis of a functional anti-phytochrome single-chain Fv protein in transgenic tobacco. Biotechnology 10:790–794

Pinkert CA, Manz J, Linton PJ, Klinman NR, Storb U (1989) Elevated PC responsive B cells and anti-PC antibody production in transgenic mice harboring anti-PC immunoglobulin genes. Vet Immun Immunopathol 23(3–4):321–332

Ritchie KA, Brinster RL, Storb U (1984) Allelic exclusion and control of endogenous immunoglobulin gene rearrangement in kappa transgenic mice. Nature 312(5994):517–520

Rogers SF, Klee HJ, Horsch RB, Fraley RT (1987) Improved vectors for plant transformation expression cassette vectors and new selectable markers. Methods Enzymol 153:253–276

Rogerson B, Hackett J Jr, Peters A, Haasch D, Storb U (1991) Mutation pattern of immunoglobulin transgenes is compatible with a model of somatic hypermutation in which targeting of the mutator is linked to the direction of DNA replication. EMBO J 10(13):4331–4341

Rothman JE (1989) Polypeptide chain binding proteins: catalysts of protein folding and related processes in cells. Cell 59(4):591–601

Rusconi S, Kohler G (1985) Transmission and expression of a specific pair of rearranged immunoglobulin mu and kappa genes in a transgenic mouse line. Nature 314:330–334

Saunders J (1989) Plant gene transfer using electrofusion and electroporation. In: Neumann E, Sowers AE, Jordan CA (eds) Electroporation and electrofusion in cell biology. Plenum, New York

Schell J (1983) Agrobacterium tumor induction. In: Shapiro J (ed) Mobile genetic elements. Academic, New York

Shorrosh BS, Dixon RA (1991) Molecular cloning of a plant putative endomembrane protein resembling vertebrate protein disulfide-isomerase and a phosphoinositidespecific phospholipase C. Proc Natl Acad Sci USA 88:10941–10945

Shorrosh BS, Dixon RA (1992) Sequence analysis and developmental expression of an alfalfa protein disulfide isomerase. Plant Mol Biol 19:319–321

Stall AM, Kroese FGM, Gadus FT, Sieckmann DG, Herzenberg LA, Herzenberg LA (1988) Rearrangement and expression of endogenous immunoglobulin genes occur in many murine B cells expressing transgenic membrane IgM. Proc Natl Acad Sci USA 85(10):3546–3550

Storb U (1990) The published data. Immunol Rev 115:253–257

Storb U, Pinkert C, Arp B, Engler P, Gollahon K, Manz J, Brady W, Brinster RL (1986) Transgenic mice with mu and kappa genes encoding antiphosphorylcholine antibodies. J Exp Med 164(2):236:627–641

Storb U, Engler P, Manz J, Gollahon K, Denis K, Lo O, Brinster R (1988) Expression of immunoglobulin genes in transgenic mice and transfected cells. Annals NY Acad Sci 546:51–56

Sturm A, Kuik AV, Vliegenthart JFG, Crispeels MJ (1987) Structure, position, and biosynthesis of the high mannose and the complex oligosaccharide side chains of the bean storage protein phaseolin. J Biol Chem 262(28):13392–13403

Underdown BJ, Schiff JM (1986) Immunoglobulin A: strategic defense initiative at the mucosal surface. Annu Rev Immunol 4:389–417

Weaver JC, Powell KT (1989) Theory of electroporation. In: Neumann E, Sowers AE, Jordan CA (eds) Electroporation and electrofusion in cell biology. Plenum, New York, pp 111–126

CHAPTER 13

Some Aspects of Monoclonal Antibody Production

R.G. RUPP

There is presently a wide range of uses for monoclonal antibodies, from research reagents in the laboratory to therapeutic products in the medical clinic. In all cases, the use to which the antibody preparation will be put must determine the manner in which the antibodies are made. Antibodies intended for therapeutic usage, for example, are subject to production restrictions that do not apply to antibodies intended solely for research purposes. Furthermore, therapeutic antibody preparations are often required in large quantities, in the gram to kilogram range, while antibodies for the research laboratory are generally needed in much smaller amounts, in the microgram to milligram range. The method selected to produce monoclonal antibodies must thus be determined by the ultimate quality and quantity desired. The purpose of this chapter is to outline my strategy for producing monoclonal antibodies – a strategy that has been used successfully to make over 60 different monoclonal products from at least that many cell lines. It is not intended as a review of the literature, since a number of excellent reviews (FEDER and TOLBERT 1985; Ho and WANG 1991; LUBINIECKI 1990; LYDERSEN 1987; ONO 1991) have already covered the field extensively.

In the initial stages of monoclonal antibody production, hybridomas are cloned in 96-well microtiter plates and, after several days in culture, the growth medium from each well is tested for the presence of antibody of the correct specificity. During this screening procedure, one also indirectly selects for cells of some minimal level of productivity, determined by the sensitivity limits of the assay used to detect the antibodies. One must not underestimate the importance of selecting a screening assay that is both sensitive and accurate. It is not unusual to develop screening assays capable of detecting antibody levels as low as $0.1\,\mu g/ml$. The primary requirement for the assay of dozens to hundreds of potentially active cell clones is a sufficient quantity of the antigen to which the monoclonal antibodies are directed. In most cases, the enzyme-linked immunosorbent assay is the technique of choice, because it can be sensitive to antigen concentrations in the nanogram to microgram range. It can also be conveniently established in the same 96-well microtiter plate format as the cloned cell populations.

Once an actively producing hybridoma culture has been selected and definitively shown to be monoclonal, the clone must be expanded into increasingly larger volumes in order to obtain sufficient numbers of cells for

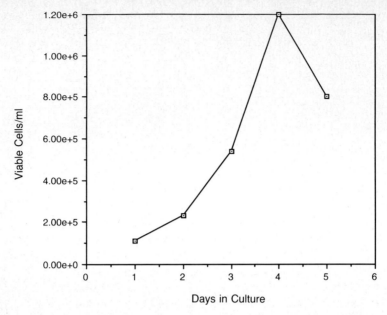

Fig. 1. Typical hybridoma growth curve. Hybridoma cells were seeded in a T-25 flask at a density of 1×10^5 cells/ml. At daily interval, the viable cells were quantitated by trypan blue exclusion and direct counting on a hemocytometer. Antibody was quantitated by enzyme-linked immunosorbent assay (ELISA)

storage in the freezer. Prompt storage minimizes the risk of loss of clonal populations due to contamination, equipment failure, or other adverse laboratory events. Cells are routinely frozen at a concentration of $5-10 \times 10^6$ cells/ml culture medium containing 10% glycerol. To obtain adequate numbers of cells for frozen storage, the clone is first removed from a well of the 96-well plate (a total volume of approximately 0.1–0.2 ml) and transferred to one well of a 12-well plate containing 1 ml medium. When the culture reaches a cell concentration of $4-5 \times 10^5$/ml or greater (cells can be quantitated by direct count or, with experience, appropriate concentrations can be approximated by visualization), the cells are transferred to a T-25 flask containing 5–7 ml medium. The process of expanding into ever increasing volumes can continue through sequentially larger T flasks. Individual hybridoma clones have population doubling times ranging from 12–36 h; murine hybridomas generally divide faster than human hybridomas. It usually takes 3–6 weeks to scale cells from a 96-well microtiter plate to the larger T flasks.

The cells can be maintained and perpetuated indefinitely in the larger T flasks, provided that care is taken to subculture before the cells exceed their maximum density and begin to die. As a rule of thumb, it is important to maintain the cell population at no less than 5×10^4 cells/ml culture medium.

Below this density, the population will not produce sufficient cellular factors to enrich the growth medium, a process essential for establishing an environment conducive to cell replication. Conversely, it is equally important to prevent the cells from exceeding a certain population density. Hybridomas display a very typical growth curve, illustated in Fig. 1. Subculturing must take place at or before the point when the dividing cells achieve maximum density. It is difficult to subculture cells effectively from the declining phase of the curve.

Many basal media adequately support the growth of hybridoma cells, provided that they are supplemented with fetal bovine serum (FBS). The usual media employed are Roswell Park Memorial Institute (RPMI) 1640, Dulbecco's modified Eagle's medium (DMEM), Iscoves, or a DMEM–HAM (developed by Dr. Richard Ham) F12 mixture (1:1), supplemented with 5%–10% FBS. It is, however, often advantageous to culture cells in a serum-free (or low-protein) medium. In general, culture of cells in serum-free medium is more successful if preceded by a gradual adaptation period. While the necessary adaptation period varies from cell line to cell line, a general method involves growing the cells for 1 week in medium supplemented with 5% FBS, followed by sequential week-long intervals in medium supplemented with 2%, 1%, 0.1%, and finally 0% FBS.

There are a number of serum-free media available commercially. The serum-free media I have found most successful, however, are based on a noncommercial basal medium developed in my laboratory and designated REM[1], the components of which are listed in Table 1. REM is a highly enriched medium which, with the addition of very few supplements, can suffice as a defined medium for many different cell lines. Table 2 lists two serum-free or low-protein media I have used for the growth of mouse, rat, and human hybridomas. These media support the growth and expression of differentiated function in other cell lines as well, including CHO, C127, and HL-60 cells. The single major difference between the two media is that REM. 1 contains bovine serum albumin, and REM. 2, human serum albumin. Supplementation with human serum albumin is preferred for production of antibodies intended for therapeutic use, since the albumin is the major contaminant of antibodies purified from cultured hybridomas.

If the established goal is to produce monoclonal antibodies in small quantities, the adaptation of the cells to serum-free medium is of minimal importance. On the other hand, if large quantities of antibody are required, as are necessary for therapeutic use, it is very important, if not essential, to adapt the cells to a defined, low-protein medium. This field is discussed by GRIFFITHS (1985) and SPIER (1988). Such adaptation has at least three major advantages. Most prominently, it significantly simplifies the process of down-

---

[1] Author's note: REM is often thought to stand for "Rupp's Enriched Medium". However, I actually named it because I developed it while working all night and in "REM" sleep.

Table 1. Components of REM basal medium

REM

Ingredient	Amount (g/l)
Group I	
NaCl	6.799646030
KCl	0.400000000
$MgSO_4$	0.097700000
$NaH_2PO_4H_2O$	0.140000000
WB/SFM Trace elements stock	0.000393300
$FeSO_4 \cdot 7H_2O$	0.000834000
$ZnSO_4 \cdot 7H_2O$	0.001400000
Group II	
L-Alanine	0.09000000
L-Arginine·NCl	0.06000000
L-Asparagine·H_2O	0.02000000
L-Aspartic Acid	0.03000000
L-Cysteine·HCl·H_2O	0.01500000
L-Cystine·2HCl	0.02000000
L-Glutamic acid	0.05000000
L-Glutamine	0.14600000
Glycine	0.05000000
L-Histidine·HCl·H_2O	0.02000000
L-Isoleucine	0.10000000
L-Leucine	0.10000000
L-Lysine·HCl	0.10000000
L-Methionine	0.03000000
L-Phenylalanine	0.05000000
L-Proline	0.03000000
L-Serine	0.04000000
L-Threonine	0.09500000
L-Tryptophan	0.01600000
L-Tyrosine $Na_2 \cdot 2H_2O$	0.10000000
L-Valine	0.09000000
Calcium chloride – $CaCl_2 \cdot 2H_2O$	0.14700000
Group III	
Dextrose	3.00000000
Ascorbic acid	0.00200000
Biotin	0.00050000
Calciferol (ergocalciferol)	0.00100000
D-Calcium pantothenate	0.00400000
Choline chloride	0.01000000
Folic acid	0.00400000
Glutathione (reduced)	0.00005000
l-Inositol	0.01000000
Lipoic acid	0.00020000
Menadione, sodium bisulfite	0.00000920
Methyl linoleate	0.00009000
Nicotinamide	0.00400000
Putrescine·2HCl	0.00020000
Pyridoxal·HCl	0.00400000
Riboflavin	0.00040000

Table 1. *Continued*

REM

Ingredient	Amount (g/l)
Thiamine·HCl	0.00400000
Sodium pyruvate	0.11000000
Vitamin E phosphate·Na$_2$	0.00001000
Vitamin A acetate	0.00010000
Vitamin B$_{12}$	0.00100000
Group IV	
AlC$_{13}$:6H$_2$O	0.00000121
AgNO$_3$	0.00000017
Ba(C$_2$H$_3$O$_2$)$_2$	0.00000255
KBr	0.00000012
CdC$_{12}$:2.5H$_2$O	0.00000228
CoC$_{12}$:6H$_2$O	0.00000238
CrC$_{12}$(anh)	0.00000016
NaF	0.00000420
GeO$_2$	0.00000053
KI	0.00000017
MnSO$_4$:5H$_2$O	0.00000024
(NH$_4$)6Mo$_7$O$_{24}$:4H$_2$O	0.00001236
NiSO$_4$:6H$_2$O	0.00000024
RbCl	0.00000121
Na$_2$SeO$_3$	0.00000129
Na$_2$SiO$_3$:9H$_2$O	0.00000028
SnC$_{12}$:2H$_2$O	0.00000226
TiCl$_4$	0.00000095
Ammonium m-vanadate	0.00000124
ZrOCl$_2$:8H$_2$O	0.00000322
Group V	
NaHCO$_3$	2.2
Group VI	
2N HCl	2 ml

Table 2. Serum-free formulations

	REM (g/l)	Sodium bicarbonate (g/l)	Insulin (g/l)	Holo-transferrin (g/l)	BSA (g/l)	HSA	L-Glutamine (mM)	Ethanolanine (g/l)	2N HCl ml/l
REM. 1	11.96	2.2	0.002	0.01	2	–	6	0.06	2
REM. 2	11.96	2.2	0.002	0.01	–	1	6	0.06	2

BSA, bovine serum albumin, fraction V; HSA, human serum albumin.

stream purification. In the absence of serum, monoclonal antibody purities in excess of 95% are easily achieved with two to three chromatographic steps, as opposed to a probable four to five chromatographic steps in the presence of serum (or high protein). Secondly, the elimination of serum confers a regulatory advantage: the presence of adventitious agents in serum may compromise the quality of antibodies intended for therapeutic use. Thirdly, cells adapted to serum-free media often show an increased cellular productivity. The desirability of this is obvious and, since this is not a well-known phenomenon, a detailed discussion of the data follows.

After culturing substantial numbers of hybridoma cell lines, results in our laboratory indicated that in most cases cells cultured in the presence of FBS produced less antibody than when cultured in our serum-free media. Similar results have been reported elsewhere (GLASSY et al. 1987). In an attempt to correlate phenotypic alterations to this change in productivity, we monitored surface antibody levels by staining the cells with a viable fluoresceinated anti-immunoglobulin G (IgG) and quantitated cell populations with a fluorescent-activated cell sorter (FACS). The application of this technique to optimization of cell cultures (i.e., selection of high producers) has been discussed recently by RUBEAI and EMERY (1993) and RUPP et al. (1989). Surface Ig production by various hybridomas was determined using two-color fluorescence FACS analysis. Cell lines were stained both with species-specific fluorescein isothiocyanate (FITC)-labeled anti-heavy chain and phycoerythrin (PE)-labeled anti-light chain reagents. A hybridoma line producing mouse IgG1, for example, was stained with goat anti-mouse IgG1/FITC and goat anti-mouse KAPPE/PE. Such two-color staining allows greater resolution of populations with differing levels of surface Ig than would be possible using a single subclass reagent.

The data can best be understood by studying the FACS results presented in Fig. 2. The viable cells with the highest intensity staining are designated the "A" group; those with little or no staining, the "B" group. The analyzer quantitates the number of cells in each group and determines the percentage of cells in each population. Figure 2A shows that in the presence of 10% FBS, only 25% of the cells displayed high levels of surface antibodies. Specific productivity of this subpopulation was found to be 18 μg antibody/ 10^6 cells per day. The cells were then switched to a serum-free formulation designated REM. 6. The components of REM. 6 are identical to those of REM. 1 (see Table 2), with the addition of a 0.1% (v/v) supplement of a commercial lipid preparation, Excyte I (Miles Laboratories). Three weeks after the transfer to serum-free medium (Fig. 2B), the percentage of cells containing surface antibodies increased to 65%; 8 weeks after transfer (Fig. 2C), the percentage was 98%. By 8 weeks, cellular productivity had increased to 28 μg antibody/10^6 cells per day. Cells were cultured in the serum-free medium for an additional 2 weeks (Fig. 2D), and the A population was then "sorted" or selected and 100% of the cells were maintained as stable high producers.

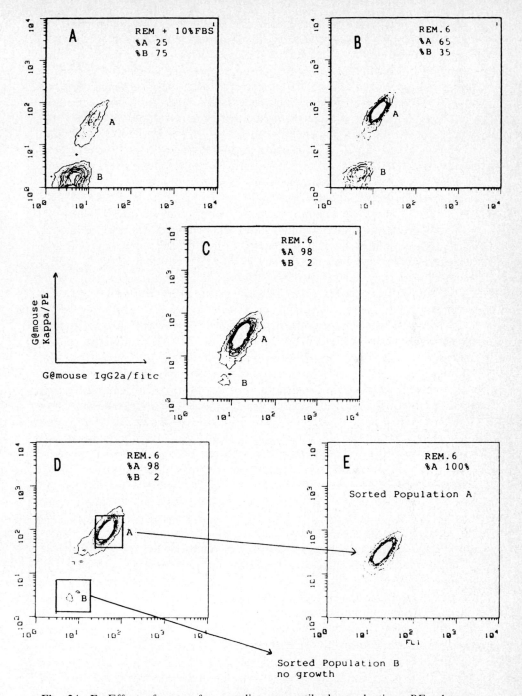

Fig. 2A–E. Effect of serum-free medium on antibody production. *PE*, phycoerythrin; *Ig*, immunoglobulin; *fitc*, fluorescein isothiocyanate

Not all serum-free media are as effective. Figure 3 summarizes an experiment in which cells were initially cultured in REM. 1 (designated Serum-Free I; Fig. 3A). The cells at this point constituted a homogeneous population producing approximately 20 μg antibody/ml. The cells were then switched to a commercially available serum-free medium (designated serum-free II). After 3 weeks in this medium, analysis revealed a bimodal population, with only 90% of the cells maintaining high productivity levels (data not shown). At 5 weeks, only 60% of the cells were in the high-productivity population (Fig. 3B). The cells were then returned to the serum-free I medium (REM. 1). After 4 weeks (Fig. 3C), high productivity cells comprised 90% of the cell population.

Methods used to estimated specific antibody production often vary from laboratory to laboratory. Some researchers simply report productivities in μg of antibody per milliliter, without adjustment for cell number or time in culture. This method, though perhaps useful for discussing data within a particular laboratory, becomes deceptive when comparing data from cell line to cell line in the general research community. To avoid such complications, we approximate specific cellular productivity by the method described in Table 3.

Cells are seeded into culture medium at a density sufficient to permit exponential growth. Samples are taken daily for determinations of cell viability and antibody concentration in the medium. Assuming that the antibody is relatively stable once secreted into the medium, the amount of antibody present in a given sample is thus proportional to the number of viable cells and their time in culture. By integrating the data, one can derive a specific productivity per cell in a given unit of time. Alternatively, one can simply sum the number of viable cells per milliliter on consecutive days to approximate the integral. This (daily) summation is referred to as the number of "accumulated" viable cells per milliliter (see column 3). The amount of antibody per milliliter on any given day is then divided by the corresponding daily number of accumulated cells per milliliter. (We standardized this figure at 10^6.) Sequential determinations of antibody/10^6 cells per day are then averaged to estimate the specific productivity. Therefore, even though the absolute amount of antibody increases daily, the specific productivity remains relatively constant. In this case (see Table 3), the cells are producing 25.1 μg antibody/10^6 cells per day. Variability in the daily estimate of specific productivity often results from not sampling the cells at the same time on consecutive days, thus resulting in varying day-to-day time intervals. This error can be eliminated by recording the exact time of sampling and calculating specific productivity as amount of antibody/10^6 cells per h.

A frequent misconception is that cells increase their specific antibody production level as they enter the stationary and subsequent death phase. This is not generally true. As can be seen from Table 3, even though the viable cell number decreases on day 5 and antibody concentration increased

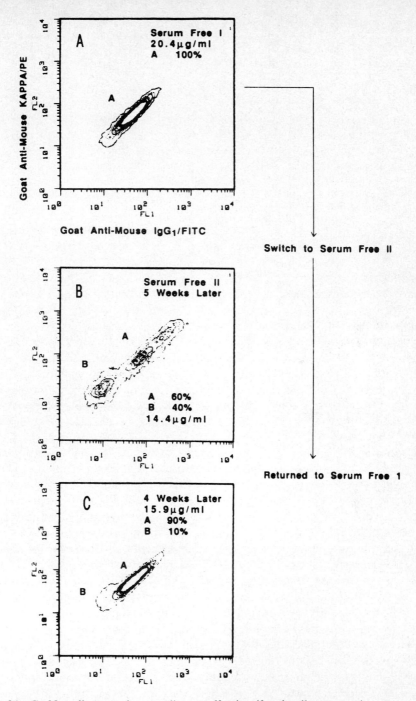

Fig. 3A–C. Not all serum-free media are effective (for details, see text)

Table 3. Working estimation of specific productivity

Days in culture	No. of viable cells/ml	Accumulation viable cells/ml	μg/Ab/ml	Specific productivity (μg Ab/ accumulated cells per day)
1	0.11×10^6	0.11×10^6	2.86	18/2
2	0.23×10^6	0.34×10^6	9.3	27.4
3	0.54×10^6	0.88×10^6	23.50	24.0
4	1.20×10^6	1.0×10^6	28.0	28.0
5	0.8×10^6	1.8×10^6	49.9	27.7
Average				25.1

Cells were cultured in T-25 flasks containing 7 ml medium. Each day 0.5 ml was received from the flask and viable cells quantitated by tripan blue exclusion on a hemocytomerter. Monoclonal antibody was estimated by enzyme-linked immunosorbent assay (ELISA).
Ao, antibody.

significantly from 28 to 49.9 μg/ml between days 4 and 5, the specific productivity remains relatively constant at 25.1 μg/10^6 cells per day. Figure 4A provides an alternative visualization of these data. When antibody and viable cell concentrations are plotted on the same graph, it can be seen that on day 5, antibody concentration increases while cell number decreases. Figure 4B, however, clearly shows that specific antibody production is relatively constant. This simple method of quantitating specific antibody productivity provides an accurate and readily comprehensible picture of the relative productivities of cell lines.

An accurate assessment of the specific productivity of cells is of particular importance for large-scale production. This information is essential for determining the efficiency of the chosen method of large-scale production and for estimating the number of cells required to produce a certain quantity of antibody in a given period of time. An average murine hybridoma, for example, will produce approximately 20 μg of antibody/10^6 cells per day. To obtain milligram quantities of antibody, murine hybridomas cultured in roller bottles, each containing 200 ml medium, can produce roughly 1–2 mg crude antibody per bottle in a 3-day culture period. (The cells can also be cultured in large T flasks, but, since the volume of medium is less in these smaller containers, production in flasks is less efficient than that in roller bottles.) Assuming a 30% yield of antibody after purification, this constitutes a net yield of only 300–600 μg of pure antibody per bottle per 3 days. Specific productivities of human hybridomas are often only 1%–10% those of murine hybridomas, with concomitantly lower overall yields. This may be sufficient for applications requiring small quantities of monoclonal antibodies.

When gram quantities of antibody are required, I recommend initially growing the cells in small (100- to 250-ml) spinner flasks. Initial propagation of the cells on this scale is important to establish growth rate and specific antibody productivity in a dynamic (stirred) environment. Often yields

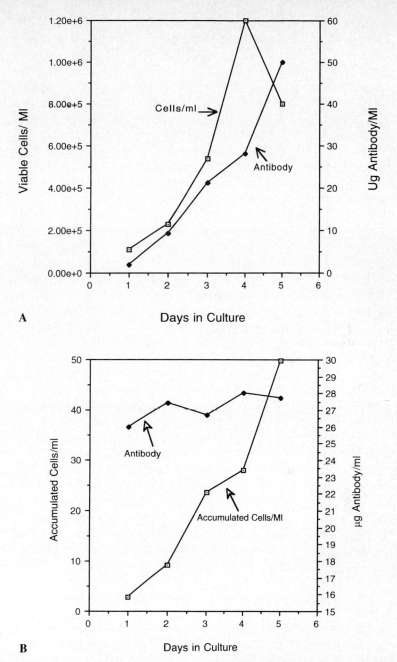

Fig. 4A,B. Antibody production. Cells and antibody were quantitated as in Fig. 1. **A** Antibody production during hybridoma growth. **B** Specific antibody production (see text for details of estimation)

change when cells are grown in stirred flasks. It is also possible to culture the cell population in fluidized bed or perfusion bioreactors, which minimize the actual movement of the cells. However, after working with many kinds of reactors, I find simple stirred flasks preferable for optimization of conditions for growth and expression of differentiated function. The simplicity and success rate possible with such vessels is highly advantageous, particularly if the goal is to obtain antibody as quickly as possible. After adapting cells to growth in small spinner (stirred) flasks, the cells can be transferred to increasingly larger vessels, beginning with 3-l spinner flasks.

The optimal volume of medium in any of these flasks is approximately one half of the total working volume; one cannot, in other words, fill the flask completely with growth medium. For optimum cell growth and productivity, one must maintain a surface-to-volume ratio of approximately 1:1. (To calculate this, determine the surface area of the medium in the vessel and divide by medium depth.) Cells grown in spinner flasks only obtain necessary oxygen through the exposed surface of the medium. If the volume of medium in the container is too great, oxygen cannot diffuse into the medium rapidly enough to support the growth of cells. Furthermore, the carbon dioxide generated by the metabolizing cells cannot escape rapidly enough, thus acidifying the medium, which may have adverse effects on cell growth and production patterns. Therefore, in a 3-l spinner flask, for example, one should usually culture cells in a medium volume of 1.5–2 l.

One 3-l spinner flask is equivalent in volume and antibody production capacity to ten roller bottles. One laboratory worker can easily maintain five to ten of these flasks at a time. Our recommended culture procedure involves maintaining the cells at a density less than maximal. If maximum population density is 10^6 cells/ml, for example, we allow the cells to reach a concentration of approximately 8×10^5 cells/ml. We then harvest 50%–75% of the medium volume, replacing it with fresh medium equal in volume to that removed, and continue culture until cell density again reaches 8×10^5/ml. This cycle of harvest and medium replenishment continues until enough medium has been collected to yield the required amount of antibody. From a hybridoma with a specific productivity of $20\,\mu$g of antibody/10^6 cells per day, five to ten 3-l spinner flasks can yield 100 mg of crude antibody per day. In 2–3 weeks, one can thus produce enough crude antibody to yield 1 g purified product.

When multiple gram to kilogram quantities of antibody are desired, I recommend the same stirred-flask technology, increasing in scale to accommodate the increased requirements. The principles of the culture are essentially the same, though in such cases the vessels are much larger, including stirred-tank bioreactors with capacities up to 10000 l, and the equipment used to handle fluids is different. The technology for growing cells in large stirred-tank reactors has proven adequate for most antibody production. While there are more expensive and much more complex cell culture systems available, the importance of using a type of bioreactor that is simple, efficient, and dependable cannot be overemphasized (OKA and RUPP 1990). A stirred-tank reactor fits this definition.

In summary, there are a few basic approaches to producing monoclonal antibodies for research or therapeutic use at either small or large scale. These are:

1. Ensure that the selected clone is indeed monoclonal.
2. Expand the clone in T flasks until sufficient quantities are available for storage in the freezer before any other work is performed. Do not risk weeks of effort to obtain the clone only to lose it to contamination or some other mishap.
3. Culture the cells in an enriched basal medium such as REM. This medium both supports the growth of cells in the presence of serum and serves as the basis for a serum-free medium. REM is highly enriched with essential vitamins, cofactors, fatty acids, and trace elements. It can be supplemented with insulin, transferrin, albumin, ethanolamine, and additional lipids to yield a functional serum-free medium. Both basal medium and serum-free formulations presented here are routinely made at pH 6.7– 7.0, since we have found that cells tend to grow and produce more effectively at low pH.
4. The calculation of specific cellular productivity is the only reliable way to determine the productivity of cell populations. It is essential that the data be presented as mass of antibody per cell per unit of time. Without this calculation, it is impossible to adequately or accurately compare effects of media, culture conditions, or other parameters from one cell line or one laboratory to the next.
5. Culturing cells in large T flasks or roller bottles is the most rapid and efficient means of producing milligram quantities of antibody from the average hybridoma, which produces approximately $10-20\,\mu g$ antibody/ 10^6 cells per day.
6. Culturing cells in spinner flasks of up to 3-l capacity at half volume in a CO_2 incubator is the most efficient means of producing antibody in gram quantities.
7. To produce multiple gram or kilogram quantities of antibody, cells should be cultured in large (50- to 10 000-l) stirred-tank bioreactors. These reactors, in simplest form, reflect the same technology used when growing cells in spinner flasks. Do not be intimidated by their size.
8. Avoid using very sophisticated bioreactors, such as hollow-fiber or perfusion systems, or any other reactors that require equally sophisticated operators and leave little room for error. In some of these bioreactors, control paradigms are so exacting that an entire culture can be obliterated if control of a single physiological parameter is lost for 5 min. If your goal is to use alternative bioreactors to study the biochemical engineering principles of culture or to investigate the effects of very high density cell culture, these reactors are effective. A traditional stirred-tank bioreactor,

however, is the system of choice for producing antibodies with least effort and cost. The primary reason for this is that investigators have been unable to develop a reactor that permits high density cell culture without using a volume of culture medium proportional to the cell number. Thus, one may obtain a very high cell density in a small bioreactor but, to maintain the population, the cells must still be perfused with a very large volume of medium. It is much more efficient to grow the cells directly in the necessarily large volume of medium in a stirred-tank bioreactor. For a more complete discussion of this, see Rupp and Oka (1990).

9. The most likely way to increase specific cellular productivity is by selecting and amplifying active clones of cells, either through genetic or physiochemical methods, rather than by developing alternative bioreactors. By selecting the appropriate serum-free medium, for example, cellular productivity can be increased by 50%–70% relative to that seen when cells are cultured in the presence of serum or in an inappropriate serum-free medium. A general discussion of methods to increase cellular productivity are discussed by Oka and Rupp (1993).

References

Feder J, Tolbert W (eds) (1985) Large-scale mammalian cell culture. Academic, London

Glassy M, Peters R, Mikhalev A (1987) Growth of human–human hybridomas in serum-free media enhances antibody secretion. In Vitro Cell Dev Biol 23: 745–781

Griffiths J (1985) Serum and growth factors in cell culture media: an introductory review. Dev Biol Standard 66:155–160

Ho C, Wang DC (eds) (1991) Animal cell bioreactors. Butterworth-Heinemann, London

Lubiniecki A (ed) (1990) Large-scale mammalian cell culture technology. Marcel Dekker, New York

Lydersen B (ed) (1987) Large scale cell culture technology. Hanser

Oka M, Rupp R (eds) (1993) Cell biology and biotechnology. Novel approaches to increased cellular productivity. Springer, Berlin Heidelberg New York

Ono D (ed) (1991) The business of biotechnolgy. Butterworth

Rubeai M, Emery A (1993) Flow cytometry in animal cell culture. Biotechnology 11:572–579

Rupp R, Tate E, Petersen L (1989) In: Spier K, Griffths J (eds) Advances in animal cell biology and technology for bioprocesses. Butterworth, London, p 129

Rupp R, Oka M (1990) Large scale annual cell culture: a biological perspecture. In: Lubiniechi A (ed) Large-scale mammalian cell culture technology. Marcel Dekker, New York

Spier R (1988) Environmental factors: medium and growth factors. Anim Cell Biotechnol 3:29–53

Section VI: Medical Applications

CHAPTER 14

Prospects for Cancer Imaging and Therapy with Radioimmunoconjugates

D.M. Goldenberg, R.D. Blumenthal, and R.M. Sharkey

A. Introduction

Immunological methods for cancer therapy have been sought for over
a century (DeLand and Goldenberg 1986), especially in an attempt to
achieve a more targeted, tumor-selective approach. This immunotherapy
has involved "active" immunity when host defenses are stimulated, such as
by immunization with nonviable tumor cells or extracts, or has been "passive," as when immune cells or antibodies are transferred from donor to
patient. Both active and passive immunotherapy strategies have received
rekindled attention in recent years, particularly with the use of anticancer
antibodies.

In the past, conventional antibodies evoked in various laboratory animals
against human cancer cells were laborious to prepare, difficult to reproduce,
and required extensive absorption to yield a high specificity and purity for
the desired antigen. Hence, these technical difficulties impeded progress
in the development of anticancer antibodies. The advent of monoclonal
antibody (MAb) production by in vitro hybridization methods in 1975
(Kohler and Milstein 1975) resolved many of these technical problems and
revitalized efforts to develop antibodies for numerous immunological applications in cancer and other diseases, such as in isotopic and nonisotopic
immunoassays, immunohistopathology, flow cytology, marrow purging,
external imaging of diseased tissues, and immunotherapy with unconjugated
antibodies. In addition, antibodies have also been conjugated with drugs,
toxins, or isotopes (Goldenberg 1993). The chapter summarizes the current
status and future prospects for the in vivo use of radiolabeled MAbs antibodies (radioimmunoconjugates) for cancer detection and therapy.

B. Nature and Pharmacology of Radioimmunoconjugates

Since the radioimmunoconjugate consists of an antibody and an isotope, and
sometimes also a linker for both, the in vivo behaviour is influenced by the
nature of each and the fidelity of the conjugate. Iodine isotopes have been
conjugated to antibodies by direct oxidation at such sites as tyrosine by
several methods, such as iodogenation, lactoperoxidase or chloramine-T
(Weadock et al. 1990). Radiometals, such as technetium 99m, indium-111,

and yttrium-90, have been conjugated to antibodies by a chelate linker (KREJCAREK and TUCKER 1976; HNATOWICH et al. 1985), or, in the case of Tc and rhenium isotopes, also by direct attachment to SH groups (HANSEN et al. 1990; GRIFFITHS et al. 1991). Different labeling methods have resulted in different degrees of conjugation stability, particularly after circulating in the body, thus fostering attempts at improving the binding without affecting the antibody's immunoreactivity. When an antibody is introduced into the body systemically, it behaves like most other proteins and faces a number of barriers and impediments that are also related to its immunological functions, thus affecting the amount of antibody delivered to a target organ or lesion. The injected antibody passes through a number of compartments, including vascular and extravascular spaces of diverse organs and tissues, as it is metabolized and excreted. Different radioconjugates and linkers behave somewhat differently, especially when different antibody forms are involved. Whole (intact) IgG molecules have a longer biological half-life than antibody fragments, and human or humanized IgGs have a longer biological half-life than rodent IgGs in humans (KHAZAELI et al. 1991). Whole IgG is metabolized principally by the liver, whereas antibody fragments show a higher metabolism by the kidneys. Iodine radioimmunconjugates are metabolized differently than radiometal conjugates, and different radiometal conjugates also behave differently from one another. For example, ^{111}In immunoconjugates show a stronger accretion to hepatocytes and certain reticuloendothelial system (RES) tissues than Tc-99m immunoconjugates, whereas the latter show a higher retention in renal tissue. The net effect of all of the barriers and metabolic processing is that in humans only a very small portion of the injected antibody, in the range of 0.0006%–1.04%, per gram of tumor usually is taken up (EPENETOS et al. 1986; SIEGEL et al. 1990). Binding of the antibody to circulating antigen or to the antigen present on normal tissue can divert the immunoconjugate from the target. A host immune response to the murine foreign antibody (human anti-mouse antibody, HAMA) can complex to the immunconjugate and hinder targeting while fostering release of the radiolabel. Poor vascularization and vascular penetration at the tumor site can also reduce the amount of antibody available for binding to tumor. In the tumor, antigen modulation, inaccessibility, and low density can affect the amount of antibody that is bound. Results from animal radioimmunotherapy experiments have shown that the amount of tumor reduction is proportional to the antigen density at the cell surface (CAPONE et al. 1984). This would suggest that use of multiple antibodies directed against different tumor-associated antigens should be injected in combination to increase targeting of the radionuclide and, hence, enhance the therapeutic results. Indeed, experimental and preliminary clinical results support this prospect (BLUMENTHAL et al. 1991). More recent theoretical and experimental considerations have suggested that efforts to increase MAb uptake in tumor should involve the use of antibodies against antigens that are highly expressed on tumor cells rather than the selection of high-affinity antibodies (SUNG et al. 1992). The organ and tissue location of the antigen-containing tumor can similarly affect antibody targeting and accretion, depending also upon the route of antibody administration. Whereas the i.v. route has been the most prevalent one, intralesional, regional, and intracavitary injections have also been undertaken for improved targeting (WAHL et al. 1988). It should be appreciated that antigens in easily accessible sites interact with circulating antibody first. These sites must thus be saturated before the antibody can pass to other compartments and, ultimately, penetrate tumor and bind to antigen in less accessible areas, as emphasized recently by KENNEL (1992). High concentration of the target tumor antigen on the tumor's cell surface or extracellular space may also impede penetration of the MAb to internal tumor sites, which may favor the use of lower affinity antibodies that may not be as highly complexed on the tumor surface as lower affinity antibodies (FUJIMORI et al. 1991).

Whether or not circulating antigen plays a significant role in antibody targeting is still not clearly understood. Although experimental animal models have suggested altered distribution properties, and in some instances a reduction in tumor uptake, clinical evidence has been less convincing. An analysis of 56 patients given ^{125}I-labeled anti-carcinoembryonic antigen (CEA) MAb followed by surgery to quantitate activity in tumors and tissues revealed no significant correlation between serum CEA and percent tumor uptake (BOXER et al. 1992). However, the same relationship may not exist for other tumor markers. For example, studies with B72.3 MAb have suggested that serum TAG-72 elevation correlates with best tumor targeting in patients (DOERR et al. 1992). Using quantitative imaging methods, we did not find any correlation between tumor uptake of a ^{131}I-labeled anti-CEA MAb and serum CEA (SHARKEY et al. 1993). In marked contrast, complexes formed between HAMA and the radiolabeled antibody are rapidly removed from the blood and deposited in the liver and spleen. Although tumors may be visualized in patients with HAMAs, there is evidence that this type of complexation will adversely affect uptake of MAb in the tumor (SAKAHARA et al. 1989). It is also important to preclude the binding of the injected MAb to circulating blood cells, because this can result in hematopoietic and systemic toxicity. In an early clinical study involving the injection of a CEA MAb which reacted with the CEA cross-reactive antigen present on human granulocytes, a marked decrease in circulating granulocytes and severe side effects were experienced (DILLMAN et al. 1984). The presence of antigen in normal tissue can also divert the antibody from tumor sites. However, if the antigen is located in a sequestered site, where access to the antibody in the circulation is limited, the presence of antigen in normal tissues should not preclude the MAb's use in tumor targeting so long as the normal tissue's antigen is not accessible to circulating antibody. Indeed, even with the presence of antigen in normal tissues, targeting to tumor may be enhanced by increasing the amount of MAb protein given, so as to saturate nontumor antigen sites. Poor vascularization and vascular penetration at the tumor site have also been cited as important parameters that affect tumor targeting. These can it part be overcome by using antibody fragments, which penetrate more quickly and distribute more uniformly in tumors than whole IgG (SUTHERLAND et al. 1987). Antibodies coupled with vasoactive compounds, such as interleukin-2 (IL-2), may open vascular barriers and improve tumor targeting (LEBERTHON et al. 1991). Likewise, inducing hyperemia in the tumor site can affect targeting and accretion (STICKNEY et al. 1987). Thus, many factors influence antibody targeting and accretion, and these factors appear to be amenable to control.

C. Radioimmunoconjugates in Detection vs Therapy

The nature of the antibody, the radionuclide attached thereto, and the scheme of administration may vary for detection as compared to therapy. Detection of a target lesion by external imaging or photoscanning requires a high count ratio, detectable by the imaging camera's crystals, in relation to that present in adjacent tissues. When scanning in a single plane is performed (planar imaging), a higher count ratio of target to background is required than when numerous sections through the body, in different planes, are made (emission tomography, also known as single-photon emission computer tomography, or SPECT). The choice of imaging radionuclide and antibody form should be compatible, particularly the physical half-life of the isotope and the biological half-life of the targeting antibody vehicle. In imaging, for example, the use of a short-lived isotope, such as Tc 99m (6 h half-life), requires a rapid targeting agent, for which reason smaller fragments, such as Fab and Fab', are suitable (GOLDENBERG et al. 1990). Isotopes of a longer half-life, such as [111]In (2.5 days), are compatible with whole IgG, especially since later imaging (3 days and beyond) appears to be desirable for ^{111}In-based agents. In therapy, an isotope of high energy and compatible physical half-life and which delivers a path-length that is suitable for the tumor size, location and type should be chosen. These should have a minor imaging (γ) energy to permit monitoring of targeting and dosimetric calculations, but pure β and α emitters, including those emitting Auger electrons, are also of promise as therapeutic nuclides (GOLDENBERG 1989). In the therapeutic mode, radioimmunoconjugates should deliver a high dose to the tumor for a long duration, thus delivering a high rad dose to tumor as compared to normal tissues. Unfortunately, these goals have been achieved only to a limited extent. At present, it appears that radioimmunotherapy will be used best against radiosensitive neoplasms and in an adjuvant setting, perhaps in combination with other therapeutic modalities and bone marrow supportive measures. As the application becomes complex, so too do the studies needed to prove efficacy of this new modality.

D. Nature and Problems of Radioimmunodetection

Although initial studies in the 1950s and later indicated the potentials of cancer imaging with radiolabeled antibodies, the anticancer antibodies and labeled products were poorly characterized, and the problems of image interpretation with this new class of agents were not adequately appreciated. In the early 1970s, we began work in human tumor xenograft models to determine the conditions and potentials of tumor targeting and imaging with radioiodinated goat antibodies to a tumor-associated antigen, CEA (GOLD and FREEDMAN 1965). The studies in the colonic carcinoma xenografts showed that: (a) the radioiodinated labels were stable, (b) there was high and specific accretion in the CEA tumors, (c) the tumors could be imaged by external photoscanning, and (d) nonspecific accretion could be demonstrated in large tumors, either with the specific or an irrelevant antibody (GOLDENBERG et al. 1974). Later studies showed that affinity purification of the polyclonal antibody improved targeting and accretion of the anti-CEA antibody in this colonic carcinoma xenograft (PRIMUS et al. 1977). Clinical studies with ^{131}I-labeled polyclonal anti-CEA antibodies ensued, and failure to image cancer was reported (REIF et al. 1974; MACH et al. 1978). In contrast, a polyclonal antibody against an undefined melanoma antigen showed evidence of tumor radiolocalization clinically (BELITSKY et al. 1978). However, whether this was due to a nonspecific accretion in highly vascular lesions or true tumor targeting was not clear. At the same time, we progressed from our encouraging animal studies with radiolabeled anti-CEA antibodies to the clinic and examined a group of patients with CEA-containing and also CEA-devoid neoplasms. The initial results showed that: (a) an antibody protein dose of 0.20 mg sufficed, (b) the antibody's radioactivity provided high counts in blood-rich organs and tissues, particularly the heart and major vessels, for several days after injection of ^{131}I-labeled whole IgG preparations, (c) tumors containing CEA away from these blood-pool regions could disclose sufficiently increased radioactivity for interpreting the tumor images, (d) non-CEA tumors could not be disclosed, and (e) circulating elevated levels of CEA did not block targeting of the radioactive anticancer antibody, although evidence of in vivo complexation was found (GOLDENBERG et al. 1978). In order to compensate for blood and interstitial nonspecific radioactivity, a dual isotope subtraction method was implemented which improved the contrast of images near blood-rich areas (GOLDENBERG et al. 1986; DELAND et al. 1980). In one case in which the tumor was excised after the radioimmunodetection (RAID) study, a radio of radioactivity in tumor to adjacent intestine of 2.5 was determined (GOLDENBERG 1979). Subsequent clinical studies involved both more patients with diverse tumor types and the use of a control, normal goat IgG preparation. The specificity of antigen targeting was confirmed, while a sensitivity of tumor imaging of 70%–90% was computed (GOLDENBERG et al. 1980).

Table 1. Potential indications for antibody imaging

Presurgical staging of disease
Postsurgical evaluation of residual disease
Confirmation of viable tumor sites revealed by other methods
Disclosure of sites of tumor elaborating a tumor-associated marker
Determination of suitability of patient for antibody-mediated therapy

---

Since these initial clinical results, many other tumor targeting antibodies, labels, and imaging protocols have been studied, as reviewed elsewhere (GOLDENBERG and LARSON 1992). Suffice it to conclude that, at the present time, cancer RAID is gaining in acceptance as an adjunct method for the disclosure of sites of cancer in diagnosed patients who are being monitored, even revealing sites of cancer that are missed by other diagnostic modalities. The potential clinical uses of RAID are given in Table 1. The disclosure of occult tumors and the confirmation of sites revealed by other, more anatomic, radiological methods appear to be the indications that are best established at present. However, evidence in support of other applications, such as in initial diagnosis, staging, and assessment of therapeutic response, is increasing (GOLDENBERG and LARSON 1992). To date, several thousand patients have been studied with radiolabeled anticancer antibodies in RAID, and usually between 60% and 90% of known lesions have been disclosed correctly. Different antibodies and their respective forms, radionuclides, and scanning methods have yielded different results, but the basic observations are fairly consistent, as summarized in Table 2. Most studies of RAID have involved colorectal and ovarian cancers, but additional tumor types are under investigation. RAID results have already contributed to better (or potentially better) patient management in about 25%–50% of patients in whom this has been measured (GOLDENBERG and LARSON 1992).

The four principal radionuclides studied for RAID are ^{131}I, ^{111}In, ^{123}I, and Tc-99m. Of these, the most desirable is Tc-99m, since it has a high intensity photon energy and is easily and inexpensively available from on-site generators. Its only limitation is a short physical half-life of 6.02 h, which therefore requires a rapid targeting agent that permits imaging within 24 h. This is achieved best with smaller antibody molecules, such as antibody fragments (bivalent and monovalent). Indeed, our experience with Fab' reagents of diverse MAbs labeled with Tc-99m indicates that early and rapid imaging can be achieved (GOLDENBERG et al. 1990).

Dehalogenation of radioiodine and sequestration of ^{111}In in reticuloendothelial organs, particularly the liver, are limitations for these isotopes. Since the liver is a major organ of metastasis, ^{111}In uptake in normal liver makes differentiation of tumor uptake of radioactivity difficult to interpret over background liver radioactivity (HNATOWICH et al. 1985). Another important consideration is the induction of host antibodies against the foreign proteins. When intact murine immunoglobulin is used, HAMAs are induced

Table 2. Current status of antibody imaging

Subject	Findings
Safety	Radioactive antibodies have been found to be safe in over 10,000 patients studied worldwide.
Sensitivity	On a tumor site basis, results between 60% and over 90% sensitivity and specificity have been reported, with the highest accuracy rates found for MAbs labeled with ^{131}I, ^{123}I, or Tc 99m.
Small tumors	Tumor as small as 0.4–0.5 cm have been disclosed with Tc-99m MAbs, especially with emission tomography, but resolution usually is in the range of 1.0–2.0 cm.
Occult tumors	Tumors missed by other methods, including CT, have been revealed by antibody imaging.
Serum antigen	Antibody imaging can be positive even before the antigen titer in the blood is elevated. Complexation with circulating antigen does not compromise antibody imaging.
HAMA	Repeated injections of animal antibodies result in human antibodies against these foreign proteins that can compromise antibody targeting.

MAbs, monoclonal antibodies; CT, computerized tomography; HAMA, human anti-mouse antibodies.

in up to 50% of patients receiving a single injection (REYNOLDS et al. 1989). Low protein doses (1 mg or less) of Fab' fragments have evoked HAMAs in less than 1% of patients after a single injection (HANSEN et al. 1992), indicating that this form will be best suited for repeated administrations. Accordingly, the development of RAID has progressed from first-generation ^{131}I-labeled whole IgGs to second-generation ^{111}In whole IgG preparations and to the current (third-generation) Tc-99m-labeled monovalent (Fab') fragments that permit imaging lesions of less than 1 cm (by SPECT methods) within a few hours of injection (GOLDENBERG et al. 1990).

E. Nature and Problems of Radioimmunotherapy

Just as radioimmunoconjugates have been targeted successfully to cancer cells for imaging, so too can they be utilized in a cancer radioimmunotherapy (RAIT). Whereas tumor imaging requires a high target to nontarget ratio of counts for optimal results, therapy with radiolabeled antibodies depends on a high concentration of tissue radioactivity for a long duration. The antitumor effects of RAIT appear to exceed the expected doses as compared to external beam therapy, which may be related to the low dose rate achieved by MAb-targeted radionuclides (KNOX et al. 1990). There is also a contribution of both nonspecific targeting to tumor, as we observed almost 20 years ago (ORDER et al. 1990), and whole-body radiation to the antitumor effects of RAIT. The range of action of radionuclides is defined predominantly by the nature of the particle (α, β, Auger) and the energy of emission. The first radionuclide used in RAIT was ^{131}I, which has had a long history in the successful treatment of thyroid cancer. It is suitable for RAIT, having an averge β energy of 0.183 MeV and a physical half-life of 8 days, while emitting γ rays for imaging and quantitation. It is also easy to bind to antibodies without affecting their immunoreactivity, and the minimal toxicity to normal tissues (hypothyroidism) and the extensive clinical experience with this isotope favor its use. However, it has shown relatively low tumor dose rates of 5 cGy/h in clinical studies (BRADY et al. 1991), and rapid excretion after antibody degradation and/or dehalogenation has been observed. Also, with ^{131}I there is excessive radiation exposure to the patient and attending staff due to the high level of high-energy γ radiation from an 81% abundant emission at 364 keV. Finally, its average β energy penetration of less than 1 mm restricts its efficacy in larger tumors. Table 3 presents a list of the more popular radionuclides for RAIT, including some advantages and disadvantages. [125]I emits Auger electrons, whose short-range energy can affect the nucleus and, thus, destroy cells once the electrons are internalized. This isotope is being used, after conjugation to MAbs, against epidermal growth factor receptor in the therapy of intracranial neoplasms (BRADY et al. 1991). Yttrium-90 is a pure β emitter with an average maximum emission of 0.937 MeV and a physical half-life of 2.7 days.

In clinical experience, a much higher radiation dose to tumor is achieved with ^{90}Y conjugated to MAbs than with ^{131}I but with increased host toxicity, since most current labeling methods result in ^{90}Y detachment and targeting to bone, resulting in bone marrow toxicity (SHARKEY et al. 1990). However, improved chelates for conjugating ^{90}Y to MAbs are in development and evaluation (DESHPANDE et al. 1990). Preclinically, the magnitude of toxicity (myelosuppression) from ^{90}Y is similar to that of ^{131}I at their maximum tolerated doses, but the duration of toxicity is about 2 weeks shorter than with [131]I (BLUMENTHAL et al. 1993). Rhenium-186 and [188]Re have gained much interest recently because of the similarity in their chemical properties to Tc-99m. [186]Re has an average β energy of 0.781 MeV, a physical half-life of 3.8 days, and emits a γ energy of 137 keV for imaging and quantitation of uptake. [188]Re has a maximum β energy of 2.116 MeV and a physical half-life of 17 h, with a γ energy component of 155 keV for imaging and quantitation of uptake. Lutetium-177 is a lanthanide β emitter (497 and 133 keV) with γ emissions of 208 and 113 keV, thus allowing imaging and quantitation of uptake. It has a physical half-life of 6.7 days, a shorter path than ^{90}Y or [188]Re, and a longer half-life than these other two therapeutic radionuclides. In addition to β emitting isotopes, α emission from bismuth-212 and astatine-211, which irradiate at very short distances, has been used in laboratory studies and appears to hold promise for the therapy of circulating tumor cells or micrometastases. Thus, each of these isotopes has advantages and disadvantages for RAIT, and it is still an important research objective to determine which one is best for each treatment situation and clinical

Table 3. Selected isotopes for antibody therapy

Isotope	Half-life	Decay mode	Mean energy distance (mm)	Advantages	Disadvantages
^{131}I	8 days	β	0.7	Inexpensive; images	Deiodination; thyroid uptake
^{90}Y	2.5 days	β	6.0	Indium chemistry	Leaches off chelate: bone uptake; no images
^{186}Re	3.5 days	β	2.0	Tc chemistry; images	Availability
^{188}Re	17 h	β	5.5	Generator-produced; Tc chemistry: images	
^{67}Cu	2.5 days	β	0.5	Images	Availability
^{211}At	7 h	α; Electron capture	0.05	Iodine chemistry; short range	No images
^{125}I	60.4 days	Electron capture (Auger)	0.02	Short range electrons	Poor images

problem. Although many radionuclides have potential for RAIT, a major problem has been the achievement of tight and stable binding to the MAb at a sufficiently high specific activity without affecting the antibody's immunoreactivity and, in turn, targeting properties. A basic problem, however, has been the very low radiation dose delivered to tumor with each systemic injection, due to the very low accretion of antibody in tumor. One consideration to compensate for this problem has been to fractionate the radiation doses delivered by RAIT. Indeed, there has been an advantage found for this method compared to single dose administration, in terms of increased therapeutic effects and/or decreased bone marrow doses, in both experimental and clinical studies (MEREDITH et al. 1992; SCHLOM et al. 1990; BUCHEGGER et al. 1989; BLUMENTHAL et al. 1989a). In contrast to other immunoconjugates, however, RAIT has the advantage of the radiation being deposited at a distance by the antibody capable of destroying cells, thus not requiring antibody uptake by each cancer cell. Depending on the form of radiation and the isotope's particular energy, the distance can be large, such as with ceretain β emitters (e.g., ^{90}Y and ^{188}Re), or short, as in the case of α emitters. This helps address the problem of tumor cell heterogeneity in the expression of target tumor antigens, which is less likely to be solved with other forms of immunoconjugates. Indeed, autoradiographic studies have shown that MAb localizes in tumor heterogeneously and can even accumulate in necrotic areas more than in viable areas (BOXER et al. 1992). This heterogeneous targeting may not only be due to variable expression of antigen in the tumor cell population, but also to variable accessibility to antibody, as discussed already above. For example, CEA-positive tumor cells near the blood vessels in an experimental tumor model were targeted better than more distant tumor cells (BOXER et al. 1992). Various methods are under investigation to increase antibody accretion in tumors, such as affecting tumor blood flow and vascular permeability, increasing antigen expression by the tumor, and use of multistep pretargeting methods (MSIRIKALE et al. 1987; CHAN et al. 1984; GREINER et al. 1987; PAGANELLI et al. 1991b). The indirect multistep targeting and therapy strategies attempt to decrease circulating time of the radionuclide, thus decreasing nonspecific targeting to the bone marrow. One method is to inject an unlabeled bispecific antibody, in which one arm of the MAb is directed against the tumor antigen and the other against the radiolabeled hapten (GOODWIN 1990). After the bispecific MAb has localized in the tumor and cleared the circulation, a radiolabeled hapten is injected and should attach to the antihapten arm of the bispecific MAb in the tumor. Variations of this approach, involving a bivalent radiolabeled hapten, have also been pursued (LEDOUSSAL et al. 1989). Still other indirect approaches have been described involving the high avidity of avidin or streptavidin for biotin (HNATOWICH et al. 1987). In one scheme, the first injection is made with an unlabeled biotinylated anticancer MAb and then radiolabeled avidin or streptavidin is given once the first agent concentrates in the tumor selectively (PAGANELLI et al. 1991b). In a three-step strategy, the nontargeted, unlabeled, biotinylated antibody is cleared from the circulation by injection of avidin, which also binds to biotin in the tumor, followed by administration of the labeled biotin that again binds at the tumor (PAGANELLI et al. 1991a). These multistep approaches appear attractive but require evaluation in well controlled trials that compare the indirect method to the direct RAIT approach under similar conditions of timing. In order to reduce circulating, nontargeted, radiolabeled antibody, we and others have merely injected a second antibody against the primary antitumor antibody and have found enhanced tumor/blood ratios, enabling an increase in the radiation dose injected and targeted to tumor, although some reduction in tumor deposition of MAb can also be found (SHARKEY et al. 1992). A simpler approach has been direct injection of the MAb into the tumor or the tumor-containing cavity (WAHL et al. 1988). ^{131}I, ^{90}Y, and ^{186}Re conjugated to antibodies have been used in these direct approaches with variable but encouraging results (RIVA et al. 1989; ORDER et al. 1986; BREITZ et al. 1992). As expected, responses have been better in patients with less extensive disease than in those with bulky tumors, and this is generally the case with all RAIT agents.

The administration of high and repeated doses of radiolabeled MAbs has been limited by the HAMA response to murine antibodies and by myelotoxicity (SIEGEL et al. 1990; GOLDENBERG 1989). The former may be avoided by the development and use of totally human or partially human ("humanized") MAbs (MORRISON 1985; STEPLEWSKI et al. 1988; HARDMAN et al. 1989), while the latter may be overcome by autologous bone marrow grafting and/or the use of marrow-stimulating cytokines, such as IL-1, IL-3, IL-7, granulocyte colony-stimulating factor (G-CSF), and granulocyte/macrophage colony-stimulating factor (GM-CSF) (MORTON et al. 1990; BLUMENTHAL et al. 1992a), which have permitted a 25%–30% increase in the radioantibody dose that can be administered in an animal model (BLUMENTHAL et al. 1992a). Preliminary results suggest that this higher dose can improve therapeutic effects against the tumor (BLUMENTHAL et al., manuscript in preparation).

F. Current Clinical Status of Radioimmunotherapy

The major problems of low MAb doses delivered to tumors and increased myelotoxicity as radiation doses are escalated have limited the exploitation of cancer RAIT, as already mentioned. It is interesting, nevertheless, that the anticancer effects of radiolabeled antibodies seem to exceed their expected tumor doses, which is probably attributable to the low dose rates achieved by RAIT (KNOX et al. 1990). Much of the recent success of cancer RAIT has involved radiosensitive neoplasms, particularly hematopoietic tumors (ROSEN et al. 1987; PRESS et al. 1989; DENARDO et al. 1988; CZUCZ-

MAN et al. 1991; KAMINSKI et al. 1991; GOLDENBERG et al. 1991; SCHWARTZ et al. 1993; SCHEINBERG et al. 1991). The majority of studies involved ^{131}I labels, particularly because of their long history of safe use, cost, and ease of radioiodination. In contrast with many other radioisotopes, free radioiodine is metabolized or excreted by the thyroid, renal organs, and gastrointestinal tract with controllable side effects, whereas some of the radiometals may accrete in the liver, gastrointestinal tract, and bone. Regressions have been reported for T and B cell lymphomas with several different antibodies using a variety of doses and schedules for achieving remissions or for induction of dose-limiting myelotoxicity. Another area of interest is the treatment of myeloid leukemia by RAIT (SCHWARTZ et al. 1993). It has been estimated that antileukemic and marrow-ablative doses of ^{131}I can be delivered with an anti-CD33 MAb, in which more than 99% of leukemia cells (about 1 kg) can be eliminated without nonhematologic toxicity (SCHEINBERG et al. 1991). This method may also be of use for ablating the marrow before marrow transplantation (SCHEINBERG et al. 1991). These results support the encouraging view that RAIT is gaining in acceptance as a promising modality in the treatment of leukemia and lymphomas, and it now needs to be determined how it will relate or be combined with other modalities in the routine treatment of these hematopoietic neoplasms.

In contrast, systemic RAIT of solid tumors has been generally disappointing for many of the reasons discussed above. A possible exception has been the use of chimeric (human/mouse) L-6 labeled with ^{131}I in the therapy of advanced breast cancer. Initial results suggest therapeutic activity related to a concomitant antibody-mediated cellular immune and/or vascular function, including high antibody accretion in tumor (DENARDO et al. 1991). High tumor accretion of antibody has also been observed in neuroblastoma treated with MAb 3F8, with excellent targeting, relatively high tumor accretion of radioactivity, and evidence of therapeutic efficacy observed (LEUNG et al. 1986). Locoregional and direct tumor administration has shown more encouraging results. For example, STEWART et al. (1989) have studied the i.p. administration of ^{131}I-labeled MAbs for the treatment of ovarian cancer and have observed responses and initial evidence of improved survival. ^{125}I labeled to a MAb against epidermal growth factor receptor has also been used in the therapy of brain neoplasms by carotid artery infusions, with encouraging clinical responses (BRADY et al. 1991). However, most of these studies have involved only modest rad doses delivered to tumor, usually less than 2000 cGy. In contrast, very high radiation doses, approaching 20 times this value, have been achieved by intratumoral injection of anti-tenescin MAb labeled with ^{131}I for brain cancers, resulting in objective and durable tumor responses (RIVA et al. 1992).

In addition to the use of ^{90}Y for RAIT of solid tumors (ORDER et al. 1986), attention has turned recently to ^{177}Lu and the rhenium isotopes ^{186}Re and ^{188}Re as preferred β emitters. As in the case of ^{131}I and ^{90}Y RAIT, preclinical studies have shown efficacy in human solid tumor xenograft models with these other radionuclides (GRIFFITHS et al. 1991; SCHLOM et al. 1991; BEAUMIER et al. 1991), and early phases of clinical trials are in progress. Unfortunately, however, the induction of HAMAs has limited administration of repeated courses, which are necessary in order to deliver therapeutic radiation doses to solid tumors. Therefore, we will need to await clinical trials with different forms of human or humanized antibodies with these different radionuclide labels before a proper assessment of the status and potential of RAIT can be made. But before this is achieved, is there a more immediate role for lower loses of RAIT in solid cancer therapy, similar to the apparent results in hematopoietic tumors? Also, can RAIT be combined with other forms of anticancer therapy for an improved result? Initial results in animal models suggest that, at least as an adjuvant therapy, RAIT can be curative at low, well tolerated radiation doses.

G. Experimental Studies of Adjuvant Radioimmunotherapy

The poor prognosis in most cancers relates principally to the relatively advanced stage of disease at the time of diagnosis. In colorectal cancer, for example, more than 50% of patients have evidence of tumor dissemination at the time of diagnosis. This fact supports the view that adjuvant therapy of cancer after surgical intervention is an important approach to improving patient survival. Radiolabeled antibodies have shown selective targeting to tumors in preclinical and clinical experiments, including improved therapeutic responses being reported in several human tumor models (LEUNG et al. 1986; GOLDENBERG et al. 1981; BADGER et al. 1985; VESSELLA et al. 1985; SHARKEY et al. 1987; BLUMENTHAL et al. 1992b). Further, animal studies have shown better responses in smaller as opposed to larger human tumor xenografts (BLUMENTHAL et al. 1989a). Also, is has been determined that higher rad doses can be delivered to smaller tumors in patients (SIEGEL et al. 1990). Therefore, it appears to be a logical conclusion that radioimmunoconjugates are of considerable potential as an adjuvant therapy of cancer. Indeed, our initial studies in animal models support this perspective (BLUMENTHAL et al. 1992b) and will be summarized here.

The first task was to establish an appropriate, and hopefully predictive, model for early disseminated cancer and to determine if RAIT could be used to extend life expectancy. In the past, most human tumor xenograft models used to study a therapeutic modality involved a subcutaneous growth site in athymic nude mice. However, since almost all human solid cancers do not present clinically as subcutaneous tumors, and because the location of the tumor may play a role in antibody targeting (BLUMENTHAL et al. 1989b), a model with tumor seeded in a visceral site would appear to be more clinically relevant. Although the liver is the principal organ for metastasis of subdiaphragmatic tumor types, such as colorectal cancer, seeding of experi-

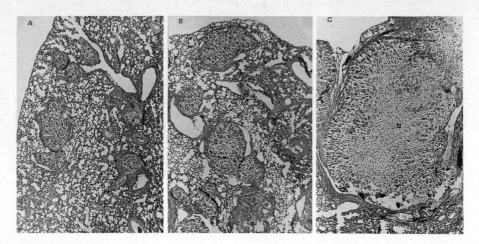

Fig. 1A–C. Microscopic tumor foci in the lungs of animals 7, 14, or 28 days (**A, B,** and **C,** respectively) after i.v. injection of GW-39 target cells. Hematoxyl in and eosin stained sections. *Arrows* indicate representative tumor nodules; *V*, viable tumor; *N*, necrotic tumor. (From Sharkey et al. 1991) ×32

mental tumors in this organ depends on either direct inoculation or indirect spread by intrasplenic or colonic injection (Giavazzi et al. 1986). These methods require surgical procedures and can present technical difficulties. Therefore, we pursued a simpler metastatic model that is representative of micrometastatic disease, namely, a human colonic carcinoma cell line. By means of intravenous injection into nude mice of a human colonic carcinoma cell line, GW-39 (Goldenberg et al. 1966), which produces CEA and other colonic tumor markers (Goldenberg and Hansen 1972), reliable and quantitative dissemination and metastasis to the lungs could be achieved (Sharkey et al. 1991). After i.v. injection of 0.03 ml of a 10% tumor cell suspension, multiple microscopic tumor colonies (10–60) that ranged in size from 50 to 200 μm in diameter resulted within 1 week. By 2 weeks, the colonies reached 0.5–1.0 mm in diameter, and as many as 100 colonies could be counted in some specimens. After 4 weeks, the colonies continued to enlarge, and many reached a diameter of 2 mm, including areas of central necrosis. Despite this central necrosis, tumor colonies continued to enlarge, with only a narrow rim of viable tumor cells seen on the colony's periphery (Sharkey et al. 1991), as shown in Fig. 1. The animals began to lose weight 3–6 weeks after tumor inoculation and died when their body weight loss exceeded 20%. The median survival time in several separate experiments ranged from 5 to 10 weeks. At death, the GW-39 human colonic tumor cells were not found in any organ other than the lungs. These results were basically similar for two other human colonic tumor cell lines, LS174T and GS-2, inoculated i.v. in nude mice, but with fewer colonies in the lungs (Sharkey et al., unpublished results). Using either CEA or mucin (CSAp)

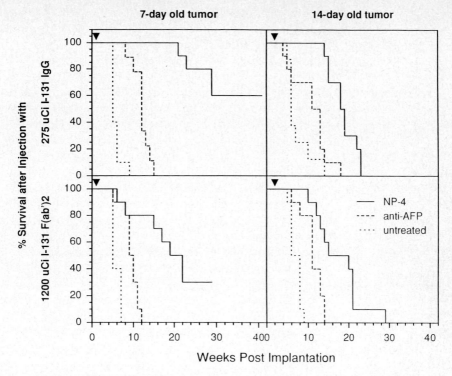

Fig. 2. Survival curves of nude mice bearing either 7- or 14-day-old GW-39 lung nodules and treated with the maximal tolerated dose of NP-4 anti-CEA or irrelevant antibody anti-AFP-7-31 IgG ($275\,\mu$Ci) or F(ab')$_2$ fragment ($1200\,\mu$Ci). Lung nodules were initiated by caudal vein injection of $30\,\mu$l of a 10% GW-39 suspension. All groups had 10 animals. *Filled triangle* indicates the time that the radioantibody was administered. (From BLUMENTHAL et al. 1992a)

antibodies labeled with ^{131}I, even as a single i.p. injection, nude mice bearing the lung metastases of the human colonic cancer model could be cured in a high percentage of cases, depending on when therapy was instituted. At early times after metastasis (7 days), higher cure rates were found than when later therapy times (14 days) were instituted (SHARKEY et al. 1991), as shown in Fig. 2. Interestingly, this RAIT protocol was more effective than a similar adjuvant therapy with 5-fluorouracil and leucovorin in this experimental model (BLUMENTHAL et al. 1992a). In conclusion, the greatest current prospect for RAIT of solid tumors appears to be for the treatment of micrometastatic cancer, as in an adjuvant setting. This view was recently supported by a mathematical analysis, especially when based on the use of extracorporeal immunoadsorption of nontargeted, circulating, radioactive MAbs, as proposed by SGOUROS (1992) and supported by ZANZONICO (1992).

H. Conclusions and Future Prospects

For over a decade, MAbs have been used in a variety of ways to detect and treat various cancer types. This has followed an even longer period in which less well characterized polyclonal antibodies (DELAND and GOLDENBERG 1986) were used. At present, a multitude of radiolabeled anticancer antibodies, involving different antibody preparations and radiolabels, have shown various degrees of clinical utility in detecting sites of cancer, even those missed by conventional radiological methods (GOLDENBERG and LARSON 1992). However, MAbs, including radioimmunoconjugates, have not yet gained a role in the therapy of cancer, although encouraging responses have been achieved in lymphomas with different radiolabeled antibody preparations. Thus, despite the absence of truly cancer-specific MAbs or target antigens, many MAbs have been shown to be relatively selective cancer agents, and it is now a subject of intensive investigation to improve the accretion of these agents into tumors in order to deliver higher therapeutic doses. Solutions to the problems of MAb immunogenicity and host myelotoxicity appear to be feasible, thus enhancing the prospects of cancer RAIT. Furthermore, RAID studies have shown that MAbs can be used safely and successfully for the improved management of cancer patients, which reinforces our enthusiasm for the development and use of similar agents for therapy.

Acknowledgements. The authors' studies have been supported in part by U.S. Public Health Service grants CA39841, CA37895, and CA49995 from the National Institutes of Health.

References

Badger CC, Krohn KA, Peterson AV, Shulman H, Bernstein ID (1985) Experimental radiotherapy of murine lymphoma with ^{131}I-labelled anti-Thy 1.1 monoclonal antibody. Cancer Res 45:1536–1544

Beaumier PL, Venkatesan P, Vanderheyden JL, Burgua WD, Kunz LL, Fritzberg AR, Abrams PG, Morgan AC Jr (1991) [186]Re radioimmunotherapy of small cell lung carcinoma xenografts in nude mice. Cancer Res 51:676–681

Belitsky P, Ghose T, Aquino J et al. (1978) Radionuclide imaging of metastasis in renal cell carcinoma patients by 131I-labeled antitumor antibody. Radiology 126:515–517

Blumenthal RD, Sharkey RM, Kashi R, Goldenberg DM (1989a) Comparison of therapeutic efficacy and host toxicity of two different ^{131}I-labelled antibodies and their fragments in the GW-39 colonic cancer xenograft model. Int J Cancer 44:292–300

Blumenthal RD, Sharkey RM, Kashi R, Natale AM, Goldenberg DM (1989b) Influence of animal host and tumor implantation site on radioantibody uptake in the GW-39 human colonic xenograft. Int J Cancer 44:1041–1047

Blumenthal RD, Kashi R, Stephens R, Sharkey RM, Goldenberg DM (1991) Improved radioimmunotherapy of colorectal cancer xenografts using antibody mixtures against carcinoembryonic antigen and colon-specific antigen-p. Cancer Immunol Immunother 32:303–310

Blumenthal RD, Sharkey RM, Goldenberg DM (1992a) Dose escalation of radio-antibody in a mouse model with the use of recombinant human interleukin-1 and granulocyte-macrophage colony-stimulating factor intervention to reduce myelosuppression. J Natl Cancer Inst 84:399–407

Blumenthal RD, Sharkey RM, Haywood L, Natale AM, Wong GY, Siegel JA, Kennel SJ, Goldenberg DM (1992b) Targeted therapy of athymic mice bearing GW-39 human colonic cancer micrometastases with ^{131}I-labeled monoclonal antibodies. Cancer Res 52:6036–6044

Blumenthal RD, Sharkey RM, Forman D, Rodriguez B, Goldenberg DM (1993) A preclinical comparison of therapy and toxicity for Mu-9 anti-CSAp IgG labeled with I-131, Y-90, or Re-188 (Abstr). J Nucl Med 34:1014

Boxer GM, Begent RHJ, Kelly AMB, Southall PJ, Blair SB, Theodorou NA, Dawson PM, Ledermann JA (1992) Factors influencing variability of localisation of antibodies to carcinoembryonic antigen (CEA) in patients with colorectal carcinoma – implications for radioimmunotherapy. Br J Cancer 65:825–831

Brady LW, Miyamoto C, Woo DV, Rackover M, Emrich J, Bender H, Dadparvar S, Steplewski Z, Koprowshi H, Black P, Lazzaro B, Nair S, McCormack T, Nieves J, Morabito M, Eshleman J (1991) Malignant astrocytomas treated with iodine-125 labeled monoclonal antibody 425 against epidermal growth factor receptor: a phase II trial. Int J Radiat Oncol Biol Phys 22:225–230

Breitz HB, Weiden PL, Vanderheyden JL, Appelbaum JW, Bjor MJ, Fer MF, Wolf SB, Ratliff BA, Seiler CA, Foisie DC, Fisher DR, Schroff RW, Fritzberg AR, Abrams PG (1992) Clinical experience with rhenium-186-labeled monoclonal antibodies for radioimmunotherapy: results of phase I trials. J Nucl Med 33:1099–1112

Buchegger F, Pfister C, Fournier K, Prevel F, Schreyer M, Carrel S, Mach J-P (1989) Ablation of human colon carcinoma in nude mice by ^{131}I-labeled mono-clonal anti-carcinomembryonic antigen antibody F(ab')$_2$ fragments. J Clin Invest 83:1449–1456

Capone PM, Papsidero LD, Chu TM (1984) Relationship between antigen density and immunotherapeutic response elicited by monoclonal antibodies against solid tumors. J Natl Cancer Inst 72:673–677

Chan RC, Babbs CF, Vetter RJ, Lamar CH (1984) Abnormal response of tumor vasculature to vasoactive drugs. J Natl Cancer Inst 72:145–150

Czuczman MS, Divgi CR, Straus DJ, Lovett DR, Garin-Chesa P, Yeh S, Feirt N, Graham M, Leibel S, Gee TS, Myers J, Pentlow K, Finn R, Schwartz M, Oettgen HF, Larson SM, Old LJ, Scheinberg DA (1991) 131-Iodine-labeled monoclonal antibody therapy of acute myelogenous leukemia and B cell lym-phoma. Antibody Immunoconj Radiopharm 4:787–793

DeLand FH, Goldenberg DM (1986) Radiolabeled antibodies: radiochemistry and clinical applications. In: Freeman LM (ed) Clinical radionuclide imaging. Grune and Stratton, Orlando, p 1915

DeLand FH, Kim EE, Simmons G, Goldenberg DM (1980) Imaging approach in radioimmunodetection. Cancer Res 40:3046–3049

DeNardo SJ, DeNardo GL, O'Grady LF et al. (1988) Treatment of B-cell malig-nancies with [131]I Lym-1 monoclonal antibodies. Int J Cancer Suppl 3:96–103

DeNardo SJ, Warhoe KA, O'Grady LF, Denardo GL, Hellstrom I, Hellstrom KE, Mills SL (1991) Radioimmunotherapy with I-131 chimeric L-6 in advanced breast cancer. In: Ceriani RL (ed) Breast epithelial antigens. Plenum, New York, p 227

Deshpande SV, DeNardo SJ, Kukis DL, Moi MK, McCall MJ, DeNardo GL, Meares CF (1990) Yttrium-90-labeled monoclonal antibody for therapy: labeling by a new macrocyclic bifunctional chelating agent. J Nucl Med 31:473–479

Dillman RO, Beauregard JC, Sobol RE, Royston I, Bartholomew RM, Hagan PS, Halpern S (1984) Lack of radioimmunodetection and complications associated with monoclonal anticarcinoembryonic antigen antibody cross-reactivity with an antigen on circulating cells. Cancer Res 44:2213–2218

Doerr RJ, Abdel-Nabi H, Krag D, Mitchell E (1992) Radiolabeled antibody imaging in the management of colorectal cancer. Results of a multicenter clinical study. Ann Surg 214:118–124

Epenetos AA, Snook D, Durbin H et al. (1986) Limitations of radiolabeled monoclonal antibodies for localization of human neoplasms. Cancer Res 46:3183–3191

Fujimori K, Fisher DR, Weinstein JN (1991) Integrated microscopic-macroscopic pharmacology of monoclonal antibody radioconjugates: the radiation dose distribution. Cancer Res 51:4821–4827

Giavazzi R, Jessup JM, Campbell DE, Walker SM, Fidler IJ (1986) Experimental nude mouse model of human colorectal cancer liver metastases. J Natl Cancer Inst 77:1303–1308

Gold P, Freedman S (1965) Demonstration of tumor-specific antigens in human colon carcinomata by immunological tolerance and absorption techniques. J Exp Med 121:439–462

Goldenberg DM (1979) CEA and other tumor-associated antigens in colon cancer diagnosis and management. In: Grundmann E (ed) Colon cancer. Fischer, Stuttgart, p 163

Goldenberg DM (1989) Future role of radiolabeled monoclonal antibodies in oncological diagnosis and therapy. Semin Nucl Med 19:332–339

Goldenberg DM (1993) Monoclonal antibodies in cancer detection and therapy. Am J Med 94:297–312

Goldenberg DM, Hansen HJ (1972) Carcinoembryonic antigen present in human colonic neoplasia serially propagated in hamsters. Science 175:1117–1118

Goldenberg DM, Larson SM (1992) Radioimmunodetection in cancer identification. J Nucl Med 33:803–814

Goldenberg DM, Witte S, Elster K (1966) GW-39: a new human tumor serially transplantable in the golden hamster Transplantation 4:610–614

Goldenberg DM, Preston DF, Primus DM, Hansen HJ (1974) Photoscan localization of GW-39 tumors in hamsters using radiolabeled anti-CEA immunoglobulin G. Cancer Res 34:1–9

Goldenberg DM, Deland F, Kim E, Bennett S, Primus FJ, Van Nagell JR Jr, Estes N, DeSimone P, Rayburn P (1978) Use of radiolabeled antibodies to carcinoembryonic antigen for detection and localization of diverse cancers by external photoscanning. N Engl J Med 298:1384–1388

Goldenberg DM, Kim EE, DeLand FH, Bennett SJ, Primus FJ (1980) Radioimmunodetection of cancer with radioactive antibodies to carcinoembryonic antigen. Cancer Res 40:2984–2992

Goldenberg DM, Gaffar SA, Bennett SJ, Beach JL (1981) Experimental radioimmunotherapy of a xenografted human colonic tumor (GW-39) producing carcinoembryonic antigen. Cancer Res 41:4354–4360

Goldenberg DM, Goldenberg H, Sharkey RM, Higginbotham-Ford E, Lee RE, Swayne LC, Burger K, Tsai D, Horowitz JA, Hall TC, Pinsky CM, Hansen HJ (1990) Clinical studies of cancer radioimmunodetection with carcinoembryonic antigen monoclonal antibody fragments labeled with ^{123}I or ^{99m}Tc. Cancer Res 50:909s–921s

Goldenberg DM, Horowitz JA, Sharkey RM et al. (1991) Targeting, dosimetry, and radioimmunotherapy of B-cell lymphomas with iodine-131-labeled LL2 monoclonal antibody. J Clin Oncol 9:548–564

Goodwin DA (1990) Strategies for antibody targeting. Antibody Immunoconj Radiopharm 4:427–434

Greiner JW, Guadagni F, Noguchi P, Pestka S, Colcher D, Fisher PB, Schlom J (1987) Recombinant interferon enhances monoclonal antibody-targeting of carcinoma lesions in vivo. Science 235:895–898

Griffiths GL, Goldenberg DM, Knapp FF, Callahan AP, Chang C-H, Hansen HJ (1991) Direct radiolabeling of monoclonal antibodies with generator-produced rhenium-188 for radioimmunotherapy: labeling and animal biodistribution studies. Cancer Res 51:4594–4602

Hansen HJ, Jones AL, Sharkey RM, Grebenau R, Blazejewski N, Kunz A, Buckley MJ, Newman ES, Ostella F, Goldenberg DM (1990) Preclinical evaluation of an "instant" ^{99m}Tc-labeling kit for antibody imaging. Cancer Res 50:7945–7985

Hansen HJ, Sharkey RM, Jones AL et al. (1992) Lack of human anti-mouse antibody (HAMA) response in patients imaged with ImmuRAID-CEA [Tc-99m] and ImmuRAID-MN3 [Tc-99m]. Antibody Immunoconj Radiopharm 5:354

Hardman N, Gill LL, Winter RFJ et al. (1989) Generation of a recombinant mousehuman chimeric monoclonal antibody directed against human carcinoembryonic antigen. Int J Cancer 44:424–433

Hnatowich DJ, Layne WW, Childs RL, Lanteigne D, Davis MA, Griffin TW, Doherty PW (1983) Radioactive labeling of antibody: a simple and efficient method. Science 220:613–615

Hnatowich DJ, Griffin TW, Kosciuczyk C et al. (1985) Pharmacokinetics of an indium-lll labeled monoclonal antibody in cancer patients. Clin Sci 26:849–858

Hnatowich DJ, Virzi F, Rusckowski M (1987) Investigations of avidin and biotin for imaging applications. J Nucl Med 28:1294–1302

Kaminski M, Zasadny K, Buchsbaum D, Wahl R (1991) 131-I anti-B1: initial clinical evaluation in B-cell lymphoma. Antibody Immunoconj Radiopharm 4:59

Kennel SJ (1992) Effects of target antigen competition on distribution of monoclonal antibody to solid tumors. Cancer Res 52:1284–1290

Khazaeli MB, Saleh MN, Liu TP, Meredith RF, Wheeler RH, Baker TS, King D, Secher D, Allen L, Rogers K, Colcher D, Schlom J, Shochat D, LoBuglio AF (1991) Pharmacokinetics and immune response of ^{131}I-chimeric mouse/human B72.3 (human γ4) monoclonal antibody in humans. Cancer Res 51:5461–5466

Knox SJ, Levy R, Miller RA, Uhland W, Schiele J, Ruehl W, Finston R, DayLollini P, Goris ML (1990) Determinants of the antitumor effect of radiolabeled monoclonal antibodies. Cancer Res 50:4935–4940

Kohler G, Milstein C (1975) Continuous culture of fused cells secreting antibody of predefined specificity. Nature 256:495–496

Krejcarek GE, Tucker KL (1976) Covalent attachment of chelating groups to macromolecules. Biochem Biophys Res Commun 77:581–585

LeBerthon B, Khwali LA, Alauddin M, Miller GK, Charak BS, Mazumder A, Epstein AL (1991) Enhanced tumor uptake of macromolecules induced by a novel vasoactive interleukin 2 immunoconjugate. Cancer Res 51:2694–2698

LeDoussal J-M, Martin M, Gautherot E, Delaage M, Barbet J (1989) In vitro and in vivo targeting of radiolabeled monovalent and divalent haptens with dual specificity monoclonal antibody conjugates: enhanced divalent hapten affinity for cell-bound antibody conjugate. J Nucl Med 30:1358–1366

Leung NKV, Landmeir B, Neely J, Nelson AD, Abramowsky C, Ellery S, Adams RB, Miraldi F (1986) Complete tumor ablation with iodine 131-radiolabeled disialoganglioside G_{D2} specific monoclonal antibody against human neuroblastoma xenografted in nude mice. J Natl Cancer Inst 77:739–745

Mach J-P, Carrel S, Merenda C, Heumann D, Rosenspier U (1978) In vivo localization of anti-CEA antibody in colon carcinoma. Can the results obtained in the nude mice model be extrapolated to the patient situation? Eur J Cancer Suppl 1:113–120

Meredith RF, Khazaeli MB, Liu T, Plott G, Wheeler RH, Russell C, Colcher D, Schlom J, Shochat D, LoBuglio AF (1992) Dose fractionation of radiolabeled antibodies in patients with metastatic colon cancer. J Nucl Med 33:1648–1653

Msirikale JS, Klein JL, Schroeder J, Order SE (1987) Radiation enhancement of radiolabeled antibody deposition in tumors. Int J Radiat Oncol Biol Phys 13: 1839–1844

Morrison SL (1985) Transfectomas provide novel chimeric antibodies. Science 229: 1202–1207

Morton BA, Beatty BG, Mison AP, Wanek PM, Beatty JD (1990) Role of bone marrow transplantation in ^{90}Y antibody therapy of colon cancer xenografts in nude mice. Cancer Res 50 Suppl:1008s–1010s

Order SE (1990) Presidential address: systemic radiotherapy – the new frontier. Int J
    Radiat Oncol Biol Phys 18:981–992
Order SE, Klein JL, Leichner PK, Frincke J, Lollo C, Carlo DJ (1986) ^{90}Y
    antiferritin- a new therapeutic radiolabeled antibody. Int J Radiat Oncol Biol
    Phys 12:277–281
Paganelli G, Magnani P, Zito F, Villa E, Sudati F, Lopalco L, Rossett C, Malcovati
    M, Chiolerio F, Seccamani E, Siccardi AG, Fazio F (1991a) Three-step mono-
    clonal antibody tumor targeting in carcinoembryonic antigen-positive patients.
    Cancer Res 51:5960–5966
Paganelli G, Malcovati M, Fazio F (1991b) Monoclonal antibody pretargetting
    techniques for tumor localization: the avidin-biotin system. Nucl Med Commun
    12:211–234
Press OW, Eary JF, Badger CC, Martin PJ, Appelbaum FR, Levy R, Miller R,
    Brown S, Nelp WB, Krohn KA, Fisher D, DeSantes K, Porter B, Kidd P,
    Thomas ED, Bernstein I (1989) Treatment of refractory non-Hodgkin's lym-
    phomas with radiolabeled MB-1 (ant-CD37) antibody. J Clin Oncol 7:1027–1038
Primus FJ, MacDonald R, Goldenberg DM, Hansen HJ (1977) Localization of GW-
    39 tumors in hamsters by affinity-purified antibody to carcinoembryonic antigen.
    Cancer Res 37:1544–1547
Reif AE, Curtis LE, Duffield R, Shauffer IA (1974) Trial of radiolabeled antibody
    localization in metastases of a patient with tumor containing carcinoembryonic
    antigen (CEA). J Surg Oncol 6:133–150
Reynolds JC, Del Vecchio S, Sakahara H et al. (1989) Anti-murine antibody re-
    sponse to mouse monoclonal antibodies: clinical findings and implications. Int J
    Rad Appl Instrum [B]16:121–125
Riva P, Lazzari S, Agostini M et al. (1990) Intracavitary radioimmunotherapy trials
    in gastrointestinal and ovarian carcinoma: pharmacokinetic, biologic and
    dosimetric problems. In: Nuclear medicine quantitative analysis in imaging and
    function. Schmidt HAE, Chambron J (eds) Shattauer, Stuttgart-New York,
    pp 586–588
Riva P, Arista C, Sturiale C, Moscatelli G, Tison V, Mariani M, Seccamani E,
    Lazzari S, Fagioli L, Franceschi A, Sarti G, Riva N, Natali PG, Zardi L,
    Scassellati GA (1992) Treatment of intracranial human glioblastoma by direct
    intratumoral administration of 131I-labelled anti-tenascin monoclonal antibody
    BC-2. Int J Cancer 51:7–13
Rosen ST, Zimmer AM, Goldman-Leikin R et al. (1987) Radioimmunodetection
    and radioimmunotherapy of cutaneous T-cell lymphoma using an 131-I-labeled
    monoclonal antibody: an Illinois Cancer Council study. J Clin Oncol 5:562–573
Sakahara H, Reynolds JC, Carrasquillo JA, Lora ME, Maloney PJ, Lotze MT,
    Larson SM, Neumann RD (1989) In vitro complex formation and biodistri-
    bution of mouse antitumor monoclonal antibody in cancer patients. J Nucl Med
    30:1311–1317
Scheinberg DA, Lovett D, Divgi CR, Graham MC, Berman E, Pentlow K, Feirt N,
    Finn RD, Clarkson ED, Gee TS et al. (1991) A phase I trial of monoclonal
    antibody M195 in acute myelogenous leukemia: specific bone marrow targeting
    and internalization of radionuclide. J Clin Oncol 9:478–490
Schlom J, Molinolo A, Simpson JF et al. (1990) Advantage of dose fracitonation
    in monoclonal antibody-targeted radioimmunotherapy. J Natl Cancer Inst 82:
    763–771
Schlom J, Siler K, Milenic DE, Egensperger D, Colcher D, Miller LS, Houchens D,
    Cheng R, Kaplan D, Goeckler W (1991) Monoclonal antibody-based therapy of
    a human tumor xenograft with a 177-lutetium-labeled immunoconjugate. Cancer
    Res 51:2889–2896
Schlom J, Siler K, Milenic DE, Egensperger D, Colcher D, Miller LS, Houchens D,
    Cheng R, Kaplan D, Goeckler W (1991) Monoclonal antibody-based therapy of
    a human tumor xenograft with a 177-lutetium-labeled immunoconjugate. Cancer
    Res 51:2889–2896

Sgouros G (1992) Plasmapheresis in radioimmunotherapy of micrometastasis: a mathematical modeling and dosimetrical analysis. J Nucl Med 33:2167–2179

Sharkey RM, Pykett MJ, Siegel JA, Alger EA, Primus FJ, Goldenberg DM (1987) Radioimmunotherapy of the GW-39 human colonic tumor xenograft with ^{131}I-labeled monoclonal antibody to carcinoembryonic antigen. Cancer Res 47:5672–5677

Sharkey RM, Motta-Hennessy C, Gansow OA, Brechbiel MW, Fand I, Griffiths GL, Jones AL, Goldenberg DM (1990) Selection of a DTPA chelate conjugate for monoclonal antibody targeting to a human colonic tumor in nude mice. Int J Cancer 46:79–85

Sharkey RM, Weadock KS, Natale A, Haywood L, Aninipot R, Blumenthal RD, Goldenberg DM (1991) Successful radioimmunotherapy for lung metastasis of human colonic cancer in nude mice. J Natl Cancer Inst 83:627–632

Sharkey RM, Boerman OC, Natale A, Pawlyk D, Monestier M, Losman MJ, Goldenberg DM (1992) Enhanced clearance of radiolabeled murine monoclonal antibody by a syngeneic anti-idiotype antibody in tumor-bearing nude mice. Int J Cancer 51:266–273

Sharkey RM, Goldenberg DM, Murthy S, Pinsky H, Vagg R, Pawlyk D, Lee RE, Siegel J, Wong GY, Gascon P, Izon DO, Vezza M, Burger K, Swayne LC, Pinsky CM, Hansen HJ (1993) Clinical evaluation of tumor targeting with a high-affinity, CEA-specific, murine monoclonal antibody, MN-14. Cancer 71: 2081–2096

Siegel JA, Pawlyk DA, Lee RE, Sasso NL, Horowitz JA, Sharkey RM, Goldenberg DM (1990) Tumor, red marrow, and organ dosimetry for ^{131}I-labeled anti-carcinoembryonic antigen monoclonal antibody. Cancer Res 50 Suppl: 1039s–1042s

Steplewski Z, Sun LK, Sherman CW et al. (1988) Biological activity of human-mouse IgG_1, IgG_2, IgG_3, and IgG_4 chimeric monoclonal antibodies with anti-tumor specificity. Proc Natl Acad Sci USA 85:4852–4956

Stewart JSW, Hird V, Sullivan M, Snook D, Epenetos AA (1989) Intraperitoneal radioimmunotherapy for ovarian cancer. Br J Obstet Gynecol 96:529–536

Stickney DR, Gridley DS, Kirk GA, Slater JM (1987) Enhancement of monoclonal antibody binding to melanoma with single dose radiation of hyperthermia. Monogr Natl Cancer Inst 3:47–52

Sung S, Shockley TR, Morrison PF, Dvorak HF, Yarmush ML, Dedrick RL (1992) Predicted and observed effects of antibody affinity and antigen density on monoclonal antibody uptake in solid tumors. Cancer Res 52:377–384

Sutherland R, Buchegger F, Schreyer M, Vacca A, Mach JP (1987) Penetration and binding of radiolabeled anti-carcinoembryonic antigen monoclonal antibodies and their antigen binding fragments in human colon multicellular tumor spheroids. Cancer Res 47:1627–1633

Vessella RC, Alvarez V, Chious RK, Rodwell J, Elgon M, Palme D, Shafe R, Lange P (1985) Radioimmunoscintigraphy and radioimmunotherapy of renal-cell carcinoma xenografts. Monogr Natl Cancer Inst 3:159–167

Wahl RL, Barrett J, Geatti O, Liebert M, Wilson ES, Fisher S, Wagner JG (1988) The intraperitoneal delivery of radiolabeled monoclonal antibodies: studies on the regional delivery advantage. Cancer Immunol Immunother 26:187–201

Weadock KS, Sharkey RM, Varga DC, Goldenberg DM (1990) Evaluation of a remote radioiodination system for radioimmunotherapy. J Nucl Med 31:508–511

Zanzonico P (1992) Radioimmunotherapy of micrometastasis: a continuing evolution (Editorial). J Nucl Med 33:2180–2183

Clinical Experience with Murine, Human and Genetically Engineered Monoclonal Antibodies

M.N. SALEH, R.M. CONRY, and A.F. LoBUGLIO

A. Introduction

Monoclonal antibodies (MAbs), by virtue of their unique antigen specificity, have emerged as important biologic reagents with wide ranging clinical utility. Over the last decade, MAbs have moved from the laboratory bench to the forefront of innovative clinical application. Early clinical trials utilized murine antibodies since these could be readily produced by immunizing mice with candidate antigens and generating hybridomas that secrete antigen-specific MAbs (KOEHLER and MILSTEIN 1975). Extensive clinical experience with murine MAbs has led to the recognition of a number of limitations associated with the use of such xenogeneic proteins. Thus, murine antibodies are immunogenic (DILLMAN et al. 1984; KHAZAELI et al. 1988) and have a short circulating half-life which often makes frequent administration necessary (MEEKER et al. 1985; LoBUGLIO et al. 1988).

Several strategies have been used to circumvent some of these limitations. The first was the utilization of human components for hybridization with the generation of human MAbs (TENG et al. 1985; COTE et al. 1986). This has been discussed in detail in Chap. 1. The generation of human MAbs is associated with a number of technical difficulties and only a few human antibodies generated by conventional methods have entered clinical trials (ZIEGLER et al. 1991; IRIE and MORTON 1986; STEIS et al. 1990).

With the advent of recombinant DNA technology, strategies to modify the murine nature of MAbs have emerged. One of these approaches involves cloning the immunoglobulin variable region genes from the murine hybridoma cells and ligating these with the constant region genes of human immunoglobulin. The resulting genetic construct can be transfected into eukaryotic cells for the production of genetically engineered chimeric mouse/human MAbs (MORRISON 1985). Chapter 5 gives a detailed overview of this technology. Such reagents were projected to be less immunogenic than their murine counterparts and to have a circulating capacity similar to human immunoglobulin. A number of such chimeric mouse/human MAbs have entered clinical trials (LoBUGLIO et al. 1989; MEREDITH et al. 1991; SALEH et al. 1992b; MEREDITH et al. 1992; BEGENT et al. 1990; MORELAND et al. 1993a).

A second genetic approach to generate MAbs that more closely resemble human antibodies has also been developed. This strategy entails grafting the

murine complementarity determining regions (CDRs) to the genetic construct for human immunoglobulin (RIECHMANN et al. 1988). Chapter 5 is
devoted to the description of this technology. Such "humanized" (CDR-
grafted) MAbs have been predicted to be less immunogenic than their
murine or chimeric counterparts and are beginning to enter into clinical
trials (HALE et al. 1988; CLENDENINN et al. 1992; KOSMAS et al. 1992).

A more recent approach goes one step further to create genetically
engineered human antibodies (HUSE et al. 1989). Human lymphocytes from
immune or nonimmune donors are used to create a library of possible V
region combinations (combinatorial library) with subsequent screening. A
selection of appropriate antigen binding F_{Ab} fragments may significantly
expand our repertoire of human monoclonals. Such agents have not reached
clinical trials as yet.

This review will discuss the limitations associated with the clinical application of murine MAbs and the clinical experience to date with human,
chimeric, and CDR-grafted MAbs.

B. Difficulties Encountered with Murine Monoclonal Reagents

I. Immunogenicity of Murine Antibodies

Clinical trials have utilized murine MAbs or their fragments for radio-
immune imaging (GOLDENBERG 1989). They have also been utilized for
therapeutic intent either as unmodified native antibodies (MEEKER et al.
1985; SEARS et al. 1982) or conjugated to radioactive isotopes (ORDER et al.
1989), cytotoxic drugs (REISFELD et al. 1989; OLDHAM 1991), or natural
toxins (VITETTA et al. 1987; BYERS et al. 1989; HERTLER and FRANKEL 1989).
These clinical trials have resulted in some degree of success but optimal
clinical application appears to be limited by the immunogenicity of the
murine protein (DILLMAN et al. 1984; KHAZAELI et al. 1988). Initial trials
using murine MAbs for the treatment of chronic lymphocytic leukemia
were associated with the development of human anti-mouse antibodies
(HAMA) in only a minority of cases (FOON et al. 1984). This, however,
appeared to be related to the underlying immunosuppression of the study
population since administration of the same antibody to patients with
cutaneous T cell lymphoma resulted in HAMA in over half of the patients
(DILLMAN et al. 1984). In general, the administration of xenogeneic murine
MAbs has resulted in frequent development of HAMA (KHAZAELI et al.
1988, 1992; SALEH et al. 1992a,b; ESTEBAN et al. 1987; VADHAN-RAJ et al.
1988). The antibody response occurs across a wide dose range of MAb with
no hint of tolerance (KHAZAELI et al. 1988; REYNOLDS et al. 1989). The
development of HAMA is more likely with repeated administration (LIND

et al. 1991). Antibody fragments appear less immunogenic than intact molecules presumably due to their short plasma circulating time and absence of the CH2 and CH3 immunogenic epitopes of the murine constant region (REYNOLDS et al. 1989). Some murine antibodies such as the anti-GD2 MAb 14G2a (SALEH et al. 1992b) and the anti-GD3 MAb R24 (VADHAN-RAJ et al. 1988; CAULFIELD et al. 1990; BAJORIN et al. 1990) appear consistently more immunogenic than others such as the anti-TAG-72 MAb B72.3 (KHAZAELI et al. 1992).

The HAMA response is generally not detected before day 8 following initial exposure and peaks between 2–6 weeks and may persist for many months (KHAZAELI et al. 1988; COURTENAY-LUCK et al. 1986). The initial exposure typically results in a primary immune response with IgM antibody followed by development of IgG antibodies (COURTENAY-LUCK et al. 1986). The specificity of HAMA is variable. In general, a pattern of early anti-constant region reactivity is followed by increasing specificity for variable region epitopes, especially following repeated exposure (SHAWLER et al. 1985; SCHROFF et al. 1985; SALEH et al. 1990). Conversely, however, we have shown that murine B72.3 can elicit an early anti-V region (anti-idiotypic) response (KHAZAELI et al. 1992). This has also been observed in renal transplant patients receiving murine OKT3 as rejection prophylaxis (CHATENOUD et al. 1986). The variability in the human immune response to murine antibodies is the subject of ongoing research.

II. Pharmacokinetics of Murine Antibodies

The pharmacokinetics of immunoglobulins differ from most therapeutic agents because of their large molecular weight and the slow equilibration between the intravascular and extravascular space. The pharmacokinetics of MAbs have been studied using radiolabeled antibodies or by quantitating the concentration of the monoclonal protein in plasma or serum using radioimmunoassay techniques.

One of the most studied murine MAbs has been MAb 17-1A, an IgG1a mouse antibody directed at the 38 kDa glycoprotein antigen expressed on a variety of adenocarcinomas (SEARS et al. 1982, 1985; LoBUGLIO et al. 1988). Using the double antigen radiometric assay, we have reported that the pharmacokinetics of 17-1A follows a one-compartment model with a mean plasma half-life ($t_{1/2}\beta$) of approximately 15 h (KHAZAELI et al. 1988). Since then, a variety of murine MAbs have been studied in humans with the $t_{1/2}\beta$ varying between 15 and 52 h (KHAZAELI et al. 1988; SALEH et al. 1992a,c; ESTEBAN et al. 1987). It is, thus, apparent that the circulating half-life of murine antibodies of all IgG subtypes is rather short and therefore repeated administrations of antibody is necessary in order to achieve sustained circulating plasma concentrations. This obviously enhances their immunogenicity and further precludes adequate circulation of the antibody due to rapid clearance by the reticuloendothelial system.

III. Clinical Efficacy of Murine Antibodies

1. In Cancer

The in vitro biologic activity of antibody usually refers to its ability to mediate antibody-dependent cellular cytotoxicity (ADCC) and/or complement-dependent lysis of antigen-positive target cells (Roitt et al. 1991). In addition, delayed clinical antitumor responses observed in some patients receiving MAbs (Koprowski et al. 1984; Vadhan-Raj et al. 1988) have been attributed to the generation of an idiotype-anti-idiotype cascade in vivo (Jerne 1974; Köhler et al. 1989). The contribution of this phenomenon to these anti-tumor responses is unclear. Immunotherapy using anti-idiotypic antibodies directed at unique immunoglobulin idiotypes expressed on malignant B cells has been associated with encouraging responses (Brown and Miller 1989; Miller et al. 1982). In addition to ADCC and complement-dependent lysis, such reagents may specifically inhibit clones that express the target idiotype. The finding that many B cell lymphomas share idiotypes may substantially enhance the practical application of this therapeutic modality (Swischer et al. 1991; Miller et al. 1989).

Some of the earliest trials employing unlabeled MAbs were conducted in patients with hematologic malignancies (reviewed Grossbard et al. 1992b; Kuzel et al. 1990). The murine antibodies were directed against cell surface antigens expressed by leukemia/lymphoma cells. In most cases, there was a prompt but transient reduction of circulating tumor cells with little impact on the disease in the bone marrow or lymph nodes. The antibodies caused significant down-modulation of the target cell surface antigens and most responses were therefore partial and short-lived (Dillman et al. 1984; Foon et al. 1984). Subsequent studies using unlabeled MAbs in solid tumors have focused primarily on colorectal carcinoma and malignant melanoma and objective clinical responses have ranged between 5% and 20% (reviewed in LoBuglio and Saleh 1992; Dimaggio et al. 1990; Schlom 1990).

Antibodies conjugated to radioactive isotopes or toxins further expand the biologic spectrum of MAbs by using the antigen binding specificity of the antibody for selective delivery to the tumor target. Administration of radio-labeled murine antibodies has resulted in favorable antitumor activity in patients with Hodgkin's disease (Lenhard et al. 1985) and non-Hodgkin's lymphoma (DeNardo et al. 1988; Goldenberg et al. 1991). Delivery of high doses (up to 600 mCi) of a ^{131}I-labeled anti-B cell MAb followed by autologous bone marrow rescue has resulted in complete responses in a majority of patients, with some patients surviving beyond 3 years (Press et al. 1989). Radioimmunoconjugates have, however, had little or no efficacy to date in solid tumors (Goldenberg 1989; Schlom 1990; Stewart et al. 1990; Knapp and Lau 1990; Dimaggio et al. 1990). Recently, very encouraging clinical responses have also been demonstrated in lymphoma patients receiving anti-B cell MAbs conjugated with blocked ricin (Grossbard et al. 1992a) or deglycosylated ricin A chain (Vitetta et al. 1991).

2. In Nonmalignant Disorders

The clinical application of murine monoclonal reagents has not been restricted to the treatment of malignant disorders. To date, only one MAb has been approved by the Food and Drug Administration (FDA) for therapeutic use. The murine MAb OKT3 is directed against the T cell receptor complex (CD3) ubiquitously expressed on mature T cells. Clinical trials of the OKT3 antibody in recipients of cadaveric renal transplants have demonstrated a superior reversal of acute rejection when compared with conventional immunosuppressive therapy. Kidney survival rates at 1 year were significantly higher for the group treated with OKT3 than for the control group which received standard immunosuppressive therapy (ORTHO MULTICENTER TRANSPLANT STUDY GROUP 1985). Therapy with OKT3, however, has been associated with a significant increase in opportunistic infections, especially cytomegalovirus. Some cases of B cell lymphomas have been linked to the chronic immunosuppression induced by such antibodies. Murine antibodies directed at the lipid A component of endotoxin have been studied for the treatment of gram-negative sepsis (GREENMAN et al. 1991). A reduction in mortality was only observed in the subset of patients with gram-negative sepsis who were not in shock. The role of such reagents and their clinical impact has yet to be defined (BONE 1991; WOLFF 1991). The CD4 subset of T cells has been implicated in the pathogenesis of a number of immunologic disorders (KUMAR et al. 1989; MORELAND et al. 1993b; KRONENBERG 1991; SINHA et al. 1990). Pilot studies using murine anti-CD4 MAbs conducted in patients with rheumatoid arthritis have yielded promising results (HORNEFF et al. 1991; REITER et al. 1991; WENDLING et al. 1991; GOLDBERG et al. 1991; HERZOG et al. 1989). In addition, anecdotal reports have suggested possible beneficial effects of anti-CD4 therapy in other autoimmune disorders (VAN DER LUBBE et al. 1991; CHOY et al. 1991; EMMRICH et al. 1991; MATHIESON et al. 1990; HAFLER et al. 1988).

Platelet aggregation plays a pivotal role in the pathogenesis of arterial vaso-occlusive processes (COLLER 1991). MAbs that block the platelet GPIIb/IIIa receptor are potent inhibitors of platelet function (COLLER et al. 1991) and phase I/II studies using 7E3, an anti-GPIIb/IIIa Fab, demonstrated profound inhibition of platelet function without causing excessive bleeding tendency (ELLIS et al. 1990).

The role of specific coagulation factors in mediating vaso-occlusion following vascular injury has led investigators to target and inhibit such coagulation factors using MAbs. A phase I study of an anti-factor VII antibody is currently ongoing at the University of Alabama at Birmingham and interim data indicate that the antibody induces an immediate, reversible and dose-dependent inhibition of factor VII activity in vivo (RUSTAGI et al. 1992).

IV. Toxicity Associated with Murine Antibodies

In general, administration of murine antibodies to patients has been associated with minimal toxicity. Acute toxicity has occasionally been associated with rapid infusion in patients with circulating target cells (shortness of breath and chest pain) which can be avoided by prolonging the duration of the antibody infusion (DILLMAN et al. 1986). Anaphylactic reactions are rare even with repeated infusions and readily reversible with minimal therapeutic intervention (LoBUGLIO et al. 1988). Despite theoretical concerns, the administration of mouse proteins in the presence of HAMA has rarely resulted in serum sickness or immune complex disease, though shortening of antibody circulation has been observed (SALEH et al. 1992). The administration of some MAb has been associated with the development of acute reversible toxicity that is felt to be related to binding of the antibody to native or cross-reactive antigens expressed on normal tissue. Thus, administration of MAb D612 which binds to a 48 kDa antigen present on malignant and normal gastrointestinal epithelium (FERNSTEN et al. 1991) induces a reversible dose-dependent secretory diarrhea (SALEH et al. 1992a), whereas administration of the murine anti-GD2 antibody 14G2a is associated with reversible neurotoxicity (SALEH et al. 1992c). Such cross-reacting toxicities have been even more dramatic with toxin conjugates (WEINER et al. 1989; GOULD et al. 1989).

C. Human Monoclonal Antibody Trials

The most extensively studied human MAb has been HA-1A, an IgM antibody directed to the lipid A component of endotoxin. Phase I trials with HA-1A were carried out in 15 nonseptic patients with cancer (KHAZAELI et al. 1990) and 34 patients with sepsis (FISHER et al. 1990) using single infusions of $50\,\mu g$ to 250 mg. No adverse reactions to HA-1A were observed and a "double antigen" radiometric assay detected no evidence of antibody response to HA-1A. The mean plasma half-life of HA-1A was 31.5 h in the nonseptic cancer patients and 15.9 h in the patients with sepsis syndrome. More recently, a multicenter double-blind trial of HA-1A in sepsis syndrome randomized patients to a single dose of 100 mg of HA-1A or placebo (ZIEGLER et al. 1991). A statistically significant ($p < 0.05$) decrease in mortality was reported among patients with gram-negative bacteremia who received HA-1A. No benefit of treatment with HA-1A was demonstrated in the patients who did not prove to have gram-negative bacteremia. All patients tolerated HA-1A well, and no anti-HA-1A antibody response was detected by double antigen assay. Among 1158 patients infused with HA-1A as of December 1991, hypersensitivity-like reactions were observed in six patients (SMITH et al. 1992). There has been no significant difference in the incidence or types of adverse events in patients receiving multiple doses of HA-1A compared with patients receiving either one dose of HA-1A or placebo.

A mixture of five IgM human MAbs against lipopolysaccharide antigens of *Pseudomonas aeruginosa*, plus a human IgG1 MAb against exotoxin A, were studied in 12 noninfected patients and eight patients with *Pseudomonas aeruginosa* bacteremia or pneumonia (SARAVOLATZ et al. 1991). The preparation was well tolerated with no evidence of an antibody response to the monoclonal reagents. Serum half-lives ranged from 34 to 99 h and infusion of the antibody preparation significantly increased serum opsonin activity.

Three neutralizing human IgG1 MAbs to cytomegalovirus (CMV) have been evaluated in phase I trials (AZUMA et al. 1991; AULITZKY et al. 1991). The first, antibody C23, was administered to 20 healthy volunteers with no evidence of an immune response by double antigen assay. The antibody had a mean plasma half-life of 24 days. Some patients received three doses at weekly intervals with no alteration of pharmacokinetics and sustained circulating levels of antibody. The other two antibodies, SDZ 89–104 and SDZ 89–109, were administered biweekly to 13 bone marrow transplant patients for 3 months. The mean serum half-lives were approximately 6 days. No adverse reactions were seen with any of the three anti-CMV human monoclonals and high serum levels were closely accompanied by increased neutralizing activity against CMV.

The first reported trial of a human MAb in cancer therapy utilized L72, an IgM antibody specific for GD2 ganglioside (IRIE and MORTON 1986). The antibody was injected directly (intralesional) into cutaneous metastatic melanoma nodules in eight patients with total doses of 3–15 mg divided into two to four treatments. They utilized a passive sheep erythrocyte agglutination assay for detection of antibody response and reported that five of eight patients gave positive assays during the course of treatment. With the exception of mild local erythema, no side effects were observed in any patient. Of the 21 nodules treated with antibody, ten completely regressed and five partially regressed. One patient with melanoma satellitosis showed complete regression with no sign of recurrence 20 months after MAb therapy.

STEIS et al. (1990) have reported on two ^{131}I-labeled human IgM MAbs (16.88 and 28A32) specific for colon cancer-associated antigens in phase I trials of metastatic colon cancer patients. A latex agglutination assay was used to monitor for anti-globulin response with no evidence of response to 16.88 in 12 patients receiving repeated injections. This assay had a high degree of "nonspecific" positivity with antibody 28A32, including five of twelve normal sera, seven of 23 colon cancer patients prior to receiving antibody, and eight of 14 patients at some time after multiple infusions of 28A32. These were all of low titer (i.e., 1/10 dilution) and did not appear to affect antibody half-life. Two patients developed localized urticarial reactions following injection of antibody 28A32, but otherwise these reagents were well tolerated. No antitumor effects were seen with either antibody. Serial scans showed tumor uptake of radioisotope in 21 of 28 patients including lesions in the liver, lung, lymph nodes, peritoneum, bone, and brain. Only rarely were tumors smaller than 2 cm in diameter imaged.

These studies suggest that human MAbs have low immunogenicity, have relatively long half-lives, and rarely produce adverse reactions in human trials. Potentially, anti-idiotype antibody responses may occur but will most likely be associated only with repeated doses and/or subcutaneous/ intratumor injections. The efficacy of human MAbs in the therapy of bacterial and viral infections and malignancies will be tested in forthcoming phase II trials.

D. Chimeric Antibody Trials

The first clinical trial of a genetically engineered chimeric antibody was reported in 1989 with the chimeric MAb 17-1A, consisting of the V regions of murine 17-1A and the constant regions of human IgG1κ (LoBUGLIO et al. 1989). Patients received single and multiple doses of 10 mg or 40 mg. The chimeric antibody was able to circulate five to six times as long as murine 17-1A. A subsequent trial with radiolabeled chimeric 17-1A in six patients (MEREDITH et al. 1991b) confirmed its improved plasma β half-life (100 h vs 19 h). The chimeric antibody was less immunogenic than murine 17-1A with only one of 16 patients demonstrating a weak anti-idiotypic immune response (LoBUGLIO et al. 1989; MEREDITH et al. 1991).

Since then, a number of chimeric antibodies have gone through phase I clinical testing. Radiolabeled chimeric anti-TAG-72 MAb B72.3 (human IgG4κ) was administered to a total of 24 patients with metastatic colon cancer (MEREDITH et al. 1992b). The pharmacokinetic data fit a two-compartment model with a mean $t_{1/2}\alpha$ of 18 h and a $t_{1/2}\beta$ of 242 h (KHAZAELI et al. 1991). Of the 24 patients, 16 developed an immune (anti-Id) response with six patients demonstrating moderate to high amounts of antibody. Two of the high responder patients developed an anamnestic response upon retreatment with chimeric B72.3 and the second treatment was associated with rapid clearance of the infused antibody. Chimeric B72.3, thus, has a four- to sixfold prolongation of circulation as compared to murine B72.3 (ESTEBAN et al. 1987) yet retains considerable immunogenicity. The maximal tolerated dose of ^{131}I-labeled-chimeric B72.3 was 36 mCi/m^2 administered as a single dose and thrombocytopenia was the dose-limiting toxicity observed (MEREDITH et al. 1992). Toxicity was not significantly ameliorated when the radiolabeled antibody was administered in split doses (MEREDITH et al. 1992a).

The chimeric anti-GD2 antibody 14.18 (human IgG1κ) has been studied in adult patients with malignant melanoma (SALEH et al. 1992b) and in children with neuroblastoma (YU et al. 1991). Toxicity consisted primarily of infusion-related abdominal pain which was similar to that observed with its murine counterpart, 14G2a (SALEH et al. 1992c). The pain resolved following completion of the antibody infusion, a phenomenon that remains unexplained since high levels of antibody persist in the circulation. Similar to the other chimeric antibodies, 14.18 has a half-life five to six times longer ($t_{1/2}\beta$ of 168 h) than that of murine 14G2a. Weak to modest anti-Id antibody responses were noted in eight of 13 patients following chimeric 14.18 therapy (SALEH et al. 1992b). This was a dramatic reduction in immunogenicity as compared to murine 14G2a which produced very high immune responses (SALEH et al. 1992c). Although no clinical responses were observed in this phase I study, patients receiving >45 mg of chimeric 14.18 had antibody detectable on tumor biopsies.

In a recently completed phase I trial, 25 patients with refractory rheumatoid arthritis received single or multiple doses of the chimeric anti-CD4 MAb M T412. There was a prompt and often sustained decline of CD4-positive T cells. Only two of the 25 patients developed a transient and low level of anti-M T412 antibodies. Significant clinical benefit as evidenced by a >50% improvement in tender joint counts was observed both at the 5 week and 6 month evaluation (MORELAND et al. 1993a).

The chimeric derivative of the anti-GPIIb/IIIa antibody 7E3 (c7E3) has undergone phase I/II testing. Whereas 20% of patients treated with m7E3 in a previous phase I study developed HAMA that were predominantly directed at the murine V region, no immune response was seen in any of the 64 patients treated with c7E3 (JORDAN et al. 1992). The chimeric antibody is currently in phase III testing in patients undergoing elective percutaneous coronary angioplasty. Interim data indicate that c7E3 blocks platelet aggregation in vivo without affecting platelet survival (SWEENEY et al. 1993). Combined therapy with c7E3, heparin, and aspirin may potentially reduce the risk of myocardial infarction following coronary angioplasty (SIMOONS et al. 1993).

The experience with chimeric antibodies to date allows some generalized observations. As predicted, chimeric antibodies survived much longer than murine antibodies but not as well as human immunoglobulin (LOBUGLIO et al. 1992; WALDMANN and SCHWAB 1965; BARTH et al. 1964). Furthermore, the degree of immunogenicity of chimeric antibodies is quite variable. Thus, chimeric 17-1A and MT412 were very weakly immunogenic in humans whereas the chimeric B72.3 antibody induced a substantial degree of antibody response, not unlike that seen with murine B72.3. Other chimeric antibodies appear intermediate between these extremes. The underlying mechanism for this variable immune response to chimeric reagents is yet to be defined and may well lie in the immunogenicity of the murine variable region sequences which are preserved in the chimeric molecule (KHAZAELI et al. 1992). The clinical efficacy of chimeric MAbs in the treatment of malignant disorders has not been tested adequately in phase II trials to estimate the advantage over murine monoclonal reagents. However, repeated administration and prolonged blood levels may prove advantageous for several of these reagents.

E. CDR-Grafted Humanized Monoclonal Antibody Trials

The second type of murine/human constructs are CDR-grafted (humanized) MAbs which utilize not only human constant regions but also human framework of the V_H and V_L regions. This reduces the murine component to the CDR regions and selected other vital residues required for CDR orientation (RIECHMANN et al. 1988). Modest information from clinical trials of these molecules is currently available.

The most widely studied of these reagents is CAMPATH-1H, a humanized IgG1 MAb specific for CD_W52 present on all lymphocytes and some monocytes. CAMPATH-1H was originally used to treat two patients with non-Hodgkin's lymphoma (NHL). Both patients demonstrated clearance of lymphoma cells from the blood and bone marrow and resolution of splenomegaly (HALE et al. 1988). A single patient with systemic vasculitis received multiple does of CAMPATH-1H in combination with a rat anti-CD4 MAb resulting in a remission lasting greater than 12 months (MATHIESON et al. 1990).

In both of these studies, no anti-globulin response was detectable by ELISA despite repeated dosing. Subsequently, eight patients with refractory rheumatoid arthritis received multiple doses of CAMPATH-1H with significant sustained clinical benefit seen in seven patients. No anti-globulin response was measurable by ELISA after one course of therapy, but three out of four patients did have an antibody response (predominantly anti-idiotype reactivity) on retreatment (ISAACS et al. 1992). Most recently, 113 patients with relapsed or refractory NHL or chronic lymphatic leukemia (CLL) received multiple doses of CAMPATH-1H (CLENDENNIN et al. 1992). Low levels of anti-CAMPATH-1H antibody were detected in 4% of patients tested and the mean terminal half-life was 28 h. Major disease responses (50% or greater reduction in disease) were seen in some patients with CLL and NHL. Circulating tumor cells appear more sensitive to CAMPATH-1H than bone marrow, nodal, or splenic disease, although major improvement in these sites has been observed. Opportunistic infections including herpes simplex, herpes zoster, pneumocystis carinii, and CMV occurred in 34 instances. In all of the above trials, acute self-limited toxicity including fever, rigors, nausea, vomiting, hypotension, and rash has been observed during or within 6 h of CAMPATH-1H infusion reflecting the lymphokine release syndrome seen with other anti-lymphocyte antibody reagents.

An IgG1 humanized anti-placental alkaline phosphatase (anti-PLAP) conjugated to ^{111}In via a macrocyclic chelating agent (DOTA) has been utilized in six patients to image a variety of carcinomas including ovarian, breast, and gastric (KOSMAS et al. 1992). ELISA assays demonstrated an anti-DOTA response without evidence of an anti-globulin response. However, the apparent dose in this report was a few hundred micrograms with relatively brief plasma circulation; so the immunogenicity of this reagent will require further study. A variety of humanized MAbs are approaching clinical trials and a more clear picture of their immunogenicity and clinical efficacy will be forthcoming.

F. Future Prospects

The structural makeup of monoclonal reagents has undergone significant genetic alteration in an effort to manipulate the circulating capacity, immunogenicity and the effector function of these molecules.

Compared to murine reagents, many, albeit not all, chimeric MAbs appear to be less immunogenic. Much, however, needs to be learned about the immunogenic epitopes of such antibodies in order to predict beforehand whether the chimeric or CDR-grafted variants of specific murine antibodies will be substantially less immunogenic and will readily permit repetitive administration.

The circulating capacity of chimeric antibodies is four to eight times longer than murine antibodies but falls short in comparison with human immunoglobulin. The role of immunoglobulin constant region glycosylation patterns may be critical in determining clearance of the protein by the reticuloendothelial system. In the case of genetically engineered antibodies, this is dictated by the expression system which is not human in origin.

The clinical efficacy of native antitumor antibodies depends substantially on the ability to achieve high and sustained levels of antibody in the tumor microvasculature and interstitium (JAIN 1987, 1988). Use of cytokines to manipulate tumor vasculature and up-regulate tumor antigen density may positively influence the balance in favor of antibody deposition onto tumor cells (FAUCI 1987; DINARELLO and MIER 1987; KIRKWOOD and ERNSTOFF 1984). Enhancing the human effector function using appropriate cytokines (ROBINSON and QUESENBERY 1990) is another modality that could improve the biologic activity of such reagents. A number of trials have been initiated to address these specific issues.

The ability of MAbs to target circulating cells (e.g., CD4 cells, platelets) or factors in plasma (e.g., endotoxin, factor VII) appears easier to achieve than targeting of malignant cells in poorly vascularized solid tumors. This may explain the increasing utility and apparent success of MAbs in the treatment of nonmalignant disorders, a phenomenon that is bound to generate new demands for these reagents.

The data generated from the clinical use of human, chimeric and CDR-grafted antibodies will certainly pave the way for more effective application of newer reagents such as those that will be generated by combinatorial libraries. The field of MAb therapy demonstrates the close-knit and interdependent relationship between basic science and clinical medicine, for what is possible in the test tube may not be readily achieved in the patient. Conversely, clinical application of novel reagents frequently leads to observations that can be best dissected and understood in the controlled setting of the laboratory.

References

Aulitzky WE, Schulz TF, Tilg H, Niederwieser D, Larcher K, Östberg L, Scriba M, Martindale J, Stern AC, Grass P, Mach M, Dierich MP, Huber C (1991) Human monoclonal antibodies neutralizing cytomegalovirus (CMV) for prophylaxis of CMV disease: report of a phase I trial in bone marrow transplant recipients. J Infect Dis 163:1344–1347

Azuma J, Kurimoto T, Tsuji S, Mochizuki N, Fujinaga S, Matsumoto Y, Masuho Y (1991) Phase I study on human monoclonal antibody against cytomegalovirus: pharmacokinetics and immunogenicity. J Immunother 10:278–285

Bajorin DF, Chapman PB, Wong G, Coit DG, Kunicka J, Dimaggio J, Cordon-Cardo C, Urmacher C, Dantes L, Templeton MA, Liu J, Oettgen HF, Houghton AN (1990) Phase I evaluation of a combination of monoclonal antibody R24 and interleukin 2 in patients with metastatic melanoma. Cancer Res 50:7490–7495

Barth W, Wochner R, Waldmann T, Fahey J (1964) Metabolism of human gamma macroglobulins. J Clin Invest 43:1036–1048

Begent RHJ, Ledermann JA, Bagshawe KD, Green AJ, Kelly AMB, Lane D, Secher DS, Dewji MR, Baker TS (1990) Chimeric B72.3 antibody for repeated radioimmunotherapy of colorectal carcinoma. Antibody, Immunoconj, Radiopharm 3:86

Bone R (1991) Monoclonal antibodies to endotoxin: new allies against sepsis? JAMA 266:1125–1126

Brown SL, Miller RA (1989) Treatment of B-cell lymphomas with anti-idiotype antibodies alone and in combination with alpha interferon. Blood 73:651–661

Byers VS, Rodvien R, Grant K, Durrant LG, Hudson KH, Baldwin RW, Scannon PJ (1989) Phase I study of monoclonal antibody-ricin A chain immunotoxin XomaZyme-791 in patients with metastatic colon cancer. Cancer Res 49:6153–6160

Caulfield MJ, Barna B, Murthy S (1990) Phase Ia-Ib trial of an anti-GD3 monoclonal antibody in combination with interferon-α in patients with malignant melanoma. J Biol Response Mod 9:319–328

Chatenoud L, Baudrihaye MF, Chkoff N, Kreis H, Goldstein G, Bach J-F (1986) Restriction of the human in vivo immune response against the mouse monoclonal antibody OKT3. J Immunol 137:830–838

Choy EHS, Chikanza IC, Kingsley GH, Panayai GS (1991) Chimeric anti-CD4 monoclonal antibody for relapsing polychondritis. Lancet 338:450

Clendeninn NJ, Nethersell ABW, Scott JE, Collier MA (1992) Phase I/II trials of CAMPATH-1H, a humanized anti-lymphocyte monoclonal antibody (MoAb), in non-Hodgkin's lymphoma (NHL) and chronic lymphocytic leukemia (CLL) (Abstr 624). Blood 80:1589

Coller BS (1991) Platelets in cardiovascular thrombosis and thrombolysis. In: Fozzard HA, Harber E, Jennings RB, Katz AM, Morgan HE (eds) The heart and cardiovascular system, 2nd edn. Raven, New York

Coller BS, Scudder LE, Beer J, Gold HK, Folts DJ, Cavagnaro J, Jordan R, Wagner C, Iuliucci J, Knight D, Ghrayeb J, Smith C, Weisman HF, Berger H (1991) Monoclonal antibodies to platelet GPIIb/IIIa as antithrombotic agents. Ann NY Acad Sci 614:193–213

Cote R, Morrissey D, Houghton A, Thomson T, Daly M, Oettgen H, Old L (1986) Specificity analysis of human monoclonal antibodies reactive with cell surface and intracellular antigens. Proc Natl Acad Sci USA 83:2959–2963

Courtenay-Luck NS, Epenetos AA, Moore R, Larche M, Pectasides D, Dhokia B, Ritter MA (1986) Development of primary and secondary immune responses to mouse monoclonal antibodies used in the diagnosis and therapy of malignant neoplasm. Cancer Res 46:6489–6493

Courtenay-Luck NS, Epenetos AA, Sivolapenko GB, Larche M, Barkans JR, Ritter MA (1988) Development of anti-idiotypic antibodies against tumor antigens and autoantigens in ovarian cancer patients treated intraperitoneally with mouse monoclonal antibodies. Lancet 2(8416):894–896

DeNardo SJ, DeNardo GL, O'Grady LF, Levy NB, Mills SL, Macey DJ, McGahan JP, Miller CH, Epstein AL (1988) Pilot studies of radioimmunotherapy of B-cell lymphoma and leukemia using ^{131}I Lym-1 monoclonal antibody. Antibody Immunoconj Radiopharm 1:17–34

Dillman RO, Beauregard JC, Halpern SE, Clutter M (1986) Toxicities and side effects associated with intravenous infusions of murine monoclonal antibodies. J Biol Response Mod 5:73–83

Dillman RO, Shawler DL, Dillman JB, Royston I (1984) Therapy of chronic lymphocytic leukemia and cutaneous T-cell lymphoma with T101 monoclonal antibody. J Clin Oncol 2:881–891

Dimaggio JJ, Scheinberg DA, Houghton AN (1990) Monoclonal antibody therapy of cancer. In: Pinedo HM, Chabner BA, Longo DL (eds) Cancer chemotherapy and biological response modifiers, annual 11. Elsevier, Amsterdam, pp 177–203

Dinarello CA, Mier JW (1987) Current concepts. Lymphokines. N Engl J Med 317:940–945

Ellis SG, Navetta GI, Tcheng JT, Weisman HF, Wang AL, Pitt B, Topol EJ (1990) Antiplatlet GPIIb/IIIa (7E3) antibody in elective PTCA: safety and inhibition of platelet function (Abstr 0755). Circulation 82:111–191

Emmrich J, Seyfarth M, Fleig WE, Emmrich F (1991) Treatment of inflammatory bowel disease with anti-CD4 monoclonal antibody. Lancet 338:570

Esteban JM, Colcher D, Sugarbaker P, Carrasquillo JA, Bryant G, Thor A, Reynolds JC, Larson SM, Schlom J (1987) Quantitative and qualitative aspects of radiolocalization in colon cancer patients of intravenously administered MoAb B72.3. Int J Cancer 39:50–59

Fauci AS (1987) Immunomodulators in clinical medicine. Ann Intern Med 106: 421–433

Fernsten PD, Primus FJ, Greiner JW, Simpson JF, Schlom J (1991) Characterization of the colorectal carcinoma-associated antigen defined by monoclonal antibody D612. Cancer Res 51:926–934

Fisher CJ, Zimmerman J, Khazaeli MB, Albertson TE, Dellinger RP, Panacek EA, Foulke GE, Dating C, Smith CR, LoBuglio AF (1990) Initial evaluation of human monoclonal anti-lipid A antibody (HA-1A) in patients with sepsis syndrome. Crit Care Med 18:1311–1315

Foon KA, Schroff RW, Bunn PA, Mayer D, Abrams PG, Fer M, Ochs J, Bottino GC, Sherwin SA, Carlo DJ (1984) Effects of monoclonal antibody therapy in patients with chronic lymphocytic leukemia. Blood 64:1085–1094

Goldberg D, Morel P, Chatenoud L, Boitard C, Menkes CJ, Bertoye P, Revillard J, Bach J (1991) Immunological effects of high dose administration of anti-CD4 antibody in rheumatoid arthritis patients. J Autoimmun 4:617–630

Goldenberg DM (1989) Future role of radiolabeled monoclonal antibodies in oncological diagnosis and therapy. Semin Nuclear Med 19:3392–3396

Goldenberg H, Lee RE, Stein R, Siegel JA, Izon DO (1991) Targeting, dosimetry and radioimmunotherapy of B-cell lymphomas with iodine-131-labeled LL2 monoclonal antibody. J Clin Oncol 9:548–564

Gould BJ, Borowitz MJ, Groves ES, Carter PW, Anthony D, Weiner LM, Frankel AE (1989) Phase I study of an anti-breast cancer immunotoxin by continuous infusion: report of a targeted toxic effect not predicted by animal studies. J Natl Cancer Inst 81:775–781

Greenman RL, Schein RMH, Martin MA, Wenzel RP, MacIntyre NR, Emmanuel G, Chmel H, Kohler RB, McCarthy M, Plouffe J, Russell JA, and the XOMA Sepsis Study Group (1991) A controlled clinical trial of E5 murine monoclonal IgM antibody to endotoxin in the treatment of gram-negative sepsis. JAMA 266:1097–1102

Grossbard ML, Freedman AS, Ritz J, Coral F, Goldmacher VS, Eliseo L, Spector N, Dear K, Lambert JM, Blättler WA (1992a) Serotherapy of B-cell neoplasms with anti-B4 blocked ricin: a phase I trial of daily bolus infusion. Blood 79: 576–585

Grossbard ML, Press OW, Appelbaum FR, Bernstein ID, Nadler LM (1992b) Monoclonal antibody-based therapies of leukemia and lymphoma. Blood 80: 863–878

Hafler DA, Ritz J, Schlossman SF, Weiner HL (1988) Anti-CD4 and anti-CD2 monoclonal antibody infusions in subjects with multiple sclerosis: immuno-suppressive effects and human anti-mouse responses. J Immunol 141:131–138

Hale G, Dyer MJ, Clark MR, Phillips JM, Marcus R, Riechmann L, Winter G, Waldmann H (1988) Remission induction in non-Hodgkin's lymphoma with reshaped human monoclonal antibody CAMPATH 1-H. Lancet 2:1394–1395

Hertler AA, Frankel AE (1989) Immunotoxins: a clinical review of their use in the treatment of malignancies. J Clin Oncol 7:1932–1942

Herzog C, Walker C, Muller W, Rieber P, Reiter C, Riethmuller G, Wassmer P, Stockinger H, Madic O, Pichler WJ (1989) Anti-CD4 antibody treatment of patients with rheumatoid arthritis. I. Effect on clinical course and circulating T-cells. J Autoimmun 2:267–282

Horneff G, Burnmester GR, Emmrich F, Kalden JR (1991) Treatment of rheumatoid arthritis with an anti-CD4 monoclonal antibody. Arthritis Rheum 34:129–140

Huse W, Sastry L, Iverson S, Kang A, Alting-Mees M, Burton D, Benkovic S, Lerner R (1989) Generation of a large combinatorial library of the immuno-globulin repertoire in phage lambda. Science 246:1275–1281

Irie RF, Morton DL (1986) Regression of cutaneous metastatic melanoma by intra-lesional injection with human monoclonal antibody to ganglioside GD2. Proc Natl Acad Sci USA 83:8694–8698

Isaacs JD, Watts RA, Hazleman BL, Hale G, Keogan MT, Cobbold SP, Waldmann H (1992) Humanized monoclonal antibody therapy for rheumatoid arthritis. Lancet 340:748–752

Jain RK (1987a) Transport of molecules in the tumor interstitium: a review. Cancer Res 47:3038–3050

Jain RK (1987b) Transport of molecules across tumor vasculature. Cancer Metast Rev 6:559–594

Jain RK (1988) Determinant of tumor blood flow: a review. Cancer Res 48:2641–2658

Jerne NK (1974) Towards a network theory of the immune system. Ann Immunol (Paris) 125c:373

Jordan RE, Knight DM, Wagner C, McAleer MF, McDonough M, Mattis JA, Coller BS, Weisman HF, Ghrayeb J (1992) A dramatic reduction of the immuno-genicity of the anti-GPIIb/IIIa monoclonal antibody, 7E3 Fab, by humanization of the murine constant domains (Abstr 1637). Circulation 86:I-411

Khazaeli MB, Saleh MN, Wheeler RH, Huster WJ, Holden H, Carrano R, LoBuglio AF (1988) Phase I trial of multiple large doses of murine monoclonal antibody CO17-1A. II. Pharmacokinetics and immune response. J Natl Cancer Inst 80:937–942

Khazaeli MB, Wheeler R, Rogers K, Teng N, Ziegler E, Haynes A, Saleh MN, Hardin JM, Bolmer S, Cornett J, Berger H, LoBuglio AF (1990) Initial evalu-ation of a human immunoglobulin M monoclonal antibody (HA-1A) in humans. J Biol Response Mod 9:178–184

Khazaeli MB, Saleh MN, Liu TP, Meredith RF, Wheeler RH, Baker TS, King D, Secher D, Allen L, Rogers K, Colcher D, Schlom J, Shochat D, LoBuglio AF (1991) Pharmacokinetics and immune response of ^{131}I-chimeric mouse/human B72.3 (human γ4) monoclonal antibody in humans. Cancer Res 51:5461–5466

Khazaeli MB, Saleh MN, Liu TP, Kaladas PM, Gilman SC, LoBuglio AF (1992) Frequent anti-V-region immune response to mouse B72.3 monoclonal antibody. J Clin Immunol 12:116–121

Kirkwood JM, Ernstoff MS (1984) Interferons in the treatment of human cancer. J Clin Oncol 4:336–351

Knapp RC, Lau CC (1990) Monoclonal antibodies and radioimmunoconjugates: keeping feasibility at center stage. J Clin Oncol 8:1938–1940

Koehler G, Milstein C (1975) Continuous culture of fused cells secreting antibody of predefined specificity. Nature 256:495–496

Köhler H, Kieber-Emmons T, Srinivhsan S, Kaveri S, Morris WJW, Muller S, Kang C-Y, Raychanduri S (1989) Short analytical review: revised immune network concepts. Clin Immunol Immunopathol 52:104

Koprowski H, Herlyn D, Lubeck M, DeFreitas E, Sears H (1984) Human anti-idiotype antibodies in cancer patients: is the modulation of the immune response beneficial for the patient? Proc Natl Acad Sci USA 81:216–219

Kosmas C, Snook D, Gooden CS, Courtenay-Luck NS, McCall MJ, Meares CF, Epenetos AA (1992) Development of humoral immune responses against a macrocyclic chelating agent (DOTA) in cancer patients receiving radioimmuno-conjugates for imaging and therapy. Cancer Res 52:904–911

Kronenberg M (1991) Self tolerance and autoimmunity. Cell 65:537

Kumar V, Kono DH, Urban JL, Hood L (1989) The T-cell receptor repertoire and autoimmune diseases. Annu Rev Immunol 7:657

Kuzel TM, Winter JN, Rosen ST, Zimmer AM (1990) Monoclonal antibody therapy of lymphoproliferative disorders. Oncology 4:77–93

Lenhard RE Jr, Order SE, Spunberg JJ, Asbell SO, Leibel SA (1985) Isotopic immunoglobulin: a new systemic therapy for advanced Hodgkin's disease. J Clin Oncol 3:1296–1300

Lind P, Lechner P, Hausmann B, Smola MG, Koeltringer P, Steindorfer P, Cesnik H, Passl R, Eber O (1991) Development of human antimouse antibodies (HAMA) after single and repeated diagnostic application of intact murine monoclonal antibodies. Antibody Immunoconj Radiopharm 4:811–818

LoBuglio AF, Saleh MN (1992) Monoclonal antibody therapy of cancer. Crit Rev Oncol Hematol 13:271–282

LoBuglio AF, Saleh MN, Lee J, Khazaeli MB, Carrano R, Holden H, Wheeler RH (1988) Phase I trial of multiple large doses of murine monoclonal antibody CO17-1A. I. Clinical aspects. J Natl Cancer Inst 80:932–936

LoBuglio AF, Wheeler RH, Trang J, Haynes A, Rogers K, Harvey EB, Sun L, Ghrayeb J, Khazaeli MB (1989) Mouse/human chimeric monoclonal antibody in man: kinetics and immune response. Proc Natl Acad Sci USA 86:4220–4224

LoBuglio AF, Liu T, Khazaeli MB (1992) Human and chimeric mouse/human monoclonal antibodies. In: Kung AHC, Baughman RA, Larrick JW (eds) Therapeutic proteins: pharmacokinetics and pharmacodynamics. Freeman, New York, pp 45–61

Mathieson DW, Cobbold SP, Hale G, Clark MR, Oliverira DBG, Lockwood CM, Waldmann H (1990) Monoclonal antibody therapy in systemic vasculitis. N Engl J Med 323:250–255

Meeker T, Lowder J, Maloney D, Miller R, Thielemans K, Warnke R, Levy R (1985) A clinical trial of anti-idiotype therapy for B cell malignancy. Blood 65:1349–1363

Meredith RF, LoBuglio AF, Plott WE, Orr RA, Brezovich IA, Russell CD, Harvey EB, Yester MV, Wagner AJ, Spencer SA, Wheeler RH, Saleh MN, Rogers KJ, Polansky A, Salter MM, Khazaeli MB (1991) Pharmacokinetics, immune response and biodistribution of iodine-131-labeled chimeric mouse/human IgG1,κ 17-1A monoclonal antibody. J Nucl Med 32:1162–1168

Meredith RF, Khazaeli MB, Liu TP, Plott G, Wheeler RH, Russell C, Colcher D, Schlom J, Shochat D, LoBuglio AF (1992a) Dose fractionation of radiolabeled antibodies in patients with metastatic colon cancer. J Nucl Med 33:1648–1653

Meredith RF, Khazaeli MB, Plott WE, Saleh MN, Liu TP, Allen LF, Russell CD, Orr RA, Colcher D, Schlom J, Shochat D, Wheeler RH, LoBuglio AF (1992b)

Phase I trial of iodine-131-chimeric B72.3 (human IgG4) in metastatic colorectal cancer. J Nucl Med 33:23–29

Miller RA, Maloney DG, Warnke R, Levy R (1982) Treatment of a B cell lymphoma with monoclonal anti-idiotypic antibody. N Engl J Med 306:517–522

Miller RA, Hart S, Samoszuk M, Coulter C, Brown S, Czerwinski D, Kelkenberg J, Royston I, Levy R (1989) Shared idiotypes expressed by human B-cell lymphomas. N Engl J Med 321:851–857

Moreland LW, Bucy RP, Tilden A, Pratt PW, LoBuglio AF, Khazaeli M, Everson MP, Daddona P, Ghrayeb J, Kilgarrif C, Sanders ME, Koopman WJ (1993a) Use of a chimeric monoclonal anti-CD4 antibody in patients with refractory rheumatoid arthritis. Arthritis Rheum 34(3):307–318

Moreland LW, Heck LW, Sullivan W, Pratt PW, Koopman WJ (1993b) New approaches to the therapy of autoimmune diseases: Rheumatoid arthritis as a paradigm. Am J Med Sci 305:40–51

Morrison SL (1985) Transfectomas provide novel chimeric antibodies. Science 229: 1202–1207

Oldham RK (1991) Custom-tailored drug immunoconjugates in cancer therapy. Mol Biother 3:148–162

Order S, Sleeper A, Stillwagon G, Klein T, Leichner P (1989) Current status of radioimmunoglobulins in the treatment of human malignancy. Oncology 3: 115–120

Ortho Multicenter Transplant Study Group (1985) A randomized clinical trial of OKT3 monoclonal antibody for acute rejection of cadaveric renal transplants. N Engl J Med 313:337–342

Press OW, Eary JF, Badger CC, Martin PJ, Appelbaum FR, Levy R, Miller R, Brown S, Nelp WB, Krohn KA, Fisher D, DeSantes K, Porter B, Kidd P, Thomas ED, Bernstein ID (1989) Treatment of refractory non-Hodgkin's lymphomas with radiolabeled MB-1 (anti CD-37) antibody. J Clin Oncol 7: 1027–1038

Reisfeld R, Yang H, Muller B, Wargalla U, Schrappe M, Wrasidlo W (1989) Promises, problems, and prospects of monoclonal antibody-drug conjugates for cancer therapy. Antibody Immunoconj Radiopharm 2:217–224

Reiter C, Kakavand B, Rieber EP, Schattenkirchner M, Riethmuller G, Kruger K (1991) Treatment of rheumatoid arthritis with monoclonal CD4 antibody M-T151: clinical results and immunopharmacologic effects in an open study, including repeated administration. Arthritis Rheum 34:525–536

Reynolds JC, Del Vecchio S, Sakahara H (1989) Anti-murine antibody response to mouse monoclonal antibodies: clinical findings and implications. Nucl Med Biol 16:121–125

Riechmann L, Clark M, Waldmann H, Winter G (1988) Reshaping human antibodies for therapy. Nature 332:323–327

Robinson BE, Quesenbery PJ (1990) Hematopoietic growth factors. Overview and clinical application (Review). Ann J Med Sci 163:237–311

Roitt I, Brostoff J, Make D (eds) (1991) Immunology, 3rd edn. Gower, London

Rustagi PK, Meluch AA, Khazaeli MB, Liu T, Morris DL, Soule HR, LoBuglio AF (1992) Clinical study of a monoclonal antibody (Fab) to factor VII, a novel anticoagulant (Abstr 655). Blood 80:166a

Saleh MN, LoBuglio AF, Wheeler RH, Rogers KJ, Haynes A, Lee JY, Khazaeli MB (1990) A phase II trial of murine monoclonal antibody 17-1A and interferon-γ: clinical and immunologic data. Cancer Immunol Immunother 32:185–190

Saleh M, Khazaeli M, Wheeler R, Allen L, Liu T, Schlom J, LoBuglio A (1992a) A phase I trial of the murine monoclonal antibody (MoAb) D612 in patient with metastatic colorectal cancer (Abstr 821). Cancer Res 53:4555–4562

Saleh MN, Khazaeli MB, Wheeler RH, Allen L, Tilden AB, Grizzle W, Reisfeld RA, Yu AL, Gillies SD, LoBuglio AF (1992b) A phase I trial of the chimeric anti-GD2 monoclonal antibody ch14.18 in patients with malignant melanoma. Hum Antibodies Hybridomas 3:19–24

Saleh MN, Khazaeli MB, Wheeler RH, Liu TP, Urist M, Miller DM, Lawson S, Dixon P, Russell CH, LoBuglio AF (1992c) A phase I trial of the murine monoclonal anti-GD2 antibody 14G2a in metastatic melanoma. Cancer Res 52:4342–4347

Saravolatz LD, Markowitz N, Collins MS, Bogdanoff D (1991) Safety, pharmacokinetics, and functional activity of human anti-*pseudomonas aeruginosa* monoclonal antibodies in septic and nonseptic patients. J Infect Dis 164:803–806

Schlom J (1990) Monoclonal antibodies: they're more and less than you think. In: Broder S (ed) Molecular foundations of oncology. Williams and Wilkins, Baltimore, pp 95–134

Schroff RW, Foon KA, Beatty SM, Oldham RK, Morgan AC Jr (1985) Human antimurine immunoglobulin responses in patients receiving monoclonal antibody therapy. Cancer Res 45:879–885

Sears HF, Atkinson B, Mattis J, Ernst C, Herlyn D, Steplewski Z, Häyry P, Koprowski H (1982) Phase I clinical trial of monoclonal antibody in treatment of gastrointestinal tumours. Lancet 1:762–765

Sears HF, Herlyn D, Steplewski Z, Koprowski H (1985) Phase II clinical trial of a murine antibody cytotoxic for gastrointestinal adenocarcinoma. Cancer Res 45:5910–5913

Shawler DL, Bartholomew RM, Smith LM, Dillman RO (1985) Human immune response to multiple injections of murine monoclonal IgG. J Immunol 135:1530–1535

Simoons ML, vd Brand M, Hoorntje JCA, de Feyter P, Zijilstra F, Heyndricks G, vd Wicken R, De Bono D, Rutsch W, Weisman HF, Schaible TF, deBoer MJ (1993) Chimeric 7E3 antiplatelet antibody Fab for treatment of refractory unstable angina: a placebo-controlled pilot study (Abstr 768-1). J Am Coll Cardiol 21:269A

Sinha AA, Lopez MT, McDevitt HO (1990) Autoimmune diseases: the failure of self tolerance. Science 248:1380–1387

Smith CR, Straube RC, Ziegler EJ (1992) A human monoclonal antibody for the treatment of gram-negative sepsis. Infect Dis Clin North Am 6:253–266

Steis RG, Carrasquillo JA, McCabe R, Bookman M, Reynolds J, Larson S, Smith J, Clark J, Dailey V, Del Vecchio S (1990) Toxicity, immunogenicity, and tumor radioimmunodetecting ability of two human monoclonal antibodies in patients with metastatic colorectal carcinoma. J Clin Oncol 8:476–490

Stewart JSW, Hird V, Snook D, Phokia B, Sivolapenko A, Hooker G, Papadimitriou JT, Rowlinson G, Sullivan M, Lambert HE, Coulter C, Mason WP, Soutter WP, Epenetos AA (1990) Intraperitoneal yttrium-90 labeled monoclonal antibody in ovarian cancer. J Clin Oncol 8:1941–1950

Sweeney J, Holme S, Heaton A, Weisman H, Whitley P, Barackman K, O'Donnell D, Herre J (1993) Infusion of a chimeric monoclonal Fab fragment (c7E3) against platelet glycoprotein IIB-IIIA potently inhibits platelet aggregation but does not affect in vivo platelet survival (Abstr 911-41). J Am Coll Cardiol 21:253A

Swischer EM, Shawler DL, Collins HA, Bustria A, Hart S, Bloomfield C, Miller RA, Royston I (1991) Expression of shared idiotypes in chronic lymphocytic leukemia and small lymphocytic lymphoma. Blood 77:1977–1982

Teng N, Kaplan H, Hebert J, Moore C, Douglas H, Wunderlich A, Braude A (1985) Protection against gram-negative bacteremia and endotoxemia with human monoclonal IgM antibodies. Proc Natl Acad Sci USA 82:1790–1794

Vadhan-Raj S, Cordon-Cardo C, Carswell E, Mintzer D, Dantis L, Duteau C, Templeton MA, Oettgen HF, Old LJ, Houghton AN (1988) Phase I trial of a mouse monoclonal antibody against GD3 ganglioside in patients with melanoma: induction of inflammatory responses at tumor sites. J Clin Oncol 6:1636–1648

Van der Lubbe P, Miltenburg AM, Breedveid FC (1991) Anti-CD4 monoclonal antibody for relapsing polychondritis. Lancet 337:1349

Vitetta ES, Fulton R, May R, Till M, Uhr J (1987) Redesigning nature's poisons to create anti-tumor reagents. Science 238:1098–1104

Vitetta ES, Stone M, Amlot P, Fay J, May R, Till M, Newman J, Clark P, Collins R, Cunningham D (1991) Phase I immunotoxin trial in patients with B-cell lymphoma. Cancer Res 51:4052–4058

Waldmann T, Schwab P (1965) IgG (7 S gamma globulin) metabolism in hypogammaglobulinemia: studies in patients with defective gamma globulin synthesis, gastrointestinal protein loss, or both. J Clin Invest 44:1523–1533

Weiner LM, O'Dwyer J, Kitson J, Comis RL, Frankel AE, Bauer RJ, Konrad MS, Groves ES (1989) Phase I evaluation of an anti-breast carcinoma monoclonal antibody 260F6-recombinant ricin A chain immunoconjugate. Cancer Res 49: 4062–4067

Wendling D, Wijdenes J, Rachadot E, Morel-Fourrior B (1991) Therapeutic use of monoclonal anti-CD4 antibody in rheumatoid arthritis. J Rheumatol 18:325–327

Wolff SM (1991) The treatment of gram-negative bacteremia and shock (Editorial). N Engl J Med 324:486–487

Yu A, Reisfeld R, Gillies S (1991) Immune response to monoclonal anti-GD2 antibody therapy. Proc Am Assoc Cancer Res 32:263

Ziegler EJ, Fisher CJ Jr, Sprung CL, Straube RC, Sadoff JC, Foulke GE, Wortel CH, Fink MP, Dellinger RP, Teng NNH, Allen IE, Berger HJ, Knatterud GL, LoBuglio AF, Smith CR, and the HA-1A Sepsis Study Group (1991) Treatment of gram-negative bacteremia and septic shock with HA-1A human monoclonal antibody against endotoxin: a randomized, double-blind, placebo-controlled trial. N Engl J Med 324:429–436

CHAPTER 16

Anti-idiotypic Monoclonal Antibodies: Novel Approach to Immunotherapy

M. CHATTERJEE, K.A. FOON, and H. KOHLER

A. Introduction

Immunoglobulin molecules possess variable regions specific for antigen recognition. The variable region is encoded by VH, D and JH genes for the heavy chains and VL and JL chains for the light chain (TONEGAWA 1983). The variable region contains determinants known as idiotypes (Ids) which are themselves immunogenic. Anti-Id antibodies can recognize idiotopes expressed by the light chain, heavy chain, or a combination of both chains (KUNKEL et al. 1963; OUDIN and MICHEL 1963). The initial studies by OUDIN and MICHEL (1963) and KUNKEL et al. (1963) indicated that an Id was unique to a small set of antibody molecules. However, the Id determinants may show a continuum of specificity ranging from more or less private to semi-public (WILLIAMSON 1976; STEVENSON and GLENNIE 1985). Shared idiotopes are attractive targets for active and passive immunotherapy approaches and may play important roles in the regulation of the antibody response.

In 1973 LINDENMANN and in 1974 JERNE proposed theories which described the immune system as a network of interacting antibodies and lymphocytes. According to this original network hypothesis, Id-anti-Id interactions regulated the immune response of a host to a given antigen. Both Ids and anti-Ids have been used to manipulate cellular and humoral immunity. Ids are distinguished by their topographical location on the immunoglobulin structures and are classified by their physical relation to the antigen binding site of the antibody (Ab1). If an anti-Id antibody (Ab2) recognizes an idiotope thereby not interfering with antigen binding, it is designated the α type (JERNE et al. 1982). If the target idiotope is close to the binding site of Ab1 so that it interferes with antigen binding, it is called the γ type (BONA and KOHLER 1984). If the Ab2 binds to the antigen-binding site (paratope) of Ab1, then this Ab2 is referred to as β type (JERNE et al. 1982; KOHLER 1984). Anti-Id antibodies of the β type are thought to express the internal image of the antigen for Ab1. They can effectively mimic the three-dimensional structures of external antigens. The original network concept predicted that only immunization with Ab2β could lead to the generation of anti-anti-Id antibodies (Ab3) that recognized the corresponding original antigen identified by the Ab1 (Fig. 1). However, recent data have shown that Ab2α and Ab2γ (KOHLER et al. 1989a) can also induce specific antibody

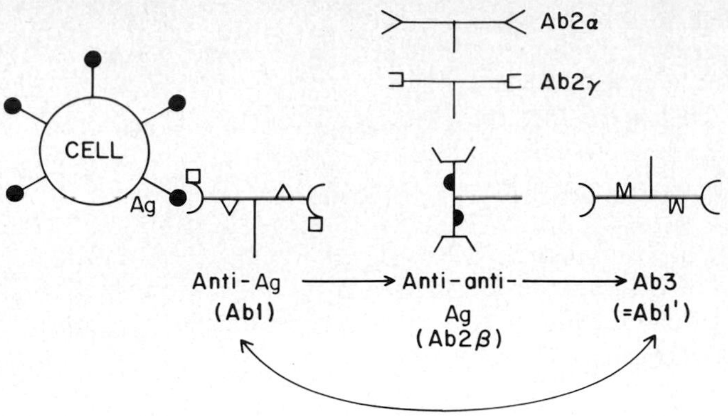

Fig. 1 Idiotype network

responses (KOHLER et al. 1989b). The induced Ab3 has Ab1-like reactivity and is called Ab1' to indicate that it might differ in its other Ids from Ab1. The cyclic nature of complimentary binding sites and idiotopes is the basis for the approach to Id vaccines (NISONOFF and LAMOYI 1981).

The idea of anti-Ids as vaccines for infectious disease has been derived from the successful preparation and characterization of anti-Id antibodies able to mimic bacterial, viral and parasitic antigens (KOHLER et al. 1989b). When appropriately manipulated, anti-Ids can serve as effective inducers of T and B cell immunity to pathogens.

B. Advantages of Anti-idiotypic Antibodies Over Conventional Vaccines

The approach to produce vaccines is entering a critical state of transition. There are serious side effects associated with certain vaccines for infectious diseases and some vaccines are limited in quantity. This has created a pressing need for safer vaccines in situations in which large segments of the population require vaccination.

In recent years, emphasis has been directed towards the use of synthetic peptides containing the antigenic or immunogenic determinants of viral surface coat proteins (SHINNICK et al. 1983). In effect, this approach presents to the immune system only a fragment of the nominal antigen, thereby alleviating the difficulties associated with antigen isolation and purification or immunization with attenuated and live viruses.

The network hypothesis (JERNE 1974) offers still another elegant concept for developing vaccines which is not based on the conventional approach of using the nominal antigenic material. These anti-Id vaccines take advantage of the fact that the repertoire of external or nominal antigens is mimicked by

Id structures on immunoglobulins and possibly on receptors and products of T cells as well. Thus, with this approach, Id-based vaccines do not contain nominal antigen nor its fragments. This excludes the possibility that Id vaccines would have the same undesired side effects which are associated with conventional vaccines.

Besides the increased safety of Id vaccines, these new kinds of antigens have other practical, economical and biological advantages over conventional vaccines. Id vaccines do not depend on the availability of large amounts of pure antigen which often is a limiting economical factor in vaccine production. By virtue of their being proteins, Id vaccines can be easily manipulated and can be coupled to potent immunogenic carriers to become T cell-dependent antigens which can receive full T help. T-dependent protein vaccines can become a decisive factor in situations in which the responding immune system is immature or suppressed. From experimental studies on animals we know that the response to T cell-dependent antigens matures earlier than the T-independent response to carbohydrate antigens and that often a genetically or acquired abnormal immune system responds better to T-dependent antigens than to T-independent antigens.

Finally, data exist showing that an acquired state of tolerance to one antigen form can be broken by using a different molecular form of the same antigenic moiety (WEIGLE 1961). This could become an important consideration in a broader context such as in the immunotherapy of cancer patients, who may be often immunodeficient or tolerant against their own tumor.

In this review, we will discuss various examples in which anti-Id antibodies are used as vaccine therapy in several models, which include acquired immune deficiency syndrome (AIDS) and malignancies. Recently, anti-Id antibodies have also been used in human therapy trials.

C. Acquired Immune Deficiency Syndrome

AIDS results from infection by the human immunodeficiency virus (HIV) of many cellular components vital for the maintenance of human immune homeostasis. The CD4 molecule is the receptor for HIV (KLATZMANN et al. 1984). AIDS can also be considered as an autoimmune disease, in which HIV mimics the "self" component and is able to bind to CD4-expressing helper-inducer T lymphocytes, thereby creating a cascade of network errors leading to the destruction of the immune system. The gp120 envelope protein of HIV is known to possess the CD4 binding domain by which HIV type 1 attaches to its target cell (McDOUGAL et al. 1986; LASKY et al. 1987). Problems in designing a vaccine and/or immunotherapy in AIDS arise from the primary target of HIV infection, i.e., the CD4 receptor-bearing cells. CD4-expressing cells fulfill important regulatory functions in immune responses, and any anti-HIV-targeted vaccine, based either on inactivated whole virus, viral recombinant subunit vaccine, or anti-Id antibodies, must circumvent side effects such as immunosuppression or the induction of autoimmune responses.

Current approaches to anti-Id vaccines include anti-Id antibodies related to viral epitopes and CD4 (ZHOU et al. 1987; ZAGURY et al. 1988; HOMSY et al. 1987). Since the CD4 binding site appears to be conserved among a variety of different HIV isolates, the induction of anti-Id antibodies bearing the internal image of the CD4 receptor appears to be a promising approach towards an Id vaccine in AIDS.

A variety of CD4 monoclonal antibodies block HIV infection of CD4-expressing cells and therefore recognize the binding epitope of HIV for the CD4 molecule (SATTENTAU et al.1986). Anti-Ids generated in mice immunized with CD4 antibodies (DALGLEISH et al. 1987) neutralized in vitro viral activity in three diverse isolates of HIV-1 and a single isolate of HIV-2. These results are important, considering that a vaccine must stimulate the production of a wide range of virus-neutralizing antibodies. Other investigators (CHANH et al. 1987) generated a murine anti-Id monoclonal antibody that mimicked the CD4 receptor and bound HIV envelope antigens.

As previously described, immunization of humans with anti-CD4 antibody preparations can be potentially harmful (DEL GUERCIO and ZANE 1987; KOFF and HOTH 1988). Anti-CD4 antibodies can block those epitopes on T cells which demonstrate helper activity (BENJAMIN and WALDMANN 1986). However, immunization of newly positive HIV patients with an intact immune system may induce an anti-Id response to the anti-CD4, which will represent a new neutralizing set of antibodies. Alternatively, in those patients with AIDS anergic to neoantigens and not responding to anti-CD4 antibody immunization, an anti-Id that mimics CD4 could be used for passive immunotherapy. Passive immunotherapy with such an antibody may be considered as a "CD4 sponge" which would bind all the HIV that is present (DALGLEISH et al. 1987; CHANH et al. 1987).

Weak neutralizing activity against different HIV isolates has been found with antibodies (Ab1) raised against HIV envelope glycoproteins (WEISS et al. 1986). However, there are recent observations that human monoclonal antibodies against HIV antigens, particularly neutralizing antibodies against HIV glycoproteins, e.g., gp120 possess virus-neutralizing activities (Ho et al. 1990). More recently, broadly neutralizing antibodies could be isolated from sera of HIV-1 infected individuals (KANG et al. 1991). These neutralizing antibodies are directed against the CD4 attachment site of gp120. Studies are underway to produce human monoclonal antibodies against HIV gp120 and gp41 to be used not only for immunotherapy or vaccination (Ho et al. 1990; GORNY et al. 1988), but also as antigen to generate anti-Ids related to HIV epitopes. Recently (KANG et al. 1992), a monoclonal antibody was raised against purified human virus neutralizing antibodies which recognizes an idiotope frequently associated with human neutralizing antibodies. The anti-Id, however, is not an internal image of the CD4 site on gp120. Instead, it recognizes human neutralizing antibodies and B cell receptors committed to produce neutralizing antibodies. This biological activity was demonstrated by immunizing nonhuman primates with this anti-Id and the induction of broadly neutralizing antibodies. Hence, this type of anti-Id can be described as an anti-clonotypic antibody. Monoclonal anti-Ids have been also used to screen sera from AIDS patients for Id-carrying anti-HIV antibodies as Id markers in the immune response of patients to HIV (MULLER et al. 1991).

D. Solid Tumors and Cutaneous T Cell Lymphoma

Active immunotherapy of cancer patients with tumor-derived material has been studied by numerous investigators with modest clinical responses (EILBER et al. 1976; MATHE et al. 1969; GUTTERMAN et al. 1974; VOGLER et al. 1978; MASTRANGELO et al. 1984; FOON et al. 1983; BERD et al. 1986; LIVINGSTON et al. 1985; TAI et al. 1985; HOOVER et al. 1984, 1985; McCUNE et al. 1981; HOLLINSHEAD et al. 1987). There exists a number of problems with using tumor material for immunization. One problem is that tumor-associated antigens are often poorly immunogenic possibly because the immune system has been tolerized (McBRIDE and HOWIE 1986; GREENE 1980; HAUBECK and KOLSCH 1982; HOWIE and McBRIDE 1982). If this is true, steps could be taken to break tolerance to tumor antigen. One approach might be to present the critical epitope in a different molecular environment to the tolerized host (WEIGLE 1961). However, this is impossible for most tumor antigens because they are poorly defined chemically and difficult to purify. The network hypothesis could offer an alternative approach (JERNE 1974).

Active immunization with tumor-specific Id vaccines has been shown to inhibit the growth of tumors in animal models (HELLSTROM and HELLSTROM 1989; STEVENSON and GORDON 1983; KAMINSKI et al. 1987). Furthermore, in some of these, the mechanisms responsible for inducing antitumor immunity have been studied in detail. We have reported on a mouse leukemia model L1210 in DBA/2 mice (RAYCHAUDHURI et al. 1986, 1987a–c), which has provided us with basic information on B and T cell induced responses using the anti-Id approach. A number of anti-Id hybridomas against monoclonal antibody to the L1210 tumor were generated. These anti-Id monoclonal antibodies induced tumor-specific delayed-type hypersensitivity, inhibition of tumor growth, generation of cytotoxic T lymphocytes and T helper cells and antitumor Ab3 antibodies. These findings are very promising since they demonstrate a cross-reaction of nominal antigen and internal image antigen for a tumor-associated antigen system at the T and B cell level. In a recent study, 100% cure of established tumors was achieved in DBA/2 mice by combining anti-Id vaccines with cyclophosphamide (CHEN et al. 1989). Similar results have been reported in a variety of murine tumor systems

(CAMPBELL et al. 1988; KENNEDY et al. 1985; NELSON et al. 1987). Cyclophosphamide (100 mg/kg), administered in combination with Id vaccines to
mice bearing 10-day-old, 1–2 cm diameter subcutaneous B cell lymphoma
(38C13), resulted in a dramatic survival benefit (CAMPBELL et al. 1988). An
anti-Id antibody was used to induce immunity to SV-40 transformed cells
(KENNEDY et al. 1985). Mice vaccinated with this anti-Id demonstrated
prolonged survival after tumor transfer. The role of Id interactions in regulating the immune response of mice to chemically induced syngeneic
sarcomas has been recently studied (NELSON et al. 1987). Treatment with
anti-Id monoclonal antibody of mice with the established sarcomas (MCA-
490 and MCA-1511) had significant anti-tumor activity.

In human clinical trials, anti-Id responses were implicated in the induction of antitumor immunity to colorectal carcinoma (KOPROWSKI et al. 1984).
In these trials, patients were passively infused with an Ab1 anti-colorectal
cancer monoclonal antibody (17-1A). A small cohort of responding patients
demonstrated Ab1' and Ab3 responses (KOPROWSKI et al. 1984). These
investigators generated polyclonal anti-17-1A (Ab2) antibodies for an active
immunization clinical trial. Thirty patients with advanced colorectal carcinoma were immunized with alum-precipitated goat Ab2 in doses between
0.5 and 4 mg per injection. All patients developed Ab3 that bound to tumor
cells and blocked Ab1 binding. Clinical responses were observed in 13 of the
30 patients treated (HERLYN et al. 1987). In another recent study it was
demonstrated that subcutaneous injection of 0.5–4 mg of a monoclonal anti-
Id generated to an Ab1 that bound a high molecular weight human
melanoma-associated antigen elicited an antitumor antibody response in
the host (MITTELMAN et al. 1990). Furthermore, reduction in the size of
metastatic lesions were observed in at least seven of the 37 immunized
patients. Repeated injection of murine anti-Id monoclonal antibody was not
associated with any side effects (MITTELMAN et al. 1990).

We have generated monoclonal Id cascades for two different human
tumor-associated antigens. The first cascade originated from a T cell
leukemia/lymphoma (BHATTACHARYA-CHATTERJEE et al. 1987, 1988) and the
other from carcinoembryonic antigen (CEA) (BHATTACHARYA-CHATTERJEE et
al. 1990). In both cascades we have generated monoclonal anti-anti-Ids
(Ab3) which bind to the original tumor-associated antigen (BHATTACHARYA-
CHATTERJEE et al. 1991).

We have initiated our first clinical trial for patients with cutaneous T cell
lymphoma. We have begun the active immunotherapy trial with a murine
monoclonal IgG1 anti-idiotype antibody (Ab2) designated 4DC6 which
mimics a unique tumor-associated cell surface glycoprotein (gp37) expressed
exclusively on T cell leukemia and lymphoma but not on normal T cells.
4DC6 was raised against the murine monoclonal antibody SN2 (Ab1) which
defines gp37. Our first patient with cutaneous T cell lymphoma (CTCL) was
immunized intradermally biweekly three times with 1 mg of alum-precipitated
4DC6. Pretherapy serum was negative for human anti-mouse antibodies

(HAMAs) and a significant HAMA response was induced after the second immunization against 4DC6. A major portion of the HAMAs was directed against the idiotype of 4DC6 as measured by ELISA against 4DC6 F(ab')2 and isotype-matched controls. The anti-anti-Id (Ab3) response increased after the third immunization and reached a plateau. The binding of radio-labeled Ab1 (SN2) to Ab2 (4DC6) was inhibited >50% by the patient's serum collected after the third immunization as compared to pretherapy serum, suggesting that the Ab1 and Ab3 share common Ids. However, the binding of labeled Ab1 to gp37-positive MOLT-4 cells was inhibited 10% – 15%, indicating that only a minor portion of the Ab3 was recognizing the same epitope as Ab1. The patient has stable disease and is being followed at monthly intervals. Interestingly, serum lactate dehydrogenase (LDH) steadily declined from 730 U/L prior to treatment to 417 U/L following the third immunization. A second patient has begun the protocol, also at the 1 mg dose. After three immunizations this patient has demonstrated significant reduction (greater than 90%) of his cutaneous tumors and has mounted a strong anti-anti-Id (Ab3) response.

While preliminary results from anti-Id vaccine trials for human cancer have demonstrated Ab3 (Ab1') responses, it is not clear whether T cell responses, including specific cytotoxic T cells to tumor cells, have been generated as reported in our animal models (RAYCHAUDHURI et al. 1987b,c). Whether the Ab3 antibodies were able to recruit human effector cells or complement to destroy tumor cells remains to be determined.

Allergic reactions to murine antibodies are a concern. Previous experience with murine antibodies in humans has not shown serious anaphylactic type reactions, although low grade fever, nausea, diarrhea and urticaria are reported (FOON 1989; DILLMAN 1989). These reactions have rarely led to discontinuation of therapy.

Anti-Id vaccines may be replaced by synthetic peptide antigens. It is possible to obtain the three-dimensional shape of Id internal antigens using computer modeling on the sequence of internal antigen hybridomas (KIEBER-EMMONS et al. 1986). The molecular modeling studies will indicate whether the internal Id antigen is expressed as a linear sequence or a complex determinant involving noncontiguous sequence contributions to the molecular architecture. After obtaining sufficient information as to the most likely peptide regions which might depict the antigenic structure, it should be possible to synthesize the relevant peptides which can mimic the structure and couple them to new synthetic adjuvants to increase the relative level of immunogenicity.

Heterogeneous expression and modulation of target antigens represent other potential problems to immunotherapy with anti-Id vaccine preparations. However, these problems are not insurmountable and they may be minimized by several complementary approaches. Heterogeneity of tumor-associated antigen expression may be addressed by utilizing cocktails of anti-Id vaccine preparations directed against multiple target antigens collectively expressed by the vast majority of tumor cells. With respect to antigenic modulations by Ab3 induced in response to Ab2β, it is known that modulation rates for different antigens may vary considerably. Hence, tumor-associated antigens with slow modulation characteristics should be chosen as targets for anti-Id active immunotherapy. Furthermore, recent studies (CARREL et al. 1985; LIAO et al. 1982) have indicated that some biological response modifiers, such as interferon, are capable of inducing the expression of tumor-associated antigens in several tumor systems. Thus, in addition to providing general stimulation to the immune system, the administration of biological response modifiers known to increase the expression of relevant tumor-associated antigens may further enhance the efficacy of anti-Id tumor vaccine application.

The anti-Id approach needs to be compared to other tumor therapies, both established and experimental. A realistic assessment of anti-Id therapy predicts that complete remission in patients with advanced disease will be unlikely. However, evidence exists that partial remission and responses can be achieved with anti-Ids. It is apparent that successful therapy of established tumors will depend on a multimodality approach. Since the immune response has a limited capacity, immunotherapeutic manipulations will probably be most effective against small tumor masses. Consequently, immunotherapy with Id vaccine would be most attractive in the adjuvant setting. Compared to chemotherapy or most lymphokine therapy, the Id vaccine approach is considerably less toxic.

E. B Cell Lymphomas and Leukemias

Another therapeutic approach with anti-Id monoclonal antibodies is in B cell malignancies. This represents a departure from the previous discussion of anti-Id Ab2 vaccines. In this case, the anti-Id (Ab2) is generated to the surface membrane immunoglobulin on the malignant B cells and is infused into patients to target directly to tumor cells. Indeed, this antigen is the closest we have come to identifying a tumor-specific antigen in humans. This specificity is based on the fact that individual B cells are committed to the synthesis of only one immunoglobulin species with a unique variable region structure (Id). Moreover, since B cell lymphomas and leukemias are clonal in nature, members of the malignant clone should express the same Ig molecule, and hence the same Id. This feature thus represents a marker by which these tumors cells can be distinguished from normal cells of the host. These facts also imply that an individual patient's tumor cell Id will be different from that of other patients, hence anti-Id antibodies must be "tailor-made" for the individual patient. Because of the highly specific nature of these antibodies, treatment with these antibodies have yielded important results regarding the ultimate potential of monoclonal antibody therapy.

The largest experience reported with anti-Id therapy is the work of Levy and coworkers. Their first attempt at this therapy was a patient with malignant lymphoma (MILLER et al. 1982). At the time of treatment, the patient had evidence of rapidly progressive systemic disease resistant to chemotherapy and interferon. Following eight continuous 6h intravenous infusions spaced over the period of a month, the patient entered a complete clinical remission for 6 years (LEVY, personal communication). The mechanisms accounting for this dramatic response are not clear. Because it was noted that the patient's antitumor response continued after the period of passive antibody administration, evidence of an anti-Id antibody response by the patient himself was investigated, but none was detected. It is still possible that indirect mechanisms could have been involved. Since the immune system may be regulated in part by networks of interactions between Ids and anti-Ids, the administered anti-Id could have triggered these types of networks of interactions which led to an antiproliferative response against the patient's tumor.

Additional patients have been treated with individually tailored anti-Id antibodies of varying antibody subclasses (MEEKER et al. 1985a). Some patients have been treated with more than one antibody (differing in isotype or epitope specificity) during the course of an individual treatment period. Significant tumor responses have been demonstrated in 50% of the patients with limited complete responses. The addition of combined therapy with interferon (BROWN et al. 1989) or chlorambucil (LEVY and MILLER 1990) may enhance responses but the numbers of patients on these trials are too small to draw critical conclusions.

There exists a number of mechanisms by which the tumor could evade the anti-Id antibody. First, some tumor secrete the immunoglobulin (Id) in sufficient quantitites to bind the anti-Id in the serum, preventing it from binding to tumor cells. This is sometimes overcome by infusing with excessive quantities of anti-Id or plasmapheresis. A second problem is that some patients (<25%) generate an immune response to the mouse antibody preventing further therapy. Another means by which the patients' tumors could evade the therapeutic effects of anti-Id antibodies is the emergence of Id variants within the tumors during the treatment (MEEKER et al. 1985b). It is now believed that somatic mutation accounts for this escape (RAFFELD et al. 1985; CARROLL et al. 1986).

Various factors have been studied to predict the response to this therapy. Included among these are the isotype of the anti-Id antibody used, the density of cell surface Id, the epitope recognized by the anti-Id antibody, the affinity of anti-Id antibody for antigen, the relative ability of the anti-Id antibody to modulate surface antigen, the direct effect of antibody on tumor cell proliferation in vitro, and the degree of T cell infiltration present in pretherapy tumor specimens. None of these factors has been positively correlated with good clinical outcome, except the number of T cells present in pretherapy tumor tissues (LOWDER et al. 1987; GARCIA et al. 1985). In the two best responding cases, the T cells actually outnumbered the tumor cells. The majority of these T cells were of the helper/inducer phenotype (CD4). Whether the anti-Id antibodies given to these patients augmented an on-going cell-mediated cytotoxic response by the host against the tumor is not clear. Certainly more observations on pretherapy T cell infiltration must be made before the actual significance and function of this finding become apparent.

Some monoclonal anti-Id antibodies made against an individual patients' tumor cells have recently been shown to cross-react with more than one patient's tumor cells. A panel of 29 anti-shared monoclonal anti-Id anti-bodies that reacted with approximately one third of cases of B cell lymphoma and chronic lymphocytic leukemia has been identified (MILLER et al. 1989; CHATTERJEE et al. 1990). These antibodies also detected rare normal B cells and minor components of serum Ig. Thus far, it is too early to determine whether anti-shared idliotype antibodies will be as effective as tailor-made anti-Id antibodies.

An active immune approach to anti-Id therapy for B-cell diseases that mimics the anti-Id internal image tumor vaccine approach is also possible. In this case, purified immunoglobulin derived from patient's tumors is used as a vaccine to immunize patients with B cell malignancies. In one study, eight of nine patients immunized with autologous tumor-derived immunoglobulin coupled to keyhole limpet hemocyanin (KLH) and mixed with adjuvant demonstrated Id-specific immune responses; either humoral, cellular or both (KWAK et al. 1992). Tumor regressions were also reported.

We have been interested in a similar approach to active anti-Id therapy in B cell diseases. We are trying to generate a murine anti-Id monoclonal antibody to an antibody designated 1D10 (GINGRICH et al. 1990) that binds to a highly restricted antigen on malignant B cells. This antigen is not found on any normal cells, including inactive and activated lymphocytes. It is, however, identified on over 75% of B cell malignancies and presents an ideal target antigen for active immunotherapy.

F. Conclusion

Immunization with anti-Id antibodies represents a novel new approach to active immunotherapy. Preclinical trials using internal image anti-Ids for infectious diseases and cancer have supported the immune network theory. Whether this will have clinical applicability in humans is currently under investigation.

Acknowledgements. The work was supported by United States Public Health Services Grants CA 47860, CA 57165 and CA 56701. We thank Michelle Barbaro and Cheryl Zuber for typing of this manuscript.

References

Benjamin RJ, Waldmann H (1986) Induction of tolerance by monoclonal antibody therapy. Nature 320:449–451

Berd D, Maguire HC Jr, Mastrangelo MJ (1986) Induction of cell-mediated immunity to autologous melanoma cells and regression of metastases after treatment with a melanoma cell vaccine preceded by cyclophosphamide. Cancer Res 46:2572–2577

Bhattacharya-Chatterjee M, Pride MW, Seon BK, Kohler H (1987) Id vaccines against human T cell acute lymphoblastic leukemia (T-ALL). I. Generation and characterization of biologically active monoclonal anti-Ids. J Immunol 139:1354–1360

Bhattacharya-Chatterjee M, Chatterjee SK, Vasile S, Seon BK, Kohler H (1988) Id vaccines against human T cell leukemia. II. Generation and characterization of a monoclonal Id cascade (Ab1, Ab2 and Ab3). J Immunol 141:1398–1403

Bhattacharya-Chatterjee M, Mukerjee S, Biddle W, Foon KA, Kohler H (1990) Murine monoclonal anti-Id antibody as a potential network antigen for human carcinoembryonic antigen. J Immunol 145:2758–2765

Bhattacharya-Chatterjee M, Foon KA, Kohler H (1991) Anti-idiotype monoclonal antibodies as vaccines for human cancer. Int Rev Immunol 7:289–302

Bona CA, Kohler H (1984) Anti-Id antibodies and internal images. In: Venter JC, Fraser CM, Lindstrom J (eds) Monoclonal and anti-Id antibodies: probes for receptor structure and function. Liss, New York, pp 141–150

Brown SL, Miller RA, Horning SJ et al. (1989) Treatment of B-cell lymphomas with anti-Id antibodies alone and in combination with alpha interferon. Blood 73:651–661

Campbell MJ, Esserman L, Levy R (1988) Immunotherapy of established murine B cell lymphoma. Combination of Id and cyclophosphamide. J Immunol 141:3227–3233

Carrel S, Schmidt-Kessen A, Giuffre L (1985) Recombinant interferon-gamma can induce the expression of HLA-DR and -DC on DR-negative melanoma cells and enhance the expression of HLA-ABC and tumor-associated antigens. Eur J Immunol 15:118–123

Carroll WL, Lowder JN, Streifer R, Warnke R, Levy S, Levy R (1986) Id variant cell populations in patients with B cell lymphoma. J Exp Med 164:1566–1560

Chanh TC, Dreesman GR, Kennedy RC (1987) Monoclonal anti-Id antibody mimics the CD4 receptor and binds human immunodeficiency virus. Proc Natl Acad Sci USA 84:3891–3895

Chatterjee M, Barcos M, Han T, Liu X, Bernstein Z, Foon KA (1990) Shared Id expression by chronic lymphocytic leukemia and B-cell lymphoma. Blood 76:1825–1829

Chen JJ, Saeki Y, Shi L, Kohler H (1989) Synergistic anti-tumor effects with combined "internal image" anti-Ids and chemotherapy. J Immunol 143:1053–1057

Dalgleish AG, Thomson BJ, Chanh TC, Malkovsky M, Kennedy RC (1987) Neutralization of HIV isolates by anti-Id antibodies which mimic the T4 (CD4) epitope: a potential AIDS vaccine. Lancet 2:1047–1050

Del Guercio P, Zane M (1987) The CD4 molecule, the human immunodefiency virus and anti-Id antibodies. Immunol Today 8:204–205

Dillman RO (1989) Monoclonal antibodies for treating cancer. Ann Intern Med 111:592–603

Eilber FR, Morton DL, Holmes EC et al. (1976) Adjuvant immunotherapy with BCG in treatment of regional lymph node metastases from malignant melanoma. N Engl J Med 294:237–240

Foon KA (1989) Biological response modifiers: the new immunotherapy. Cancer Res 49:1621–1639

Foon KA, Smalley RV, Riggs CH, Gale RP (1983) The role of immunotherapy in acute myelogenous leukemia. Arch Intern Med 143:1726–1731

Garcia CF, Lowder J, Meeker TC, Bindl J, Levy R, Warnke RA (1985) Differences in "host infiltrates" among lymphoma patients treated with anti-Id antibodies: Correlation with treatment response. J Immunol 135:4252–4260

Gingrich RD, Dahle CE, Hoskins KF, Senneff MJ (1990) Identification and characterization of a new surface membrane Ag found predominantly on malignant B lymphocytes. Blood 75(12):2375–2387

Gorny MK, Gianakakos V, Zolla-Pazner S (1988) Generation of human monoclonal antibodies to HIV. IV Int. Conf. AIDS, Book 2: Program and Abstracts, p 74

Greene MI (1980) The genetic and cellular basis of regulation of the immune response to tumor antigens. Contemp Top 11:81–116

Gutterman JU, Hersh EM, Rodriguez B et al. (1974) Chemoimmunotherapy of adult acute leukaemia: prolongation of remission in myeloblastic leukaemia with BCG. Lancet 2:1405–1409

Haubeck HD, Kolsch E (1982) Regulation of immune responses against the syngeneic ADJ-PC-5 plasmacytoma in BALB-c mice. III. Induction of specific T suppressor cells to the BALB/c plasmacytoma. ADJ-PC-5 during early stages of tumorigenesis. Immunology 47:503–510

Hellstrom KE, Hellstrom I (1989) Anti-idiotypic tumor vaccines. Int Rev Immunol 4:337–346

Herlyn D, Wettendorf M, Schmoll E, Hopoulos D, Schedel I, Dreikhausen U, Raab R, Ross AH, Jaksche H, Scriba M, Koprowski H (1987) Anti-Id immunization of cancer patients: modulation of immune response. Proc Natl Acad Sci USA 84:8055–8059

Ho DD, McKeating JA, Xiling L, Moudgil T, Daar ES, SUN N-C, Robinson JE (1991) Conformational epitope on gp 120 important in CD4 binding and human immunodeficiency virus Type 1 neutralization identified by a human monoclonal antibody. J Virology 65:489–493

Hollinshead A, Stewart TH, Takita M, Dalbow M, Concannon J (1987) Adjuvant specific lung cancer immunotherapy trials. Tumor-associated antigens. Cancer 60:1249–1262

Homsy J, Steimer K, Kaslow R (1987) Towards an AIDS vaccine: challenges and prospects. Immunol Today 8:193–196

Hoover HC Jr, Surdyke M, Dangel RB, Peters LC, Hanna MG Jr (1984) Delayed cutaneous hypersensitivity to autologous tumor cells in colorectal cancer patients immunized with an autologous tumor cell: bacillus Calmette-Guerin vaccine. Cancer Res 44:1671–1676

Hoover HC Jr, Surdyke MG, Dangel RB, Peters LC, Hanna MG Jr (1985) Prospectively randomized trial of adjuvant active-specific immunotherapy for human colorectal cancer. Cancer 55:1236–1243

Howie S, McBride WH (1982) Tumor-specific T helper activity can be abrogated by two distinct suppressor cell mechanisms. Eur J Immunol 12:671–675

Jerne NK (1974) Towards a network theory of the immune system. Ann Immunol (Paris) 125C:373–389

Jerne NK, Roland J, Cazenave PA (1982) Recurrent Ids and internal images. EMBO J 1:243–247

Kaminski MS, Kitamura K, Maloney DG, Levy R (1987) Id vaccination against murine B cell lymphoma. Inhibition of tumor immunity by free Id protein. J Immunol 138:1289–1296

Kang CY, Nara P, Chamat S, Caralli V, Ryskamp T, Haigwood N, Newman R, Kohler H (1991) Evidence of non- V3 specific neutralizing antibodies in HIV-1 infected humans which interfere with gp120/CD4 binding. Proc Natl Acad Sci USA 88:6171–6175

Kang CY, Nara P, Morrow WJ, Ho D, Kohler H (1992) Monoclonal anti-clonotypic antibody elicit broadly neutralizing antibodies in monkeys. Proc Natl Acad Sci USA 89:2546–2550

Kennedy RC, Dreesman GR, Butel JS, Lanford RE (1985) Suppression of in vivo tumor formation induced by simian virus 40-transformed cells in mice receiving anti-Id antibodies. J Exp Med 161:1432–1439

Kieber-Emmons T, Ward RE, Raychaudhuri S, Rein R, Kohler H (1986) Rational design and application of Id vaccines. Int Rev Immunol 1:1–26

Klatzmann D, Champagne E, Chamaret S, Gruest J, Guetard D, Hercend T, Gluckman JC, Montagnier L (1984) T-lymphocyte T4 molecule behaves as the receptor for human retrovirus. Nature 312:767–768

Koff WC, Hoth DF (1988) Development and testing of AIDS vaccines. Science 241:426–432

Kohler H (1984) The immune network revisited. In: Kohler H, Urbain J, Cazenave P (eds) Idiotypy in biology and medicine. Academic, New York, pp 3–28

Kohler H, Kieber-Emmons T, Srinivasan S et al. (1989a) Revised immune network concepts. Clin Immunol Immunopathol 52:104–116

Kohler H, Kaveri S, Kieber-Emmons T, Marrow WJW, Muller S, Raychaudhuri S (1989b) Overview of idiotypic networks and the nature of molecular mimicry. Methods Enzymol 178:3–35

Koprowski H, Herlyn D, Lubeck M, DeFreitas E, Sears HF (1984) Human anti-Id antibodies in cancer patients: is the modulation of the immune response beneficial for the patient? Proc Natl Acad Sci USA 81:216–219

Kunkel HG, Mannik M, Williams RC (1963) Individual antigenic specificity of isolated antibodies. Science 140:1218–1219

Kwak LW, Campbell MJ, Czerwinski DK, Hart S, Miller RA, Levy R (1992) Induction of immune responses in patients with B-cell lymphoma against the surface immunoglobulin idiotype expressed by their tumors. N Engl J Med 327:1209–1238

Lasky LA, Nakamura G, Smith DH, Fennie C, Shimasaki C, Patzer E, Berman P, Gregory T, Capon DJ (1987) Delineation of a region of the human immunodeficiency virus type I gp120 glycoprotein critical for interaction with the CD4 receptor. Cell 50:975–985

Levy R, Miller RA (1990) Therapy of lymphoma directed at idiotypes. Monogr Natl Cancer Inst 10:61–68

Liao SK, Kwong PC, Khosravi M, Dent PB (1982) Enhanced expression of melanoma-associated antigen monoclonal antigens and β2-microglobulin on cultured human melanoma cells by interferon. J Natl Cancer Inst 68:19–25

Lindenmann J (1973) Speculations on Ids and hombodies. Ann Immunol (Paris) 124:171–184

Livingston PO, Kaelin K, Pinsky CM, Oettgen HF, Old LJ (1985) The serologic response of patients with stage II melanoma to allogeneic melanoma cell vaccines. Cancer 56:2194–2200

Lowder JN, Meeker TC, Campbell M, Garcia CF, Gralow J, Miller RA, Warnke R, Levy R (1987) Studies on B lymphoid tumors treated with monoclonal anti-Id antibodies: correlation with clinical responses. Blood 69:199–210

Mastrangelo MJ, Berd D, Maguire HC Jr (1984) Current condition and prognosis of tumor immunotherapy: a second opinion. Cancer Treat Rep 68:207–219

Mathe G, Amiel JL, Schwarzenberg L et al. (1969) Active immunotherapy for acute lymphoblastic leukaemia. Lancet 1:1697–1699

McBride WH, Howie SE (1986) Induction of tolerance to a murine fibrosarcoma in the two zones of dosage – the involvement of suppressor cells. Br J Cancer 53:707–711

McCune CS, Schapira DV, Henshaw EC (1981) Specific immunotherapy of advanced renal carcinoma. Evidence for the polyclonality of metastasis. Cancer 47:1984–1987

McDougal JS, Kennedy MS, Sligh JM, Cort SP, Mawle A, Nicholson JK (1986) Binding of HTLV-III/LAV to T4 + T cells by a complex of the 110K viral protein and the T4 molecule. Science 231:382–385

Meeker TC, Lowder J, Maloney DG, Miller RA, Thielemans K, Warnke R, Levy R (1985a) A clinical trial of anti-Id therapy of B cell malignancy. Blood 65:1349–1363

Meeker TC, Lowder J, Cleary ML, Stewart S, Warnke R, Sklar J, Levy R (1985b) Emergence of Id variants during treatment of B-cell lymphoma with anti-Id antibodies. N Engl J Med 312:1658–1665

Miller RA, Maloney DG, Warnke R, Levy R (1982) Treatment of B-cell lymphoma with monoclonal anti-Id antibody. N Engl J Med 306:517–522

Miller RA, Hart S, Samoszuk M, Coulter C, Brown S, Czerwinski D, Kelkenberg J, Royston I, Levy R (1989) Shared Ids expressed by human B cell lymphomas. N Engl J Med 321:851–857

Mittelman S, Chen ZJ, Kageshita T, Yang H, Yamada M, Baskind P, Goldberg N, Puccio C, Ahmed T, Arlin Z, Ferrone S (1990) Active specific immunotherapy in patients with melanoma. J Clin Invest 86:2136–2144

Muller S, Wang HT, Kaveri S, Chattoppadhyay S, Kohler H (1991) Generation and specificity of monoclonal anti-Id antibodies against human HIV-specific antibodies. J Immunol 147:933–941

Nelson KA, George E, Swenson C, Forstrom JW, Hellstrom KE (1987) Immunotherapy of murine sarcomas with auto-anti-anti-Id monoclonal antibodies which bind to tumor-specific T cells. J Immunol 139:2110–2117

Nisonoff A, Lamoyi E (1981) Implications of the presence of an internal image of the antigen in anti-Id antibodies: possible application to vaccine production. Clin Immunol Immunopathol 21:397–406

Oudin J, Michel M (1963) A new allotype form of rabbit serum gammaglobulins, apparently associated with antibody function and specificity. C R Acad Sci (Paris) 257:805–808

Raffeld M, Neckers L, Longo DL, Cossman J (1985) Spontaneous alteration of Id in a monoclonal B-cell lymphoma. Escape from detection by anti-Id. N Engl J Med 312:1653–1658

Raychaudhuri S, Saeki Y, Fuji H, Kohler H (1986) Tumor-specific Id vaccines. I. Generation and characterization of internal image tumor antigen. J Immunol 137:1743–1749

Raychaudhuri S, Saeki Y, Chen JJ, Kohler H (1987a) Tumor specific Id vaccines. II. Analysis of the tumor-related network response induced by the tumor and by internal image antigens (Ab2β). J Immunol 139:271–278

Raychaudhuri S, Saeki Y, Chen JJ, Kohler H (1987b) Tumor-specific Id vaccines. III. Induction of T helper cells by anti-Id and tumor cells. J Immunol 139:2096–2102

Raychaudhuri S, Saeki Y, Chen JJ, Iribe H, Fuji H, Kohler H (1987c) Tumor-specific Id vaccine. IV. Analysis of the Id network in tumor immunity. J Immunol 139:3902–3910

Sattentau QJ, Dalgleish AG, Weiss RA, Beverley PC (1986) Epitopes of the CD4 antigen and HIV infection. Science 234:1120–1123

Shinnick TM, Sutcliffe JG, Green N, Lerner RA (1983) Synthetic peptide immunogens as vaccines. Annu Rev Microbiol 37:425–446

Stevenson FK, Gordon J (1983) Immunization of Id Ig protects against development of B lymphocytic leukemia, but emerging tumor cells can evade antibody attack by modulation. J Immunol 130:970

Stevenson GT, Glennie MJ (1985) Surface Ig of B-lymphocytic tumours as a therapeutic target. Cancer Surv 4:213–244

Tai T, Cahan LD, Tsuchida T, Saxton RE, Irie RF, Morton DL (1985) Immunosensitivity of melanoma-associated gangliosides in cancer patients. Int J Cancer 35:607–612

Tonegawa S (1983) Somatic generation of antibody diversity. Nature 302:575–581

Vogler W, Bartolucci AA, Omura GA et al. (1978) A randomized clinical trial of remission induction, consolidation and chemo-immunotherapy maintenance in adults acute myeloblastic leukemia. Cancer Immunol Immunother 3:163–170

Weigle WO (1961) The immune response of rabbits tolerant to bovine serum albumin to the injection of other heterologous serum albumins. J Exp Med 114:111–125

Weiss RA, Clapham PR, Weber JN, Dalgleish AG, Lasky LA, Berman PW (1986) Variable and conserved neutralization antigens of human immunodeficiency virus. Nature 346:572

Williamson AR (1976) The biological origin of antibody diversity. Annu Rev Biochem 45:467

Zagury D, Bernard J, Cheynier R, Desportes I, Leonard R, Fouchard M, Reveil B, Ittele D, Lurhuma Z, Mbayo K et al. (1988) A group specific anamnestic immune reaction against HIV-1 induced by a candidate vaccine against AIDS. Nature 332:728–731

Zhou EM, Chanh TC, Dreesman GR, Kanda P, Kennedy RC (1987) Immune response to human immunodeficiency virus. In vivo administration of anti-Id induces an anti-gp160 response specific for a synthetic peptide. J Immunol 139:2950–2956

Subject Index

Springer-Verlag
and the Environment

We at Springer-Verlag firmly believe that an international science publisher has a special obligation to the environment, and our corporate policies consistently reflect this conviction.

We also expect our business partners – paper mills, printers, packaging manufacturers, etc. – to commit themselves to using environmentally friendly materials and production processes.

The paper in this book is made from low- or no-chlorine pulp and is acid free, in conformance with international standards for paper permanency.

Printing: Saladruck, Berlin
Binding: Buchbinderei Lüderitz & Bauer, Berlin